The Amphibians and Reptiles of Missouri

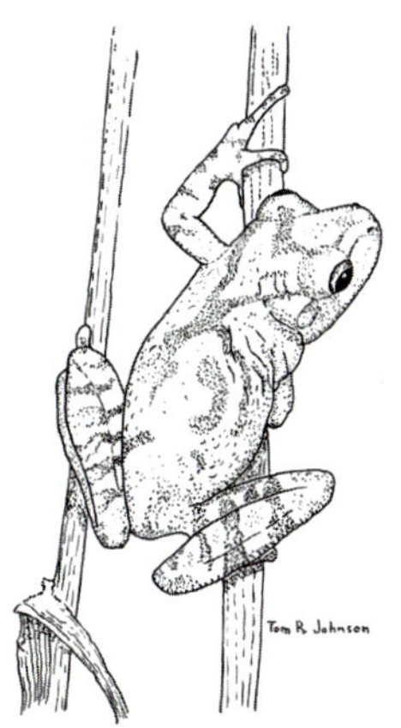

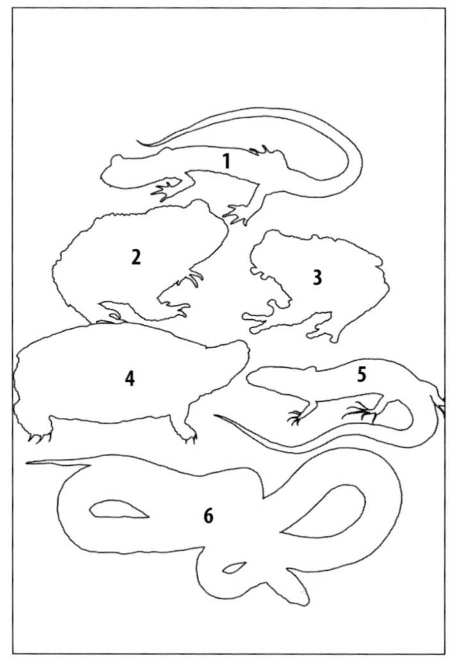

Front Cover:
A sampling of Missouri species;
watercolors by Tom R. Johnson

1. Cave Salamander (*Eurycea lucifuga*)
2. American Toad (*Anaxyrus americanus*)
3. Gray Treefrog (*Hyla versicolor*)
4. Northern Map Turtle (*Graptemys geographica*)
5. Common Five-lined Skink (*Plestiodon fasciatus*)
6. Prairie Kingsnake (*Lampropeltis calligaster*)

The Amphibians and Reptiles of Missouri

by Jeffrey T. Briggler and Tom R. Johnson

Revised and Expanded Third Edition

Edited by Larry Archer

Designed by Les Fortenberry

Illustrations by Tom R. Johnson

ISBN 978-1-73229442-4

Published by the Missouri Department of Conservation
P.O. Box 180, Jefferson City, MO 65102-0180

Contents

(continued)

(contents continued)

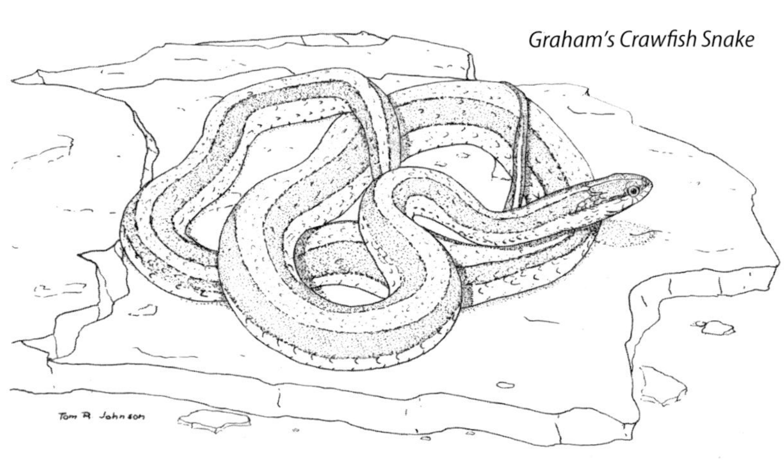

Graham's Crawfish Snake

Dedication

As senior author of the third edition of *The Amphibians and Reptiles of Missouri*, I whole-heartedly dedicate this book to Douglas A. James, PhD (1925–2018), University of Arkansas, Fayetteville. Dr. James contributed greatly to the field of ornithology and was committed to the conservation of many flora and fauna species, including herpetofauna. He was an outstanding naturalist, advisor, and mentor for my graduate studies, and a friend.

—Jeffrey T. Briggler

As co-author of this edition, I respectfully submit my dedication to Julius Hurter Sr. (1842–1916) and Paul Anderson (1914–1962). Both men contributed a great deal of knowledge about Missouri's herpetofauna through their tireless statewide fieldwork, observations, studies, and publications.

—Tom R. Johnson

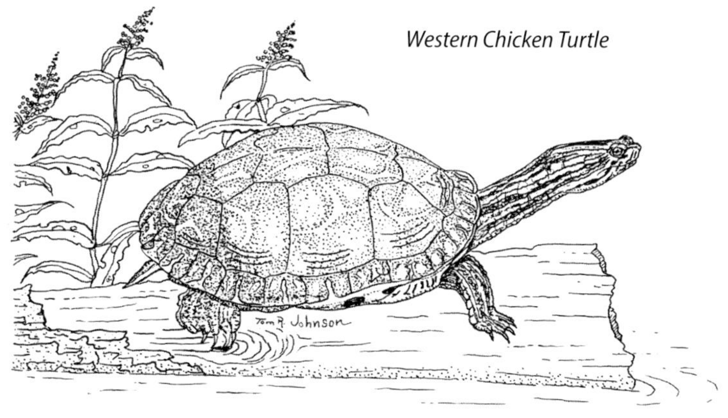

Western Chicken Turtle

Acknowledgments

We wish to thank our colleagues and supervisors in the Missouri Department of Conservation for their support throughout our years together. They have provided a professional atmosphere, continuous support, encouragement, and interest in this project.

This book was greatly improved by the expertise of the editor, Larry Archer, and designer, Les Fortenberry. Larry carefully edited and ensured consistency of the text to reduce the number of errors as humanly possible. Thank you, Larry, for a very readable and accurate publication. A special thanks to Les for an improved book layout from the attractive cover to the colorful and updated species accounts. We would also like to thank Nichole LeClair Terrill, Cliff White, and Susan Ferber for their contributions to the publication process. Finally, we would like to thank Phillip Pitts Sr. for his numerous contributions, from initial and final proofreading to reviewing of the literature citations.

Updated county records were generously supplied by Richard Daniel and Brian Edmond of the Missouri Herpetological Atlas Project. Many of these county records were provided by numerous biologists, researchers, and interested citizens through the years. Such records, published in the Missouri Herpetological Association Newsletter, were compiled by Robert Powell prior to 1986 and by Richard Daniel and Brian Edmond to 2020. Their valuable assistance is most appreciated and added greatly to the clarity and accuracy of this book.

Numerous people have helped us with various aspects of the herpetology of Missouri over the years, and it is inevitable that we have forgotten some of them. This by no means diminishes our gratitude but simply indicates a weakness of the human mind. We offer thanks to the following individuals who helped us in a variety of capacities, from providing literature, photographs, and natural history data to field assistance and many other kindnesses (in alphabetical order): Scott Ballard, Melissa Bax, Kurt A. Buhlmann, Dorothy Butler, Beth Churchwell, Shelly Colatskie, Charles J. Cole, Roger Conant, Trisha Crabill, Richard Daniel, Robert DiStefano, Lauren E. Brown, Francis Durbian, David A. Easterla, Brian Edmond, Gerry Emich, Jeff Ettling, Cynthia A. Evans, Richard S. Funk, Carl Gerhardt, Brian Greene, Karen Goellner, Melanie and Tyler Green, Jim W. Grace, A.J. Hendershott, John Hess, Charles H. Hoessle, Scott House, Stan Hudson, Clay Jensen, Michael Jones, Donald A. Kangas, Arnold Kluge, Robert L. Krager, Jeffrey W. Lang, Louise Langbein, Day Ligon, Glenn Manning, Alicia Mathis, John Miller, Julie Miller, Mark Mills, Don Moll, Tom Nagel, Max. A. Nickerson, Krista Noel, D. Novinger, Mark Pelton, Christopher A. Phillips, Katherine and Adam Pogatshnik, Robert Powell, Bill Resetarits, Daren Riedle, Rhonda Rimer, Larry Rizzo, Sharon M. Sanborn, Bruce Schuette, Richard A. Seigel, Raymond D. Semlitsch, Owen J. Sexton, G. Silverton, Donald D. Smith, Janet Sternburg, Michael Sutton, Richard H. Thom, Stanley E. Trauth, David Urich, Brandon Vandalsem, Mark Wanner, Claire and Tony Weiss, Darrin Welchert, and Robert F. Wilkinson, Jr.

The senior author would like to thank his friends and family, especially his wife, Malissa Briggler, for their support, encouragement, and understanding of completing such an endeavor. Finally, we wish to express our sincere gratitude to the Missouri Department of Conservation for publishing this third edition of *The Amphibians and Reptiles of Missouri.*

Introduction

This book is about the amphibians and reptiles occurring in Missouri. The study of these creatures is known as herpetology, which is a branch of zoology that specializes in both amphibians and reptiles. The term herpetology (Greek: herpeto- "animals that creep" and -logos "to study") means the study of animals that "creep." However, amphibians and reptiles are in separate animal classes: Amphibians are in the class Amphibia, and reptiles are in the class Reptilia. The two classes were first described together as a group by an Englishman, John Ray, in a systematic work called *Synopsis Methodica Animalium Quadrupedum et Serpentini Generis* (1693). He combined the two classes of animals because of their similarity of structures of the heart. Ray also was the first author to use the term "species" in a biological publication. As a group, the salamanders, toads, frogs, turtles, lizards, and snakes are the least known and understood of all the vertebrate animals in this region. Few Missourians come in direct contact with most of these animals: Amphibians and reptiles have secretive habits and are inactive during the winter. Fear and myth still surround many amphibians and reptiles, due largely to a lack of knowledge about them. We hope this book will reduce negative attitudes and replace them with understanding, tolerance, and an appreciation of amphibians and reptiles as a natural part of outdoor Missouri. An informed public is vital to a successful wildlife conservation program, and we hope that a variety of individuals—from the casual amateur naturalist to the serious student and professional biologist—will benefit from this book. Scientists recognize more than 7,300 species of amphibians and about 10,700 species of reptiles living on earth, with approximately 600 of these occurring within the continental United States. At present, 116 species (with an additional 11 subspecies or geographic races) of amphibians and reptiles are native to Missouri. In addition, two species of non-native lizards have become established in Missouri. Of this total, only six species (venomous snakes) are considered dangerous to humans. Most of Missouri's amphibians and reptiles are benign creatures of little direct economic consequence to people but add immeasurably to the complexity, diversity, and esthetics of this region.

Southern Leopard Frog

A Brief History of Missouri Herpetology

***Julius Hurter, Sr.** (1842–1916). Photo provided by Ron Goellner, St. Louis Zoological Park, from the photo collection of the St. Louis Naturalist Club.*

***Paul Anderson** (1914–1962). Photo by Bert W. Landfried.*

The first major contributions to the knowledge of Missouri's amphibians and reptiles were made by Julius Hurter Sr. (1842–1916), a Swiss-born engineer. Around the turn of the 20th century, Hurter published several papers and one book about Missouri amphibians and reptiles (see literature cited).

Hurter was born in Schaffhausen, a city located along the Rhine River in extreme northern Switzerland. Upon graduation from high school in 1858, he became a millwright apprentice and later a draftsman in Kriens, a small city in central Switzerland. He spent two years as a mechanical engineer in Paris, then emigrated to the United States in 1865. He lived briefly in Minnesota and Chicago before settling in St. Louis in 1866. He became a mechanical engineer and chief draftsman at a foundry and remained there until 1906. Although his vocation was mechanical engineering, Hurter had a strong avocation for the study of natural history. He spent many days observing and collecting Missouri birds (the study skins he gathered are housed at the St. Louis Science Center).

Hurter's fieldwork eventually led him to observe and gather amphibians and reptiles in many areas of Missouri. Many of his specimens are in the permanent collection at the National Museum of Natural History in Washington, D.C. Hurter's book, *Herpetology of Missouri* (1911), was the only major reference available on Missouri amphibians and reptiles for decades. The Hurter's spadefoot, *Scaphiopus hurterii*, was named in his honor (Conant 1991). Hurter died in St. Louis on December 6, 1916, at the age of 74.

Paul Anderson (1914–1962) was another contributor to the knowledge of Missouri's amphibians and reptiles, and he also lacked a formal education in herpetology. A resident of Independence, Anderson had a lifelong interest in reptiles; he was self-taught and became recognized as an authority on Missouri herpetofauna by many professional herpetologists. He worked at a petroleum refinery for 28 years but spent his free time collecting, preserving, and cataloging thousands of Missouri amphibians and reptiles.

Anderson wrote 19 technical papers, many of them published by the Chicago Academy of Science, and corresponded with many professional herpetologists and several budding zoologists. While in his 20s, he became interested in writing a book on the herpetology of Missouri; most of the specimens he collected and observed in captivity were for the book.

Anderson was assisted in the field by Philip Evans. Howard K. Gloyd, while director of the Chicago Academy of Science, was a valuable source of information and encouragement. Anderson's wife, Zelma, typed the manuscript for his Missouri book. Anderson's untimely death in 1962 at the age of 48 came when the reptile section was nearly ready to go to press. The University of Missouri-Columbia published it as *The Reptiles of Missouri* in 1965. Anderson's storehouse of information on Missouri amphibians, however, remained in his field notes. A major part of Anderson's preserved amphibian and reptile collection was donated to the University of Kansas Museum of Natural History after his death. The Chicago Academy of Science gained many specimens over the years through its close association with Anderson.

Tom R. Johnson was another major contributor to the herpetofauna of Missouri. He grew up in central Wisconsin and received his Bachelor of Science degree in zoology from the University of Wisconsin-Stevens Point in 1970. He moved to St. Louis and became a reptile keeper at the St. Louis Zoo in 1973. Johnson wrote his first major publication, *The Amphibians of Missouri*, in his spare time. *The Amphibians of Missouri* was illustrated with Johnson's black and white photographs, as well as numerous pen and ink drawings, and was published in 1977 by the University of Kansas Museum of Natural History. This work took four years to complete and was published shortly after Johnson departed the St. Louis Zoo in 1977 to become the Missouri Department of Conservation's (MDC) first state herpetologist.

In 1987, MDC published Johnson's *The Amphibians and Reptiles of Missouri*. A revision of his amphibian book, *The Amphibians and Reptiles of Missouri* included new information on Missouri's native reptiles and was the first work on Missouri's herpetofauna to be illustrated with color photographs. It also featured numerous pen and ink drawings by Johnson. After serving 23 years as the state herpetologist, Johnson retired in 1998 but continued to work for MDC two additional years to complete the second edition.

MDC published the second edition of *The Amphibians and Reptiles of Missouri* in 2000. This edition provided a thorough update on the knowledge and taxonomy of amphibians and reptiles in Missouri. Like the 1987 edition, the second edition was illustrated with color photographs, numerous drawings, and watercolor paintings by Johnson. In addition to his three books on the herpetofauna of Missouri, Johnson's photographs and illustrations have appeared in numerous publications, including the third edition of *A Field Guide to Reptiles and Amphibians of Eastern and Central North America* by Roger Conant and Joseph T. Collins (1991).

Johnson, along with Ron Goellner and Robert Powell, founded the Missouri Herpetological Association in 1988. The organization's focus is to share herpetological information among academics and conservation-minded individuals in Missouri. This organization publishes an annual newsletter that includes abstracts of herpetology research, natural history notes, county records, and other herpetology related information (**mha.moherp.org**). Major contributors to organizing annual meetings, archiving county records, and editing the annual newsletters from 1988 to present include Jeffrey T. Briggler, Richard Daniel, Brian Edmond, Tom R. Johnson, and Robert Powell.

In 1997, Richard Daniel and Brian Edmond initiated the Missouri Herpetological Atlas Project, making herpetofauna records and detailed distribution maps for Missouri amphibians and reptiles easily updated and accessible online (**www.atlas.moherp.org**).

Natural Divisions of Missouri

This information is furnished to familiarize the reader with the natural divisions or ecological units of the state. These physical features strongly influence the kinds of plants and animals found in Missouri (Nelson 2005; Nigh and Schroeder 2002).

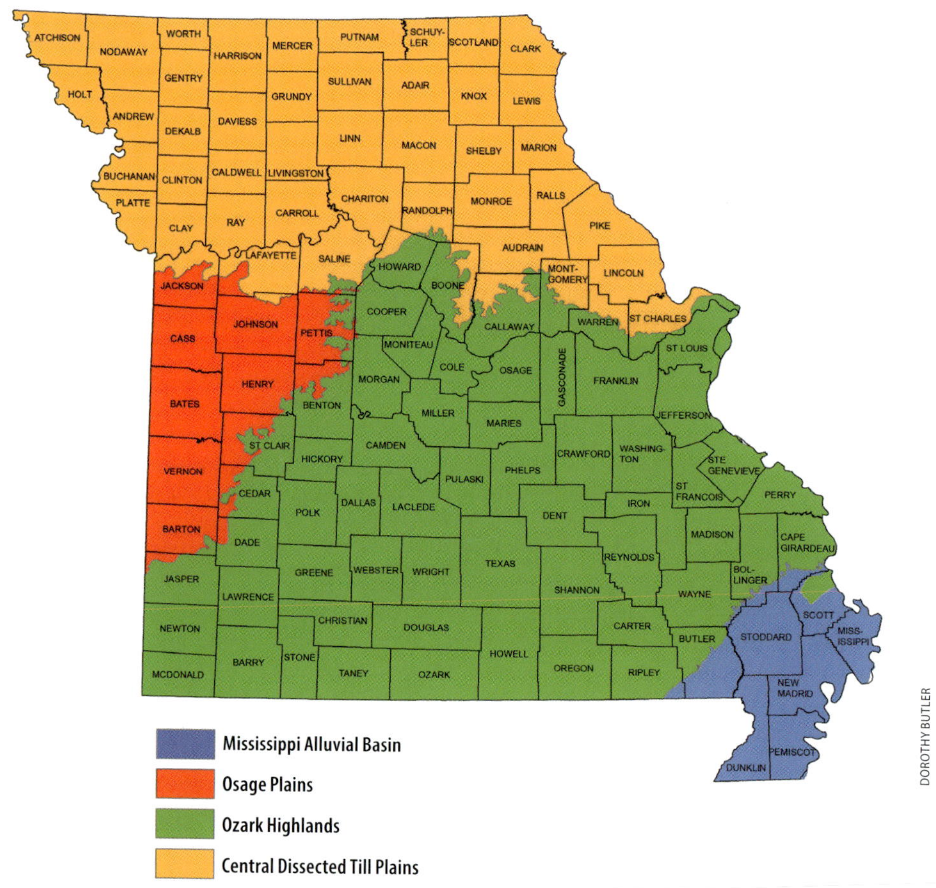

NOPPADOL PAOTHONG

Open rolling hills are characteristic of the Central Dissected Till Plains in northern Missouri.

Central Dissected Till Plains

The physical features of this region reflect the southernmost movement and eventual retreat of the Kansan period of glaciation, which took place about 400,000 years ago. This ecological unit was formerly known as the Glacial Plains (Thom and Wilson 1980). The rolling hills and broad, flat valleys were caused by the leveling effect of the glaciers. Soils developed from glacial till and loess deposits. Marshes, native prairie, and upland deciduous forests were dominant plant communities prior to European settlement. Most of the region has been altered by agriculture. The Lincoln Hills along the Mississippi River have limestone bluffs and steep hills of till or loess soils, which likely escaped glaciation. Mean annual precipitation ranges from 86 to 99 cm (34 to 39 in.; Nigh and Schroeder 2002). Some of the amphibians associated with the Central Dissected Till Plains are the eastern tiger salamander (*Ambystoma tigrinum*), Great Plains toad (*Anaxyrus cognatus*), plains spadefoot (*Spea bombifrons*), northern crawfish frog (*Lithobates areolatus circulosus*) and plains leopard frog (*Lithobates blairi*). Associated reptiles include the yellow mud turtle (*Kinosternon flavescens*), western painted turtle (*Chrysemys picta bellii*), Blanding's turtle (*Emydoidea blandingii*), Kirtland's snake (*Clonophis kirtlandii*), eastern foxsnake (*Pantherophis vulpinus*), western foxsnake (*Pantherophis ramspotti*), plains gartersnake (*Thamnophis radix*) and prairie massasauga (*Sistrurus tergeminus tergeminus*).

JIM RATHERT

Deciduous forest, rough terrain, and river valleys are characteristic of the Ozark Highlands.

Ozark Highlands

This unit was formerly recognized as both the Ozarks and Ozark Border (Thom and Wilson 1980). This region is characterized by rough terrain, thin soils, limestone bluffs, steep-sided sandstone canyons, springs, caves, fens, clear streams, and rocky glades. The Ozark Highlands make up about 45 percent of the state. Most of the forest cover is oak and hickory with some maple. Shortleaf pine is a common tree species in the southeastern section of the region. The eastern section of the Missouri Ozarks has Precambrian (igneous) bedrock and steep hills that are known as the St. Francois Mountains. This area has the highest elevation point in the state—Taum Sauk Mountain at 1,772 feet above sea level. The forest products industry has been important here, and the U.S. Forest Service manages several large areas. Major sections of the Ozark forest, however, have been cleared and seeded in cool-season grasses for pasture. The average annual precipitation ranges from 102 to 122 cm (40 to 48 in.; Nigh and Schroeder 2002). Amphibian and reptile species indicative of this region include the eastern hellbender (*Cryptobranchus alleganiensis alleganiensis*), Ozark hellbender (*Cryptobranchus a. bishopi*), ringed salamander (*Ambystoma annulatum*), spotted salamander (*Ambystoma maculatum*), grotto salamander (*Eurycea spelaea*), Ozark zigzag salamander (*Plethodon angusticlavius*), pickerel frog (*Lithobates palustris*), wood frog (*Lithobates sylvaticus*), northern map turtle (*Graptemys geographica*), three-toed box turtle (*Terrapene carolina triunguis*), eastern collared lizard (*Crotaphytus collaris*), southern coal skink (*Plestiodon anthracinus pluvialis*), broad-headed skink (*Plestiodon laticeps*), rough earthsnake (*Haldea striatula*), red-bellied snake (*Storeria occipitomaculata*), and western pygmy rattlesnake (*Sistrurus miliarius streckeri*).

JIM RATHERT

Most of the state's remaining native tallgrass prairies are located in the Osage Plains region of west-central Missouri.

Osage Plains

This unglaciated region in the west-central section of Missouri is characterized by rolling hills, prairie streams, and tallgrass prairie plants. Most of Missouri's last native tallgrass prairie is in this area. The average annual precipitation ranges from 99 to 107 cm (39 to 42 in.; Nigh and Schroeder 2002). Amphibians in this region include the boreal chorus frog (*Pseudacris maculata*), northern crawfish frog (*Lithobates areolatus circulosus*), and southern leopard frog (*Lithobates sphenocephalus*). Reptiles associated with the Osage Plains are the ornate box turtle (*Terrapene ornata*), Great Plains skink (*Plestiodon obsoletus*), western slender glass lizard (*Ophisaurus attenuatus attenuatus*), prairie kingsnake (*Lampropeltis calligaster*), bullsnake (*Pituophis catenifer sayi*), and plain-bellied watersnake (*Nerodia erythrogaster*).

Native cypress swamps of southeastern Missouri provide important habitat for many amphibians and reptiles but most have been eliminated.

Mississippi Alluvial Basin

This region was formerly recognized as the Mississippi Lowlands (Thom and Wilson 1980). The Bootheel region, located in Missouri's southeastern corner, is mostly a broad, flat alluvial plain. The area has an average elevation of 325 feet above sea level, and the state's the lowest point (230 feet above sea level) is located in Dunklin County. A ridge of low hills of sand, gravel, and bedrock—known as Crowley's Ridge—runs diagonally across the lowlands from northern Dunklin County to northern Scott County. Most of the area was bald cypress, tupelo swamp, and bottomland hardwood forest prior to European settlement. Also, there were sand prairies and savannas on terraces. Extensive lumbering and drainage programs during the last century have converted most of the area to agriculture. Average annual precipitation ranges from 122 to 132 cm (48 to 52 in.; Nigh and Schroeder 2002). Some amphibians and reptiles typical in this region include mole salamander (*Ambystoma talpoideum*), three-toed amphiuma (*Amphiuma tridactylum*), eastern spadefoot (*Scaphiopus holbrookii*), green treefrog (*Hyla cinerea*), Illinois chorus frog (*Pseudacris illinoensis*), Mississippi mud turtle (*Kinosternon subrubrum hippocrepis*), southern painted turtle (*Chrysemys dorsalis*), western mudsnake (*Farancia abacura reinwardtii*), dusty hog-nosed snake (*Heterodon gloydi*), broad-banded watersnake (*Nerodia fasciata confluens*), and northern cottonmouth (*Agkistrodon piscivorus*).

Explanation of Family and Species Accounts

Accounts

The 22 families of amphibians and reptiles native to Missouri and two additional families for established nonnative species are presented in the phylogeny sequence used by Pough et al. (2018). These accounts are furnished to acquaint the reader with the general characteristics common to each family. Some species that occur outside of Missouri are discussed to give the reader an idea of the relationships and distributions of amphibians and reptiles on a broader scale. A typical species from each family is illustrated at the beginning of each family account.

Common and Scientific Names

The common and scientific names used throughout this book are those currently accepted by most authorities. These names follow the recommended usage by Crother (2017) in *Scientific and Standard English Names of Amphibians and Reptiles of North America North of Mexico*, with *Comments Regarding Confidence in Our Understanding*, published by the Society for the Study of Amphibians and Reptiles. If there has been a recent change in any name, a comment to that effect will be in the remarks section of the species account. To help standardize amphibian and reptile nomenclature, usage of the names contained in this book is recommended. Although changes in taxonomy are often controversial, we agree with the recommendation of Frost et al. (1992) that a regional field guide to amphibians and reptiles, such as this book, is not the proper place to argue for or against taxonomic issues. The genera in each family are arranged in alphabetical order, and the species are arranged in alphabetical order within each genus, except for the placement of the earthsnakes and massasaugas. We chose to place the rough earthsnake (*Haldea striatula*) next to the western smooth earthsnake (*Virginia valeriae elegans*) for ease of comparison. For the same reason, we place eastern massasauga (*Sistrurus catenatus*) next to the prairie massassauga (*Sistrurus tergeminus tergeminus*).

For an understanding of the scientific naming of animals, it is helpful to refer to a biology or general zoology textbook.

Description

The description of species outlines characteristics to help identify each species; it includes general shape, a brief description of color and pattern and, when appropriate, a discussion of external characteristics that help separate the sexes. The size of each species is given in both metric and English units. Total length is given for salamanders, lizards, and snakes (i.e., length from snout to end of tail). Sizes of toads and frogs are given in snout-to-vent length (i.e., from the tip of the nose to the vent). Turtle sizes are of the carapace or upper shell length. Maximum sizes of United States specimens are those reported by Powell et al. (2016) and other authors and are not necessarily the maximum lengths that these animals are capable of reaching in the wild.

Habits and Habitat

Aspects of the natural history of each species are discussed. These include habitat preferences, daily and seasonal activity, food and specific behavioral traits.

Breeding

The reproductive biology of Missouri's amphibians and reptiles is discussed in this section: courtship, mating, egg-laying, descriptions of eggs or newborn young, and births or hatching. Also included are the duration of the breeding season and, in the case of toads and frogs, a description of their breeding calls. Much of this information has been gleaned from published papers in technical journals, gray literature papers (reports, magazine articles, etc.), and personal observation and unpublished data of the authors.

Subspecies

This section is offered only in those species accounts where more than one subspecies or geographic race occurs in Missouri. Descriptions used to distinguish the subspecies are presented.

Remarks

This section provides special information such as taxonomic changes or problems, hybridization, rare or endangered species status, population status, or comments on unusual behavior.

Distribution and Maps

Each species account has a brief description of the species' range in Missouri. A map of Missouri showing county lines illustrates the presumed range of each species and/or subspecies (horizontal and diagonal lines) and the known county records (black dots). Sections of the state where subspecies are thought to intergrade are shown by overlapping shading. An open circle in a county indicates a field observation that was published in the second edition, but with no available museum voucher specimen. Many county records and observations were published in the Missouri Herpetological Association Newsletter and are updated annually in the Missouri Herpetological Atlas Project. The general distribution of each species outside of Missouri is also furnished.

Photographs

Photographs have been selected to show each animal in a natural setting while clearly illustrating individual characteristics. Each species account has one photograph to assist in identification, plus one or more photographs to illustrate a subspecies, show a calling male toad or frog, eggs or larvae, hatchlings, habitat, etc. Photo credit is provided for each image.

For readers who wish to learn about the distribution of amphibians and reptiles outside the confines of Missouri, we suggest referring to *A Field Guide to Reptiles and Amphibians of Eastern and Central North America* by Robert Powell, Roger Conant, and Joseph T. Collins (2016).

Checklist of Missouri Amphibians and Reptiles

Class Amphibia

Order Caudata – Salamander

Family Cryptobranchidae – Hellbenders and Giant Salamanders

Eastern Hellbender .... *Cryptobranchus alleganiensis alleganiensis* (Daudin)
Ozark Hellbender .... *Cryptobranchus alleganiensis bishopi* Grobman

Family Sirenidae – Sirens

Western Lesser Siren .... *Siren intermedia nettingi* Goin

Family Salamandridae – Newts

Central Newt .... *Notophthalmus viridescens louisianensis* (Wolterstorff)

Family Ambystomatidae – Mole Salamanders

Ringed Salamander .... *Ambystoma annulatum* Cope
Spotted Salamander .... *Ambystoma maculatum* (Shaw)
Marbled Salamander .... *Ambystoma opacum* (Gravenhorst)
Mole Salamander .... *Ambystoma talpoideum* (Holbrook)
Small-mouthed Salamander .... *Ambystoma texanum* (Matthes)
Eastern Tiger Salamander .... *Ambystoma tigrinum* (Green)

Family Proteidae – Mudpuppies and Olm

Common Mudpuppy .... *Necturus maculosus maculosus* (Rafinesque)
Red River Mudpuppy .... *Necturus maculosus louisianensis* Viosca

Family Amphiumidae – Amphiumas

Three-toed Amphiuma .... *Amphiuma tridactylum* Cuvier

Family Plethodontidae – Lungless Salamanders

Eastern Long-tailed Salamander .... *Eurycea longicauda longicauda* (Green)
Dark-sided Salamander .... *Eurycea longicauda melanopleura* (Cope)
Cave Salamander .... *Eurycea lucifuga* Rafinesque
Grotto Salamander .... *Eurycea spelaea* Stejneger
Oklahoma Salamander .... *Eurycea tynerensis* Moore and Hughes
Four-toed Salamander .... *Hemidactylium scutatum* (Temminck and Schlegel *in* Von Siebold)
Western Slimy Salamander .... *Plethodon albagula* Grobman
Ozark Zigzag Salamander .... *Plethodon angusticlavius* Grobman
Southern Red-backed Salamander .... *Plethodon serratus* Grobman

Order Anura – Toads and Frogs

Family Scaphiopodidae – North American Spadefoots

Eastern Spadefoot .... *Scaphiopus holbrookii* (Harlan)
Plains Spadefoot .... *Spea bombifrons* (Cope)

Family Bufonidae – True Toads

Eastern American Toad .... *Anaxyrus americanus americanus* (Holbrook)
Dwarf American Toad .... *Anaxyrus americanus charlesmithi* (Bragg)
Great Plains Toad .... *Anaxyrus cognatus* (Say *in* James)
Fowler's Toad .... *Anaxyrus fowleri* (Hinckley)
Rocky Mountain Toad .... *Anaxyrus woodhousii woodhousii* Girard

Family Hylidae – Cricket Frogs, Treefrogs, and Chorus Frogs

Blanchard's Cricket Frog .... *Acris blanchardi* Harper
Cope's Gray Treefrog .... *Hyla chrysoscelis* Cope
Green Treefrog .... *Hyla cinerea* (Schneider)
Gray Treefrog .... *Hyla versicolor* LeConte
Spring Peeper .... *Pseudacris crucifer* (Wied-Neuwied)
Upland Chorus Frog .... *Pseudacris feriarum* (Baird)
Cajun Chorus Frog .... *Pseudacris fouquettei* Lemmon, Lemmon, Collins, and Cannatella
Illinois Chorus Frog .... *Pseudacris illinoensis* Smith
Boreal Chorus Frog .... *Pseudacris maculata* (Agassiz)

Family Microhylidae – Narrow-mouthed Toads

Eastern Narrow-mouthed Toad .... *Gastrophryne carolinensis* (Holbrook)
Western Narrow-mouthed Toad .... *Gastrophryne olivacea* (Hallowell)

Family Ranidae – True Frogs

Northern Crawfish Frog .... *Lithobates areolatus circulosus* (Rice and Davis)
Plains Leopard Frog .... *Lithobates blairi* (Mecham, Littlejohn, Oldham, Brown, and Brown)
American Bullfrog .... *Lithobates catesbeianus* (Shaw)
Green Frog .... *Lithobates clamitans* (Latreille)
Pickerel Frog .... *Lithobates palustris* (LeConte)
Northern Leopard Frog .... *Lithobates pipiens* (Schreber)
Southern Leopard Frog .... *Lithobates sphenocephalus* (Cope)
Wood Frog .... *Lithobates sylvaticus* (LeConte)

Class Reptilia

Order Testudines – Turtles

Family Chelydridae – Snapping Turtles
Eastern Snapping Turtle .... *Chelydra serpentina* (Linnaeus)
Alligator Snapping Turtle .... *Macrochelys temminckii* (Harlan)

Family Kinosternidae – American Mud and Musk Turtles
Yellow Mud Turtle .... *Kinosternon flavescens* (Agassiz)
Mississippi Mud Turtle .... *Kinosternon subrubrum hippocrepis* Gray
Eastern Musk Turtle .... *Sternotherus odoratus* (Latreille *in* Sonnini and Latreille)

Family Emydidae – Pond and Box Turtles
Southern Painted Turtle .... *Chrysemys dorsalis* Agassiz
Western Painted Turtle .... *Chrysemys picta bellii* (Gray)
Western Chicken Turtle .... *Deirochelys reticularia miaria* Schwartz
Blanding's Turtle .... *Emydoidea blandingii* (Holbrook)
Northern Map Turtle .... *Graptemys geographica* (LeSueur)
Ouachita Map Turtle .... *Graptemys ouachitensis* Cagle
Mississippi Map Turtle .... *Graptemys pseudogeographica kohnii* (Baur)
Northern False Map Turtle .... *Graptemys pseudogeographica pseudogeographica* (Gray)
Eastern River Cooter .... *Pseudemys concinna concinna* (LeConte)
Three-toed Box Turtle .... *Terrapene carolina triunguis* (Agassiz)
Ornate Box Turtle .... *Terrapene ornata* (Agassiz)
Red-eared Slider .... *Trachemys scripta elegans* (Wied-Neuwied)

Family Trionychidae – Softshells
Midland Smooth Softshell .... *Apalone mutica mutica* (LeSueur)
Eastern Spiny Softshell .... *Apalone spinifera spinifera* (LeSueur)

Order Squamata – Lizards and Snakes

Suborder Sauria – Lizards

Family Geckkonidae – True Geckos
Mediterranean Gecko .... *Hemidactylus turcicus* (Linnaeus) [Introduced]

Family Crotaphytidae – Collared and Leopard Lizards
Eastern Collared Lizard .... *Crotaphytus collaris* (Say *in* James)

Family Phrynosomatidae – Spiny and Sand Lizards
Texas Horned Lizard .... *Phrynosoma cornutum* (Harlan)
Prairie Lizard .... *Sceloporus consobrinus* Baird and Girard

Family Scincidae – Skinks
Southern Coal Skink .... *Plestiodon anthracinus pluvialis* (Cope)
Common Five-lined Skink .... *Plestiodon fasciatus* (Linnaeus)
Broad-headed Skink .... *Plestiodon laticeps* (Schneider)
Great Plains Skink .... *Plestiodon obsoletus* Baird and Girard
Southern Prairie Skink .... *Plestiodon septentrionalis obtusirostris* (Bocourt)
Northern Prairie Skink .... *Plestiodon septentrionalis septentrionalis* Baird
Little Brown Skink .... *Scincella lateralis* (Say *in* James)

Family Teiidae – Racerunners, Whiptails, and their Relatives

Eastern Six-lined Racerunner .... *Aspidoscelis sexlineata sexlineata* (Linnaeus)
Prairie Racerunner .... *Aspidoscelis sexlineata viridis* (Lowe)

Family Lacertidae – Wall Lizards and Relatives

Italian Wall Lizard .... *Podarcis siculus* (Rafinesque) [Introduced]

Family Anguidae – Glass and Alligator Lizards, and their Relatives

Western Slender Glass Lizard .... *Ophisaurus attenuatus attenuatus* Cope

Suborder Serpentes – Snakes

Family Colubridae – Typical Snakes

Western Wormsnake .... *Carphophis vermis* (Kennicott)
Northern Scarletsnake .... *Cemophora coccinea copei* Jan
Kirtland's Snake .... *Clonophis kirtlandii* (Kennicott)
Eastern Yellow-bellied Racer .... *Coluber constrictor flaviventris* Say
Southern Black Racer .... *Coluber constrictor priapus* Dunn and Wood
Eastern Coachwhip .... *Coluber flagellum flagellum* Shaw
Prairie Ring-necked Snake .... *Diadophis punctatus arnyi* Kennicott
Mississippi Ring-necked Snake .... *Diadophis punctatus stictogenys* Cope
Western Mudsnake .... *Farancia abacura reinwardtii* Schlegel
Dusty Hog-nosed Snake .... *Heterodon gloydi* Edgren
Plains Hog-nosed Snake .... *Heterodon nasicus* Baird and Girard
Eastern Hog-nosed Snake .... *Heterodon platirhinos* Latreille
Prairie Kingsnake .... *Lampropeltis calligaster* (Harlan)
Speckled Kingsnake .... *Lampropeltis holbrooki* Stejneger
Eastern Black Kingsnake .... *Lampropeltis nigra* (Yarrow)
Eastern Milksnake .... *Lampropeltis triangulum* (Lacépède)
Mississippi Green Watersnake .... *Nerodia cyclopion* (Duméril, Bibron and Duméril)
Plain-bellied Watersnake .... *Nerodia erythrogaster* (Forster)
Broad-banded Watersnake .... *Nerodia fasciata confluens* (Blanchard)
Northern Diamond-backed Watersnake .... *Nerodia rhombifer rhombifer* (Hallowell)
Midland Watersnake .... *Nerodia sipedon pleuralis* (Cope)
Northern Watersnake .... *Nerodia sipedon sipedon* (Linnaeus)
Northern Rough Greensnake .... *Opheodrys aestivus aestivus* (Linnaeus)
Smooth Greensnake .... *Opheodrys vernalis* (Harlan)
Great Plains Ratsnake .... *Pantherophis emoryi* (Baird and Girard)
Western Ratsnake .... *Pantherophis obsoletus* (Say)
Western Foxsnake .... *Pantherophis ramspotti*
Crother, White, Savage, Eckstut, Graham, and Gardner
Eastern Foxsnake .... *Pantherophis vulpinus* (Baird and Girard)
Bullsnake .... *Pituophis catenifer sayi* (Schlegel)
Graham's Crawfish Snake .... *Regina grahamii* Baird and Girard
Variable Groundsnake .... *Sonora semiannulata semiannulata* Baird and Girard
Dekay's Brownsnake .... *Storeria dekayi* (Holbrook)
Red-bellied Snake .... *Storeria occipitomaculata* (Storer)
Flat-headed Snake .... *Tantilla gracilis* Baird and Girard
Orange-striped Ribbonsnake .... *Thamnophis proximus proximus* (Say)
Plains Gartersnake .... *Thamnophis radix* (Baird and Girard)
Red-sided Gartersnake .... *Thamnophis sirtalis parietalis* (Say)

Eastern Gartersnake .... *Thamnophis sirtalis sirtalis* (Linnaeus)
Lined Snake .... *Tropidoclonion lineatum* (Hallowell)
Rough Earthsnake .... *Haldea striatula* (Linnaeus)
Western Smooth Earthsnake .... *Virginia valeriae elegans* Kennicott

Family Viperidae – Vipers and Pit Vipers [Venomous]
Eastern Copperhead .... *Agkistrodon contortrix* (Linnaeus)
Northern Cottonmouth .... *Agkistrodon piscivorus* (Lacépède)
Timber Rattlesnake .... *Crotalus horridus* Linnaeus
Eastern Massasauga .... *Sistrurus catenatus* (Rafinesque)
Prairie Massasauga .... *Sistrurus tergeminus tergeminus* (Say)
Western Pygmy Rattlesnake .... *Sistrurus miliarius streckeri* Gloyd

Amphibian and Reptile Keys

How to use the keys

The illustrated keys that follow are furnished to help the reader identify Missouri's adult amphibians and reptiles. These keys have been designed to identify living animals, not preserved specimens. Tadpoles, salamander larvae, and some young adults of many Missouri species cannot be identified using these keys.

Keys used to identify plants or animals are organized into a series of numbered couplets that offer a choice of characteristics (a and b) in each couplet. By choosing the description that best fits the animal being identified, and by continuing to refer to the next couplet numbered at the end of the appropriate characteristics, the reader is eventually directed to a couplet that ends in the name of the animal. Illustrations of important characteristics are furnished and labeled. The page numbers, in parentheses at the end of the species name, refer to the page where each species account begins.

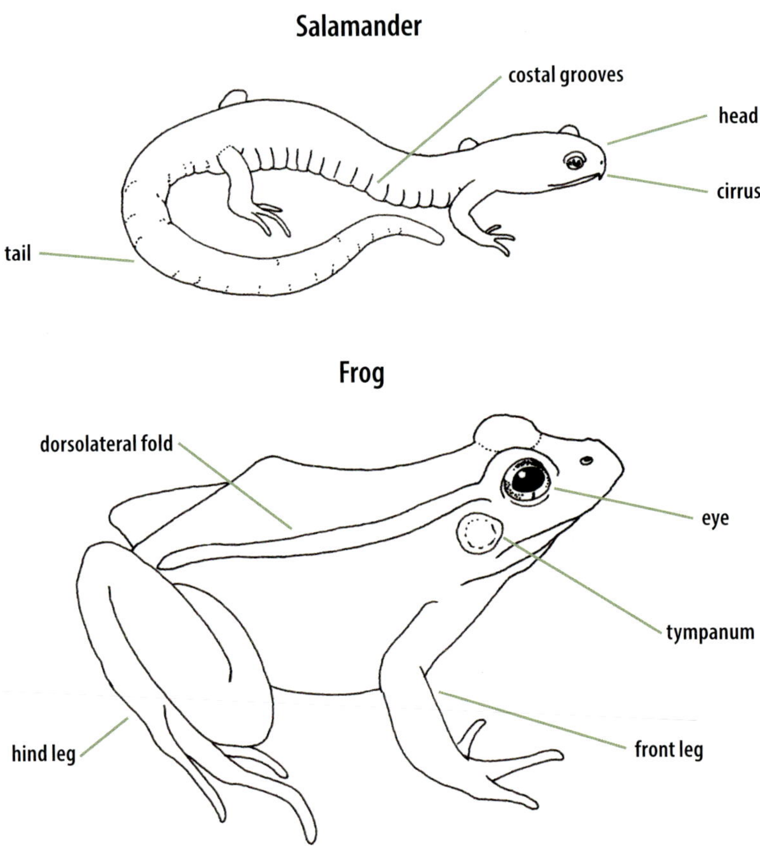

Fig. 1. *Generalized drawing of a typical salamander and frog.*

Key to the Salamanders of Missouri			
1a	Body eel-like; with or without gills; limbs reduced or hind limbs absent		2
1b	Body not eel-like; with or without gills; limbs well developed		3
2a	Only forelimbs present; with gills; tail compressed vertically; color dull brown; no opening present on either side of the rear of the head	Western Lesser Siren, *Siren intermedia nettingi* (Fig. 2A; p. 52)	
2b	Fore- and hind limbs present, tiny with three toes on each limb; no gills; tail not compressed; color dark gray brown to black; opening present on either side of the rear of head	Three-toed Amphiuma, *Amphiuma tridactylum* (Fig. 2C; p. 81)	
3a	Both fore- and hind limbs with four toes		4
3b	Forelimbs with four toes; hind limbs with five toes		5
4a	Large feathery gills; tail compressed; color tan to dull gray brown; numerous dusky or dark spots over dorsum and tail	Common Mudpuppy and Red River Mudpuppy, *Necturus maculosus* (Fig. 2B; p. 77)	
4b	No gills; tail rounded and constricted at base; white belly with black spots	Four-toed Salamander, *Hemidactylium scutatum* (p. 97)	
5a	Costal grooves absent		6
5b	Costal grooves present		8
6a	Adult size less than 150 mm (6 in.); head small, body rounded; color olive-brown dorsum with two rows of tiny red spots, belly yellow with numerous black spots; lives in ponds and swamps	Central Newt, *Notophthalmus viridescens louisianensis* (p. 55)	
6b	Adult size considerably greater than 150 mm (6 in.); head wide and flattened, body with longitudinal wrinkles; lives in Ozark rivers (Fig 2D)		7

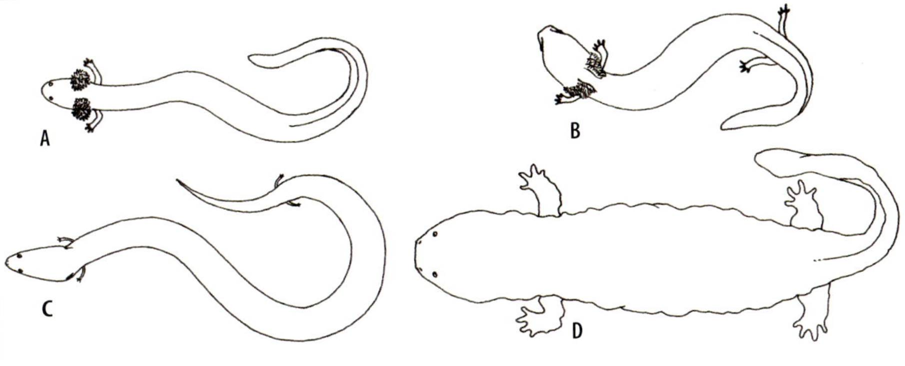

A B C D

Fig. 2. *Outline drawing of aquatic salamanders: (A) siren, (B) mudpuppy, (C) three-toed amphiuma, and (D) hellbender.*

Key to the Salamanders of Missouri			
7a	Adult size 330–580 mm (13 to 22.8 in.); color gray to reddish-brown dorsum; body uniform in color with few to no dark markings; occurs in north and northeast flowing rivers in the northern Missouri Ozarks	Eastern Hellbender, *Crytobranchus alleganiensis alleganiensis* (p. 45)	
7b	Adult size 270 to 540 mm (10.6 to 21.3 in.); color gray-brown to olive-green dorsum; body with large dark markings and blotching; occurs in south flowing rivers in southern Missouri Ozarks	Ozark Hellbender, *Cryptobranchus alleganiensis bishopi* (p. 48)	
8a	Tail thick, same length as body or shorter; naso-labial grooves absent		9
8b	Tail slender, as long or longer than body; naso-labial groove present		14
9a	Tail as long as head and body length; dorsum with grayish flecks or crossbands; costal grooves 14 to 15		10
9b	Tail as long as or shorter than head and body length; dorsum with flecks, round spots, cross bars or large blotches; costal grooves 10 to 14		11
10a	Body and tail plain or with numerous gray flecks; no rings	Small-mouthed Salamander, *Ambystoma texanum* (p. 70)	
10b	Body and tail with narrow, transverse white or yellow rings	Ringed Salamander, *Ambystoma annulatum* (p. 59)	
11a	Transverse gray or silvery bars; costal grooves 11 to 12	Marbled Salamander, *Ambystoma opacum* (p. 65)	
11b	No transverse bars; uniform dark color or with spots or blotches; costal grooves 10 to 14		12
12a	No yellow spots or blotches; with or without faded lichenlike gray flecks; costal grooves 10 to 11	Mole Salamander, *Ambystoma talpoideum* (p. 68)	
12b	Yellow spots or large yellowish-white blotches over body and tail; no lichenlike gray flecks; costal grooves 11 to 14		13
13a	Dorsum with two rows of round, yellow spots; white flecks on sides of body; costal grooves 11 to 13	Spotted Salamander, *Ambystoma maculatum* (p. 62)	
13b	Dorsum with olive-brown, yellow, or occasionally white irregular blotches; no white flecks on sides of body; costal grooves 11 to 14	Eastern Tiger Salamander, *Ambystoma tigrinum* (p. 73)	
14a	Tail long and slender; with dorsal and ventral fins; with or without gills; costal grooves 19 to 21	Oklahoma Salamander, *Eurycea tynerensis* (p. 94)	

Key to the Salamanders of Missouri			
14b	Tail rounded or slightly compressed, without fins; no gills; costal grooves 13 to 20		15
15a	Ground color brown or black, covered with flecks of white; red, orange, or yellow dorsal stripe (sometimes absent)		16
15b	Ground color tan, pink, yellow or orange; no white flecks or dorsal stripe; with or without dark spots		17
16a	Color jet black; dorsum of head, body, and tail with numerous white or yellowish-white flecks; no dorsal stripe; belly gray; costal grooves 14 to 15	Western Slimy Salamander, *Plethodon albagula* (p. 100)	
16b	Ground color brown; dorsum of body and tail with red or orange stripe; belly with black and white mottling		18
17a	Ground color light tan to pink; no markings; eyes sometimes covered with skin; tail rounded; costal grooves 16 to 19	Grotto Salamander, *Eurycea spelaea* (p. 91)	
17b	Color greenish yellow, orange or red orange; dark spots on head and body; spots or bars on tail, tail long and slender; costal grooves 13 to 14		19
18a	Dorsal stripe uniform in width, not widest at hind limbs; stripe edge serrated or saw toothed; serration points correspond to costal grooves; costal grooves 18 to 19	Southern Red-backed Salamander, *Plethodon serratus* (Fig. 3A; p. 106)	
18b	Dorsal stripe narrow and broken or lobed; stripe widest at hind limbs; costal grooves 17 to 19	Ozark Zigzag Salamander, *Plethodon angusticlavius* (Fig. 3B; p. 103)	

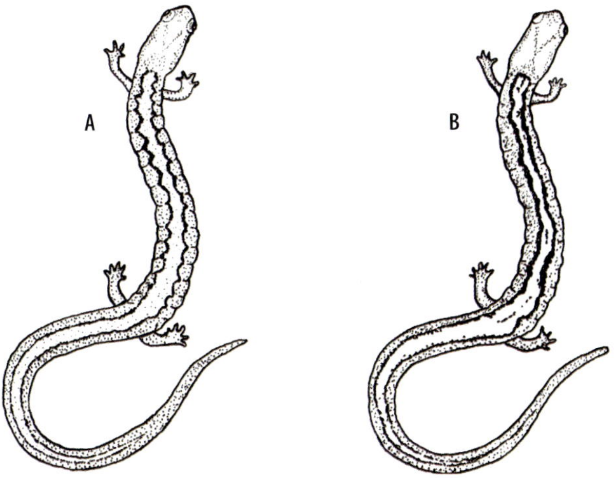

Fig. 3. *Dorsal view of red-backed salamanders: (A) southern red-backed salamander and (B) Ozark zigzag salamander.*

Key to the Salamanders of Missouri			
19a	General color greenish yellow, yellow or orange yellow; dark irregular spots on dorsum; dark markings on tail in the shape of bars or fused to form a dark lateral line on body and tail	Eastern Long-tailed Salamander and Dark-sided Salamander, *Eurycea longicauda* (p. 85)	
19b	General color bright orange or red orange; numerous small black spots on body and tail; no dark lateral line on sides of body and tail	Cave Salamander, *Eurycea lucifuga* (p. 88)	

Key to the Toads and Frogs of Missouri

1a	One or two large spades on heel of hind feet; parotoid glands behind eyes; little webbing between toes of hind feet; skin relatively dry		2
1b	No spades on heel of hind feet; no parotoid glands behind eyes; weak to strongly webbed toes on hind feet; skin relatively moist		6

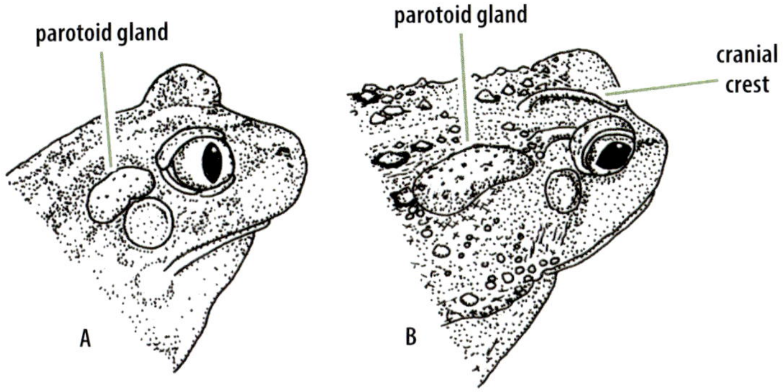

Fig. 4. *Side view of heads of (A) spadefoot and (B) true toad.*

2a	Paratoid glands small and round; eye with vertical pupil when exposed to strong light (Fig. 4A); no bony cranial crest on head; only one spade per hind foot (Fig. 5A)		3
2b	Parotoid glands large; eye with horizontal or round pupil (Fig. 4B); bony cranial crest on head; one large and one small spade per hind foot (Fig. 5B); numerous "warts"		4

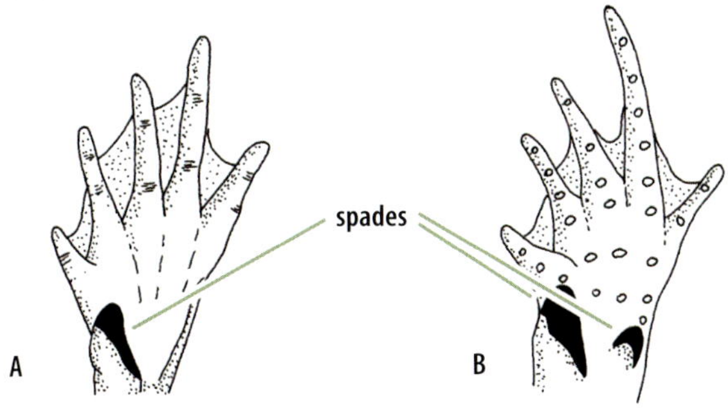

Fig. 5. *Underside of hind feet of (A) spadefoot and (B) true toad showing spades.*

3a	Pronounced boss between eyes; spade on hind feet short and rounded; general color tan to greenish tan; red "warts"	Plains Spadefoot, *Spea bombifrons* (p. 114)	
3b	No boss between eyes; spade elongated and sickle-shaped; general color dull yellow, tan, or brown; two or four yellowish longitudinal lines on dorsum; "warts" may be orange	Eastern Spadefoot, *Scaphiopus holbrookii* (p. 111)	

Key to the Toads and Frogs of Missouri			
4a	Parotoid glands oblong and connected to cranial crest; general color light gray to greenish gray; three or more "warts" per large dorsal spot; belly white or with one dark spot on chest		5
4b	Parotoid glands kidney-shaped, either separate from cranial crest or connected with a short spur; general color light brown to reddish brown; three or fewer "warts" per large dorsal spot; chest spotted with dusky markings	Eastern and Dwarf American Toads, *Anaxyrus americanus* (p. 118)	
5a	Raised boss between eyes; large, paired dorsal blotches greenish brown bordered by white	Great Plains Toad, *Anaxyrus cognatus* (p. 121)	
5b	No raised boss between eyes; usually with a single chest spot	Fowler's Toad and Rocky Mountain Toad, *Anaxyrus fowleri* and *A. w. woodhousii* (p. 124 and p. 126, respectively)	
6a	Tips of toes on fore- and hind feet expanded to form adhesive toe pads, which may be very small or large; if toe pads not present, with a V-shaped or small triangular marking between eyes; no dorsolateral fold		7
6b	Tips of toes on fore- and hind feet without adhesive toe pads; no V-shaped marking or small triangle between eyes; with or without dorsolateral fold; if without dorsolateral fold, may have a transverse fold of skin behind head		12
7a	Tips of toes with adhesive toe pads; no V-shaped or small triangular marking between eyes		8
7b	Tips of toes without adhesive toe pads; V-shaped or small triangular marking between eyes		9
8a	Toe pads small; dorsal markings in the form of long stripes or lines of spots	Upland Chorus Frog, Cajun Chorus Frog, and Boreal Chorus Frog, *Pseudacris feriarum*, *P. fouquettei,* and *P. maculata* (p. 144, p. 147 and p. 152, respectively)	
8b	Toe pads medium to large		10
9a	V-shaped marking between eyes; front legs large and muscular; no alternating light and dark bars on upper lip; a dark spot beneath each eye; no black stripe along inner thighs	Illinois Chorus Frog, *Pseudacris illinoensis* (p. 149)	

Key to the Toads and Frogs of Missouri			
9b	Dark triangle-shaped marking between eyes; forelegs small and not muscular; alternating light and dark bars on upper lip; no dark spot below each eye; rough-edged black stripe along each inner thigh	Blanchard's Cricket Frog, *Acris blanchardi* (p. 130)	
10a	Toe pads prominent; no white spot beneath each eye; color yellow green to emerald green; yellow or white line extending from snout to about midway to hind legs; dorsum may have a few small white or golden spots	Green Treefrog, *Hyla cinerea* (p. 138)	
10b	Toe pads small to large; with or without white spot below eyes; color gray, tan, green, or greenish gray; no light lateral line; dorsum without white or golden spots		11
11a	Toe pads small; no white spot below eyes; color pinkish tan to grayish tan; dark line between eyes; X-shaped dorsal marking; no bright orange-yellow color on underside of thighs	Spring Peeper, *Pseudacris crucifer* (p. 141)	
11b	Toe pads prominent; large white spot below eyes; color may be light gray to green, with dark irregular blotches on dorsum; no dark line between eyes; underside of thighs bright orange yellow	Cope's Gray Treefrog and Gray Treefrog, *Hyla chrysoscelis* and *H. versicolor* (p. 133)	
12a	Transverse fold of skin behind head; no tympanum; no dorsolateral fold; head small (less than ¼ snout-to-vent length)		13
12b	No transverse fold of skin behind head; tympanum prominent; with or without dorsolateral folds; head large (greater than ¼ snout-to-vent length)		14
13a	Color uniform dark brown, two wide tan stripes along each side of dorsum; belly heavily marked with dark pigment	Eastern Narrow-mouthed Toad, *Gastrophryne carolinensis* (p. 156)	
13b	Color uniform gray to olive tan; no markings or stripes on dorsum; belly light gray or creamy yellow without mottling	Western Narrow-mouthed Toad, *Gastrophryne olivacea* (p. 159)	
14a	Large dark masklike marking on sides of head extending from snout through eye and nearly to forelegs; with distinct dorsolateral folds; general color pinkish tan or grayish brown; no prominent dorsal markings	Wood Frog, *Lithobates sylvaticus* (p. 185)	
14b	No large dark masklike markings on sides of head; with or without dorsolateral folds; general color tan, brown, or greenish brown; with or without large dorsal spots or blotches		15

Key to the Toads and Frogs of Missouri			
15a	No light line along upper lip; tympanum same size or larger than eye; dorsal spots few and indistinct; with or without dorsolateral folds, but if present, never extending full length of body		16
15b	With or without light line along upper lip; tympanum same size as eye; dorsal spots or blotches distinct; dorsolateral fold prominent, extending full length of body		17
16a	No dorsolateral folds; prominent tympanal fold of skin from eye to shoulder; upper lip green; dorsal spots usually obscure if present	American Bullfrog, *Lithobates catesbeianus* (p. 168)	
16b	Dorsolateral folds extending halfway to hind legs; tympanal fold of skin from eye to shoulder not prominent; head greenish, with a few small dorsal spots	Green Frog, *Lithobates clamitans* (p. 172)	
17a	No light line on upper lip; dorsal spots numerous and closely set, interspaced with reticulations of dark brown or black	Northern Crawfish Frog, *Lithobates areolatus circulosus* (p. 162)	
17b	Light line on upper lip; dorsal spots scattered or in two distinct rows down the dorsum, not interspaced with reticulations of dark brown or black		18
18a	Dorsolateral folds wide and moderately raised; dorsal spots large and weakly to strongly ringed with white		19
18b	Dorsolateral folds narrow and distinctly raised; dorsal spots small to medium and not ringed with white		20
19a	Dorsal spots somewhat square or rectangular, set in two longitudinal rows, faintly ringed with white; dorsal ground color tan, never green; groin area and underside of hind legs bright yellow	Pickerel Frog, *Lithobates palustris* (Fig. 6A; p. 176)	
19b	Dorsal spots round, not set in two longitudinal rows, strongly ringed with white; dorsal ground color often green; groin area and underside of hind legs white or pale green	Northern Leopard Frog, *Lithobates pipiens* (Fig. 6B; p. 179)	
20a	Dorsolateral folds broken posteriorly and displaced toward midline; snout rounded, often with a dark spot; dorsal spots small, round, closely spaced, and often greenish brown; dorsal ground color tan to light brown, never green	Plains Leopard Frog, *Lithobates blairi* (Fig. 6C; p. 165)	

Key to the Toads and Frogs of Missouri			
20b	Dorsolateral folds continuous, but, if broken posteriorly, broken sections not displaced toward midline; snout pointed, usually without a dark spot; dorsal spots small to medium, rounded and/or oblong, well spaced and brown to black; dorsal ground color usually green, or brown with some green	Southern Leopard Frog, *Lithobates sphenocephalus* (Fig. 6D; p. 182)	

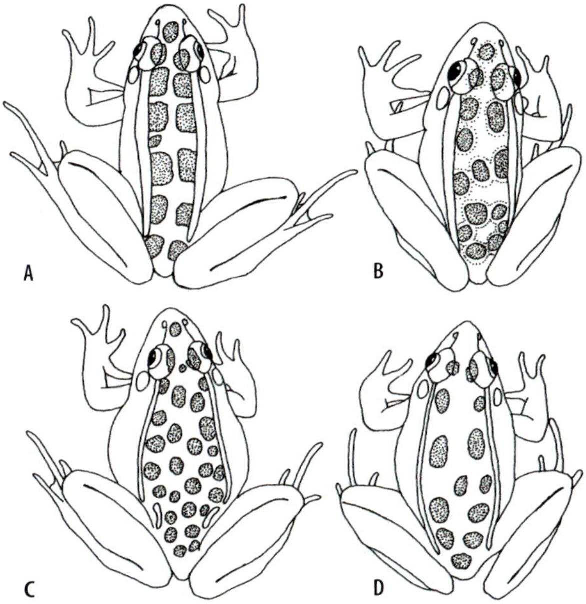

A

B

C

D

Fig. 6. *Dorsal view of (A) pickerel frog, (B) northern leopard frog, (C) plains leopard frog, and (D) southern leopard frog.*

Key to the Turtles of Missouri			
1a	Carapace flat, soft, leathery, and without scales; snout tube shaped		2
1b	Carapace dome shaped, hard, and covered with large scales; snout not tube shaped		3
2a	Anterior edge of carapace smooth, without bumps or spines; nostrils without projections from septum	Midland Smooth Softshell, *Apalone m. mutica* (p. 253)	
2b	Anterior edge of carapace rough, with short bumps or spines; nostrils with a projection on each side of septum	Eastern Spiny Softshell, *Apalone s. spinifera* (p. 256)	
3a	Tail long and at least half the length of carapace		4
3b	Tail short and less than one-third the length of carapace		5
4a	Top of head lacking large plates; beak moderately hooked; carapace without supramarginal scales	Eastern Snapping Turtle, *Chelydra serpentina* (Fig. 7A; p. 192)	
4b	Top of head with large plates; beak strongly hooked; carapace with supramarginal plates	Alligator Snapping Turtle, *Macrochelys temminckii* (Fig. 7B; p. 196)	

Fig. 7. *Dorsal view of (A) eastern snapping turtle and (B) alligator snapping turtle.*

Key to the Turtles of Missouri			
5a	Plastron with 11 plates (Fig. 8A), pectoral plates not connected to marginal plates or carapace; carapace smooth, dome shaped, posterior marginal plates not serrated		6
5b	Plastron with 12 plates (Fig. 8B), pectoral plates connected to marginal plates of carapace; carapace smooth or with a mid-dorsal ridge, dome-shaped or low, posterior marginal plates slightly to heavily serrated		8

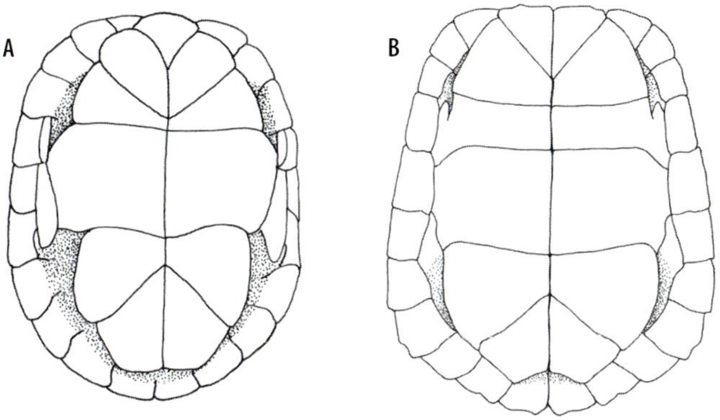

A

B

Fig. 8. *Plastron with (A) 11 plates and (B) with 12 plates.*

6a	Plastron narrow, with one hinge, but incapable of closing; ninth marginal plate along edge of carapace same size as eighth marginal; head with two distinct, yellow lateral lines	Eastern Musk Turtle, *Sternotherus odoratus* (p. 207)	
6b	Plastron moderately wide, with two distinct hinges, capable of closing slightly; ninth marginal plate along edge of carapace same size or much higher than eighth; tail with horny, clawlike tip; head without distinct yellow, lateral lines		7
7a	Ninth marginal plate same size as eighth; head with two irregular rows of lateral, yellow markings; chin not yellow	Mississippi Mud Turtle, *Kinosternon subrubrum hippocrepis* (p. 204)	
7b	Ninth marginal plate higher than eighth; head gray, without two rows of yellow lateral markings, chin yellow	Yellow Mud Turtle, *Kinosternon flavescens* (p. 201)	
8a	First vertebral plate in broad contact with nuchal and four marginals		9
8b	First vertebral plate in broad contact with nuchal and only two marginals		10

Key to the Turtles of Missouri			
9a	Carapace moderately dome shaped, dark brown with numerous small, yellow spots; neck without yellow lines; plastron with transverse hinge that does not allow shell to completely close	Blanding's Turtle, *Emydoidea blandingii* (p. 220)	
9b	Carapace slightly dome shaped, patterned with light-colored network of lines on dark brown ground color; neck strongly patterned with yellow lines; plastron not hinged	Western Chicken Turtle, *Deirochelys reticularia miaria* (p. 217)	
10a	Plastron with transverse hinge that allows both anterior and posterior halves to completely close against carapace		11
10b	Plastron without transverse hinge and immovable		13
11a	Each rear foot usually with three claws; plastron a uniform brownish yellow; carapace without keel or ridge	Three-toed Box Turtle, *Terrapene carolina triunguis* (p. 238)	
11b	Each rear foot usually with four claws; plastron boldly marked with radiating yellow and brown lines; carapace with keel or ridge		12
12a	Plastron dark and without radiating yellow lines, carapace without a keel or ridge	Eastern Box Turtle, *Terrapene c. carolina* (p. 238)	
12b	Plastron boldly marked with radiating yellow and brown lines; carapace with keel or ridge	Ornate Box Turtle, *Terrapene ornata* (p. 244)	
13a	Carapace rounded and smooth, posterior edge not serrated, first marginal extends to or beyond ventral; numerous red markings along marginal plates		14
13b	Carapace with or without a dorsal ridge, posterior edge moderate to strongly serrated, first marginal does not extend to or beyond ventral; no red color along marginal plates		15
14a	Plastron with intricate dark markings that follow plate seams outward from center; carapace without a mid-dorsal stripe	Western Painted Turtle, *Chrysemys picta bellii* (p. 214)	
14b	Plastron immaculate yellow; carapace with a distinct orange-red mid-dorsal stripe	Southern Painted Turtle, *Chrysemys dorsalis* (p. 211)	
15a	Carapace without a mid-dorsal keel; head with large, red, lateral markings, or with thin, yellow, longitudinal lines and no yellow spot behind each eye		16

Key to the Turtles of Missouri			
15b	Carapace with a low or prominent mid-dorsal keel; head with a yellow triangular, crescent-shaped or large and globular marking behind each eye; with at least some yellow transverse or vertical lines		17
16a	Plastron yellow and boldly marked with 12 dark-brown smudges; head usually with a wide reddish stripe behind each eye (adult males melanistic)	Red-eared Slider, *Trachemys scripta elegans* (p. 248)	
16b	Plastron yellow, with a few faint smudges anteriorly or without markings; head with numerous thin, yellow, lateral lines, widest yellow line begins below eye through angle of jaw (Fig. 9)	Eastern River Cooter, *Pseudemys c. concinna* (p. 235)	

Fig. 9. *Head of eastern river cooter.*

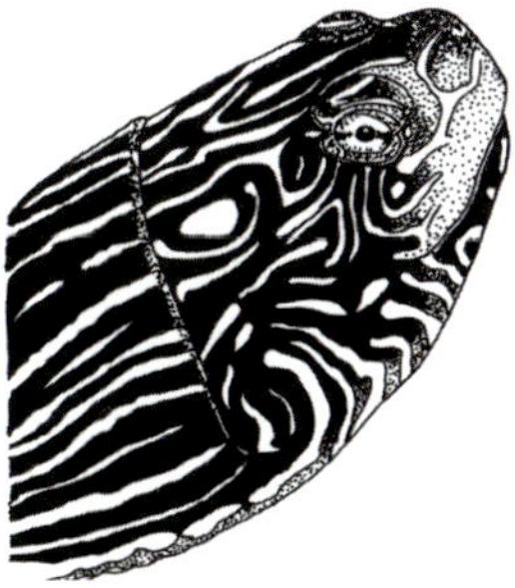

Fig. 10. *Head of northern map turtle.*

17a	Carapace with low mid-dorsal ridge; yellow marking behind eye small and triangular (Fig. 10)	Northern Map Turtle, *Graptemys geographica* (p. 223)	
17b	Carapace with medium to high mid-dorsal ridge; yellow marking behind eye crescent-shaped or globular		18
18a	Carapace with medium mid-dorsal ridge; yellow marking behind eye is crescent-shaped; thin yellow lines of the neck may or may not reach the eye; no yellow spots below eye (Fig. 11A and 11B)	Northern False Map Turtle and Mississippi Map Turtle, *Graptemys pseudogeographica* (p. 230)	
18b	Carapace with high mid-dorsal ridge; yellow marking behind eye globular-shaped; a few thin, yellow lines of neck touch each eye; one or two round or oblong, yellow spots below each eye (Fig. 11C)	Ouachita Map Turtle, *Graptemys ouachitensis* (p. 227)	

Key to the Turtles of Missouri

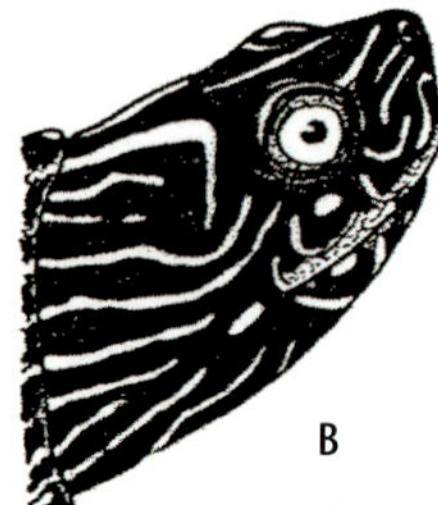

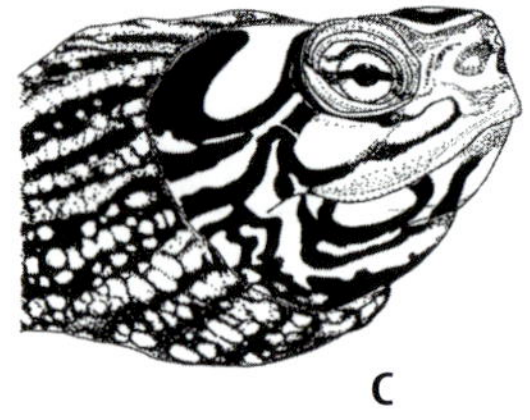

Fig. 11. *Head of (A) Mississippi map turtle, (B) northern false map turtle, and (C) Ouachita map turtle.*

Key to the Lizards of Missouri			
1a	Four limbs present		2
1b	Limbs absent	Western Slender Glass Lizard, *Ophisaurus a. attenuatus* (p. 295)	3
2a	Dorsal and lateral scales large		3
2b	Dorsal and lateral scales small and granular		4
3a	Large, rough scales on body, limbs, and tail		5
3b	Dorsal and lateral scales same size as ventrals; scales smooth and shiny		6
4a	Head large; dorsal and lateral scales small and granular; ventral scales hexagonal and in more than 15 rows across venter; a distinct black collar on neck	Eastern Collared Lizard, *Crotaphytus collaris* (p. 261)	
4b	Head long and narrow; dorsal and lateral scales small and granular; ventral scales large, transverse, and in fewer than 15 rows across venter; six or seven longitudinal light stripes on body; no distinct black collar on neck	Six-lined and Prairie Racerunners, *Aspidoscelis sexlineata* (p. 291)	
5a	Back of head with large, pointed scales in the form of spines or "horns"; tail very short	Texas Horned Lizard, *Phrynosoma cornutum* (p. 266)	
5b	Back of head without spines; tail about same length as body; scales on body, limbs, head, and tail strongly keeled; grayish brown to rusty brown	Prairie Lizard, *Sceloporus consobrinus* (p. 269)	
6a	Frontal scale V-shaped; no supranasal scales present; lower eyelids with a transparent "window"; size small (maximum snout-to-vent length 57 mm; 2.2 in.)	Little Brown Skink, *Scincella lateralis* (p. 287)	
6b	Frontal scale rectangular shaped; supranasal scales present; lower eyelids without a transparent "window"		7
7a	Dorsal and lateral scales form parallel rows with the long axis of body (Fig. 12A)		8
7b	Dorsal and lateral scale rows in oblique pattern to long axis of body (Fig. 12B); tan ground color with numerous black spots in longitudinal rows forming stripes	Great Plains Skink, *Plestiodon obsoletus* (p. 281)	

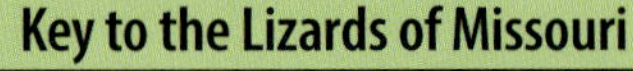

Key to the Lizards of Missouri

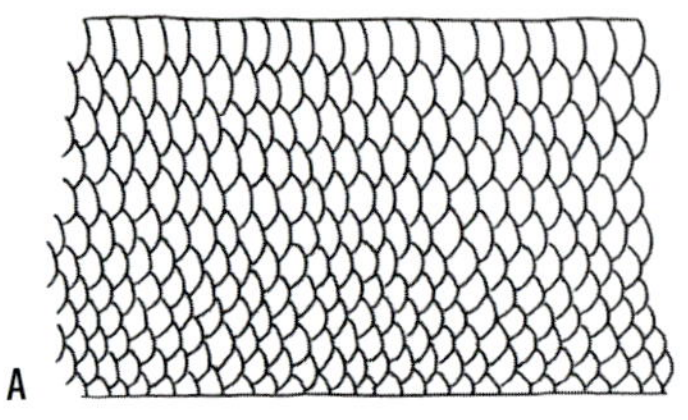

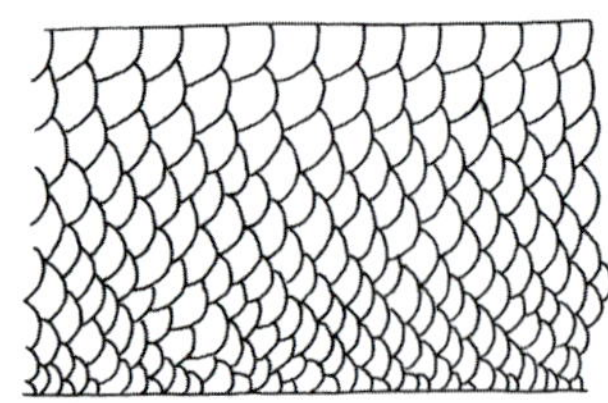

Fig. 12. *(A) Common five-lined skink showing parallel body scale rows and (B) Great Plains skink showing oblique body scale rows.*

8a	Chin with one postmental scale (Fig. 13A); no light lines on head; dark lateral stripe along body black	Southern Coal Skink, *Plestiodon anthracinus* (p. 273)	
8b	Two postmental scales present (Fig. 13B)		9
9a	No postnasal scale (Fig. 14); dorsal light line on head never forked; seven light stripes on body bordered by dark stripes; widest dark stripe extends onto tail; complete tail nearly 1½ times the length of head and body	Northern and Southern Prairie Skinks, *Plestiodon septentrionalis* (p. 284)	
9b	Postnasal scales present; may or may not have light lines on head		10
10a	Eight or nine upper labials (Fig. 15A); no postlabial scales, or one or two small postlabials present; large lizard with a maximum snout-to-vent length over 85 mm (3.3 in.)	Broad-headed Skink, *Plestiodon laticeps* (p. 278)	
10b	Seven upper labials (Fig. 15B); two large postlabial scales present; maximum snout vent length under 85 mm (3.3 in.)	Common Five-lined Skink, *Plestiodon fasciatus* (p. 275)	

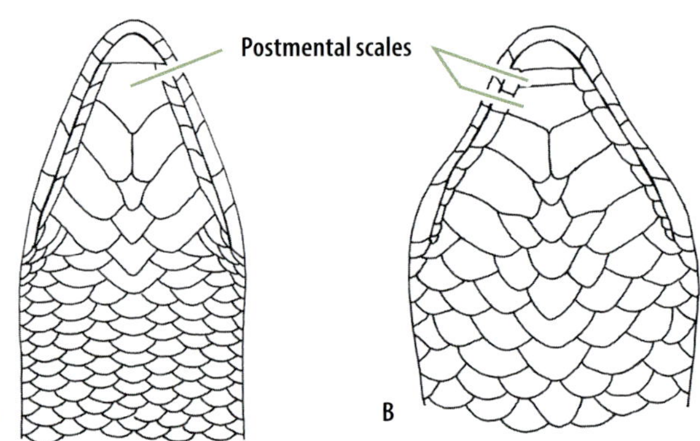

Fig. 13. *(A) Underside of southern coal skink head showing one postmental scale and (B) of common five-lined skink showing two postmental scales.*

Key to the Lizards of Missouri

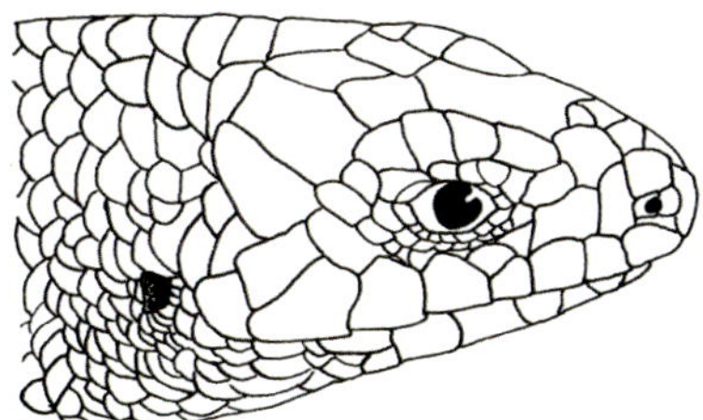

Fig. 14. *Head of northern and southern prairie skink showing the absence of a postnasal scale.*

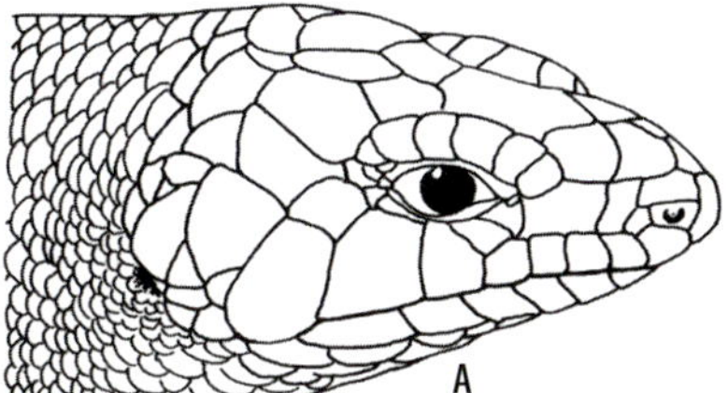

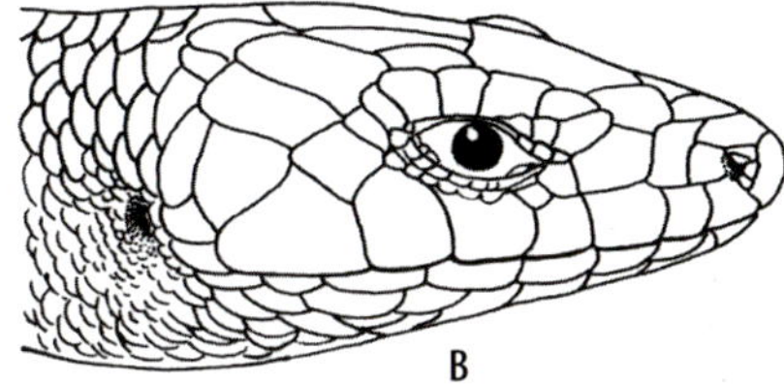

Fig. 15. *(A) Lateral view of broad-headed skink head with eight labial scales and (B) common five-lined skink with seven labial scales.*

Key to the Snakes of Missouri

1a	A facial pit present between nostril and eye on each side of head; eyes with elliptical, vertical pupils (Fig. 16A); ventral surface of tail covered with one row of scales		2
1b	No facial pit between nostril and eye on each side of head; eyes with round pupils (Fig. 16B); ventral surface of tail covered with two rows of scales		7

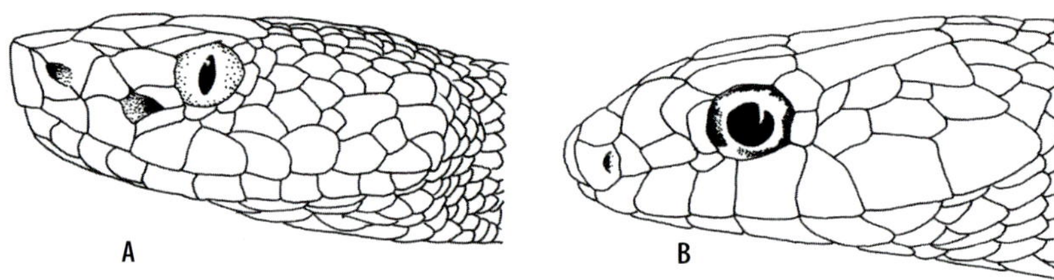

Fig. 16. *(A) Head of pit viper showing facial pit and elliptical, vertical pupil of eye and (B) head of colubrid snake without facial pit; pupil is round.*

2a	Tail ends with a rattle		3
2b	Tail does not end with a rattle		6
3a	Head covered with small scales along with supraocular scales; dorsal markings are crossbands or chevrons; a rust-colored dorsal stripe normally present; tail black; rattle size large	Timber Rattlesnake, *Crotalus horridus* (p. 429)	
3b	Head covered with nine large scales; round, dark blotches along dorsum; with or without rust-colored dorsal stripe; tail banded; rattle size small to medium		4
4a	Dorsal blotches small, irregular shaped, dark brown; ground color purple gray; a rust-colored dorsal stripe present; rattle size tiny	Western Pygmy Rattlesnake, *Sistrurus miliarius streckeri* (p. 440)	
4b	Dorsal blotches rounded, dark gray, brown, or black, closely spaced; ground color gray or brownish gray; no rust colored dorsal stripe; rattle size medium		5
5a	Color is gray or brownish gray with 29 to 39 (average 34) dark brown or black dorsal blotches; belly is chiefly black or dark gray with irregular light markings; occurs from St. Louis area and north along Mississippi River in Missouri	Eastern Massasauga, *Sistrurus catenatus* (p. 433)	

Key to the Snakes of Missouri			
5b	Color may be light to dark gray or gray brown with 34 to 50 (average 40) dark brown or black dorsal blotches; belly is normally light gray with darker gray mottling; occurs in north-central and northwestern Missouri	Prairie Massasauga, *Sistrurus t. tergeminus* (p. 436)	
6a	Distinct, dark brown, hourglass-shaped dorsal markings over a pinkish-tan ground color; head without black stripe behind each eye; no white line along upper labials	Eastern Copperhead, *Agkistrodon contortrix* (p. 421)	
6b	Dorsal markings mostly obscured by dark brown or black; head with black stripe behind each eye; a white line present along upper labials	Northern Cottonmouth, *Agkistrodon piscivorus* (p. 425)	
7a	Rostral scale rounded, not pointed or projecting forward, not upturned and without a medial keel		10
7b	Rostral scale pointed, turned upward		8
8a	Rostral scale slightly upturned; ventral coloration a mottled gray, underside of tail lighter than belly (Fig. 17A); dorsal scale rows at midbody number 25 (Fig. 18C)	Eastern Hog-nosed Snake, *Heterodon platirhinos* (p. 329)	
8b	Rostral scale sharply upturned; ventral coloration is black with small light patches, often edged in yellow, underside of tail is same coloration as belly (Fig. 17B); dorsal scale rows at midbody number 23 (Fig 18C);		9
9a	Occurs in southeastern Missouri	Dusty Hog-nosed Snake, *Heterodon gloydi* (p. 323)	
9b	Occurs in northwestern Missouri	Plains Hog-nosed Snake, *Heterodon nasicus* (p. 326)	

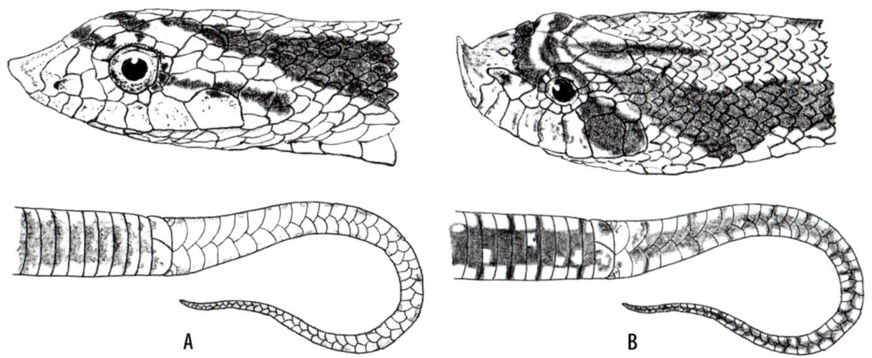

Fig. 17. *Hog-nosed snakes head profiles and ventral patterns (A) eastern hog-nosed snake and (B) dusty hog-nosed snake and plains hog-nosed snake.*

Key to the Snakes of Missouri

10a	Dorsal scales and at least some lateral scales weakly or strongly keeled (Fig. 18A)		11
10b	Dorsal and lateral scales smooth, without keels (Fig. 18B)		31

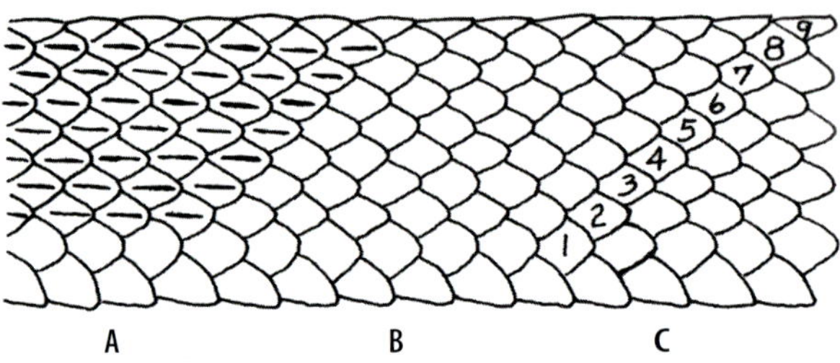

Fig. 18. *Diagram of snake body scales showing (A) keeled scales, (B) smooth scales, and (C) method of counting scale rows.*

11a	Anal plate not divided (Fig. 19A)		12
11b	Anal plate divided (Fig. 19B)		16

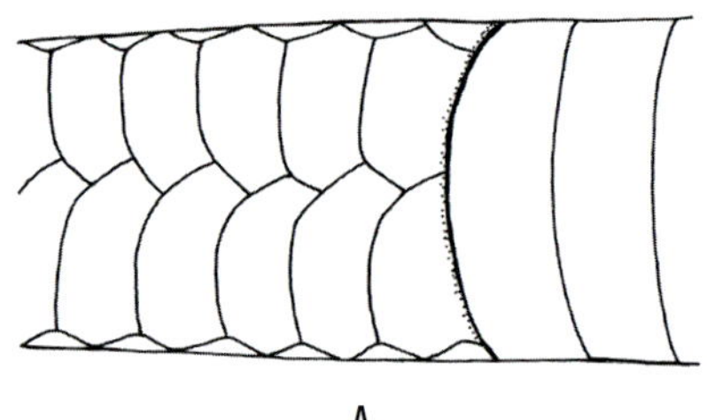

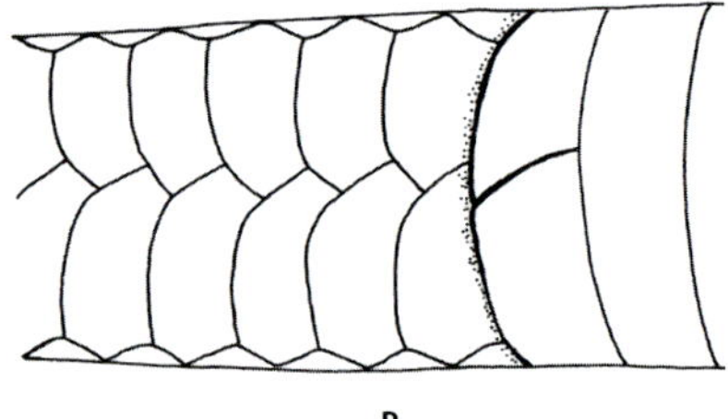

Fig. 19. *Snake anal plate (A) not divided and (B) divided.*

12a	Rostral scale narrow and high, curving upward to tip of snout; four prefrontals present; dorsal scale rows at midbody number 25 or more; ventral scales number 200 or more	Bullsnake, *Pituophis catenifer sayi* (p. 383)	
12b	Rostral scale wide and low, not curving upward to tip of snout; two prefrontals present; dorsal scale rows at midbody number 24 or fewer; ventral scales number 190 or less		13
13a	Venter with two rows of black spots down the center; supralabial scales number six or less	Lined Snake, *Tropidoclonion lineatum* (p. 411)	
13b	Venter with no black spots, or black spots present along lateral edges; supralabial scales number seven or more		14

Key to the Snakes of Missouri			
14a	Lateral light stripe includes scale rows two and three anteriorly (Fig. 20A); supralabial scales with narrow black bars; interspace between lateral body scales sometimes red	Eastern and Red-sided Gartersnakes, *Thamnophis sirtalis* (p. 407)	
14b	Lateral stripe includes scale rows three and four (Fig. 20B)		15

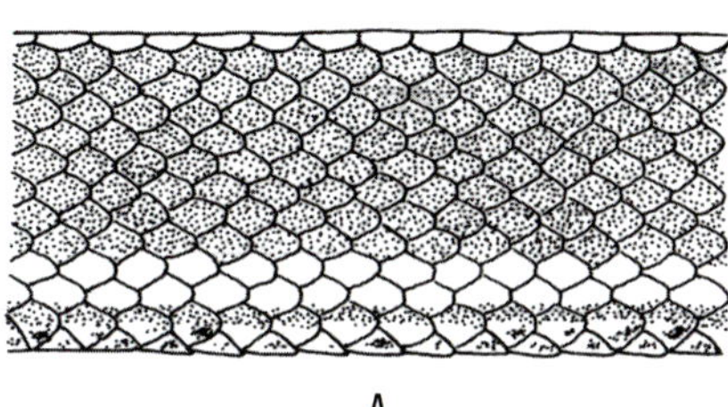

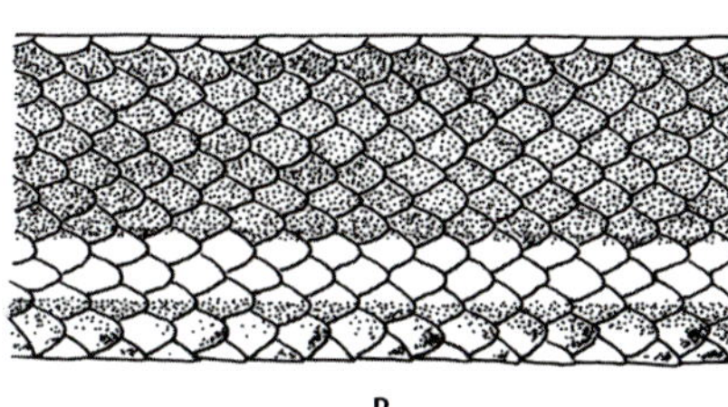

Fig. 20. *(A) Lateral light stripe on scale rows 2 and 3 and (B) lateral light stripe on scale rows 3 and 4.*

15a	Supralabial scales white or pale green, without black bars; space between mid-dorsal light stripe and lateral light stripe uniformly dark; dorsal scale rows at mid-body number 19 or less; tail length usually more than 27 percent of total length	Orange-striped Ribbonsnake, *Thamnophis p. proximus* (p. 401)	
15b	Supralabial scales light in color, boldly marked with black bars; space between mid-dorsal light stripe and lateral light stripe marked with alternating round, dark spots; dorsal scale rows at mid-body usually number 21; tail length usually less than 27 percent of total length	Plains Gartersnakes, *Thamnophis radix* (p. 404)	
16a	Dorsal scale rows at mid-body number 19 or fewer		17
16b	Dorsal scale rows at mid-body number 23 or more		22
17a	Dorsal scale rows at mid-body number 15 to 17		18
17b	Dorsal scale rows at mid-body number 19; light lateral stripe on scale rows one to three, along entire length of body; venter yellow and bordered by rows of small dark markings	Graham's Crawfish Snake, *Regina grahamii* (p. 386)	
18a	Dorsal scale rows at mid-body number 15; one or more preocular scales present; supralabial scales number five or six; belly usually red-orange	Red-bellied Snake, *Storeria occipitomaculata* (p. 395)	

Key to the Snakes of Missouri			
18b	Dorsal scale rows at mid-body number 17; no more than one preocular; belly not red-orange		19
19a	Preocular scales present		20
19b	Preocular scales absent		21
20a	Light-tan, mid-dorsal stripe, usually bordered by small black dots; venter cream colored or pale pink with small, dark dots along each side	Dekay's brownsnakes, *Storeria dekayi* (p. 392)	
20b	No dorsal stripe; uniform color with no markings; dorsal green; venter yellow	Northern Rough Greensnake, *Opheodrys a. aestivus* (p. 363)	
21a	Supralabial scales number five, one postocular scale present; one inter-nasal scale on snout; dorsal scales strongly keeled	Rough Earthsnake, *Haldea striatula* (Fig. 21A; p. 414)	
21b	Supralabial scales number six; two postocular scales present; usually two internasal scales on snout; dorsal scales weakly keeled	Western Smooth Earthsnake, *Virginia valeriae elegans* (Fig. 21B; p. 417)	

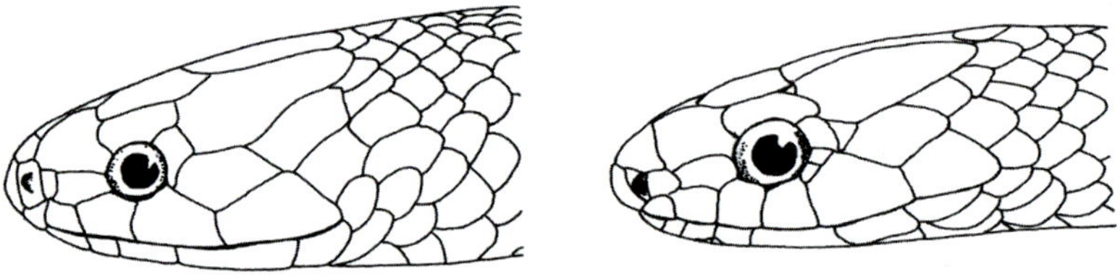

Fig. 21. *(A) Head of rough earthsnake with five labial scales and (B) head of western smooth earthsnake with six labial scales.*

22a	Ventral scales number 159 or fewer; dorsal scales strongly keeled		23
22b	Ventral scales number 190 or more; dorsal scales weakly keeled		28
23a	Dorsal scale rows number 25 or fewer		24
23b	Dorsal scale rows number 27 or more		27
24a	Dorsal scale rows number 19; dorsal color is red brown to gray brown with four alternating, round spots; pink to red belly with stippling along each side	Kirtland's Snake, *Clonophis kirtlandii* (p. 307)	
24b	Dorsal scale rows number 20 to 25		25
25a	Venter boldly patterned with irregular dark markings		26

Key to the Snakes of Missouri			
25b	Venter yellow and without markings, or with some light-brown mottling; dorsal coloration uniformly dark brown to nearly black, or with obscure bands or blotches of brown	Plain-bellied watersnakes, *Nerodia erythrogaster* (Fig. 22A; p. 350)	
26a	Dorsal bands and blotches number 20 or less; ventral coloration yellow with irregular brown or reddish-brown markings that may cover parts of two or three scales, or may merge to form large ventral blotches	Broad-banded Watersnake, *Nerodia fasciata confluens* (Fig. 22B; p. 353)	
26b	Dorsal bands and blotches number more than 20; ventral coloration light yellow with irregularly spaced half-moons or spots of orange, red, brown, or black	Northern and Midland Watersnakes, *Nerodia sipedon* (Fig. 22C; p. 359)	
27a	One or two subocular scales separating eye from supralabials; dorsal scale rows at midbody number 27 to 29; dorsal coloration dark olive brown; venter mottled brown with two or three yellow half-moons per scale	Mississippi Green Watersnake, *Nerodia cyclopion* (Fig 22D; p. 347)	
27b	No subocular scales between eye and supralabials; dorsal scale rows at midbody number 25 to 29; dorsal coloration yellow brown with narrow, dark-brown, dorsal spots alternating, but connected to, narrow lateral spots; venter yellow with small half-moon-shaped dark markings	Northern Diamond-backed Watersnake, *Nerodia r. rhombifer* (Fig. 22E; p. 356)	

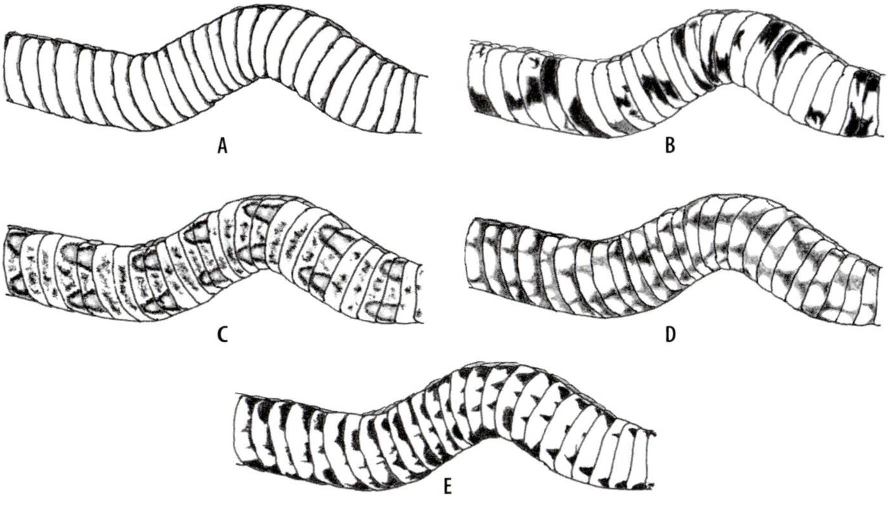

A

B

C

D

E

Fig. 22. *Watersnakes belly patterns (A) plain-bellied watersnake, (B) broad-banded watersnake, (C) northern/midland watersnake, (D) Mississippi green watersnake, and (E) northern diamond-backed watersnake.*

Key to the Snakes of Missouri			
28a	Ventral scales number 220 or more; dorsal coloration uniform black or dark brown or with a faint pattern of 28 to 38 dark blotches; head dark, no marking between eyes; ventral coloration black-and-white checkerboard pattern anteriorly, gray mottling posteriorly	Western Ratsnake, *Pantherophis obsoletus* (p. 373)	
28b	Ventral scales number less than 220; head light, with or without markings		29
29a	Head gray or brownish gray with reddish-brown spear-point marking between eyes; ventral coloration white with numerous small square, dark-gray markings; dark-gray stripes along venter of tail	Great Plains Ratsnake, *Pantherophis emoryi* (p.370)	
29b	Head orange to orange brown with no spear-point marking between eyes; ventral coloration yellow with large square, dark brown markings		30
30a	Averages 37 dorsal blotches; occurs in northwestern Missouri	Western Foxsnake, *Pantherophis ramspotti* (p. 377)	
30b	Average of 43 dorsal blotches; occurs in northeastern Missouri	Eastern Foxsnake, *Pantherophis vulpinus* (p. 380)	
31a	Anal plate divided (Fig. 19B)		32
31b	Anal plate not divided (Fig. 19A)		39
32a	Light-colored ring around neck contrasting with dark brown or black dorsal color; venter yellow to orange with small black spots	Prairie and Mississippi Ring-necked Snakes, *Diadophis punctatus* (p. 317)	
32b	No light-colored ring around neck		33
33a	Dorsal scale rows at mid-body number 15 or less		34
33b	Dorsal scale rows at mid-body number 17 or more		37
34a	Dorsal scale rows at mid-body number 13; brownish-purple dorsal color in strong contrast to pink venter	Western Wormsnake, *Carphophis vermis* (p. 301)	
34b	Dorsal scale rows at mid-body number 15		35
35a	Venter cream colored; venter of tail with two broken rows of gray or black spots; dorsal color tan, orange or reddish brown, with or without saddle-shaped dark markings	Variable Groundsnake, *Sonora s. semiannulata* (p. 389)	
35b	Venter yellow or salmon pink; no gray or black spots along venter of tail		36

Key to the Snakes of Missouri			
36a	Venter yellow; dorsal color bright green; head same color as body	Smooth Greensnake, *Opheodrys vernalis* (p. 368)	
36b	Venter salmon pink; dorsal color tan; head usually darker than body	Flat-headed Snake, *Tantilla gracilis* (p. 398)	
37a	Dorsal scale rows at mid-body number 19; venter with red bands over black; tip of tail with pointed terminal scale	Western Mudsnake, *Farancia abacura reinwardtii* (p. 320)	
37b	Dorsal scale rows at mid-body number 17; venter without light and dark bands		38
38a	Dorsal coloration dark brown to black anterior ⅔ of body, light tan to reddish posterior ⅓; 13 or fewer scale rows at end of body; venter coloration pale orange brown, anterior ventral scales edged in gray	Eastern Coachwhip, *Coluber f. flagellum* (p. 314)	
38b	Dorsal coloration uniform and varies from black to blue green; 15 scale rows at end of body; venter cream colored to yellow with some small random brown spots	Eastern Yellow-bellied and Southern Black Racers, *Coluber constrictor* (p. 310)	
39a	Dorsal coloration tan or cream with distinct red or orange-red blotches edged in black		40
39b	Dorsal coloration black with yellow spots or tan with brown blotches		41
40a	Snout pointed, venter white and without markings	Northern Scarletsnake, *Cemophora coccinea copei* (p. 304)	
40b	Snout blunt, venter white with gray or black square or rectangular markings	Eastern Milksnake, *Lampropeltis triangulum* (p. 343)	
41a	Dorsum tan or brown with 45 to 80 mid-dorsal brown or dark-brown blotches edged in black; dorsal scale rows at mid-body number 25 to 27	Prairie Kingsnake, *Lampropeltis calligaster* (p. 334)	
41b	Dorsum black; each scale generally with a yellow, white or cream spot; dorsal scale rows at mid-body number 19 to 21		42
42a	Dorsal scales are boldly patterned with one white or light-yellow spot that causes the snake to appear speckled	Speckled Kingsnake, *Lampropeltis holbrooki* (p. 337)	
42b	Doral scales are faint pattern with small yellow to cream speckles or with faint, white or cream chainlike markings along the sides	Eastern Black Kingsnake, *Lampropeltis nigra* (p. 340)	

Part I—**AMPHIBIANS**

(Class Amphibia)

Salamanders, Toads, and Frogs

Missouri has 44 species of amphibians, with an additional four subspecies or geographic races. Herpetologists recognize 7,303 species of amphibians, with more being discovered each year (Pough et al. 2018). These animals are placed in three major groups: salamanders (Order Caudata), toads and frogs (Order Anura), and caecilians (Order Gymnophiona). Caecilians—long, limbless, wormlike animals—are represented by about 200 species; because they live only in the tropics, they are not treated in this book. Amphibians have played a vital role in the evolution of the land vertebrates. This took place during the late Devonian Period, about 400 million years ago, when some freshwater, lobed-finned fish began moving overland to escape the drying of freshwater swamps, leading to the rise of early amphibians. Some amphibians became established on land (if only during part of their lives) and eventually advanced to the point where they no longer needed to return to the swamps to breed and lay eggs. Some of these animals gave rise to the first reptiles. Salamanders, toads, and frogs generally have moist skin; none has scales or claws. They require freshwater or a damp environment in which to live. Most of Missouri's amphibians must return to water

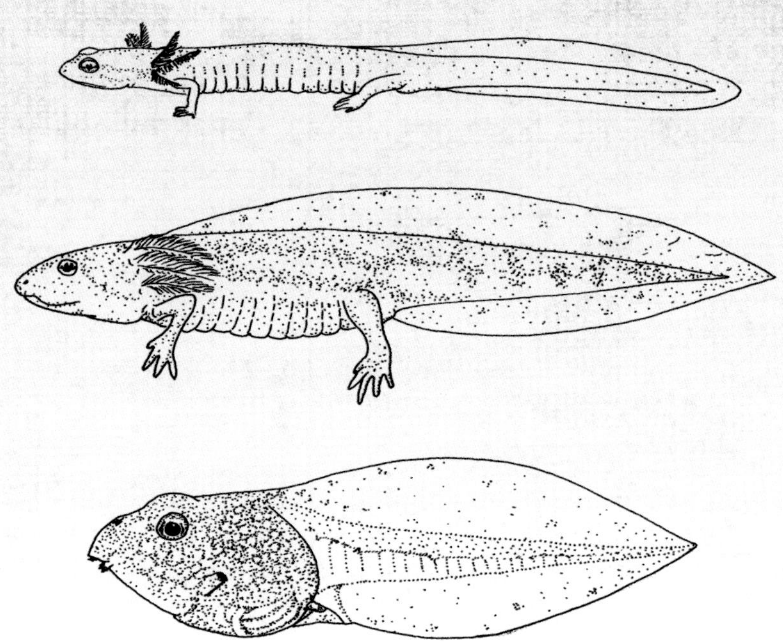

A typical stream-type salamander larva (top) has small gills and a low tail fin. A typical pond-type salamander larva (center) has larger gills and a high tail fin. A typical tadpole (bottom).

to reproduce; some species are totally aquatic, others lay their eggs in moist places on land. Amphibians are ectothermic—or "cold-blooded"—animals, meaning that they are unable to internally control their body temperature. Their skin secretions are toxic or irritating to the mucus membranes of our eyes; thorough washing is recommended after handling amphibians. All Missouri amphibians lay eggs, either in the water or in moist areas. The eggs of toads and frogs hatch into tadpoles, which have covered gills. Most tadpoles eat aquatic plants, especially algae. Most salamander eggs hatch into aquatic larvae that have exposed gills and eat various small aquatic invertebrates. Adult salamanders may be distinguished from adult toads and frogs by the presence of a tail. The hind legs of toads and frogs are much larger than those of salamanders. Also, almost all salamanders are voiceless; male toads and frogs produce a breeding call.

(Order Caudata)

Salamanders

Salamanders consist of about 665 extant species in 66 genera and are found on every continent except Antarctica and Australia (Pough et al. 2018). North and South America have more salamander species than the rest of the world combined (Powell et al. 2016). They are most diverse in North America, representing nine out of 10 families. Many salamander species are restricted to the moist, forested regions of the United States, with many species occurring in the Appalachian Mountains and Interior Highlands of central and eastern United States. Most salamander species in Missouri are found in the Ozark Highlands and bottomland forests of eastern and southern Missouri. Life history requirements for salamander species in Missouri are diverse. A few species are aquatic for their entire life, others depend on water for part of their life, especially breeding, and others only require moist habitats. Reproductive strategies are very diverse. The more ancient species, such as hellbenders, have external fertilization with males spraying sperm onto eggs extruded from the females. Most species of salamanders have internal fertilization with females picking up a sperm packet, provided by males, within her cloaca and depositing eggs through the sperm for fertilization. Many of these species rely on aquatic habitats (e.g., permanent or temporary wetlands, streams) to deposit their eggs and allow the hatching larvae to forage and grow in this aquatic environment (e.g., mole salamanders and brook salamanders). The larvae will eventually become small salamanders, with many spending most of their lives on land. Other species are only dependent on moist environments for reproduction. *Plethodon* species deposit eggs in moist areas, such as rotting logs, cave walls, and other subterranean environments. The embryos develop directly within the egg membranes and hatch into tiny replicates of the adult. There are 200 or more salamander species in the United States, with 19 species and three additional subspecies, consisting of seven families and nine genera found in Missouri.

Family Cryptobranchidae

Hellbenders and Giant Salamanders

This family of permanently aquatic salamanders contains two genera, *Andrias* and *Cryptobranchus* (Pough et al. 2018). *Andrias* is strictly Asiatic, with two living forms, the Chinese and Japanese giant salamanders (*A. davidianus* and *A. japonicus*, respectively). The giant salamander is the largest living amphibian in the world, with specimens growing to 5 feet in total length. The other genus, *Cryptobranchus*, is restricted to the United States. Two subspecies, the eastern hellbender (*C. alleganiensis alleganiensis*) and the Ozark hellbender (*C. alleganiensis bishopi*), are recognized. Although primarily a species of eastern North America, the hellbender's range includes Missouri. The Ozark hellbender is endemic to southern Missouri and a small part of northern Arkansas. Missouri is the only state that contains both subspecies. The current taxonomy of hellbenders will likely be changed soon with the elevation of additional species and subspecies.

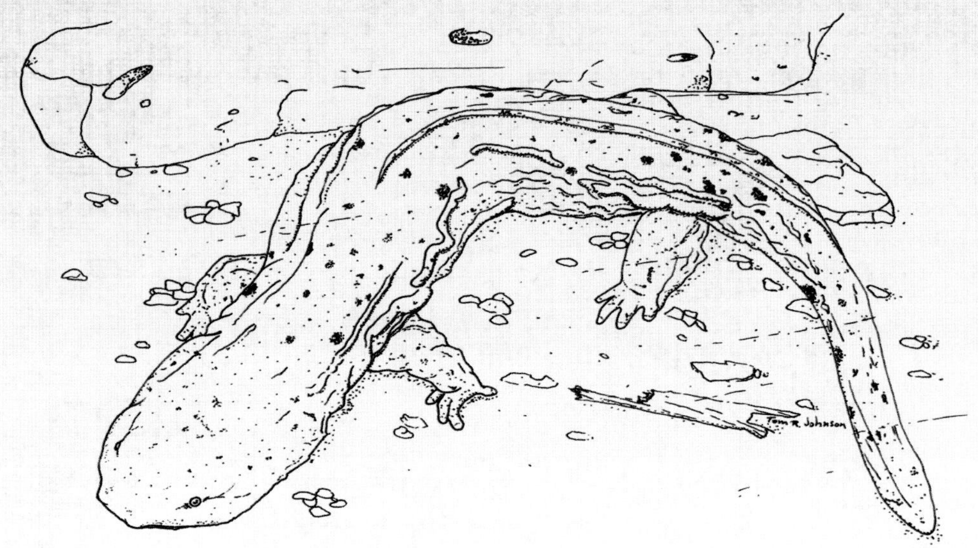

Eastern Hellbender

Eastern Hellbender

Cryptobranchus alleganiensis alleganiensis (Daudin)

JEFF BRIGGLER

Adult eastern hellbender from Pulaski County.

Description: The eastern hellbender is a large, entirely aquatic salamander. Its head is broad and flat, with small, lidless eyes. The sides of the body have soft, pronounced folds of skin. The legs also have large flaps of skin. The tail is flattened and rudderlike. A gill opening is present on each side of the head. Body color varies from red brown to dull gray brown. Brown to black irregular spots are often present on juveniles, but adults are typically uniform in color. Chin and lower lip are usually lacking in dark markings or spots. The belly is a uniform dark tan or gray brown and is lighter in color than the rest of the body; however, during the breeding season the belly can be orange.

Adult eastern hellbenders range in total length from 330 to 580 mm (13 to 22.8 in.) but have been known to reach 740 mm (29.1 in.; Powell et al. 2016).

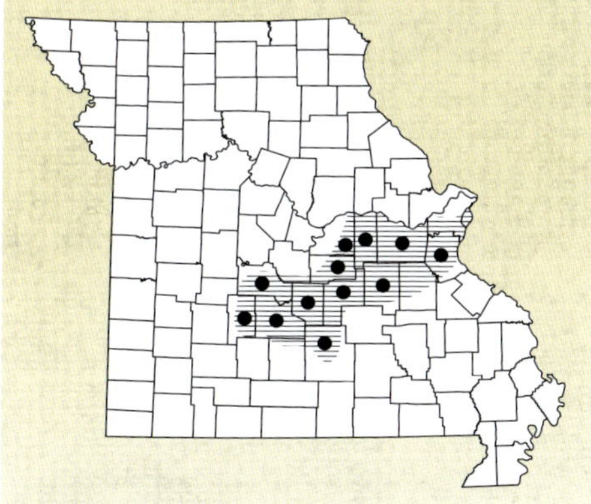

Distribution: Missouri: Northern Ozark Highlands in spring-fed rivers that drain north into the Missouri River and Meramec River drainages. North America: Southern New York southward to the western Carolinas, northern Georgia, Alabama, and Mississippi, and west to central Missouri (Powell et al. 2016).

Habits and Habitat: The hellbender is well suited for an aquatic existence. Its flat head and streamlined shape reduce water resistance. This salamander makes its home under flat rocks and within bedrock crevices in large permanent streams and rivers. It is a slow swimmer and often moves by walking along the bottom. It is normally a solitary animal and prefers cover (Nickerson and Mays 1973; Peterson and Wilkinson 1996). In a study of hellbender movements along a section of the Niangua River in southern Missouri, females had an average home range of 28 m^2 (301 ft^2); males had an average home range of 81 m^2 (872 ft^2; Peterson and Wilkinson 1996). Hellbenders can live 25 to 30 years in the wild (Peterson et al. 1983). Several animals marked in the early 1980s in Missouri rivers have been recently recaptured, proving that some individuals can indeed live over 30 years in the wild (J. Briggler unpubl. data). In addition, these marked animals were still present in the same area where they were tagged. Additional natural history information can be found in Peterson et al. 1988.

The food of adult eastern hellbenders consists primarily of crayfish, small fish, and occasionally snails, worms, and aquatic insects (Peterson et al. 1989). It is also known to eat scavenged material. On two occasions adult hellbenders captured in the Gasconade River have regurgitated fish parts (head and backbone) from filleted fish (J. Briggler pers. obs.). Captive eastern hellbenders will eat crayfish, minnows, tadpoles, frogs, small mice, earthworms, and sliced beef liver. In the wild, larvae are known to feed on various aquatic insects (Pitt and Nickerson 2006).

Breeding: Eastern hellbenders breed mainly in September in Missouri; they are one of the few salamanders in Missouri that fertilize their eggs externally. Hellbenders will typically congregate in small groups (6–12 adults) around a breeding site for courtship and subsequent mating. Animals can locate each other by following pheromones (olfactory signals) that females produce during the breeding season (Peterson 1988). Nesting sites are excavated depressions under large rocks, within bedrock crevices, and occasionally under logs or within mud banks in rivers (Alexander 1927; Bishop 1941; Peterson 1988; J. Briggler unpubl. data). The female hellbender enters or is guided into a nesting chamber by the attending male. She will deposit cream to light-yellow eggs that are about 6 mm (0.24 in.) in diameter in long strands, like a string of beads. The male fertilizes the eggs by spraying sperm within the nesting chamber. The number of eggs found in 14 nests in Missouri ranged from 186 to 403, with an average of 284 eggs (J. Briggler unpubl. data). The male hellbender remains with the eggs and is known to defend them against other hellbenders and other potential predators (Smith 1907; Bishop 1941; Peterson 1988). The number of eggs found in a nest is considerably less than the number of ovarian eggs found inside the female. A study on the Niangua River by Topping and Ingersol (1981) found ovarian egg counts in a female to range from 169 to 645 and that larger females produce a greater number of eggs. Approximately 24 percent of the eggs within a female are retained and eventually resorbed. Eggs are known to hatch in about 45 days (Peterson 1988); however, this can vary by a week or more depending on water temperature. The newly hatched larvae, averaging 27.5 mm (1.1 in.) in total length, have large amounts of yolk in their abdomen, and in captivity are known to move around in short, jerky swimming movements. The yolk furnishes nourishment to the young hellbender for up to three months (Smith 1912). Once larvae begin to eat other food, their color becomes nearly black. Eastern hellbender larvae initially have external gills but lose the gills after about 1.5 years when they reach a length of 100 to 130 mm (3.9 to 5.1 in.; Bishop 1943; J. Briggler unpubl. data). The duration of the male guarding the nest with eggs and subsequent hatchlings is poorly documented and thought to vary a great deal. Recent observations of nests in Missouri suggest most larvae depart the nest chamber in March or April (J. Briggler unpubl. data). Females are reported to be sexually mature at a total length of 380 mm (15 in.) and at six to eight years of age (Taber et al. 1975). Males typically become sexually mature at approximately 300 mm (11.8 in.) total length and at five to six years of age (Taber et al. 1975).

TOM R. JOHNSON

JEFF BRIGGLER

Three-week-old hatchling eastern hellbender from the Niangua River, Dallas County (above). Freshly laid eggs of the eastern hellbender from the Big Piney River, Texas County (left).

Remarks: Eastern hellbenders are often caught by anglers in streams and rivers of the Ozark Highlands, and are often thought to be dangerous, poisonous, and even deadly. Quite the contrary, the hellbender is a harmless amphibian that seldom, if ever, attempts to bite. If caught on hook-and-line, individuals should be released unharmed.

Eastern hellbenders in Missouri have experienced a marked decline since the 1970s (80 percent decline) with larger, mature individuals mostly observed today and few observations of juvenile individuals (Wheeler et al. 2003; J. Briggler unpubl. data). For example, surveys for hellbenders in the upper Meramec River from mid-1980s to early 1990s resulted in capturing over 500 hellbenders. Recent surveys in the 2000s have resulted in only 17 hellbenders within this same section of river. Due to the dramatic population decline, the Missouri Department of Conservation listed the eastern hellbender as state endangered in 2003, and in 2021 the U.S. Fish & Wildlife Service listed it as federally endangered. Hellbenders are habitat specialists that depend on consistent levels of flow, temperature, and dissolved oxygen, as well as large rocky habitat for shelter. Habitat alteration and degradation resulting from excessive sediment input from the surrounding landscape are a major threat to this species, as well as degraded water quality, illegal collecting, predation, and disease. Increased efforts to collect eggs from the wild and rear young at the Saint Louis Zoo are occurring to bolster wild populations and ensure this animal remains a part of Missouri's biodiversity.

Recent unpublished genetic information shows that the eastern hellbender and the Ozark hellbender should both be elevated to full species status. In addition, there appears to be several distinct lineages throughout the North American range of the eastern hellbender resulting in the possibility of additional distinct species. It is likely the disjunct population of the eastern hellbender in Missouri will be classified as one of the distinct species or subspecies as information is published.

Ozark Hellbender

Cryptobranchus alleganiensis bishopi Grobman

JEFF BRIGGLER

Adult Ozark hellbender from Shannon County.

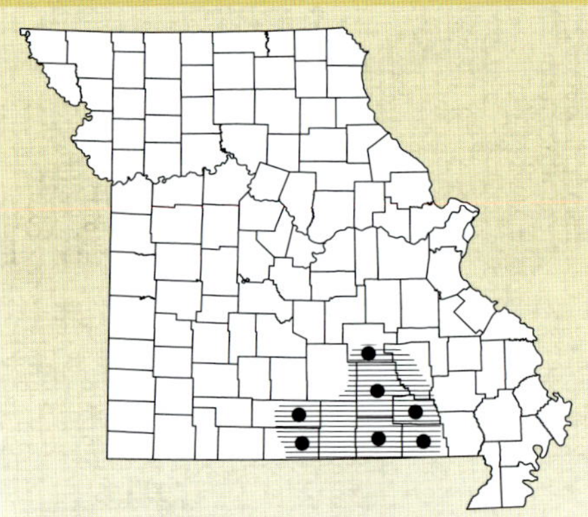

Distribution: Missouri: Restricted to the southern Ozark Highlands in spring-fed sections of the Black River system and the North Fork of the White River system. North America: Only found in southern Missouri and adjacent northern Arkansas (Powell et al. 2016).

Description: The Ozark hellbender is a large, permanently aquatic salamander that has a broad and flattened head with small, lidless eyes. The sides of the body, and often the legs, have pronounced wrinkled folds of skin. The tail is vertically flattened and rudderlike. A gill opening, although often hidden, is located on each side of the head. Body color varies from gray brown to olive green with large dark markings and blotching in both juveniles and adults. Black or dusky markings over a large part of the lip and chin are often present. The belly is a uniform dark tan or gray brown and is lighter in color than the rest of the body.

Adult Ozark hellbenders range in total length from 270 to 540 mm (10.6 to 21.3 in.) but have been known to reach 570 mm (22.4 in.; Trauth et al. 2004).

Habits and Habitat: Like the eastern hellbender, the Ozark hellbender is well adapted to riverine life. Its flattened head allows this salamander to crawl under

JEFF BRIGGLER

JEFF BRIGGLER

Ozark hellbenders live and breed in rocky rivers, such as this example in Ozark County (above). Developing Ozark hellbender eggs from the North Fork of White River in Ozark County (left).

large rocks or within bedrock crevices along the stream bottom. Its streamline body reduces water resistance, allowing easier movement while walking on the river bottom. Numerous wrinkly folds of skin along its sides provide increased surface area for respiration. Ozark hellbenders prefer rivers cooled by natural springs that contain large amounts of limestone rock for shelter. Some of the largest known populations of hellbenders were historically found in southern Missouri streams. Nickerson and Mays (1973) estimated between 341 and 573 hellbenders lived within 1 km (0.6 mi) of stream with one hellbender per 8–10 m^2 (86–108 ft^2). Additional natural history information can be found in Nickerson and Mays 1973. Because of their nocturnal and secretive behavior, these salamanders are seldom observed by the casual visitor of Ozark streams.

In a study in Missouri (Nickerson and Mays 1973), hellbenders were found to feed mainly on crayfish, although small fish, snails, and insects were occasionally eaten. Cannibalism of eggs by both males and females is known to occur also (Nickerson and Mays 1973; J. Briggler unpubl. data). Little is known on food choice by the larval hellbender. In the wild, it is assumed to feed on various aquatic insects similar to the eastern hellbender. In captivity, larval hellbenders are known to eat blackworms, bloodworms, *mysis* crustaceans, brine shrimp, and ghost shrimp (M. Wanner pers. comm.).

Breeding: Courtship and mating of the Ozark hellbender in Missouri takes place mainly in October. During this time, the hellbender increases its activity to move to breeding sites and may be seen walking on the bottom of the river. Males either occupy a nesting site or move into the nesting site just prior to mating. Such nesting sites are excavated depressions under large rocks or within bedrock crevices along the river bottom (Nickerson and Mays 1973; J. Briggler unpubl. data). Males remain posed at the entrance of the nesting chamber waiting for gravid females. The female will enter the nesting chamber and mating will occur with the female depositing eggs and the male fertilizing the eggs by spraying sperm within the nesting chamber. Egg laying is completed within two to three days (Nickerson and Mays 1973; Smith 1907), and the attending male remains with the eggs to protect them from being eaten by other hellbenders and predators (Bishop 1941; Peterson 1988; Smith 1907). After the eggs are laid, the male hellbender increases tail fanning and rocking of the body back and forth to increase oxygen supply to the eggs and himself (Bishop 1943; Settle et al. 2018). The eggs are cream to light yellow and 5 to 7 mm (0.2 to 0.3 in.) in diameter, which is considered very large compared to other salamanders. The eggs are laid in long strands similar to a string of beads and each is surrounded by a large gelatinous envelope that is approximately three times the diameter of the egg (Nickerson and Mays 1973). The number of eggs found in 37 nests in Missouri ranged from 82 to 348 with an average of 217 eggs (J. Briggler unpubl. data). It is not unusual for two or three females to deposit eggs within one nest chamber (J. Briggler unpubl. data). Larger females will produce more eggs than smaller females (Topping and Ingersol 1981). Eggs typically hatch in about 45 days (Peterson 1988); however, this can vary by a couple weeks depending on water temperature. The hatching of the embryos through the gelatinous casing typically takes place over a 7- to 14-day period with the majority occurring in a 2- to 3-day timeframe (J. Briggler unpubl. data). The newly hatched, gilled larvae are 25 to 30 mm (1 to 1.2 in.) in total length and have conspicuous, yellow yolk sacs that furnish nourishment to the young hellbender for up to three months (Nickerson and Mays 1973). Ozark hellbender larvae lose their external gills after about 1.5 to 2 years at a total length of 100 to 130 mm (3.9 to 5.1 in.; Nickerson and Mays 1973; J. Briggler unpubl. data). Information related to nest guarding by the attending male hellbender is relatively unknown, but observations in Missouri show that most larvae are absent from nesting chambers by April (J. Briggler unpubl. data), and it is assumed they move into the gravel substrate or under smaller rocks. Male and female Ozark hellbenders typically become sexually mature at a slightly smaller size and a slightly younger age compared to the eastern hellbenders (Nickerson and May 1973; Peterson et al. 1988; Taber et al. 1975; J. Briggler unpubl. data).

Remarks: Recent surveys of Missouri's populations of Ozark hellbenders have shown a 70 percent decline, with little to no evidence of recruitment of young individuals (Wheeler et al. 2003; J. Briggler unpubl. data). The specimens observed were larger, older adults. Because of these findings, the Missouri Department of Conservation classified the Ozark hellbender as state endangered in 2003. In 2011 the U.S. Fish & Wildlife Service listed it as a federally endangered species (USFWS 2011). The primary threats to this salamander are habitat alteration and degradation, over collecting, disease (mainly amphibian chytrid fungus), predation, and degraded water quality. Hellbenders are long-lived, slow-to-mature amphibians that seldom venture far within the river; therefore, changes in substrate conditions where increased sediment eliminates rocky habitat should be avoided. Because of the significant population decline of the Ozark hellbenders and the threats they face, hellbenders are being zoo bred and reared at the Saint Louis Zoo. The animals raised at the Saint Louis Zoo are being released back into their native river to ensure the Ozark hellbender remains a part of the Missouri landscape (Ettling et al. 2017).

Recent unpublished genetic information shows that the eastern hellbender and the Ozark hellbender should both be elevated to full species status. Additional genetic information suggests that two or possibly three lineages occur within the Ozark hellbender range in Missouri and Arkansas. It is likely that distinct subspecies will be classified as more information becomes available.

Family Sirenidae

Sirens

This family of permanently aquatic salamanders is restricted to North America. There are four species in two genera: *Pseudobranchus* (two species) and *Siren* (two species; Pough et al. 2018). All members of this family have external gills, are somewhat eel-like in appearance, and have only forelimbs. The dwarf sirens (*Pseudobranchus*) are small, slender salamanders that live in the extreme southeastern United States. The lesser siren and the greater siren (genus *Siren*) are much larger than *Pseudobranchus* and occur throughout the southeastern and south-central United States and the Gulf Coast of Mexico. The greater siren (*Siren lacertina*) is one of the longest salamanders in the United States with a maximum length of over 90 cm (3 ft). In Missouri, the family is represented by the western lesser siren (*Siren intermedia nettingi*).

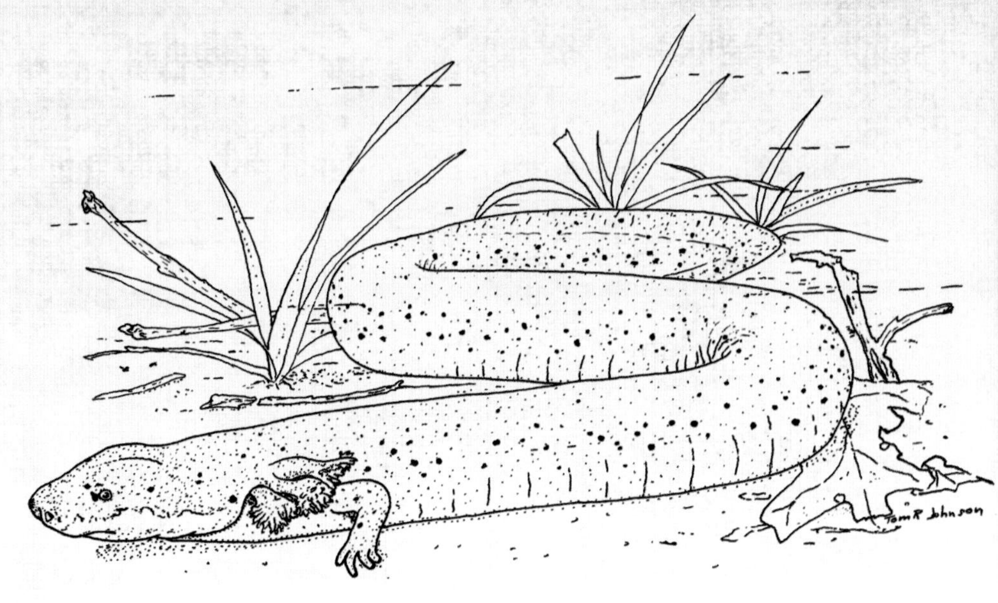

Western Lesser Siren

Western Lesser Siren

Siren intermedia nettingi Goin

TOM R. JOHNSON

Adult western lesser siren.

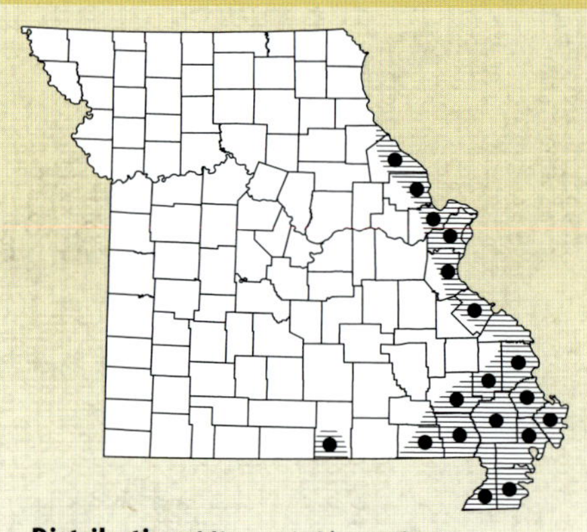

Distribution: Missouri: Along the eastern edge of the state, near the Mississippi River, and in southeastern Missouri. North America: East Texas, all of Louisiana and Mississippi, northward through the Mississippi and Ohio River drainages to central Illinois, western Indiana, and southeastern Michigan (Powell et al. 2016).

Description: The western lesser siren is characterized by external gills, small eyes, small forelimbs with four toes, and a lack of hind limbs. The general shape of this animal is similar to an eel. The three pairs of external gills are red or gray red and have a bushy appearance. Color varies from dark gray to brown to almost black. The belly is lighter than the back. Tiny dark-brown or black flecks or spots are usually scattered over the back. Costal grooves vary from 31 to 38. There is no apparent external difference between the sexes except that males are larger than females and typically have larger heads.

Adult western lesser sirens range in total length from 180 to 407 mm (7.1 to 16 in.) but have been known to exceed 500 mm (19.7 in.; Powell et al. 2016; Trauth et al. 2004).

Habits and Habitat: This aquatic salamander lives in a variety of permanent to semipermanent wetlands, including sluggish streams, ditches, ponds, sloughs, and

TOM R. JOHNSON

Head of a western lesser siren.

swamps. By day it remains hidden under clumps of aquatic plants and submerged roots or branches but becomes active at night to search for food. Its diet includes small crayfish, aquatic insects, snails, and worms. Altig (1967) found that this species can apparently obtain food by filter feeding through bottom material and in aquatic vegetation. He reported that small, aquatic crustaceans (e.g., water fleas, pill bugs, scuds, seed shrimp, etc.) may account for up to 87 percent of the total number of food items eaten. Predators, such as minks and great egrets, have been observed feeding upon sirens in Stoddard County (Frese and Britzke 2001).

A study in a managed wetland in southeastern Missouri showed that population density can be high, ranging from 1.4 to 2.2 individuals per 1 m^2 (10.8 ft^2; Frese et al. 2003). Home range averaged 95 m^2 (1,023 ft^2) but varied greatly, with some animals moving only 1 m^2 (10.8 ft^2) and others moving 232 m^2 (2,497 ft^2; Frese et al. 2003).

If the pond or slough where a western lesser siren lives begins to dry up, the animal will burrow into the bottom mud. As the mud begins to dry, the siren's skin glands produce a thick mucus layer that becomes a parchmentlike cocoon covering the entire body except the head. This covering prevents the salamander from drying too much and allows it to estivate for several months until rains again fill the pond (Gehlbach et al. 1973).

Breeding: Until recently, courtship, mating, and mode of reproduction have not been observed in this species. Reinhard et al. (2013) observed courtship among a pair of sirens swimming in circles with tail undulation and rubbing of sides. In the spring, typically March and April, a female lays approximately 100 to over 350 eggs (Gehlbach and Kennedy 1978; Godley 1983) in small pockets in the bottom mud of a pond or ditch or in aquatic vegetation. The male fertilizes the eggs by spraying sperm externally on the eggs within the nesting area (Reinhard et al. 2013). The male protects the nest by biting other sirens that get too close. He also provides paternal care to the eggs with increased aeration from constantly moving them with tail fanning and keeping the nest cleaned (Reinhard et al. 2013). Eggs hatch in 6 to 10 weeks and the young are about 11 mm (0.4 in.) in total length with stripes along the body and red patches on the head (Godley 1983). Maturity is reached after two years of development.

Remarks: Western lesser sirens are known to produce two kinds of sounds that may be used as means of communication between individuals. These include clicking sounds made by an individual when approached by another siren and a "*yelp*" sound made when captured (Gehlbach and Walker 1970).

Sirens produce a large amount of mucus on their skin. Add to this the ability to wriggle and squirm, and handling becomes next to impossible. These salamanders are not known to bite and are completely harmless to humans.

Family Salamandridae

Newts

This salamander family has 107 species in 21 genera distributed in Europe, western Asia, northwestern Africa, and North America (Pough et al. 2018). In North America, it is represented by two genera: *Taricha*, with species on the West Coast, and *Notophthalmus*, with species occurring from the East Coast to the Great Plains. Unlike most salamanders, which have smooth skin, newts have rough, almost bumpy skin (eft stage only). Adults of the West Coast newts (*Taricha*) spend their lives on land in mosses and under logs, while *Notophthalmus* adults are usually aquatic. The central newt (*Notophthalmus viridescens louisianensis*) is the only representative of this family in Missouri.

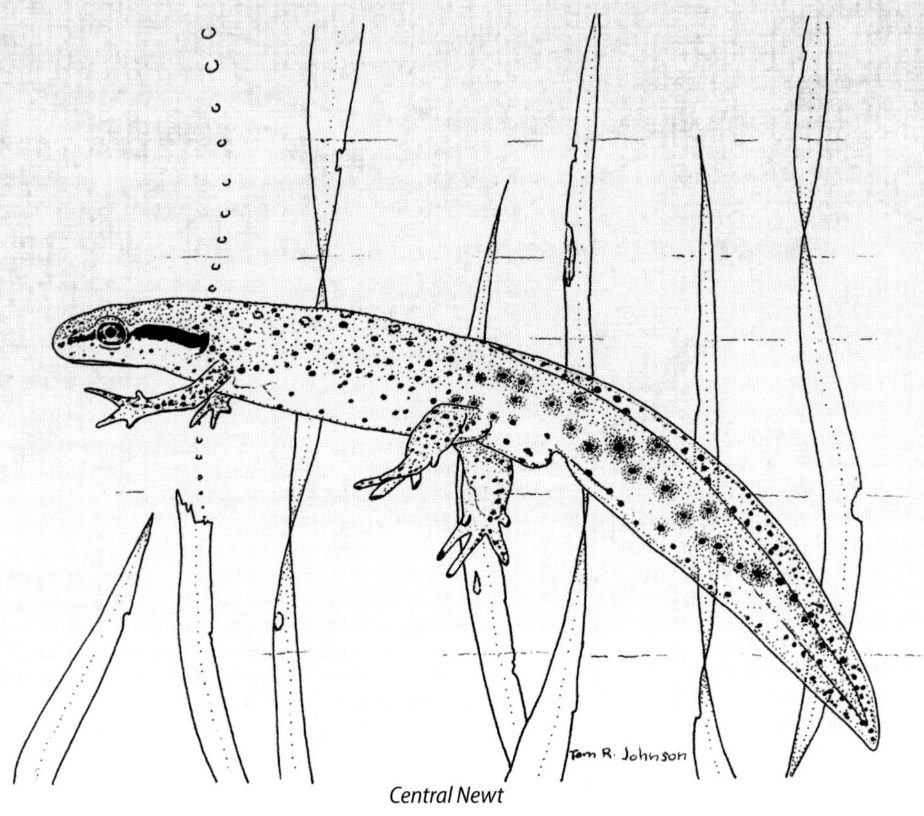

Central Newt

Central Newt

Notophthalmus viridescens louisianensis (Wolterstorff)

JEFF BRIGGLER

Adult male Central Newt from Callaway County.

Description: The adult central newt is a small, slender aquatic salamander without costal grooves or external gills. It has an olive-brown back and a bright orange-yellow belly. The dark color of the back and the yellow of the belly are distinctly separated along the sides of the body. A number of very small red spots ringed with black may be present along the back on both sides of the spine. Usually the entire body is covered with numerous small black spots that may be somewhat larger on the belly than on the back. A dark-brown or black line is present from the nostril through the eye to the forelimbs. The eyes are often orange yellow.

During the breeding season, adult males can be distinguished by the presence of very high fins on the tail, and large, swollen hind limbs with patches of black cornification along the inner thighs, as well as cornified toe tips and enlarged cloacal lips; females appear heavy bodied. The eft or middle stage of this salamander lives on land. Its dull, brown to red-brown or cream color,

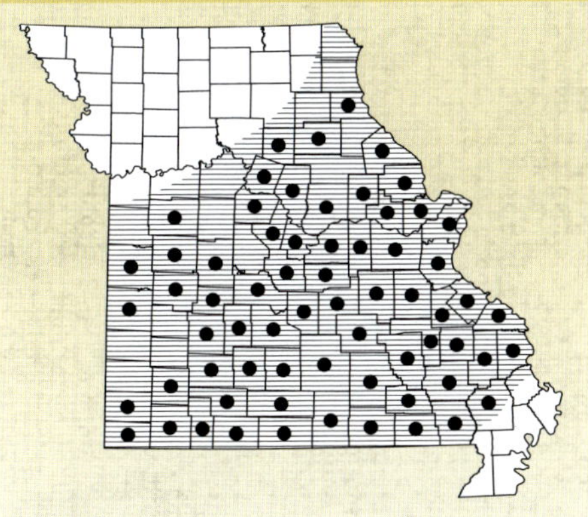

Distribution: Missouri: Throughout the southern two-thirds of the state except for extreme southeastern and the prairie regions of western Missouri. North America: Atlantic states west to Minnesota, south to eastern Texas and east through Florida (Powell et al. 2016).

rounded tail, rough skin and lack of any secondary sexual characteristics distinguish the eft stage from the adult stage.

Adult central newts range in total length from 57 to 122 mm (2.2 to 4.8 in.) but have been known to reach up to 140 mm (5.5 in.; Powell et al. 2016). Efts range in length from 35 to 86 mm (1.4 to 3.4 in.).

Habits and Habitat: Adult central newts live in woodland ponds, swamps, and occasionally water-filled ditches. They are seldom numerous in ponds that harbor fish or that lack aquatic plants. The efts take shelter under logs, rocks, or piles of dead leaves in wooded areas and may travel far from their natal ponds. A small, undisturbed pond may contain a surprisingly high number of adult newts. On one night, 102 central newts were observed in a fishless, 602 m^2 (6,480 ft^2) woodland pond in northwest Arkansas (J. Briggler unpubl. data). They are active by day or night and may be observed from shore as they swim near the surface, come up for air, or remain still in midwater. Newts remain active throughout the year and have been observed in winter swimming under the ice (Conant and Collins 1991). Newts can navigate by use of olfaction and a light-dependent magnetic compass. When displaced from their home pond, they will orient and travel toward the pond (Phillips et al. 1995).

Efts and adult newts have few natural predators because they produce toxic skin secretions that make them taste bad (Brodie et al. 1974). A study of skin toxins of the eastern subspecies documented that efts are up to ten times more toxic than aquatic adults (Brodie 1968). Also, it is well documented in Missouri's studies that newts can detect predator and prey by use of cues (i.e., visual, chemical) and adjust their behavior to increase survivorship (Mathis 2003; Mathis and Vincent 2000).

Central newts are carnivorous and will eat almost any animal food available. The food of adult newts consists mainly of worms, small mollusks, crustaceans, insects, leeches, amphibian eggs, salamander larvae, and small tadpoles (Beatty 1973). The terrestrial efts feed on small insects and tiny snails found under logs and rocks. The aquatic larvae eat smaller aquatic invertebrates.

Breeding: The central newt has a more complex courtship and life cycle than most other Missouri salamanders, generally involving four stages: egg, aquatic larva, terrestrial juvenile (eft), and aquatic adult. Breeding takes place from late March through early May. A male will swim after a female, mount her back, clasp her body with his hind limbs, and then will fan his tail toward the female. Tail fanning may help to propel a sexually stimulating odor secreted from the male's cloaca toward the female; this induces the female to allow the courtship to continue to completion. Eventually the female will begin fanning her tail, which induces the male to begin swimming about. At this time, the male's cloaca becomes more swollen; he leaves the female, and, moving forward, begins more rapid undulations. The female follows him; if she touches his tail with her head, the male can then complete his courtship act by depositing a spermatophore. The female then moves over the spot where the sperm packet was deposited and picks it up with her cloacal lips.

After the eggs are internally fertilized, they are laid singly on leaves and stems of aquatic plants and hatch three to five weeks later. A single female may lay 200 to 375 eggs from April through June (Beatty 1973; Bishop 1943); it may take several weeks for a female to complete egg laying. The gilled, pond-type larvae average 7 to 9 mm (0.3 to 0.4 in.) in length at hatching. They generally remain in the water until mid-July or early September, but some larvae do not leave the pond until October (Hocking et al. 2008). They then metamorphose into rough-skinned, land-dwelling efts without gills. These small salamanders generally live two to four years on land, hiding in leaf litter, under logs and rocks, or in rotten stumps. Here they find shelter, moisture, and food. Occasionally efts are found on rocky glades and in caves (J. Briggler unpubl. data). After the eft stage, these young newts return to a pond

TOM R. JOHNSON
JEFF BRIGGLER

Central newt larva from Pulaski County (above). Central newt eft from Callaway County (left).

or swamp to live out the rest of their lives as aquatic adults. If the adults are forced to leave a drying pond, they probably travel to a deeper pond in the area or remain on land near their original pond while awaiting rainfall. Gilled adult central newts have been found in southern Illinois (Brandon and Bremer 1966) and Massachusetts (Healy 1970), but this condition has not been reported in Missouri. Most adult newts are generally between three and eight years of age, but some individuals can likely live as long as 13 years (Petranka 1998).

Remarks: Central newts are known to have a high rate of infection with amphibian chytrid fungus (*Batrachochytrium dendrobatidis*) in Missouri (J. Briggler unpubl. data). This fungus infects the skin of salamanders and frogs, and it is known to cause massive die-offs of many species around the world. However, newt populations in Missouri appear to be stable even with carrying this fungus. Continued monitoring of newt populations in Missouri for this fungus and other introduced pathogens is necessary to ensure that newts and other amphibians remain common throughout Missouri's landscape.

The presence of fishless woodland ponds, swamps, and small sloughs is vital to the existence of this species. Efts require downed logs, brush piles, and other forest floor debris to survive their sojourn on land. Concerned landowners and land managers are encouraged to provide or protect these habitat elements.

Family Ambystomatidae

Mole Salamanders

Members of this family (32 species in the genus *Ambystoma*) live throughout most of the United States and parts of Canada and Mexico (Pough et al. 2018). There are six species of *Ambystoma* in our state. Mole salamanders are thus named because they spend most of their time underground. The majority of Missouri species breed in early spring, courting and depositing their eggs in shallow ponds. Exceptions to this breeding behavior are discussed in the species accounts. The gilled larvae usually transform from aquatic to land animals in summer, but some may spend the winter as larvae and transform the following summer. Outside of the breeding season, adults spend most of their time in the soil (often in burrows made by small mammals) or under logs and rocks. Salamanders in this family are often active at night, especially after a heavy rain.

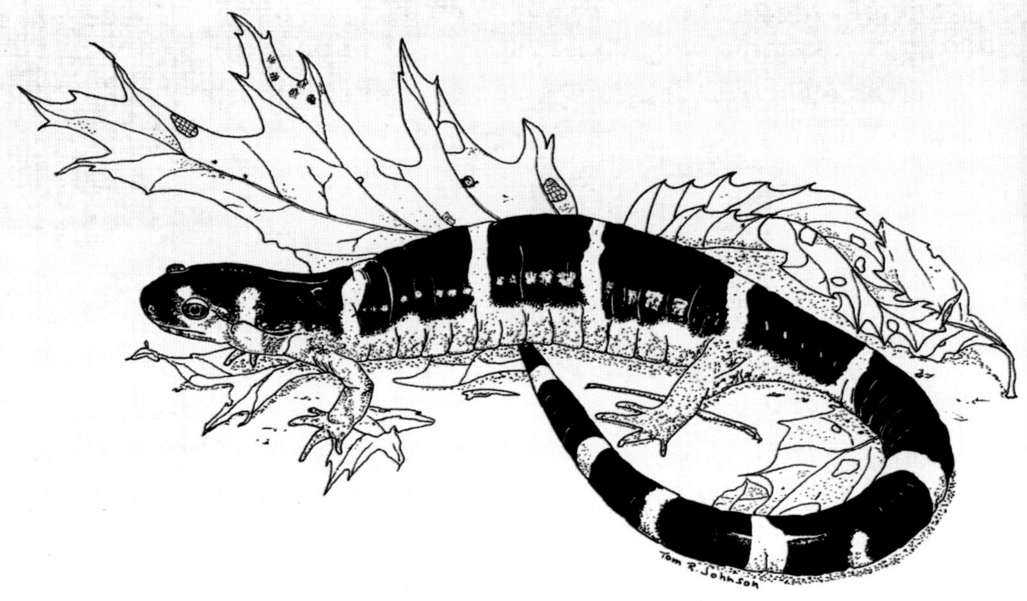

Ringed Salamander

Ringed Salamander

Ambystoma annulatum Cope

JEFF BRIGGLER

Adult ringed salamander from Stone County

Description: The ringed salamander is a slender and elongated salamander, usually with 15 costal grooves between the limbs. The head and neck are small and somewhat elongated compared to other members of the genus. The dorsal ground color ranges from grayish black to black; the belly is normally slate gray to buff yellow. A series of bold, narrow white to yellow colored rings usually extends over the back but may be broken at the midline. The rings never completely encircle the body.

Adult ringed salamanders range in total length from 140 to 180 mm (5.5 to 7.1 in.) but have been known to reach 238 mm (9.4 in.; Powell et al. 2016).

Habits and Habitat: Because of the secretive nature of this salamander, little is known about its habits. Ringed salamanders are fossorial and probably live under logs and rocks or in burrows made by small mammals, seldom venturing into the open and preferring heavily forested areas during

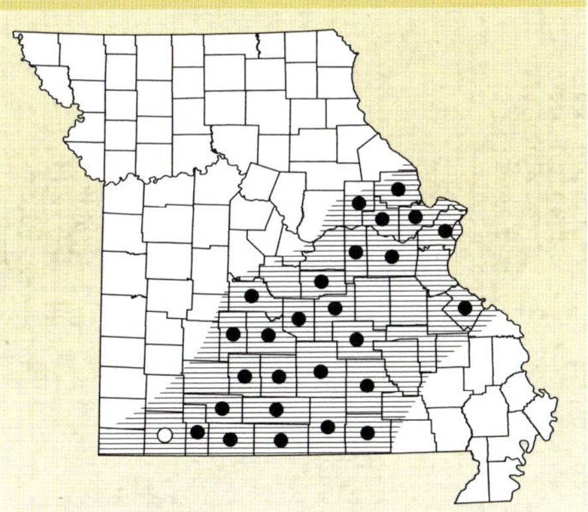

Distribution: Missouri: Southwestern and central Missouri Ozarks, and in the river hills of the Missouri River in the eastern section of the state. North America: An Ozark Highland and Ouachita Mountain endemic that occurs only in extreme eastern Oklahoma, northern and western Arkansas, and southern and eastern Missouri (Powell et al. 2016).

Late larval stage of a ringed salamander from Warren County (above). Freshly laid eggs of a ringed salamander from a fishless pond in Warren County (right).

much of the year (Dowling 1956). They migrate to ponds in the autumn for breeding with population size ranging from as few as 319 to as high as 1,971 individuals (Semlitsch et al. 2014a). Healthy populations of the ringed salamander are found in large, intact forest with numerous fishless ponds (Crawford et al. 2017; Semlitsch et al. 2014a). Although unusual and likely underreported, ringed salamanders, especially juveniles, have been observed climbing small saplings, grass clumps, stumps, and tree trunks to a maximum height of 152 cm (59.8 in.; Osbourn et al. 2012; T. Johnson pers. obs.).

There is little information available on the diet of juvenile and adult ringed salamanders, with only earthworms being documented (Hutcherson et al. 1989), but food in the wild probably includes insects and land snails. In captivity, this salamander feeds on earthworms and small crickets (J. Briggler pers. obs.). Larvae are known to feed on a variety of aquatic organisms such as water fleas, mosquito larvae, snails, earthworms, frog eggs, etc. (Hutcherson et al. 1989; Nyman et al. 1993).

Breeding: The ringed salamander breeds in autumn, primarily in September and October (Hocking et al. 2008; Hutcherson et al. 1989; Semlitsch et al. 2014a). However, winter breeding has been observed on occasion (Trauth et al. 1989a; J. Briggler unpubl. data). The salamanders are stimulated by heavy rains and cooler temperatures to travel by night to fishless, woodland ponds, where they may congregate by the hundreds (Briggler et al. 2004; Trapp 1956). The males arrive at the ponds before females and can be distinguished by their slender bodies and swollen cloacae. Spotila and Beumer (1970) and Spotila (1976) studied the ringed salamander in northwestern Arkansas. They observed the courtship of this species to be similar to the liebesspiel ("love play"; Bishop 1941) of the spotted salamander (*Ambystoma maculatum*).

Courtship commences soon after females begin to arrive at the breeding pond, with 2 or 3 to 25 males courting each gravid female. The spermatophore has a wide base and a short stalk that supports the sperm cap. Egg laying begins on the night of courtship or the next night, and egg laying is completed within a couple of days. Each female will lay a number of eggs either singly, in small strings, or in loose masses, with each clump containing generally 4 to 31 eggs (Trapp 1959). The eggs are laid on submerged branches, or aquatic plant stems, or on the bottom of shallow ponds. In a population of ringed salamanders studied in Stone County, 16 females contained between 300 and 500 eggs inside the body (Hutcherson et al. 1989). The eggs begin hatching in 9 to 16 days depending on water temperature. Newly hatched larvae are approximately 11 to 12 mm in length (0.4 to 0.5 in.; Semlitsch et al. 2014a). The larval period may last 6 to over 8 months and transformation to juveniles occurs mainly in May or early June (Hocking et al. 2008; Hutcherson et al. 1989; Semlitsch et al. 2014a). Over 99 percent of embryos and larvae die prior to hatching or during transformation (Peterson et al. 1991; Semlitsch et al. 2014a). An additional study of the Stone County population found that some ringed salamander larvae became cannibalistic when there was an extended breeding season, with larvae from early egg laying eating newly hatched larvae near the end of the breeding season. Also, cannibals transformed into larger juveniles than noncannibalistic larvae (Nyman et al. 1993). Fish greatly affect breeding and larval survivorship (Crawford et al. 2017), and even small mosquitofish and fathead minnows prey on early stages of ringed salamanders (Drake et al. 2014). Most juveniles become reproductively mature within two to three years of age (Semlitsch et al. 2014a). Most of the adult population likely lives less than 10 years, but estimates indicate that some individuals may live up to 18 years (Winter and Wilkinson 1994).

Remarks: The ringed salamander is listed as a species of conservation concern in Missouri because of its restricted range in highland forests in Missouri, Oklahoma, and Arkansas, patchy distribution, and limited information on basic biology. Management of this species should focus on creating and maintaining fishless, semipermanent to permanent ponds needed for breeding and larval growth within large areas of intact forests. Fragmentation of the forest by conversion into open lands and roads can lead to defined migration patterns of this species to reach breeding ponds. A study along a 400 m section of a paved county road observed over 1,000 ringed salamanders crossing from a forested ravine to breed in a small pond located in a nearby pasture (Briggler et al. 2004). Automobile traffic unfortunately had an impact by running over these inconspicuous salamanders on wet roads at night. Mortality was especially high when rainy nights occurred during heavy traffic nights on weekends (Briggler et al. 2004). As the forest is removed and road traffic increases, local populations will decline or disappear across the landscape.

Spotted Salamander

Ambystoma maculatum (Shaw)

Adult spotted salamander from Warren County.

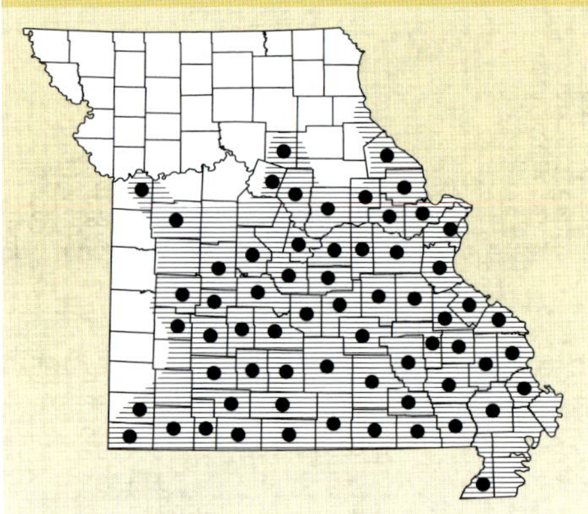

Distribution: Missouri: Throughout the forested habitats in the southern two-thirds of the state, except uncommon in the eastern part of the Bootheel. North America: Southeastern Canada to Georgia, west to eastern Texas and Oklahoma, and north to northwestern Wisconsin and southern Ontario (Powell et al. 2016).

Description: The spotted salamander has a background color of slate black, with a dark-gray belly. There are two irregular rows of rounded yellow spots from the head onto the tail. The number of yellow spots may range from 17 to 78 (Pierce and Shayevitz 1982). Some specimens collected in Missouri lacked most or all yellow spots. Spots on the head may be bright orange. Sides of the head, neck, and body usually have small white flecks. Specimens may have 11 to 13 costal grooves.

Adult spotted salamanders range in total length from 112 to 197 mm (4.4 to 7.8 in.) but have been known to reach 248 mm (9.8 in.; Powell et al. 2016).

Habits and Habitat: This species lives in damp, hardwood forests in the vicinity of shallow ponds, usually hidden under logs or rocks, inside piles of dead leaves or in burrows of other small animals. Spotted salamanders venture forth at night in search of earthworms, insects, spiders, and land snails and are often seen crossing roads on

JEFF BRIGGLER

TOM R. JOHNSON

Fishless, breeding pool used by spotted salamanders from Stoddard County (above). Spotted salamander eggs from Warren County (left).

warm, rainy nights in late winter or early spring. A radio-telemetry study in south-central New York state showed that spotted salamanders spend most of their time during the active season in shallow burrows made by short-tailed shrews and overwinter in the deeper burrows of white-footed mice (Madison 1997). The population size of spotted salamanders is influenced by the availability of small mammal burrows (Regosin et al. 2003). Spotted salamanders can be quite abundant in breeding ponds with numbers ranging from as few as 202 to as high as 1,083 individuals (Semlitsch and Anderson 2016). Although unusual, spotted salamanders, especially juveniles, have been observed climbing grass clumps and tree trunks, to an observed maximum height of 1 m (39.4 in.; Osbourn et al. 2012).

Breeding: Spotted salamanders breed in the late winter to early spring. Rainfall and warming of the soil after freezing winter temperatures are needed to stimulate breeding migrations (Sexton et al. 1990). During the first warm rains in late February to mid-March, they begin to congregate in shallow, fishless, woodland ponds to court and lay eggs. There is a strong possibility that this species uses its sense of smell (olfactory orientation) to help locate suitable breeding ponds (McGregor and Teska 1989) and terrestrial habitat (Rittenhouse et al. 2004).

In studies in St. Louis County (Sexton et al. 1986) and Warren County (Semlitsch and Anderson 2016) breeding migration of spotted salamanders was followed for many years. More males than females were found at ponds in these studies. Certain areas of a pond, especially the area of the pond connected to wooded valleys, were used as entrance and exit points (Sexton et al. 1986). Males appeared to be more consistent users of these pathways than females. The salamanders studied traveled up to 172 m (564 ft) to the pond (Sexton et al. 1986). Another study (Phillips and Sexton 1989) found that both sexes have a strong ability to orient themselves to a specific woodland pond during breeding migrations. It is speculated that this helps prevent random wandering over the forest floor to locate suitable breeding ponds or pools.

During courtship two or more salamanders go through a series of movements somewhat like a nuptial dance (Arnold 1976; Johnson 1983). Aquatic courtship, known as liebesspiel ("love play"; Bishop 1941), takes place at night in shallow water. The males walk and swim around, nudging the females. A male may swing his head from side to side near a female or nudge her at the base of her tail. Eventually, the female will move forward and pick up a packet of sperm left by one of the males. The sperm are located on the top of a clear stalk of jelly, known as a spermatophore, which is about 6 mm (0.24 in.) high. These stalks of jelly may nearly cover a pond bottom and can be found the next day as evidence of the salamanders' breeding frenzy the previous night. The female picks up the sperm packet with her cloaca and stores it in a chamber (spermatheca) in the roof of the cloaca. The eggs are fertilized as they are laid in a mass on submerged branches or aquatic plants, usually within one or two days after courtship. A female spotted salamander generally lays two to four egg masses, each usually between 80 and 110 mm (3.2 and 4.3 in.) across and produces a total of 200 to 400 eggs. Each egg mass may contain from 20 to over 300 eggs, but averages 149 eggs (Trauth et al. 1995).

Green, single-celled algae (*Oophila amblystomatis*) usually grow in the egg masses of spotted salamanders (Gatz 1973). Recently, Kerney et al. (2011) documented that the algae actually grow directly on the early stages of the developing embryos within the egg capsule. Mode of transmission of the algae to the embryo is still somewhat confusing, but it is likely obtained from water in ponds or passed down from the parents. The developing egg and egg capsule takes on a light-green color due to the presence of the algae, but the color is not visible to the naked eye until the algae have had a week or more to grow. This is viewed as a type of symbiotic relationship; it is assumed that the algae provide oxygen to the embryos and the developing larvae provide carbon dioxide and waste products that can be used by the algae. However, the benefits of this association for the algae and salamander are still somewhat complex (Burns et al. 2017).

The gilled larvae hatch in about four to six weeks, average 13 mm (0.5 in.) in length, and remain in the water to feed. Most larvae depart the pond from mid-June through early September (Hocking et al. 2008), but some larvae may overwinter in the pond and metamorphose during the following spring. Many individuals will return to the same pond to breed year after year. In a Missouri study, only a very small number (average 0.5 percent) of the larvae depart the pond due to high mortality (Semlitsch and Anderson 2016). Most juveniles become reproductively mature in two or more years (Semlitsch and Anderson 2016) and have been reported to live over 20 years in captivity (Pope 1928; 1937).

Remarks: A genetic study of spotted salamanders in several midwestern states showed that this species may not have moved into Missouri until around 5,000 years ago, long after the retreat of the Wisconsin glacier (Phillips 1994).

This salamander is often not seen by casual observers due to its activity on rainy nights but is considered quite abundant in Missouri, especially in the Ozarks. Keeping this species common in Missouri will require small, fishless ponds necessary for breeding and larval growth within large intact forest for adults to live. Preserving a 200 to 500 m (656 to 1,640 ft) buffer of mature forest around breeding wetlands is important for maintaining healthy populations of the spotted salamander.

Marbled Salamander

Ambystoma opacum (Gravenhorst)

JEFF BRIGGLER

Adult marbled salamander from Bollinger County.

Description: This small, stout salamander has silvery, white, or gray, saddle-shaped markings on its body from head to tail. These crossbands vary in shape and also in color, from silver white to gray. The back is generally shiny jet black while the belly is plain black. Adult male marbled salamanders have white or silver crossbands; adult females have light- to dull-gray crossbands. Individuals may have 11 or 12 costal grooves.

Adult marbled salamanders range in total length from 90 to 107 mm (3.5 to 4.2 in.) but have been known to reach 136 mm (5.4 in.; Powell et al. 2016).

Habits and Habitat: The marbled salamander is secretive—it spends most of its time under rocks, logs, or forest debris. It prefers forested areas but also may occur in open, sandy woods and on rocky, dry hillsides. Food probably consists of insects, worms, slugs, and other invertebrates found under

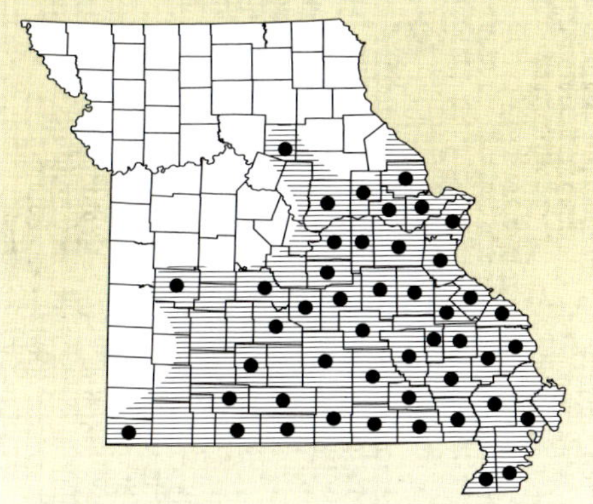

Distribution: Missouri: The Mississippi Alluvial Basin, the Ozark Highlands, the forested river hills of the Missouri River in eastern Missouri, and an isolated report from St. Clair County. North America: New England to northern Florida, west to eastern Texas and Oklahoma, and north to isolated populations in Michigan and northern Indiana (Powell et al. 2016).

Adult female marbled salamander with a freshly laid clutch of 175 eggs from Warren County (above). Fishless, woodland pond used for breeding by marbled salamanders from Carter County (right).

TOM R. JOHNSON

JEFF BRIGGLER

rocks or logs. These salamanders are seldom encountered except during the autumn breeding season.

Breeding: In this species, courtship and spermatophore deposition take place on land near fishless, woodland ponds or swamps. However, a study in South Carolina showed that close to half (31 to 49 percent) of females had been inseminated before reaching their breeding pond. This showed that, in some populations, courtship occurs before males and females reach their breeding pond (Krenz and Scott 1994). Fertilization is internal (eggs are fertilized as they move through the cloaca), and each female may lay 50 to 200 eggs (Bishop 1943) with an average clutch size of 107 eggs from an Arkansas study (Trauth et al. 1989b). We have found six females on the margins of a small pond in Warren County in late September to early October attending from 96 to 131 eggs per female. A gravid female also was found and retained in captivity, and she eventually laid 175 fertilized eggs. The eggs are laid in small depressions under logs, in leaf litter, or under vegetation at the edge of reduced ponds or dry beds of temporary ponds (Petranka 1998). In a population in Warren County, the majority of nests were found along the shallowest parts of breeding ponds; however, nest site selection within a pond basin can be variable, as marbled salamanders often show a preference to nest under grass clumps (Figiel and Semlitsch 1995).

The female stays coiled around the eggs until they are covered by rising water levels caused by autumn rains. Brooding behavior by the female probably assists with retaining moisture to the eggs, reducing fungal outbreaks, and protecting them from predators (Petranka 1998). However, in a study of this species in North Carolina, up to one-quarter of the nests were not brooded by a female, although nests were more likely to survive to hatching if guarded by a female. The North Carolina study also found that marbled salamander nests with attending females had more eggs than unattended nests (92 vs. 76 eggs; Petranka 1990). A study by Croshaw and Scott (2005) further supports that nest guarding by the female increases nest success. It appears that nests attended by females are protected from some predators, such as millipedes that are known to eat eggs of marbled salamanders (Mitchell et al. 1996).

Females remaining with the eggs through hatching are likely to brood eggs between two weeks and several months depending on rainfall that inundates the pond and the eggs (Petranka 1998). Embryos can develop within the egg capsule to within a few days of hatching while awaiting fall rains (Trauth et al. 2004). The larvae hatch and develop in the pond over the winter, and metamorphosis takes place in late spring into summer. At the time of transformation, the larvae tend to average approximately 50 mm (2 in.) in total length (Minton 2001; Trauth et al. 1989b). Recently transformed juveniles are black with numerous scattered light flecks on the body. After one to two months, the adult pattern is more pronounced (Petranka 1998).

Remarks: The marbled salamander and many other Missouri amphibians require small, fishless, woodland ponds for reproduction. Ponds and pools with a gentle sloping bank are especially important for this species. Their breeding habitat can be enhanced by providing logs along the edge of the water for egg-laying niches.

Mole Salamander

Ambystoma talpoideum (Holbrook)

Adult mole salamander from Ripley County.

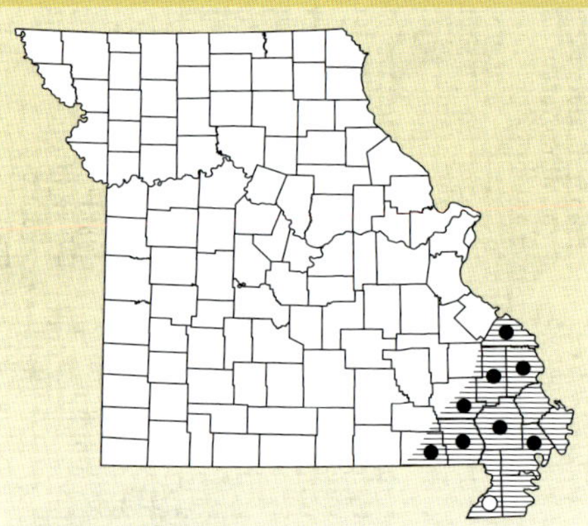

Distribution: Missouri: Restricted to the Mississippi Alluvial Basin and adjacent Black River Hills Border of the Ozark Highlands in southeastern Missouri. North America: Southern Virginia south to the northern third of Florida, west to extreme eastern Texas and southeastern Oklahoma, north to southeastern Missouri and extreme southern Illinois with disjunct populations in Tennessee, Mississippi, Georgia, and North Carolina (Powell et al. 2016).

Description: A mole salamander has a large, broad head, small body and tail, and large limbs. There are 10 or 11 costal grooves. It is usually dull gray or brown, and on most individuals, there are light-gray to blue-gray flecks over most of the body, limbs, and tail.

Adult mole salamanders range in total length from 75 to 100 mm (3 to 3.9 in.) but have been known to reach 132 mm (5.2 in.; Daniel 2011; Powell et al. 2016).

Habits and Habitat: This species lives primarily in lowland forests in association with marbled and small-mouthed salamanders but also occurs in adjacent upland hardwood forests. It finds shelter under logs, leaf litter, and in the soil. Mole salamanders are seldom encountered because they rarely venture above ground except during the breeding season. A variety of small insects, worms, and land snails make up their diet.

Breeding: Mole salamanders breed in leaf-littered, ephemeral or semipermanent

Late larval stage of mole salamander from Wayne County (above). Fishless, ephemeral pool created by a tree fall where mole salamanders breed from Bollinger County (left).

pools, ditches, and woodland, fishless ponds. Adults move on rainy nights through the forest to breeding ponds and pools during late autumn or early winter, where breeding takes place between December and February. Courtship and mating occurs in the water. Eggs are fertilized as they pass through the cloaca. In a Louisiana population study, females produced 226 to 401 eggs (Shoop 1960). The eggs are loosely attached to submerged twigs or leaves in small clumps containing 4 to 20 eggs (Bishop 1943). However, a study by Trauth et al. (1995) found an average of 41 eggs per mass. Of mole salamanders studied in northwestern Louisiana, 68 percent of the population breed annually and another 18 percent breed every other year, and some specimens were documented to be at least eight years of age (Raymond and Hardy 1990). The larval stage generally lasts three to four months (Smith 1961), and transformation occurs from late April into mid-June throughout much of its range. In a Missouri study, most larvae overwintered and metamorphosed in May, approximately 15 months after hatching (Russell 2004). Populations with neotenic or extended larval periods, however, have been found in several states (Semlitsch 1987; Trauth et al. 1993), including Missouri (Russell 2004). A study in South Carolina showed that mole salamander larvae will develop faster if their pond or pool begins to dry (Semlitsch and Wilbur 1988). Newly transformed mole salamanders may be 50 to 65 mm (1.9 to 2.6 in.; Bishop 1943) in total length.

Remarks: The mole salamander was first reported in Missouri by Easterla and Gregory (1967). This is a species of conservation concern in Missouri. The mole salamander requires fishless wetlands to breed and lowland forests to survive. This habitat has been greatly reduced in southeastern Missouri by the clearing of forests and draining of the lands, and undoubtedly populations have been reduced. Protection of the remaining fishless wetlands and forest, as well as the construction of wetlands and planting of trees, will better secure the future of this species in Missouri.

Small-mouthed Salamander

Ambystoma texanum (Matthes)

JEFF BRIGGLER

Adult small-mouthed salamander from Bollinger County.

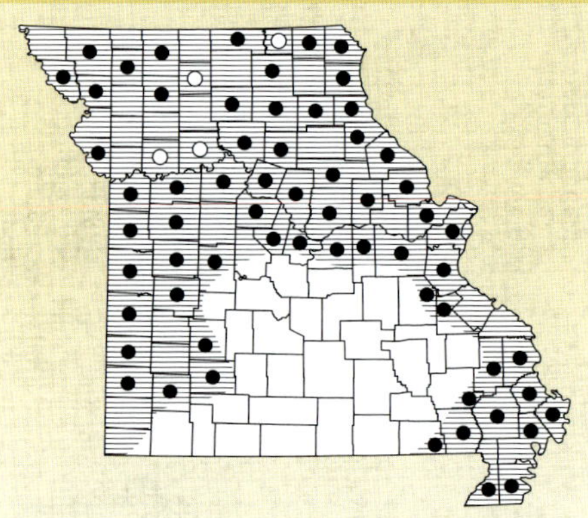

Distribution: Missouri: Throughout much of the state except for most of the Ozark Highlands. North America: Southeastern Michigan south to Alabama, west to the eastern half of Texas and north to extreme southeastern Nebraska and southern Iowa (Powell et al. 2016).

Description: The small-mouthed salamander is medium sized, has a small head and mouth, and is usually dark gray to black or dark brown. The body, limbs, and tail may be mottled with small, irregular flecks of tan, grayish blue or gray. The belly is usually dark gray to black, but small flecks may be present. There are 13 to 15 costal grooves.

Adult small-mouthed salamanders range in total length from 100 to 140 mm (3.9 to 5.5 in.) but have been known to reach 191 mm (7.5 in.; Powell et al. 2016; Redmer 1995).

Habits and Habitat: Small-mouthed salamanders live under rocks, rotten logs, piles of dead leaves, or in burrows in the soil. They typically live in bottomland and floodplain forests but can be found on rocky hillsides, woodlands, prairies, and even farmlands. In Missouri, they have been observed living in mole burrows (J. Bursewicz pers. comm.) and crayfish burrows (J. Briggler pers. obs.; R. Daniel pers.

JEFF BRIGGLER

JEFF BRIGGLER

Ephemeral pool used for breeding by small-mouthed salamanders from Callaway County (above). Small-mouthed salamander egg masses from Callaway County (left).

comm.). Three adults were found under a pile of old plywood within a hay barn in late winter (J. Briggler pers. obs.).

As with other species in this genus, the food of small-mouthed salamanders consists of earthworms, slugs, moths, centipedes, beetles, and insects (Minton 2001; Owen and Juterbock 2013). When threatened by a predator, small-mouthed salamanders exhibit defensive posturing consisting of lowering the head, curling the body, and raising and waving the tail, which exudes a sticky, white secretion that is noxious (Brodie 1977; Minton 2001; Petranka 1998).

Breeding: In Missouri, small-mouthed salamanders typically breed from late February to early April. They emerge from underground wintering locations en masse on rainy nights and migrate to fishless wetlands. Breeding locations can be found both in wooded or grassland habitats with preference to breed in shallow, temporary wetlands (e.g., pools, roadside ditches, flooded fields), but they will utilize more permanent wetlands, such as farm ponds, borrow pits, and swamps (LeClere 2013; Petranka 1998; Trauth et al. 2004). Occasionally eggs are deposited on twigs or the bottom of sluggish streams in headwater tributaries (Petranka 1998; J. Briggler pers. obs.).

Courtship involves a male nudging a female with his snout, then moving or swimming over and under the female (Petranka 1982a). Wyman (1971) reported an observation of male *A. texanum* courtship in an Illinois population that included the clasping (amplexus) of females during courtship, but such behavior has not been observed in other studies (Garton 1972; Keen 1975; Petranka 1998). Males deposit spermatophores on submerged twigs, leaves, or on the pond bottom, and a female follows and picks up the sperm located on the top of the spermatophore with her cloaca (Garton 1972; Keen 1975).

Each gravid female may deposit 300 to over 800 eggs (Plummer 1977) with an average clutch size of 545 ova in an Arkansas study (Trauth et al. 1990). Small-mouthed salamander eggs are laid singly, in small, loose clusters or sausage-shaped masses generally consisting of 6 to 30 eggs (containing an average of 13 eggs; Trauth et al. 1990) that are attached to submerged twigs or plant material (Petranka 1982b). Depending on water temperature, eggs may take two to eight weeks to hatch into gilled, pond-type larvae. The larvae remain in the wetland for two to four months feeding upon numerous types of aquatic invertebrates (McWilliams and Bachmann 1989). A study by Ryan (2007) showed that completed metamorphosis of larvae increased with longer wetland hydroperiod and drying of wetlands that are less than 2.5 months could lower the number of larvae surviving. Larvae, approximately 50 to 60 mm (2 to 2.4 in.) in total length, typically move onto land from late May through July (Trauth et al. 2004), and most of the population remains within 50 to 60 m (164 to 197 ft) of their hatching location (Parmelee 1993; Williams 1973).

Remarks: Easterla (1970) reported finding an albino small-mouthed salamander in Stoddard County. The small-mouthed salamander is likely common throughout much of its range, but is rarely observed due to its secretive habits. Likely local extinctions have occurred due to land conversion and destruction of shallow, temporary wetlands. Keeping this species common in Missouri is dependent upon protection and construction of temporary wetlands that are fishless within the natural terrestrial habitat.

Eastern Tiger Salamander

Ambystoma tigrinum (Green)

JEFF BRIGGLER

Adult eastern tiger salamander from Howell County.

Description: The eastern tiger salamander is a dark, medium to large salamander with yellow or olive blotches over the head, body, and tail. The ground color is black or dark brown. The large spots or blotches vary greatly in size and shape; blotch color ranges from bright yellow to dull olive brown. The belly is dark gray or black with yellow to olive-yellow mottling. There are 11 to 14 costal grooves. As with all members of the genus *Ambystoma*, males usually have longer tails than females, and during the breeding season have a swollen cloaca.

Adult eastern tiger salamanders range in total length from 180 to 210 mm (7.1 to 8.3 in.) but have been known to reach 330 mm (13 in.; Powell et al. 2016).

Habits and Habitat: Eastern tiger salamanders live in a wide variety of habitats, including woodlands, savannas, swamps, prairies, old fields (near farm ponds), and are occasionally observed in and around croplands (LeClere 2013; Steen et al. 2006; Trauth et al. 2004).

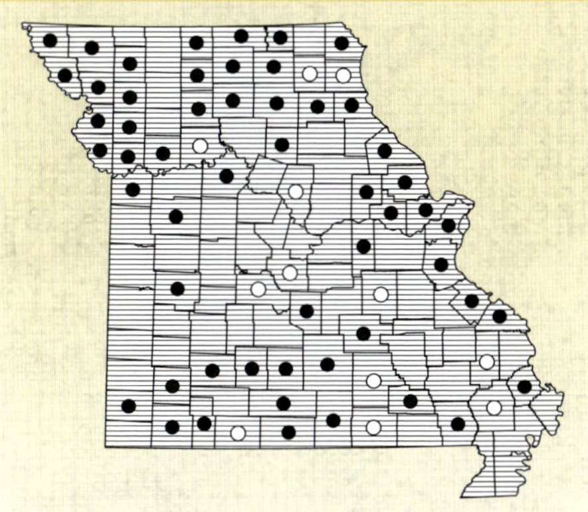

Distribution: Missouri: Presumed to occur statewide with few to no records in the Osage Plains of southwest and Mississippi Alluvial Basin of southeast. Most common in the northern half of the state, especially northwestern. Small, isolated populations occur throughout the Ozark Highlands in the former savanna and prairie regions. North America: Southeastern Manitoba south to northern Arkansas and eastern Texas, east to northern Florida and north to New York. Mostly absent from the Appalachian Highlands (Powell et al. 2016).

Fishless, prairie pool where eastern tiger salamanders breed from Linn County (above). Eastern tiger salamander larva from Linn County (right).

JEFF BRIGGLER

TOM R. JOHNSON

In Missouri, they are most commonly encountered in grassland habitat and are occasionally found in wells, basements, and root cellars. They spend most of their time in burrows or under logs and are active only at night. Eastern tiger salamanders become migratory during autumn rains, moving to ponds where breeding will take place in the upcoming winter or early spring. Unlike many of the *Ambystoma* salamanders, eastern tiger salamanders actively dig their own burrows in the soil (Gruberg and Stirling 1972; Semlitsch 1983), although they can live within crevices and burrows of small animals (Collins et al. 2010).

Prey of eastern tiger salamanders includes any animal small enough for them to swallow, ranging from small invertebrates to young mice, small frogs and snakes, and even other salamanders (Collins et al. 2010; Petranka 1998). Common foods include earthworms, insects, spiders, slugs, and snails.

Breeding: Little is known on the biology of this species in Missouri. Tiger salamanders, especially males, begin to move to temporary wetlands, fishless ponds, or marshes during rainy autumn weather. Reports of tiger salamanders by Missouri's residents occur after rains, mainly in November and December, as the salamanders are observed near homes or crossing roads. Females also may move toward ponds in autumn, but it appears that much of their migration is in early spring. Courtship and egg laying take place in the water, generally between February and April in Missouri, with most observations of adults at breeding wetlands in March. In Holt County, hundreds of tiger salamanders were observed crossing a highway in late March en route to a breeding site (Easterla 1972). However, Seale (1980) documented breeding in the St. Louis area in February. Breeding has been reported in January and February in Arkansas (Trauth et al. 1990) and from late February through April in Illinois and Iowa (Brandon and Bremer 1967; LeClere 2013). Courtship usually involves males and females rubbing together, with much tail thrashing and some nipping. Eventually the male moves away and the female follows, keeping her head close to the cloaca of the male. He then deposits a spermatophore on the pond bottom, which the female quickly picks up with her cloaca, and her eggs are fertilized as they pass through the cloaca.

Each female may lay up to 1,000 eggs (Collins et al. 2010), which are deposited in small clusters or masses containing 23 to 110 eggs (averaging 52 eggs per mass; Bishop 1943). However, Trauth et al. (2004) reported two egg masses in Arkansas containing 96 and 165 eggs. The eggs generally hatch within several weeks but may take up to 40–50 days if water temperatures are cold (Petranka 1998; Trauth et al. 1990). The recently hatched pond-type, gilled larvae are approximately 14–17 mm (0.6–0.7 in.) long and grow rapidly throughout the spring and summer, feeding on a variety of aquatic invertebrates, other amphibian larvae, and even each other. Transformation to a land-dwelling subadult takes place from June through August in Missouri when larvae reach 60–75 mm (2.4–3 in.) snout-to-vent length (Seale 1980). Large larvae between 90 and 110 mm (3.5–4.3 in.) in total length that likely were close to metamorphosis were observed in temporary wetlands in southern Missouri (Dent County) in mid-May and in northwestern Missouri (Buchanan County) in late June. Neoteny (the condition of gilled adults) has been reported for this species throughout its range but to date has not been conclusively found in Missouri. However, a larva measuring 127 mm (5 in.) total length was collected in March in Lewis County that must have at least overwintered (Shulse 1994). Tiger salamanders are known to live over 10 years in captivity with a longevity record of 20 years and 6 months (Snider and Bowler 1992). A larva collected from a small, fishless pond in southwest Missouri lived 12 years and 7 months in captivity (J. Briggler unpubl. data).

Eastern tiger salamanders are known to travel on land a considerable distance from breeding wetlands. Radio-tracking by Madison and Farrand (1998) in New York showed individuals can migrate up to 286 m (938 ft) with an average movement of 60.5 m (198.5 ft) from a wetland. Another radio-tracking study conducted in Georgia showed a maximum movement of 255 m (837 ft) from the wetland with an average of 110 m (361 ft; Steen et al. 2006).

Remarks: The larvae of this species are often erroneously called waterdogs. The term waterdog is a common name for the mudpuppy. At one time, larval tiger salamanders could be sold and used for fish bait in Missouri, but this market was eliminated because many bait animals carry infectious diseases that would negatively impact the native amphibians once established.

The eastern tiger salamander is listed as a species of conservation concern in Missouri because of loss and fragmentation of native prairies and savanna habitat, loss of breeding wetlands, patchy distribution, and small population size. Constructing and maintaining shallow, fishless wetlands for breeding is vital to the long-term persistence of this species in Missouri.

Family Proteidae

Mudpuppies and Olm

This family contains only two genera with six species (Pough et al. 2018). The first, *Proteus*, contains only one species, *P. anguinus*, the European olm. It is a slender, white, blind cave salamander, reaching a total length of 250 mm (9.8 in.). This species has bright-red gills and three toes on the front limbs and two toes on hind limbs; it inhabits cave streams in western Balkans and northern Italy (Pough et al. 2018).

The other genus of this family is *Necturus*, which consists of five species that are exclusively North American. They are commonly known as mudpuppies or waterdogs; together their ranges cover most of the eastern United States. These salamanders are totally aquatic. They have permanent gills and can live in a variety of habitats, including streams, rivers, sloughs, and lakes. In larger lakes in the northern part of its range, the mudpuppy has been netted at a depth of 27 m (88.6 ft; Pough et al. 2018). *Necturus maculosus* is the only species represented in Missouri and consists of two subspecies.

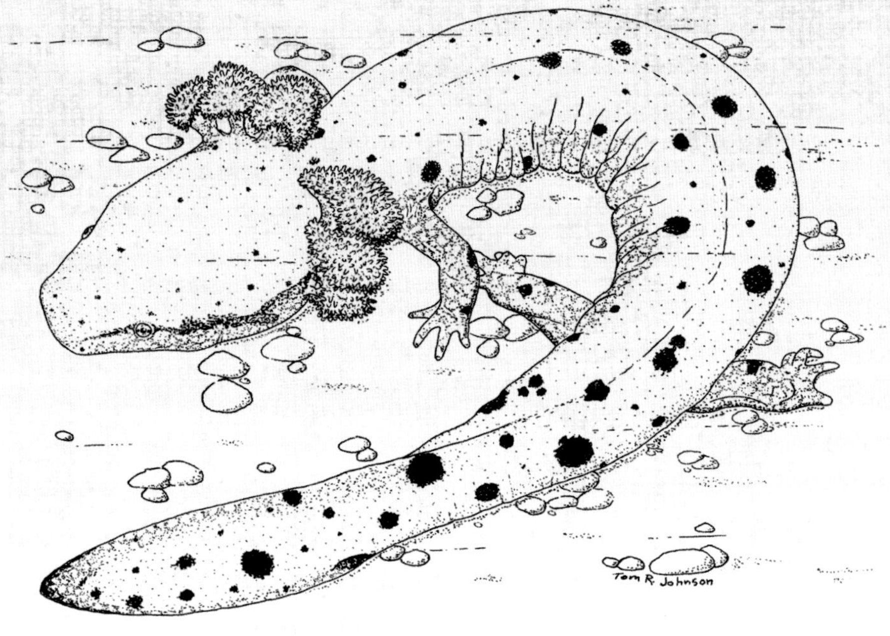

Common Mudpuppy

Common Mudpuppy

Necturus maculosus maculosus (Rafinesque)

JEFF BRIGGLER

Adult common mudpuppy from Gasconade River, Pulaski County.

Description: The common mudpuppy is a permanently aquatic species. It has a gray-brown back and a pale gray belly. Most of the body has numerous small, irregular dark-brown to black spots that usually appear on the belly. Behind the head are plumes of red gills. These gills vary in size, depending on the oxygen content of the salamander's aquatic habitat. There are four toes on both the fore and hind limbs. The eyes are small and lack eyelids. All *Necturus* species are neotenic gilled adults.

Adult common mudpuppies range in length from 200 to 330 mm (7.9 to 13 in.) but have been known to reach 486 mm (19.1 in.; Powell et al. 2016).

Habits and Habitat: In Missouri, this species lives in large creeks, rivers, and reservoirs. They can be found in muddy or gravel bottoms of shallow rivers and streams, and in deep pools in lakes up to 27 m (88.6 ft) deep (Reigle 1967). Mudpuppies are usually inactive during the day; they

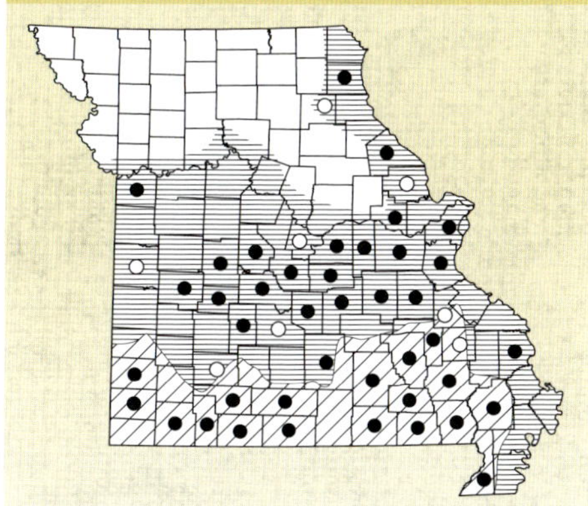

Distribution: Missouri: Throughout most of Missouri except the northwestern, northcentral and southern parts of the state (horizontal lines). It is replaced by the Red River mudpuppy in the extreme southern part of the state (diagonal lines). North America: Southern New York and Delaware, north to southern Quebec, west to southern Manitoba, south to eastern Kansas and Oklahoma, and northern Louisiana, northeast to states within the Tennessee and Ohio River drainages (Powell et al. 2016).

Larval stage of a mudpuppy from Pulaski County (above). Eggs of the mudpuppy are attached singly to the underside of a submerged rocks, such as this location in Crawford County (right).

JEFF BRIGGLER FR

JEFF BRIGGLER

usually remain hidden under submerged logs, rocks, boulders, debris piles, or tree roots. Seasonally, mudpuppies remain active throughout the year. We have received reports of these salamanders being caught on hook-and-line in Missouri's Lake of the Ozarks, Osage River, and other reservoirs in winter and early spring. They can be found by the hundreds within locks on the upper Mississippi River in northeast Missouri (J. Briggler pers. obs.).

They feed on any aquatic animals small enough to be captured and swallowed, including crayfish, mollusks, small fish, worms, and aquatic insects and their larvae (Harris 1959; Petranka 1998). They will also consume fresh carrion that they can readily locate by a great sense of smell (Harding 1997; J. Briggler unpubl. data).

Breeding: Detailed information of the breeding habits of the mudpuppy is lacking in Missouri, but the species has been studied elsewhere. Mating takes place in the fall and winter months; fertilization is internal, but the eggs are not laid until the following spring or summer. The female is known to excavate a depression beneath rocks, logs, boards, or man-made objects for nesting (Craig et al. 2015; Petranka 1998). The female will attach the eggs singly to the underside of the cover within the nesting chamber. Generally, between 30 and 200 eggs are laid by each female. A common mudpuppy nest with 87 eggs in early development was found on July 2, 2008, on the underside of a large rock in the Big Piney River, Pulaski County (J. Briggler unpubl. data). A Red River mudpuppy nest containing 42 well-developed eggs was found on May 30, 2012, under a large rock in the Current River, Carter County (J. Briggler unpubl. data). The female guards the eggs from predators through hatching (Bishop 1941), and in both nests documented in Missouri an adult mudpuppy was present. Incubation of the eggs is estimated to be 30 to 70 days depending on water temperature (Shoop 1965). Hatchlings have prominent, yellow yolk sacs and are about 22 mm (0.9 in.) long (Bishop

JEFF BRIGGLER

Head of an adult Red River mudpuppy from Eleven Point River, Oregon County.

1943). Nine recently hatched Red River mudpuppy larvae measuring 19–23 mm (0.7–0.9 in.) were found under a rock with an attending adult on July 14, 2008, in Bryant Creek, Douglas County (J. Briggler pers. comm.). Mudpuppy larvae have gills, a tail fin of medium height, a dark-brown mid-dorsal stripe bordered by pale-yellow lines and a broad, dark-brown stripe on each side from the head to the tip of the tail. Guarding of the eggs by an adult mudpuppy through hatching is well known, but only recently has guarding of the larval stage months after hatching been documented (Hime et al. 2014). On September 17, 2014, three nests with an average of 20 larvae (range 7–31) were discovered under rocks on the Big Piney River, Pulaski County (J. Briggler pers. obs.). All three nests had an attending adult and, based upon a larvae size of 40–45 mm (1.6–1.8 in.) in total length, egg deposition likely occurred in early to mid-summer. Larvae reach sexual maturity in four to six years at about 200 mm (7.9 in.; Collins et al. 2010; Trauth et al. 2004). Mudpuppies are long-lived species and are known to live over 30 years (Matson 2005).

Subspecies: There are two subspecies of *Necturus maculosus* in Missouri: the common mudpuppy (*N. m. maculosus*) described above and the Red River mudpuppy (*N. m. louisianensis* Viosca). The latter may be distinguished from the common mudpuppy by a lighter gray-brown or red-brown ground color. Dark spots on the upper part of the body are more distinct and more numerous. The belly has a wide, light, unspotted area down the center, which may be light gray with edges of pale pink. The Red River mudpuppy is somewhat smaller than the common mudpuppy; adults range from 180 to 230 mm (7.1 to 9.1 in.) and have a maximum length of about 300 mm (11.8 in.; Powell et al. 2016). The Red River mudpuppy was elevated to full species status by Collins (1991) because of the assumption that there is no contact between the two geographic races due to watershed separations for interbreeding. We choose to retain the subspecific name for this salamander until more taxonomical information is published.

Remarks: Mudpuppies are the only host for the larvae of the salamander mussel (*Simpsonaias ambigua*), an endangered species in Missouri. The small, dark larvae (glochidia) attach themselves to the mudpuppy's gills as external parasites. As an adult, this freshwater mussel averages 25–38 mm (1–1.5 in.) in length. Its range in Missouri is restricted to a few isolated locations within the Meramec River watershed (McMurray et al. 2012).

Mudpuppies are harmless to humans and to natural fish populations and are an integral part of the aquatic fauna of Missouri. Anglers often catch this species on baited hook-and-line or in minnow traps; if caught, mudpuppies should be released unharmed.

Family Amphiumidae

Amphiumas

This strictly North American family is represented by only one genus, *Amphiuma*, consisting of three recognized species: the one-toed amphiuma (*Amphiuma pholeter*), the two-toed amphiuma (*A. means*) and the three-toed amphiuma (*A. tridactylum*; Pough et al. 2018). Their combined ranges cover the southeastern and southern United States from southeastern Virginia to eastern Texas. The three-toed amphiuma ranges northward into southeastern Missouri. Amphiumas are the longest salamanders in North America and have been known to reach over 1.06 m (41.7 in.; Powell et al. 2016). They have tiny, almost useless limbs, small eyes (lacking eyelids), and smooth skin.

Three-toed Amphiuma

Three-toed Amphiuma

Amphiuma tridactylum Cuvier

TOM R. JOHNSON

Adult three-toed amphiuma.

Description: The three-toed amphiuma is a completely aquatic salamander shaped like a long cylinder with a somewhat pointed head; it has tiny, gray, lidless eyes and very small fore and hind limbs. Each limb has three very small toes. The dorsal color is dark brown or black; the belly is lighter brown or gray. Adults do not have gills, but a gill slit is present on each side of the head. Amphiumas have lungs and must breathe air at the surface of the water. The throat has a dark patch. The sexes are difficult to tell apart. Cunningham and Trautwein (1996) reported that the inside of an adult female's cloaca is black; the inside of an adult male's cloaca is pink or light gray. In a population of amphiumas studied in southeastern Missouri, the majority of specimens were found to be missing toes and/or limbs (Cunningham and Trautwein 1996).

Adult three-toed amphiumas range in total length from 457 to 760 mm (18 to 29.9 in.) but have been known to reach 1.06 m (41.7 in.; Powell et al. 2016).

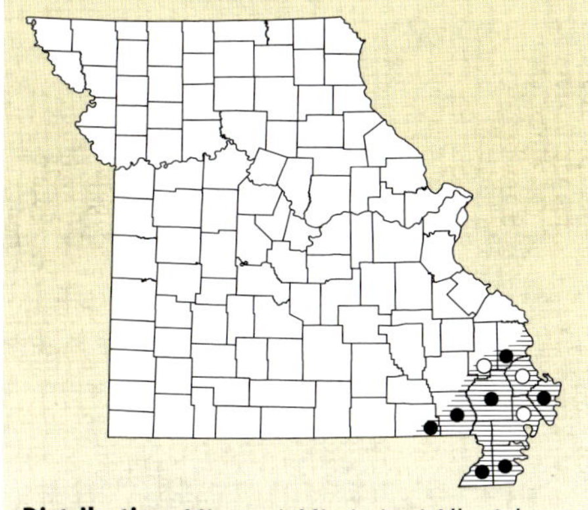

Distribution: Missouri: Mississippi Alluvial Basin of southeastern Missouri. North America: Southeastern Missouri and western edge of Kentucky and Tennessee, south through eastern and southern Arkansas, and eastern Texas and east to western Alabama (Powell et al. 2016).

Three-toed amphiumas live and breed in natural cypress swamps like this one in Butler County (above). Head of three-toed amphiuma (right).

Habits and Habitat: The three-toed amphiuma makes its home in semipermanent to permanent habitats, such as ditches, sloughs, sluggish streams, and swamps. In Missouri, cypress swamps are the favored haunt of this species. These animals are difficult to collect in the wild because they are alert, fast, slippery, and may bite viciously. They are a secretive species that spend the daylight hours buried in muck or hiding in holes or under submerged roots, debris, or aquatic plants. Their heads and necks may be exposed, and periodically they must come to the surface for air. During drying conditions, amphiumas can burrow in soft muck or utilize crayfish burrows to escape desiccation and can live for many months without food (Mount 1975; Smith and Secor 2017; Trauth et al. 2004). They can crawl overland on nights with heavy rains, and one individual was known to move at least 396 m (1,299.2 ft) from its original site of capture (Cagle 1948).

Amphiumas venture forth at night in search of small fish, crayfish, tadpoles, snails, aquatic insects, earthworms, and other aquatic animals (Bounty and Carr 2017). In southern Louisiana, an adult three-toed amphiuma ate a two-inch common snapping turtle,

Chelydra serpentina (Fontenot and Fontenot 1989). Large adults are more active on dark, warm nights; young or smaller amphiumas may be more active on cool, moonlit nights (Cunningham and Trautwein 1996). The western mudsnake (*Farancia abacura reinwardtii*) is known to prey on amphiumas.

Breeding: Very little is known about the breeding habits of the three-toed amphiuma in Missouri. In areas where it has been studied, the three-toad amphiuma is known to breed in late summer and early autumn. Courtship apparently is unusual in that the females court males (Porter 1972). Animals rapidly swim in a spiral fashion twisting and turning throughout the water column. Eventually, the female will rub her snout along the side of the male and position herself to allow sperm transfer between cloacae (Baker et al. 1947). Females can store sperm for at least six to eight months (Kreeger 1942).

A female generally lays an average of 200 eggs (Salthe 1973), but there is considerable variation throughout the species range. In Louisiana, Cagle (1948) reported an average clutch size of 98 eggs from 26 females while Fontenot (1999) had an average clutch size of 201 eggs (range 44–282 eggs) with larger females producing more eggs. Trauth et al. (1990) counted 80 and 127 ovarian eggs in two Arkansas specimens. Females generally produce a clutch of eggs biennially (Petranka 1998).

The large, yolky eggs, measuring 6.5 mm (0.3 in.) in diameter, are laid on land in a rosary-like string, usually under a rotten log near water, and the female stays with the eggs until they hatch in about four to five months (Fontenot 1999). Once water from autumn rains covers the eggs, they complete development and hatch. Most reports of females attending eggs occur between August and November with generally earlier dates in the northern part of the species range. It is likely that egg laying occurs during the late summer months in Missouri. The gilled larvae are 63 to 75 mm (2.5 to 3 in.) long. A study of this species in southern Alabama found that newly hatched amphiumas take air from the surface of the water and resorb their gills within three weeks of hatching. Apparently, they have well developed lungs at the time of hatching (Ultsch and Arceneaux 1988). Sexual maturity is likely reached within three to four years. Adults have been reported to live more than 12 years (Niemiller and Reynolds 2011).

Remarks: Local common names for this salamander are congo eel, blue eel, and ditch eel, although this species is not a fish. The three-toed amphiuma is a species of conservation concern in Missouri. The cypress swamps of southeastern Missouri are important to the survival of this interesting amphibian and should be protected. Anglers who catch amphiumas on hook-and-line should cut their line and release them unharmed.

Family Plethodontidae

Lungless Salamanders

This large family of salamanders is represented by 27 genera containing around 443 species (Pough et al. 2018). The family probably originated in the southern Appalachian Mountains of the eastern United States. Plethodontid salamanders occur over the eastern half of North America, the West Coast, and into Mexico, Central America, and northern South America. A few species also occur in southern Europe and South Korea (Pough et al. 2018).

This family is represented in Missouri by three genera with eight species and one additional subspecies. The adults lack lungs and most lack gills; the oxygen they require is taken from their environment through the skin and the mucous membrane of the mouth. One characteristic exclusive to this family is the presence of a groove in the skin running from each nostril down to the lip. In some species, associated with this groove (called the naso-labial groove) is a projection of skin that extends the groove below the upper lip. These projections are called cirri and are more pronounced on adult males. The groove and cirri may be associated with the sense of smell, specifically for detecting pheromones. Lungless salamanders live in a wide variety of moist habitats. Woodlands, springs, caves, cold streams, and rock outcrops with seepage areas are habitats commonly associated with this group.

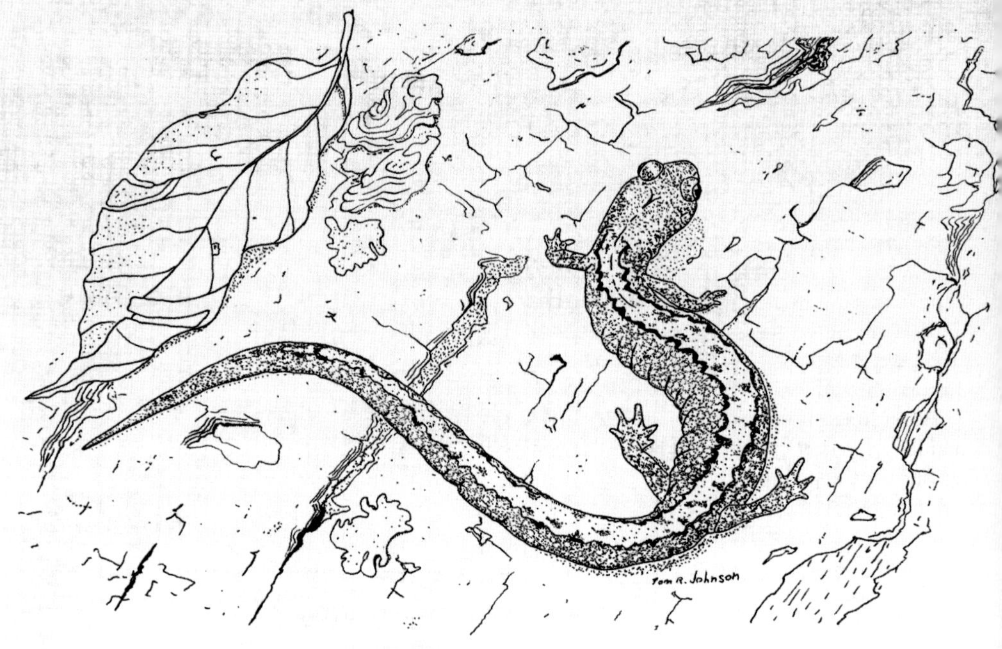

Southern Red-backed Salamander

Eastern Long-tailed Salamander

Eurycea longicauda longicauda (Green)

TOM R. JOHNSON

Adult eastern long-tailed salamander from Scott County.

Description: The eastern long-tailed salamander is a medium-sized, slender salamander with a long tail. It is usually yellow but may vary from green yellow to orange yellow. The belly is plain yellow. Dark-brown or black markings and spots occur along the back and sides. Prominent vertical bars are present on the tail. There are 13 or 14 costal grooves. Young of this species have fewer and smaller dark markings and a proportionately shorter tail.

Adult long-tailed salamanders range in total length from 92 to 159 mm (3.6 to 6.3 in.) but have been known to reach 197 mm (7.8 in.; Powell et al. 2016).

Habits and Habitat: The long-tailed salamander is nocturnal but may emerge from hiding during the day after a heavy rain. It usually lives under rocks or logs near streams, springs, and seepages in forested areas; it also frequently occurs in caves. In May 2003, 122 eastern long-tailed salamanders were observed in an hour while

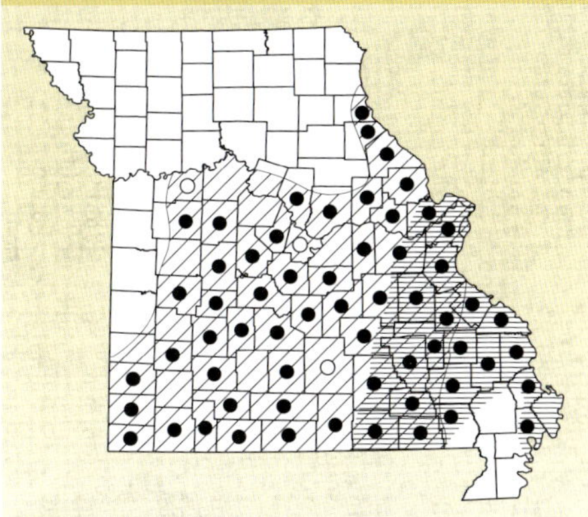

Distribution: Missouri: Restricted to southeastern Missouri except the Mississippi Alluvial Basin (horizontal lines); it is replaced by the dark-sided salamander throughout most of southern and western Missouri (diagonal lines). Approximately a two-county wide intergradation zone occurs between the two subspecies. North America: Southern New York state south to northern Virginia, west to extreme southeastern Kansas, eastern Oklahoma and east to northern Alabama (Powell et al. 2016).

JEFF BRIGGLER

JEFF BRIGGLER

Adult dark-sided salamander from St. Francois County (above). Dark-sided salamanders live mainly under rocks along small streams, such as this example in Christian County (right).

searching under rocks along a stream drainage in St. Genevieve County (J. Briggler pers. obs.). These salamanders are quite agile and can escape predators by using their tails for quick jumps. Tail waving is used to distract a predator's attention away from the salamander's head. They also can twist off their tail if it is grasped by a predator. Individuals without tails or with partially regrown tails are not uncommon.

Food consists of various small arthropods, especially ostracods and dipterans (Rudolph 1978).

Breeding: Courtship and breeding take place in or near springs, within caves, and in or near cool, rocky creeks between November and early March (Ireland 1974). The eggs are fertilized internally, and egg laying occurs mainly between December and March in Ozark populations in Arkansas (Ireland 1974). The eggs are laid in small clumps or in a single row. They are most frequently encountered in caves. In Missouri, eggs were found in caves in December and January. The eggs were attached to small rocks in the cave stream or along the margins of small rimstone pools (J. Briggler pers. obs.). Eggs are also laid outside of caves beneath rocks or in shallow water of springs or creeks. Due to females depositing small clusters of eggs in several locations, it is difficult to estimate the exact number of

TOM R. JOHNSON

A young dark-sided salamander larva.

eggs deposited by a female. However, clutch size is likely from 60 to 100 eggs per female (Hutchison 1956; Minton 2001; Mohr 1943). The eggs hatch in 4 to 10 weeks depending on water temperature, and hatchlings are about 10 mm (0.4 in.) snout-to-vent length (Ireland 1974). The larvae are dark colored, stream type, with streamlined bodies and short well-developed gills. They generally transform into juveniles in about seven months (Ireland 1974), which is usually by late summer, but some might overwinter and transform the following year (Rudolph 1978). Most individuals reach adulthood in one to two years and have a snout-to-vent length from 31 to 44 mm (1.2 to 1.7 in.).

Subspecies: There are two subspecies of long-tailed salamanders in Missouri: the eastern long-tailed salamander, *Eurycea longicauda longicauda* (Green), described above, and the dark-sided salamander, *Eurycea longicauda melanopleura* (Cope). The dark-sided salamander has large amounts of dark pigment along the sides of the body from the head onto the tail and larger and more numerous dark spots on the back than the long-tailed salamander. The sides are often spotted with white flecks. Vertical bars on the tail are more irregular in shape and may form dark vermiculations. Ground color may vary from yellow green to yellow brown. The belly is dull yellow with numerous dark flecks. The dark-sided salamander is slightly smaller than the long-tailed salamander.

Remarks: Intergradation between these subspecies is known from many areas of southeastern Missouri (see map). Many *Eurycea* salamanders observed within this intergradation zone are difficult to determine by appearance. Morphological and genetic analyses by Beasley (2010) and Potter (2008) of *Eurycea* salamanders from caves further confirmed intergradation among these two subspecies but also suggested hybridization by the long-tailed salamander with the cave salamander (*Eurycea lucifuga*). Additional research is needed to better understand the interactions among these *Eurycea* species.

The long-tailed salamander is widespread and common throughout most of the forested habitat of the state. Maintaining forested habitats and protection of cave systems and their water quality will keep this species common in Missouri.

Cave Salamander

Eurycea lucifuga Rafinesque

JEFF BRIGGLER

Adult cave salamander from Howell County.

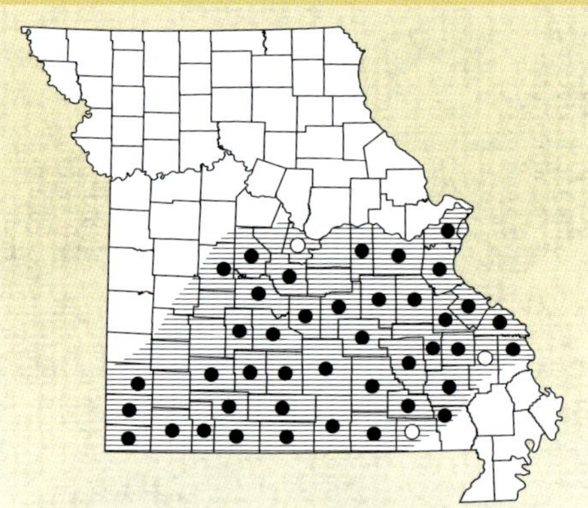

Distribution: Missouri: Throughout most of the southern half of Missouri, except for the Mississippi Alluvial Basin and Osage Plains. North America: Southern Indiana and northern West Virginia south to northern Georgia and Alabama, northwest to northern Arkansas, southern Missouri, and extreme eastern Oklahoma and southeastern Kansas (Powell et al. 2016).

Description: The cave salamander is a medium-sized salamander with a long tail. It is normally bright orange but can vary from yellow brown to orange red. Distinct dark-brown or black spots cover most of the body. The belly is usually yellow orange and without spots. There are 13 or 14 costal grooves. The end of the tail is often black. Young cave salamanders are yellow and have short tails. Adult females are often larger than males; males have more prominent cirri. Males also can be distinguished from females by the presence of a swollen cloaca.

Adult cave salamanders range in total length from 100 to 152 mm (3.9 to 6 in.) but have been known to reach 181 mm (7.1 in.; Powell et al. 2016).

Habits and Habitat: The range of this species is confined to areas having limestone outcrops. Although they mostly occur in caves, the cave salamander also can live in wooded areas, along rocky streams and springs, under rocks on glades during the spring,

JEFF BRIGGLER

JEFF BRIGGLER

Cave salamanders can live in and around cave entrances or deep within a wet cave, such as this cave opening in Crawford County (above). Cave salamanders attach eggs to rocks in cave streams, such as this example from Shannon County (left).

and even in wells and spring-fed swamps (Hutchison 1958). Cave-dwelling individuals mainly live in the twilight zone (dimly lit area beyond a cave entrance) but also occur beyond the entrance in areas of permanent darkness. They are good climbers and can cling to walls with their wet bodies. Their long tails help support them on stalactites and stalagmites. In habitats other than caves, this species is nocturnal and spends daylight hours under rocks or rotten logs. After a heavy rain, cave salamanders often rest on rocks or boulders during the day. Several times we have found adult cave salamanders under flat rocks on rocky, open hillsides during the spring, when the soil was quite moist. If pursued, these salamanders will jump and scamper away with remarkable agility. A cave salamander will wave its tail in an effort to distract a predator's attention away from its head.

The food of this species consists of a variety of small arthropods, especially isopods, ostracods, and dipterans (Rudolph 1978; Smith 1948).

Breeding: Courtship and mating likely occur in the summer or early autumn (Organ 1968; Smith 1948) prior to egg laying from late summer into winter. A study of a Perry County cave by Ringia and Lips (2007) found the majority of egg laying occurred from August to October. However, eggs of cave salamanders have also been documented from November

RICK THOM

Increasingly cave salamanders are being observed lacking much of their usual coloration, such as this example from Crawford County.

to February in the Ozarks (Myers 1958a; Niemiller et al. 2009; Ringia and Lips 2007). Cave salamanders are known to have a prolonged egg laying period throughout their range that appears to correspond to low stream flow in cave systems. Egg laying mainly occurs in underground habitats, especially in caves with streams, although eggs can be found in springs or rocky streams outside of caves. A female cave salamander generally lays between 60 and 120 immaculate white eggs with an average of 74 (Trauth et al. 1990). Eggs are generally laid singly or in clusters up to 31 (Niemiller et al. 2009; Ringia and Lips 2007) on the bottom or along the sides of rocks in streams or small rimstone pools. The gilled stream-type larvae hatch in about 10 to 20 days at a total length of 9–11 mm (0.3–0.4 in.) and most live 12–15 months in the water. When ready to leave the water, the young may be up to 55 mm (2.2 in.) in total length and reach adulthood within an additional one to two years.

Remarks: There have been several reports of unique color patterns of cave salamanders throughout their Missouri range. Patterns range from heavily black pigmented individuals to individuals that are lacking almost all pigment. Additional research is needed to determine if hybridization is occurring with other *Eurycea* species (see remark section in the long-tailed salamander species account), or if individuals are slowly becoming a cave-obligate species.

The cave salamander is considered relatively common throughout most of the karst, forested habitat of the state. Caves and associated subterranean habitats are important to the long-term persistence of many salamanders by providing ideal egg-laying sites and shelter from hot, dry summer months. As with all animals living in caves, increased human disturbance can negatively influence cave salamanders. The ecological balance in a cave is extremely fragile and any disturbance could be harmful to this balance.

Grotto Salamander

Eurycea spelaea Stejneger

JEFF BRIGGLER

Adult grotto salamander from Shannon County.

Description: Adult grotto salamanders range from tan white to pink white. These troglobitic salamanders lack gills as adults. They are partly or completely blind. The head is rather wide and flat; the eyes are small; and the tail is long, slender, and finless. There are 16 to 19 costal grooves. Males can be distinguished from females by the presence of fleshy projections on the upper lip (cirri) and a round mental gland on the chin. In addition, males have larger, swollen cloacal lips. Adults of both sexes have reduced eyes that are covered or partially covered by a fusion of the eyelids. Their eyes may appear sunken into the head.

The larvae of this species have gills, functional eyes, and broad tail fins. Larvae have more pigment than adults and are brown to dark gray. Dark pigmentation may form spots or streaks along the sides and tail.

Adult grotto salamanders range in total length from 75 to 121 mm (3 to 4.8 in.) but have been known to reach 135 mm (5.3 in.; Powell et al. 2016).

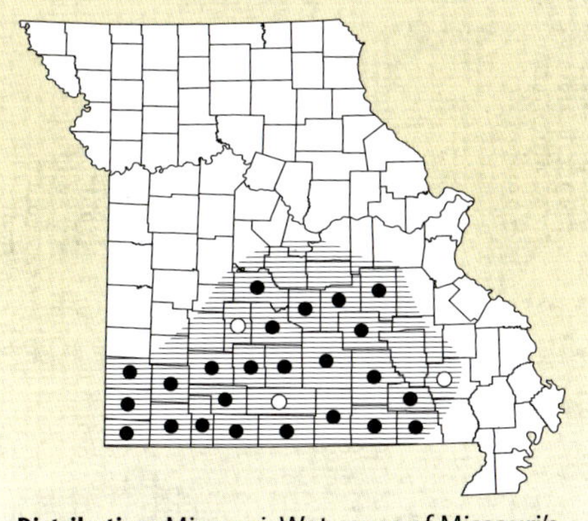

Distribution: Missouri: Wet caves of Missouri's Ozark Highlands. North America: Endemic to the Ozark Highlands and occurs in one county in extreme southeastern Kansas, in northeastern Oklahoma, northern Arkansas and the Ozarks of Missouri (Powell et al. 2016).

JEFF BRIGGLER

JEFF BRIGGLER

Grotto salamander larva from Wright County (above). Grotto salamanders live on cave walls and rimstone pools deep within a cave, like this example in Shannon County (right).

Habits and Habitat: This blind salamander living in Missouri is quite unusual because the adult is a true cave-adapted species (troglobites). The adults live in total darkness and require caves with a spring or stream running through the cave while the larvae, which possess functional eyes, can occur in small streams and springs outside of caves. Grotto salamanders typically occur in greater abundance in caves that have a large number of bats (likely due to a plentiful supply of invertebrates that feed upon bat guano; Brandon 1971; Hendricks and Kezer 1958). Hendricks and Kezer (1958) observed 106 adult grotto salamanders during a single survey in a Laclede County cave. Population density ranged from 0.04 to 0.12 individuals per 1 m^2 in a cave in Oklahoma (Fenolio et al. 2014).

The food of adults is mainly small aquatic and terrestrial invertebrates, as well as other *Eurycea* larvae (Smith 1948; Brandon 1971).

Breeding: This species is not well studied due mainly to its secretive habits in caves. Much of our knowledge of their reproductive biology is based upon examination of collected specimens and seasonal development of reproductive parts (e.g., eggs inside females, cirri and mental gland in males), but courtship, egg laying, and embryo development has not been well documented in the wild.

It appears that adults mate during the late spring into the summer months due to an influx of nutrients (bat guano) that support this period of the greatest food supply (Brandon 1971; Fenolio and Graening 2009; Hendricks and Kezer 1958), and females appear to oviposit within one to four months after mating (Petranka 1998; Rudolph 1978; Trauth et al. 1990). Egg laying has not been observed in the wild. Fertilization is internal (as with most salamanders), and the eggs are probably attached to stones in or near water in caves. A female that was induced laid 13 eggs in captivity that were 2–2.2 mm (0.08–0.09 in.) in diameter (Barden and Kezer 1944; Kezer 1952). Hatchlings generally average 17 mm (0.7 in.) in total length (Brandon 1970). The larvae are stream type and normally inhabit cave streams, although they are found in springs or streams that flow out of caves. During this time, they range from 36 to 56 mm (1.4 to 2.2 in.) in snout-to-vent length (Brandon 1966; Fenolio et al. 2014). Larvae eat many types of small, freshwater aquatic invertebrates, such as amphipods, isopods, copepods, etc. (Rudolph 1978), but larvae occasionally are known to consume bat guano to supplement their diet (Fenolio et al. 2006). Larvae generally transform into juveniles around 85–96 mm (3.3–3.8 in.) in total length (Brandon 1965; Brandon 1970). Brandon (1971) documented that the larvae may take two to three years to transform into adults in Missouri, but an Oklahoma study found that it can take up to six years (Fenolio et al. 2014). Fenolio et al. (2014) estimated a life span of at least nine years in the wild, and one grotto salamander has been reported to live for 9.9 years in captivity (Snider and Bowler 1992).

Remarks: The grotto salamander is listed as a species of conservation concern because of its extremely restricted range in karst, highland forests of the Ozarks, relatively low population size, affinity to cave ecosystems, and limited information on basic biology. Because of the delicate balance of cave ecosystems, grotto salamanders and their cave and spring habitats should be protected from human disturbance, overcollecting, and water pollution. The type locality for this species is in Barry County.

There appear to be three lineages suggesting distinct species within the distribution of the species (Phillips et al. 2017). Missouri has two distinct lineages with recommendations to elevate to full species status. The northern lineage (*Eurycea nerea*) that occurs mainly in the Salem Plateau of central and southern Missouri, and a western lineage (*Eurycea spelaea*) that occurs in the Springfield Plateau of southwestern Missouri (Phillips et al. 2017).

Oklahoma Salamander

Eurycea tynerensis Moore and Hughes

JEFF BRIGGLER

Adult neotenic Oklahoma salamander from Christian County.

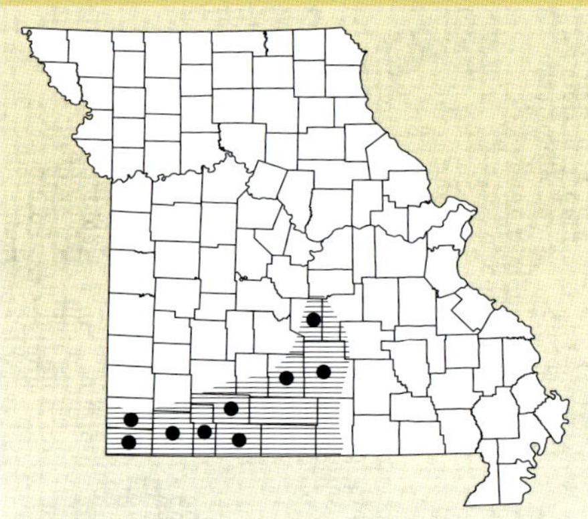

Distribution: Missouri: Occurs in the central and southwestern portions of the Ozark Highlands. North America: Eastern edge of Oklahoma, southwestern Missouri and northern Arkansas (Powell et al. 2016).

Description: The Oklahoma salamander is a small, dark, slender salamander, generally yellowish tan to dark brown or gray. The dorsal stripe may be golden or yellow tan to brown and is bordered by dark-brown lines. There may be a medial row of dark-brown chevrons along the dorsum corresponding to each of the 19 to 21 costal grooves. Small white flecks (iridophores) are present along the sides and tail. The belly is gray but also may be somewhat yellow. Most commonly, the gilled larvae and the neotenic (retain gills) adults are observed. The back and sides are cream or light tan covered with extensive gray stippling, with two lines of small white spots along the sides, and a pale belly. Poorly developed dorsal and ventral fins are present on the tail.

Adult Oklahoma salamanders range in total length from 44 to 83 mm (1.7 to 3.3 in.) but have been known to reach 102 mm (4 in.; Daniel 2011; Powell et al. 2016).

JEFF BRIGGLER

Oklahoma salamanders live and breed in clear, shallow streams such as this example in Christian County.

Habits and Habitat: Neoteny is common in this species. Most of Missouri's population is permanently aquatic, living in small, clear, cherty, gravel-bottom creeks, streams, and springs. When surface water is present, individuals are found in the interstitial spaces among loose gravel or under rocks (Tumlison et al. 1990a). During dry periods and presumably during the winter, it follows the water level down into the creek gravel bed; using this type of aquatic habitat allows it to remain active all year. Although extremely rare in Missouri, transformed individuals hide under rocks and logs in or near cave springs or cold, clear streams. Most often encountered in twilight zones (dimly lit area beyond cave entrances) and leafy seepage areas of the Ozark Highlands. When outside caves, these salamanders are nocturnal and may venture away from streams on rainy nights. Population densities in Missouri range from 0.21 to 2.2 salamanders per 1 m^2 (Dundee 1958).

Small arthropods are the primary food of both adults and larvae, especially ostracods, isopods, dipterans, and ephemeropterans (Rudolph 1978; Tumlison et al. 1990b). Dodd (1980) reported on laboratory observations of the feeding behavior of this species. He observed a salamander approach a prey item (pieces of tubifex worms), slowly and deliberately arch its pectoral region, then snap or suck up the prey with a rapid movement.

Breeding: Little is known about the reproduction of this species in Missouri. In general, fertilization is internal and the eggs are attached to the undersurface of rocks or within the opening spaces in the gravel substrate of springs and streams (Dundee 1965; Spotila and Ireland 1970; Trauth et al. 1990). The reproductive season varies considerably within the range of the species due to temperature variation among breeding sites; more thermally stable sites (e.g., cave streams and springs) have prolonged reproductive seasons compared to widely fluctuating temperatures of surface streams. Generally, breeding may last from September into July with many populations laying eggs during the fall to late spring (Cline et al. 1989; Ireland 1976; Trauth et al. 1990). Females lay 3 to 21 eggs with cream-colored yolk. The incubation period is about 30 days (Ireland 1976). Recently hatched larvae, typically 9–10 mm (0.4 in.) in snout-to-vent length, are gilled stream type. Larvae either metamorphose

JEFF BRIGGLER

Fully transformed Oklahoma salamanders, such as this example from Christian County, are rarely seen.

into adults or retain the gills and mature as neotenic individuals. Metamorphosis generally occurs at 33–48 mm (1.3–1.9 in.) snout-to-vent length and five to eight months. Neotenic individuals may reach a larger size than metamorphosed individuals (Dundee 1965). Sexual maturity is likely variable based upon growth rates of individuals. Males can reach maturity during or soon after transformation while females may take up to two years (Dundee 1965; Ireland 1976).

Remarks: The taxonomic history of this salamander has been confusing for many years. The *Eurycea multiplicata* complex consisted of two surface-dwelling subspecies, the many-ribbed salamander (*Eurycea multiplicata multiplicata*) in the Ouachitas of Arkansas and Oklahoma, and the gray-bellied salamander (*Eurycea multiplicata griseogaster*) in the Ozarks of Missouri, Arkansas, and Oklahoma (Petranka 1998). In addition, the Oklahoma salamander was considered a distinct species of this complex as a neotenic salamander living in gravel-bottom creeks although its range overlaps the gray-bellied salamander (Petranka 1998). It was once recognized that Missouri had both the gray-bellied salamander and the Oklahoma salamander (Johnson 1987) and such separation in species was based primarily upon measurements of head and trunk (Tumlison et al. 1990c). Based upon unpublished new evidence of the relatedness of these two species, Johnson (2000) recognized that these two salamanders were the same species and chose to recognize the gray-bellied salamander because the correct scientific name had yet to be published in a professional journal. Therefore, the Oklahoma salamander was no longer recognized in the state. More recent genetic work by Bonett and Chippindale (2004) concluded that the gray-bellied salamander and the Oklahoma salamander were the same species, and the scientific name of this salamander would be *Eurycea tynerensis* (Oklahoma Salamander) instead of the gray-bellied salamander (*Eurycea multiplicata griseogaster*). The gray-bellied salamander is no longer recognized in the state.

The Oklahoma salamander has a restricted distribution in Missouri, Oklahoma, and Arkansas but appears to be relatively secure in Missouri. However, its requirements for clean creeks, springs, and an abundance of clean gravel make it vulnerable to water pollution and habitat loss. Siltation (filling of interstitial spaces among gravel), followed by gravel removal, are the primary threats for this species (Cline and Tumlison 2001).

Four-toed Salamander

Hemidactylium scutatum (Temminck and Schlegel *in* Von Siebold)

TOM R. JOHNSON

Adult four-toed salamander from Reynolds County, photographed on a mirror to show the belly.

Description: The four-toed salamander is a small, delicate salamander with a thick, round tail and four toes on both fore and hind limbs. The snout appears short and blunt. General color is yellowish tan to brown on the back with many faint, irregular black spots. Sides of the body are grayish brown with black stippling, and the belly is pure white with numerous large, irregular black spots. The tail is distinctly constricted near its base. There are 13 to 14 costal grooves. Males of this species are smaller, slimmer, and have longer tails than females (Bishop 1943). Also, the upper lip overhangs the lower jaw in males but is almost straight in females.

Adult four-toed salamanders range in length from 51 to 90 mm (2 to 3.5 in.) but have been known to reach 113 mm (4.4 in.; Powell et al. 2016).

Habits and Habitat: This salamander is commonly associated with sphagnum (peat) bogs across most of its range. In Missouri,

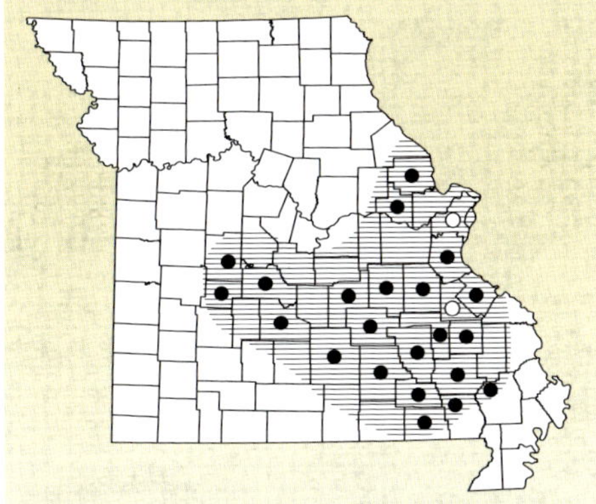

Distribution: Missouri: Eastern half of the Missouri Ozarks, including the St. Francois Mountains. Small populations are also scattered north of the Missouri River in Warren and Lincoln counties. North America: Minnesota, southeastern Canada and eastern Maine south to northern Georgia and eastern Alabama; numerous isolated populations occur in New Brunswick and Nova Scotia, Indiana, Kentucky, Illinois, Missouri, Arkansas, Oklahoma, Louisiana, Mississippi, and northwestern Florida (Powell et al. 2016).

Four-toed salamanders live and breed in overhanging vegetation, especially moss, along intermittent streams, for example this location in Phelps County (above). Female four-toed salamander with egg clutch in a clump of moss, Iron County (right).

JEFF BRIGGLER

TOM R. JOHNSON

however, the four-toed salamander occurs in mosses along heavily forested, headwater streams and spring-fed creeks associated with sandstone or igneous (Precambrian) bedrock, in and along the margins of fens, and also in and near natural sinkhole ponds. Occasionally, individuals are found along margins of man-made ponds (Hocking et al. 2008; J. Briggler unpubl. data). Away from egg-laying sites, individuals live under rotten logs, in leaf litter, or under rocks in seepage areas. This species preys on a variety of small arthropods and mollusks. If captured, a four-toad salamander can easily break off its tail and escape, or occasionally coil its body and tail with head tucked to the inside and limbs tight against the body (Herman 2013).

Breeding: Four-toed salamanders breed in autumn. Sperm are stored in a chamber (spermatheca) inside the cloaca of females. Females move to a creek, ephemeral pool, or sinkhole pond in the spring (usually during the first two weeks in April), soon after ending their winter dormancy. Eggs are fertilized as they pass through the female's cloaca. About 30 eggs are laid in a protected pocket of moss, usually overhanging the water. The eggs are attached to stems of grasses and sedges or roots of mosses and average 8.5 cm (3.3 in.) above the waterline (Bishop 1943; Herman 2013). Trauth et al. (1990) reported an average of 42 ovarian eggs (range 27–57) per female in Arkansas. Most females will remain with eggs until just before hatching. The presence of a female attending a clutch of eggs increases embryonic survival (Harris and Gill 1980), and likely the skin secretion by the female imparts a resistance to fungal infection of the eggs (Banning et al. 2008).

TOM R. JOHNSON

Four-toed salamander larva from Iron County.

A study of egg brooding in this species showed that females do not protect eggs from possible egg predators such as ground beetles, centipedes, or newts. In laboratory trials 20 different predators were placed in containers with egg-brooding female four-toed salamanders. None of the females showed any defensive behavior to the predators and some eggs were eaten by predators (Carreno and Harris 1998). However, Hess and Harris (2000) suggested that eggs might be unpalatable to certain ground beetles.

Communal nesting (two or more complements of eggs in one site) may occur, especially where there are few ideal nesting sites. Breitenbach (1982) reported that communal nesting in this species was observed in 12 percent of the nests he examined. He also found that the majority of females deserted their nest just prior to egg hatching. Communal nesting has been observed in Missouri. Communal nests of 60 or more eggs have been observed on several occasions in Iron County (T. Johnson unpubl. data). Most of the nests observed were along small, fishless creeks in thick mats of mosses, *Thuidium delicatulum* var. *delicatulum* or *Climacium americanus* (P. Redfearn pers. comm.). Easterla (1971) documented 15 clusters of eggs with attending adults within mosses overhanging water along small streams in early April in Wayne County. A study of brooding females showed that the communal nesting behavior is an indication of predator dilution since females do not attempt to protect their eggs from possible predators (Carreno and Harris 1998).

Eggs generally begin to hatch in about four weeks, and the gilled larvae, typically 11–15 mm (0.4–0.6 in.) in total length, find their way to water. The larvae have a high caudal fin or keel that begins on the back and extends down the entire length of the tail. The amount of time a larva remains in the water, 21 to 42 days, is short compared to other plethodontid salamanders. A study in northern Virginia showed that larvae developing in small, woodland, ephemeral pools may metamorphose in less than 40 days. Those developing in streams take nearly twice that long (Berger-Bishop and Harris 1996). After metamorphosing, the juveniles, which average 20 mm (0.8 in.) become terrestrial. They may take over two years to reach sexual maturity (Bishop 1943). Maximum longevity of this species is unknown, but Harris (2005) suggests at least nine years.

Remarks: This species is listed as a species of conservation concern in Missouri due to relatively low numbers, patchy distribution, and affinity to unique wetland habitats associated with moss. A number of new populations, however, have been discovered in Missouri in recent years, and it is likely the species is relatively secure in Missouri.

The four-toed salamander is considered a glacial relict with many disjunct populations throughout its range. Recent genetic analysis throughout the range of the species showed that there are six divergent lineages that may warrant full species or subspecies status in the future (Herman and Bouzat 2016).

Western Slimy Salamander

Plethodon albagula Grobman

JEFF BRIGGLER

Adult western slimy salamander from Madison County.

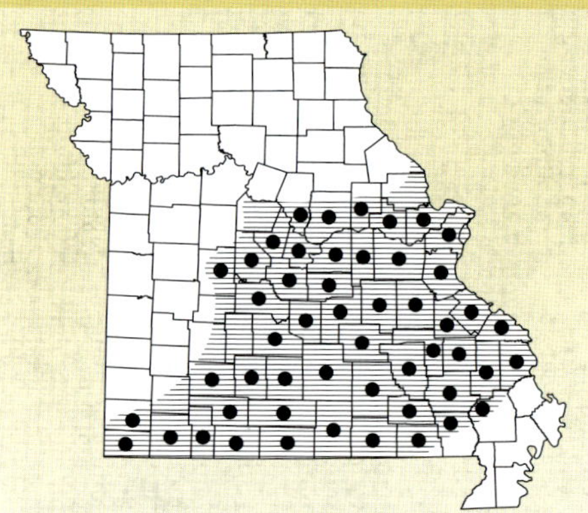

Distribution: Missouri: Throughout the Ozark Highlands and Lincoln Hills north of the Missouri River. North America: Southern Missouri, northern and western Arkansas, and eastern Oklahoma. Disjunct populations in southern and eastern Texas (Powell et al. 2016).

Description: The western slimy salamander is a black to blue-black, medium-sized woodland salamander with a long, rounded tail and numerous silver flecks irregularly distributed over the head, back, limbs, and tail. The chin and belly are dark gray. There are usually 16 costal grooves but can range from 14 to 16. Males can be distinguished from females by the presence of a light-colored swelling (mental gland) under the chin during the breeding season. The western slimy salamander is one of 13 in a group of closely related species comprising the *Plethodon glutinosus* complex (Highton et al. 1989).

This is the largest plethodontid salamander in Missouri. Adult slimy salamanders range in total length from 121 to 172 mm (4.8 to 6.8 in.) but a specimen in the slimy salamander complex was documented at 206 mm (8.1 in.; Powell et al. 2016).

JEFF BRIGGLER

BRANDON VANDALSEM

Western slimy salamanders live under rocks and logs along forest slopes, such as this location in Phelps County (above). A female western slimy salamander attending her developing eggs found in an Ozark County cave (left).

Habits and Habitat: Slimy salamanders commonly live under rocks or logs in damp ravines and moist wooded hillsides. A study in Warren County revealed that they are most abundant in dense-canopy ravine habitat with high moisture and low sun exposure (Peterman and Semlitsch 2013). They are most active on the surface during cooler, wet conditions in the spring and autumn, but during the hot summer months they are difficult to find. During dry summer weather they retreat underground, especially in cooler, moist caves (Briggler and Prather 2006) or burrow into large piles of leaf litter to find a damp place to live. They venture out of hiding at night and activity is greatest after heavy rains. This species is frequently found in caves, especially the twilight zone, throughout the Ozark Highlands (Briggler and Prather 2006; Myers 1958b). Western slimy salamanders are a common species within the forested landscape of south-central Missouri with density as high as 213 salamanders per hectare (86.2 per acre; Herbeck and Larsen 1999).

This salamander is appropriately named as its skin glands secrete a thick, very sticky substance that adheres to human skin like glue. It causes dust, dirt, or bits of dead leaves to stick to one's hands and is difficult to remove.

They are generalist feeders and eat a wide variety of small arthropods, especially ants and beetles, and worms (Milanovich et al. 2008; Petranka 1998).

Breeding: The reproductive biology of the western slimy salamander has been fairly well studied in Missouri. About 8 to over 15 eggs are laid by females in underground burrows, under rotten logs, in abandoned mines, and in caves. In studies in Arkansas, female slimy salamanders entered a mine and deposited eggs from late August into September (Milanovich et al. 2006; Trauth et al. 2006). A total of 372 clutches were observed between 1982 and 2004 in this mine with an average of 10 eggs per female (Milanovich et al. 2006). There are several documented reports of brooding females with eggs in caves throughout the Missouri Ozarks (Missouri Cave Database 2019). Initial egg laying has occurred as early as August 6 and appears to continue through September with a range of 8 to 11 eggs for five clutches (Missouri Cave Database 2019; Wells and Gordon 1958). The eggs are attached to a stalk in a grape-like cluster and suspended from the ceiling of a small cavity. Like all others of the genus *Plethodon*, females remain with the eggs during the incubation period that is about two or maybe three months. Brooding females generally position their bodies to physically touch the eggs. Such postures include coiling body or tail around the eggs, or the touching of the eggs with the head or trunk region (Trauth et al. 2006). Likely brooding postures by the female serve to deter predators, retain moisture for developing eggs, and reduce fungal outbreaks on eggs. As with all members of the *Plethodon* genus, there is no aquatic larval stage; the hatchlings resemble adults but have proportionately shorter tails. Most eggs hatch by late December in Missouri. Wells and Gordon (1958) found two clutches of well-developed eggs with brooding females in a Crawford County cave in October. Most of the embryos hatched within the next couple of days. Hatchlings averaged about 31 mm (1.2 in.) snout-to-vent length.

Although some females return to the same crevice to lay eggs each year, most of the population likely exhibits the typical biennial reproduction cycle (Milanovich et al. 2006; Trauth et al. 2006). A study in Texas reported that reproductive maturity is generally one to two years with females maturing earlier than males (Taylor et al. 2015) and between 55 and 72 mm (2.2 and 2.8 in.) in snout-to-vent length (Trauth et al. 2004).

Remarks: Plethodontid salamanders are an important part of the ground-dwelling forest community. Because these salamanders can be quite abundant, they are the primary predator of many invertebrates and are food for larger vertebrates. Direct mortality of slimy salamanders by fire is likely extremely low due to utilizing protective structures (e.g., rocks and aging logs), but surface-dwelling slimy salamanders are seven times less likely to be found following a fire (O'Donnell et al. 2016). Supposedly they retreat below ground to find more favorable environments until the surface conditions are more hospitable. It is important that landowners and land managers understand the significance of fallen logs on the forest floor. Missouri's three species of woodland salamanders (genus *Plethodon*) require rotten logs for their survival. Fallen logs located along steep, forested hillsides in the southern half of the state are especially valuable to these salamanders and other wildlife.

Ozark Zigzag Salamander

Plethodon angusticlavius Grobman

JEFF BRIGGLER

Adult Ozark zigzag salamander from Christian County.

Description: The Ozark zigzag salamander is a small, dark, slender, woodland salamander with a narrow, somewhat lobed mid-dorsal stripe. The dorsal stripe usually has irregular or wavy edges and may range from yellow to yellow orange, orange, or red (Briggler and Puckette 2003). Dark-brown or black pigment may invade the dorsal stripe causing it to appear lobed, or may cover a large part of the stripe. Normally the width of the dorsal stripe is less than one-third the width of the body; it is widest near the hind limbs. Some individuals may lack a dorsal stripe. The belly has white and black mottling. The sides of the body are generally dark gray or brownish gray with some orange or red and small white flecks. There are 17 to 19 costal grooves. The sexes are difficult to distinguish, but males are typically smaller and slimmer than females.

Adult Ozark zigzag salamanders range in total length from 60 to 98 mm (2.4 to 3.9 in.).

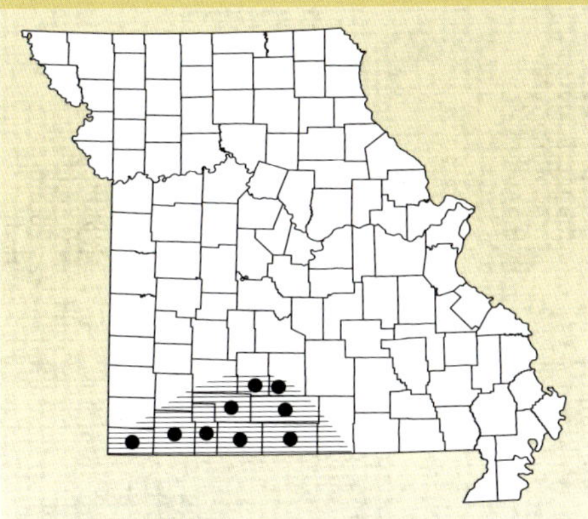

Distribution: Missouri: South and southwestern portion of the Missouri Ozarks. North America: Southwestern Missouri, northern and northwestern Arkansas and extreme northeastern Oklahoma (Powell et al. 2016).

Ozark zigzag salamanders live under rocks, logs, and moist leaves along forested slopes, such as this example in Christian County (above). Developing eggs of the Ozark zigzag salamander from northwest Arkansas (right).

JEFF BRIGGLER

JEFF BRIGGLER

Habits and Habitats: Ozark zigzag salamanders inhabit upland forested areas. We have observed them living in cooler and damper habitats compared to other plethodontid salamanders occurring in Missouri. They live mainly in or under rotten logs, rocks, and leaf litter in seepages near small streams and on steep hillsides, as well as in or near caves (Briggler and Prather 2006; Myers 1958b). They are easier to encounter on the surface during cooler, wet conditions between September and May. During the dry, summer months they are difficult to find and likely take refuge in deep underground environments (Meshaka and Trauth 1995; Wilkinson et al. 1993). Like many plethodontid salamanders, Ozark zigzag salamanders communicate among themselves and establish territorial boundaries by use of chemical odors (Dalton and Mathis 2014).

Their diet consists of very small arthropods, such as flies, mites, and ants (Britton 1981).

Breeding: The breeding biology of this species has not been well studied in Missouri, but much is known from studies in northern Arkansas. In Arkansas, mating occurs from January through May (Meshaka and Trauth 1995). In Missouri, courtship and breeding likely take place during the autumn, winter, and early spring. Little is known regarding

JEFF BRIGGLER

Occasionally, yellow-striped individuals are observed, such as this one from Barry County.

egg-laying sites with the only documented case in a sandstone fracture cave in northwestern Arkansas (Briggler and Puckette 2003). Eggs are likely deposited in deep underground cavities and crevices or in other cool damp niches that are inaccessible by humans. Clutch size ranges from two to eight eggs (average 4.3) per female with larger females depositing more eggs (Briggler and Puckette 2003). Females lay their eggs primarily in June and early July, and they remain with the eggs until they hatch in mid-August and early September (Briggler and Puckette 2003). Such brooding behavior by the female presumably assists with retaining egg moisture, reducing fungal outbreaks, and protection from predation. It is likely that females reproduce biennially, based on different females utilizing egg-laying sites within specific crevices over time (Briggler and Puckette 2003). As with all members of the genus *Plethodon*, the young go through their complete development in the egg and hatch into tiny replicas of the adults. Newly hatched Ozark zigzag salamanders have proportionately shorter tails than adults and average 22 mm (0.9 in.) in total length. Sexual maturity is reached when individuals are two to three years of age and greater than 30 mm (1.2 in.) in snout-to-vent length (Meshaka and Trauth 1995; Wilkinson et al. 1993).

Remarks: This endemic salamander of the southern Ozark Highlands is considered quite common in damp hardwood forests with rocky substrates and rotting logs. Keeping this species common in Missouri will require protection of intact upland forests with seeps, small creeks, and karst habitat.

Southern Red-backed Salamander

Plethodon serratus Grobman

JEFF BRIGGLER

Adult southern red-backed salamander from St. Francois County.

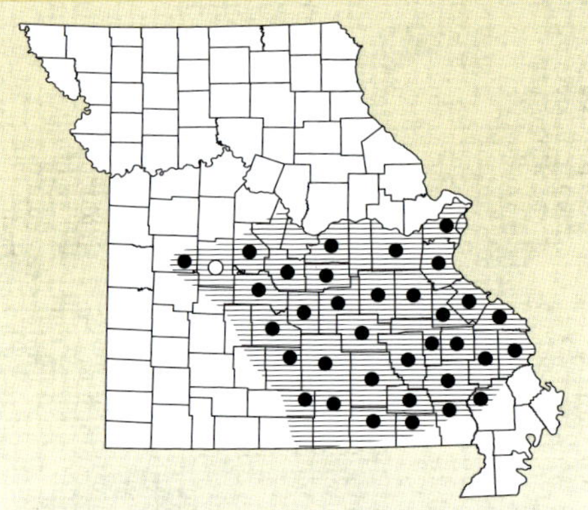

Distribution: Missouri: Throughout the northern and southeastern Ozarks. North America: Isolated populations in extreme eastern Tennessee, the southwestern corner of North Carolina, northwestern Georgia, northeastern Alabama, central Louisiana, west-central Arkansas and extreme southeastern Oklahoma and southern Missouri (Powell et al. 2016).

Description: The southern red-backed salamander is a small, dark, slender salamander with a long, rounded tail. A distinct, narrow, red or orange mid-dorsal stripe with saw-toothed edges that correspond with the costal grooves is usually present. Specimens lacking the red dorsal stripe have been found in Missouri. Although very unusual, a grey striped and a white striped individual have been observed (Drake and O'Donnell 2014). The sides are brownish gray with some red pigment. The belly is covered with gray mottling. There are 18 or 19 costal grooves. Sexes are difficult to distinguish, but males are normally smaller and slimmer than females.

Adult southern red-backed salamanders range in total length from 81 to 105 mm (3.2 to 4.1 in.; Powell et al. 2016) but have been known to reach 107 mm (4.2 in.; Daniel 2011).

Habits and Habitat: This terrestrial salamander commonly lives in upland forests where it

Southern red-backed salamanders live under rocks and logs along forest slopes, such as this location in Shannon County (above). An adult and six-month-old hatchling southern red-backed salamander from Crawford County (left).

hides under rocks, damp leaves, clumps of moss, and rotten logs. A study in a forested area in Dent County captured southern red-backed salamanders mostly in leaf litter (72 percent), especially the day following rainfall, as compared to woody cover and rocks (O'Donnell et al. 2014). They are most active on the surface during cooler, wet conditions between September and May, but they are difficult to find in the summer. During dry parts of the summer, we have found southern red-backed salamanders near seepages, near springs, or inside thick leaf litter in ravines. If moisture is not present on the surface, many salamanders retreat underground via bedrock crevices or utilize burrows created by other animals (e.g., beetles, worms). Drake et al. (2012) found southern red-backed salamanders to utilize abandoned burrows created by the emergence of periodical cicadas in Dent County. They are also found in the twilight zone of Missouri caves (the dimly lit area beyond a cave entrance; Myers 1958b; J. Briggler pers. obs.).

This species eats ants, beetles, spiders, termites, earthworms, and small snails.

Breeding: Courtship and mating likely take place between December and March in Missouri (Herbeck and Semlitsch 2000). Eggs are laid during late May or June (Herbeck and

Semlitsch 2000) and presumably the eggs are attached to a thin stalk suspended from the top of a cavity in underground burrows, under rocks, clumps of moss, or rotten logs similar to many other *Plethodon* species (Camp 1988; Trauth et al. 2004). Usually six to seven eggs are produced; fertilization occurs as they pass through the cloaca. Herbeck and Semlitsch (2000) found an average of 6.3 (range 4 to 10) eggs inside of gravid females, and reproduction among females was every other year. Females remain with the eggs until they hatch, which is sometime between late July and August. Hatchlings begin to emerge on the surface in September and October and are small replicas of adults. They average 17 mm (0.7 in.) with a range from 15 to 20 mm (0.6 to 0.8 in.) in snout-to-vent length (Herbeck and Semlitsch 2000). Sexual maturity is reached in Missouri when individuals are at least two years of age and greater than 35 mm (1.4 in.) in snout-to-vent length (Herbeck and Semlitsch 2000).

Remarks: Before the mid-1970s, the red-backed salamander of the eastern Missouri Ozarks was identified as the eastern red-backed salamander (*Plethodon cinereus*) by most authors (Highton 1962; Johnson and Bader 1974; Conant 1975). Investigations by Highton and Webster (1976) and Highton and Larson (1979) showed that the Missouri population is related to the Arkansas population (*P. serratus*), as well as to populations of red-backed salamanders in several southeastern and southern states. Missouri's population was reclassified as the southern red-backed salamander (*Plethodon serratus*). There are several disjunct populations of *Plethodon serratus*, and recent genetic work by Newman and Austin (2016) and Thesing et. al. (2016) found five divergent groups that likely warrant full species status, which includes the population in south-central Missouri.

The two species of red-backed salamanders (southern red-backed salamander and Ozark zigzag salamander) that occur in Missouri may be easily confused. The southern red-backed salamander's dorsal stripe is usually uniform in width and has serrated edges. The dorsal stripe of the Ozark zigzag salamander is usually very thin (less than one-third of the body width); it may be broken up into lobes and is always widest near the hind limbs. These two species may occur together in several Missouri counties, but the possibility of hybridization between the two forms is remote (R. Highton pers. comm.).

Southern red-backed salamanders are one of the most abundant and dominant species within the forested landscape of south-central Missouri with density estimates of 7,300 to 12,900 salamanders per hectare (2,954 to 5,221 per acre; Semlitsch et al. 2014b). They are most abundant in undisturbed, mature forest and rarely seen in newly regenerated forests (Herbeck and Larsen 1999). Protection and management of hardwood forests will help keep this salamander common in Missouri.

(Order Anura)

Toads and Frogs

Anurans consist of more than 6,400 extant species in 539 genera and are found worldwide, except where limited by extremely cold or arid conditions (Pough et al. 2018). They are most diverse in the tropical regions of the world. Toads and frogs can be found in forested, grassland, and even arid habitats. Many species prefer moist areas, especially near wetlands. However, they are better adapted to dry environments compared to salamanders. In Missouri, toad and frog species can be found in every habitat type with a good number of species occurring statewide.

Males are known for their calls to attract mates during the breeding season. Breeding season will vary among species with some calling in late winter, many calling in spring, and yet others calling in early summer in Missouri. Their life histories and reproductive modes are very diverse worldwide; many anurans deposit eggs in permanent or temporary bodies of water, others lay eggs in foam nests in leaves or trees, and yet others carry eggs on their bodies. All of Missouri's toads and frogs deposit eggs in aquatic habitats (e.g., streams, lakes, ponds, temporary wetlands, etc.). Their eggs hatch into tadpoles or polliwogs that feed mainly on plant material in the wetland. Metamorphosis can vary considerably among species; taking only several weeks for spadefoots and over a year for bullfrogs. Newly transformed young anurans look similar to the adults. There are about 100 species of anurans native to the United States. The 25 species and one additional subspecies of toads and frogs known to occur in Missouri consist of five families and eight genera.

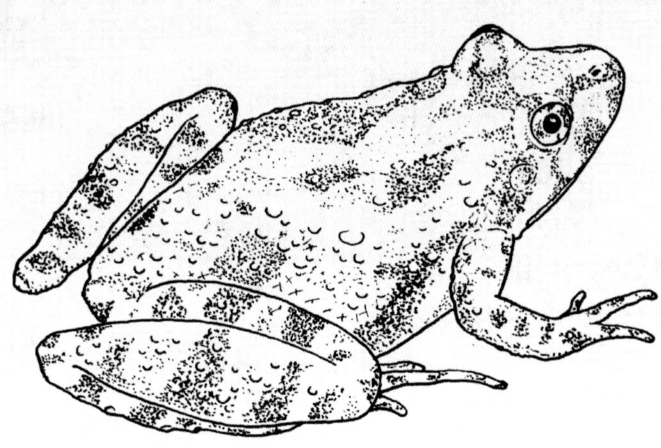

Blanchard's Cricket Frog

Family Scaphiopodidae

North American Spadefoots

In North America, this family is represented by two genera, *Scaphiopus* and *Spea*, with several species, two of which occur in Missouri (Pough et al. 2018). Spadefoots, often called spadefoot toads, are not true toads (Family Bufonidae) but resemble them in general appearance and habits. The name spadefoot is derived from a special tubercle on the hind feet that is spade shaped and used to dig into soil; spadefoots are highly fossorial, and explosive breeders in ephemeral pools. They have smoother skin than true toads (genus *Anaxyrus*) and the pupils of their eyes are vertical.

Eastern Spadefoot

Eastern Spadefoot

Scaphiopus holbrookii (Harlan)

JEFF BRIGGLER

Adult eastern spadefoot from New Madrid County.

Description: The eastern spadefoot is a stout, toad-like amphibian with large protruding eyes, vertically elliptical pupils, short legs, and large feet. Small, inconspicuous parotoid glands are present. There is no boss (a raised area) between the eyes. The general ground color is light brown to yellow brown. Head, back, and upper parts of legs are mottled with dark brown. The amount of dark brown on the dorsum may be great enough to form two or three light yellow-brown longitudinal stripes in a pattern resembling a lyre. The belly is pale white to gray. The inner surface of each hind foot has a sickle-shaped spur or spade. Males can be distinguished during the breeding season by their swollen front limbs, dark-gray throat, and the presence of a dark, horny covering on the inner surface of each front limb.

Adult eastern spadefoots range in snout-to-vent length from 44 to 57 mm (1.7 to 2.2 in.) but have been known to reach 78 mm (3.1 in.; Powell et al. 2016).

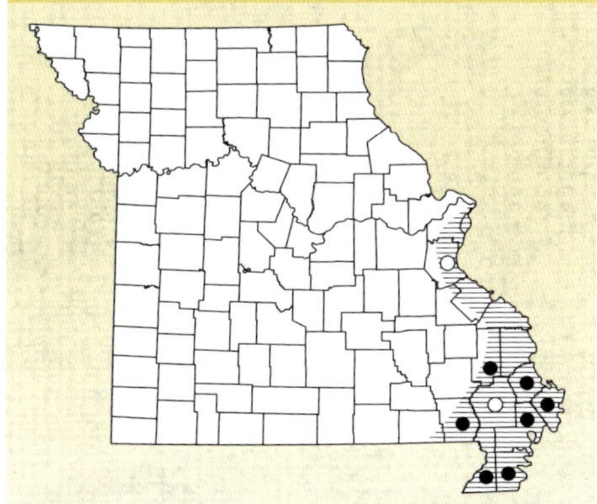

Distribution: Missouri: Eastern counties along the Mississippi River and in southeastern Missouri. North America: Southern New England, eastern Virginia, North Carolina, South Carolina, Georgia, Alabama, and Mississippi, most of Florida, southern Ohio, Indiana and Illinois, Kentucky and Tennessee, southeastern Missouri and eastern Arkansas (Powell et al. 2016).

Eastern spadefoot eggs attached to vegetation in Mississippi County (above). Calling male eastern spadefoot from Mississippi County (right).

Habits and Habitat: Eastern spadefoots spend most of their time in burrows dug with their hind feet. They are nocturnal and become active on warm, damp, or rainy nights. They occasionally occur in wooded areas but seem to prefer open fields where loose sand and soil facilitate burrowing. Spadefoots are most common within the sand prairies and sand savannas of southeastern Missouri. They are mostly observed during the breeding season and seldom seen at other times. They can be quite abundant within favorable habitat with estimates as high as 631 individuals per hectare (255.4 per acre) in a Florida study (Pearson 1957). Spadefoots limit activities to a small area with an average home range of 9.9 m^2 (106.6 ft^2) in a Florida population (Pearson 1957). One individual was found to inhabit the same burrow for over four years (Pearson 1955).

Their diet includes a variety of invertebrates, especially ants, beetles, flies, grasshoppers, crickets, termites, and worms.

Breeding: Breeding takes place after heavy rains, usually between March and June in southeastern Missouri. In some years, however, warm torrential rains in February cause early breeding, and it is likely to breed in the summer after heavy rains. Heavy rains (over 25 mm or 1 in.) are needed to stimulate breeding aggregations. If heavy rains do not occur, spadefoots will not breed during that season. Most breeding takes place after sunset, but daytime chorusing may occasionally be heard during favorable weather. Eastern spadefoots breed in temporarily flooded fields, ditches, or other fishless, temporary pools. In Missouri, a breeding chorus generally ranges from 20 to 40 individuals calling at a site. However, it is not uncommon to hear a very large chorus during favorable weather conditions. On the evening of March 13, 2019, a breeding chorus in Dunklin County was estimated to be greater than 1,000 individuals (J. Briggler pers. obs.). This area received from 3.8 to 7.6 cm (1.5 to 3 in.) of rainfall during the previous week, initiating some breeding. An additional 3.8 to 5.1 cm (1.5 to 2 in.) of rainfall fell that evening, stimulating additional widespread breeding. During

JEFF BRIGGLER

Eastern spadefoots breed in shallow, ephemeral pools, such as this roadside ditch in Mississippi County.

this time, eastern spadefoots were calling in every county with sandy habitat in southeastern Missouri, and thousands were observed crossing roads. They are explosive breeders, and very few choruses were heard the following night.

Spadefoots use a form of amplexus known as pelvic amplexus, where a male grasps a female just anterior of the pelvic area. During amplexus the eggs are fertilized by the male as they are laid. Eggs are brown or black above with a light-cream to white color below. The tangled eggs are laid in loose, wide strands or bands measuring between 25 and 75 mm (1 and 3 in.) wide and 25 and 300 mm (1 and 11.8 in.) long (Dodd 2013b) and are attached to submerged vegetation or twigs. Individual females produced 1,000 to over 2,500 eggs in Alabama (Mount 1975). Female clutch sizes ranged from 1,961 to 4,847 eggs in Arkansas (Trauth et al. 1990; Trauth and Holt 1993). Hatching takes place in 1 to 15 days (depending on the water temperature). Prior to transformation, the tadpoles' body and tail musculature are dark brown to shiny, brassy brown with numerous small, brassy-to-gold flecks (iridophores). The tail fins are clear. Tadpoles collected on April 16 in Mississippi County were 19 to 27 mm (0.7 and 1.1 in.) long. Tadpoles may transform between two and five plus weeks after hatching from eggs (depending on water temperature). Most juveniles become reproductively mature within two years of metamorphosis, and adults have lived in captivity for over 12 years (Snider and Bowler 1992).

The call of the male eastern spadefoot is a quick series of coarse "*wank, wank, wank*" sounds repeated every few seconds. Wright and Wright (1949) describe their call as sounding like a young crow.

Remarks: The eastern spadefoot is listed as a species of conservation concern in Missouri because of its restricted range, the destruction or modification of natural sand prairies, and limited knowledge on basic biology. Extensive agriculture practices in the sand prairie, as well as urban development, have destroyed and modified many of the natural ephemeral pools where spadefoots breed. Increased survey effort over the past 15 years has documented that spadefoots are more common than once believed in southeastern Missouri. Management of this species should focus on maintaining fishless, temporary wetlands for breeding and tadpole growth within former sand prairie habitat.

Plains Spadefoot

Spea bombifrons (Cope)

Adult plains spadefoot from Callaway County.

Distribution: Missouri: Counties along the Missouri River. North America: Southern Alberta and Saskatchewan, south to north and west Texas, New Mexico, northeastern Arizona, and northern Mexico, with isolated populations in Arkansas and the southern tip of Texas (Powell et al. 2016).

Description: The plains spadefoot can be distinguished from the eastern spadefoot by the presence of a boss (a raised area) extending forward from the eyes and by the wedge-shaped spade at the base of each hind foot. The plains spadefoot ranges from gray to tan gray to brown; some green may be mixed into the background color. A number of irregular dark-brown markings are usually present on the back and hind legs. There may be two or four faint light stripes on the back. The belly is white. Small round "warts," often red, are usually on the back and sides. Males can be distinguished from females by their dark throat and swollen forelegs.

Adult plains spadefoots range in snout-to-vent length from 38 to 51 mm (1.5 to 2 in.) but have been known to reach 64 mm (2.5 in.; Powell et al. 2016).

Habits and Habitat: This species is at home on the Great Plains, where it lives in grassland or open floodplain environments. By day

JEFF BRIGGLER

JEFF BRIGGLER

Plains spadefoots breed in shallow, ephemeral pools along the Missouri River floodplain, such as this example in Callaway County (above). Plains spadefoots have a spade-shaped tubercle on each hind foot used for digging into the soil. (left).

it hides in burrows, usually in sandy soil. It emerges at night to forage and breed, especially after heavy rains. It digs its own burrows backward using the wedge-shaped spade. In Missouri, this amphibian lives in the loose, sandy soils of the Missouri River floodplain. In studies from other states, most adults remain within 400 m (1,312 ft) of breeding wetlands although some individuals have been reported to travel 2.3 km (1.4 mi), and they can travel at least 1 km (0.6 mi) to a breeding site within a single night (Landreth and Christensen 1971; Werner et al. 2004).

Earthworms and a variety of insects make up the bulk of the plains spadefoot's diet. A specimen from Jackson County was found to have eaten a mole cricket (Smith and Powell 1993).

Breeding: The plains spadefoot is an explosive breeder that appears suddenly in great numbers after warm, heavy rains. Breeding in Missouri usually takes place from early April through August, although some calling activity occurs in late March in warmer years (Femmer 1978; J. Briggler unpubl. data). A breeding population of several hundred were heard in June 1993

TOM R. JOHNSON

Calling male plains spadefoot from Callaway County.

in a flooded Holt County field (Bufalino and Easterla 1996). Breeding may be completed in a few nights. Males congregate at temporary pools in flooded fields or roadside ditches with depths varying from 15 cm to 100 cm (5.9 to 39.4 in.; Femmer 1978). Males primarily call while floating in the water, and if a female approaches a calling male, he will clasp her waist. The female can produce up to 2,000 eggs; they are fertilized by the male as they are laid. The eggs usually are attached to submerged vegetation in small masses of 10 to 250 (Collins et al. 2010). The rate of hatching and transformation is regulated by water temperature, amount of dissolved oxygen, and food supply. If conditions are favorable, the eggs hatch in a few days. Metamorphosis can vary from as few as 13 days to 30 days depending on water temperature. Femmer (1978) studied a population of spadefoots in Howard County that had a tadpole development time of 17–20 days for eggs laid in late June. Spadefoot tadpoles are mostly omnivorous, but some become carnivorous. If the temporary pool in which they hatch evaporates too fast, tadpoles become overcrowded and sometimes become cannibalistic.

The call of the breeding plains spadefoot is an extended rasping or nasal "*garvank*" repeated at intervals of one-half to one second.

Remarks: Spadefoots have a greater tolerance to desiccation than other amphibians native to North America. They may lose up to half of their water content and still survive. A study of plains spadefoots in South Dakota showed that young specimens can protect themselves from tissue damage due to freezing during extremely cold periods (Swanson and Graves 1995). Adults are known to dig up to 90 cm (35.4 in.) deep to avoid freezing temperatures (Black 1970).

Spadefoots are known to produce skin secretions that can be noxious to other animals. When handling a spadefoot, it is best to avoid touching your face or eyes until you have washed your hands.

Family Bufonidae

True Toads

Members of this family live nearly worldwide, except for New Zealand, Australia, Madagascar, and the polar regions. Worldwide, the family comprises 47 genera with more than 590 species (Pough et al. 2018). There are 19 species of true toads in the United States, and all are members of the genus *Anaxyrus*. Missouri has four species of true toads with one additional subspecies. Toads have dry skin compared to frogs. They have no teeth, lack extensive webbing on the hind feet, and have large parotoid glands behind their eyes. Numerous "warts" over most of their bodies produce toxic skin secretions that are irritating to a predator's mucous membrane. At least for the species of toads in Missouri, females are larger than males. All species of toads in Missouri are primarily nocturnal, seeking shelter during the day among piles of dead leaves, under rocks and logs, or in loose soil. These amphibians are well known for their consumption of large numbers of insects. During summer nights, toads often catch and eat insects as they fall to the ground under outdoor lights. Contrary to popular belief, toads do not cause warts on humans.

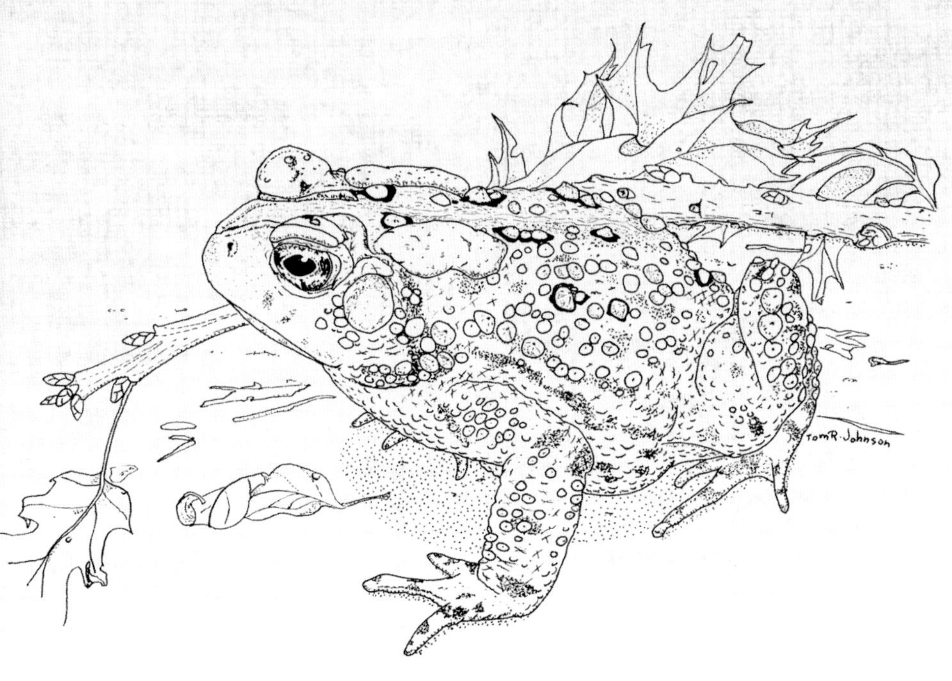

Eastern American Toad

Eastern American Toad

Anaxyrus americanus americanus (Holbrook)

Adult American toad from Callaway County.

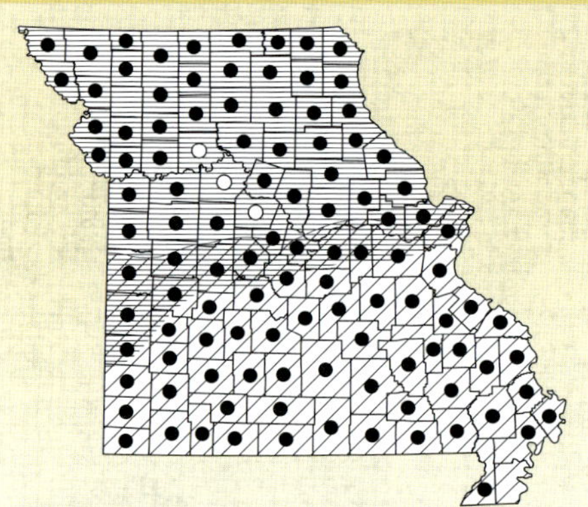

Distribution: Missouri: Occurs in the northern and western half of the state (horizontal lines), while the dwarf American toad occurs in the southern and southeastern half (diagonal lines). North America: The American toad ranges over most of the eastern half of North America from Labrador and Manitoba, south to Georgia and west to eastern Texas and Oklahoma (Powell et al. 2016).

Description: The eastern American toad is a medium-sized toad with kidney-shaped parotoid glands behind the eyes. The bony cranial ridges between the eyes usually do not touch the parotoid glands but may be connected by a small spur. Large black or dark-brown spots on the back usually encircle one to two "warts." A narrow, light stripe down the back may be present. The general color may be gray, green gray, and various shades of brown. The belly is white with dark-gray mottling. Adult males can be distinguished from adult females during the breeding season by swollen thumbs, an enlarged, horny pad on the inside of each foreleg, and a dark-gray throat. Females are larger in head and body size than males.

Adult American toads range in snout-to-vent length from 51 to 90 mm (2 to 3.5 in.) but have been known to reach 155 mm (6.1 in.; Powell et al. 2016).

Habits and Habitat: The American toad can be found in about any habitat in Missouri, but

JEFF BRIGGLER

Calling male American toad from Johnson County (above). Freshly deposited American toad eggs on top of green algae in Cole County (left).

it prefers rocky, wooded areas and often lives along the edge of hardwood forests. American toads are also found in disturbed habitats, such as urban (e.g., backyards, parks, gardens), agricultural, and mined sites. This species resides more in upland habitats compared to the equally common Rocky Mountain toad and Fowler's toad, which occur more often in lowlands and river floodplains. American toads hide during the day under rocks and logs where there is loose moist dirt or burrow into a depression where dead leaves have accumulated. They are known to travel long distances of about 0.5 km (0.3 mi) to over 6 km (3.7 mi) with an average home range of 717 m^2 (7,718 ft^2) in a Maryland study (see review in Dodd 2013a; Forester et al. 2006).

Like most toads, the American toad becomes active at dusk; it feeds at night on earthworms and insects.

Breeding: In Missouri, male American toads begin calling on warm nights during mid-March or early April. Breeding generally peaks from late April into May, with calling activity occurring into July. Breeding sites include both temporary pools (ditches, road ruts, borrow pits, flooded fields) and permanent ponds (sinkhole ponds, man-made ponds), and slow, shallow streams. The number of toads calling at a breeding site will vary annually; some years a site might have over 50 toads and the following year only 5 to 10 individuals. The male grasps the female behind her forelegs; while they float on the water, the female begins laying eggs in long, double strands. As the eggs emerge they are fertilized by the male. One female may lay from 2,000 to over 20,000 eggs. The eggs are deposited on the bottom in shallow water or entwined in aquatic plants, twigs, and branches. A study by Trauth et al. (1990) in Arkansas reported an average number of eggs per female of 4,701 (range 1,840–13,982). The eggs hatch in about one week into tiny black tadpoles. The tadpoles often are found in dense aggregations in shallow water near the shore. A mutualistic relationship between algae and salamander eggs has been well known, but only recently has such a relationship been documented in tadpoles. Tumlison and Trauth (2006) first documented the relationship between a green algae (*Chlorogonium*) and toad tadpoles in Arkansas. American toad tadpoles collected from Clark County contained green algal patches on their skin that documented *Chlorogonium* algae on tadpoles in Missouri (Drake et al. 2007).

The tadpoles generally remain in the water from late June into July, at which time they metamorphose into small toadlets 7 to 12 mm (0.3 to 0.5 in.) in length (Wright and Wright 1949). A study in central Missouri reported that most breeding occurred in May with metamorphoses occurring from July through August (Hocking et al. 2008). Even though American toads can lay a large number of eggs, there is no guarantee that the tadpoles will survive to metamorphosis. A study in Missouri estimated that 200,000 eggs hatched in a pond over a two-year period resulted in no tadpoles surviving due to high predation (Seale 1980). However, hundreds of toadlets that transformed from an abandoned Callaway County sewage lagoon during the summer of 2016 were observed actively feeding on insects in a grassy backyard (J. Briggler pers. obs.). American toads are known to live about four to five years.

The call of the male American toad is a sustained high-pitched extended trill, lasting 6 to 30 seconds.

Subspecies: There are two subspecies of *Anaxyrus americanus* in Missouri: the eastern American toad, *Anaxyrus a. americanus* (Holbrook) as described above, and the dwarf American toad, *Anaxyrus a. charlesmithi* (Bragg).

The dwarf American toad is generally smaller than the eastern American toad with an average snout-to-vent length of 51 mm (2 in.). The ground color can be red brown and the size and number of dark spots on the back are reduced or absent. The belly is cream colored with a small number of dark-gray spots on the breast. The call of a male *A. a. charlesmithi* is similar to that of the eastern American toad but slightly higher in pitch.

After examining a number of specimens of both subspecies of American toads from many parts of Missouri, it is apparent that there is a wide band of intergradation between these two subspecies along a line from St. Louis to northern Joplin (see map).

Remarks: The two subspecies of the American toad in Missouri have been known to hybridize with the Fowler's toad (*Anaxyrus fowleri*) and the Rocky Mountain toad (*Anaxyrus woodhousii woodhousii*) in some parts of the state. The offspring of this crossbreeding are very difficult to identify because they often have characteristics of both species. Generally, hybridization occurs when there is an overlap of breeding activity or where human activity has caused extensive habitat disturbance.

Great Plains Toad

Anaxyrus cognatus (Say *in* James)

JEFF BRIGGLER

Adult Great Plains toad from Saline County.

Description: The Great Plains toad is a medium-sized toad with large dark, paired blotches on the back and sides of the body. Each blotch is usually encircled with white or light tan and contains many "warts." Blotch color may vary from gray to brown, dark green, green, or yellow. A light, narrow stripe may run down the back of some individuals. The parotoid glands are kidney shaped and are connected to the bony cranial crest. The bony crests unite between the eyes and form a raised boss on the snout. The belly is cream colored with little or no gray spotting. Adult males are smaller than adult females. A dark throat and the presence of horny pads on the inside of each forelimb distinguishes males from females during the breeding season.

Adult Great Plains toads range in snout-to-vent length from 48 to 90 mm (1.9 to 3.5 in.) but have been known to reach 115 mm (4.5 in.; Powell et al. 2016).

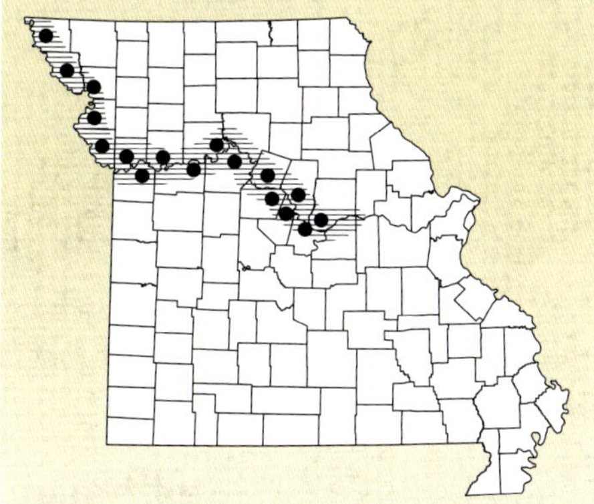

Distribution: Missouri: Restricted to the floodplain of the Missouri River from the northwestern corner to the central part of the state. North America: Great Plains and arid southwestern United States and northern Mexico (Powell et al. 2016).

Habits and Habitat: This species has a limited distribution in Missouri; its natural history

JIM RATHERT
TOM R. JOHNSON

Great Plains toads live in the sandy soils and breed in the shallow, flooded fields along the floodplain of the Missouri River (above). Calling male Great Plains toad from Callaway County (right).

has not been studied well in this state. In other states, it occurs in mixed grass and short-grass prairies and frequents open floodplains. In Missouri, it only occurs in the loose, sandy soils of the open floodplains along the Missouri River valley from central to northwest Missouri and avoids forested areas and upland prairies. However, thousands of individuals, mainly metamorphs, were observed on roads and yards while escaping Missouri River flooding in the summer of 2019 (J. Briggler pers. obs.). As with most species of toads, the Great Plains toad hides in underground burrows by day but emerges at night to feed. It uses deeper burrows to escape both hot and cold temperatures. Great Plains toads can burrow or use mammal burrows between 74 and 104 cm (29.1 and 40.9 in.) below the surface during the extreme weather months (Ewert 1969; Graves and Krupa 2005).

Ants, termites, beetles and other insects make up the bulk of its diet.

Breeding: Individuals have been found on the surface from April through early October with breeding choruses in Missouri documented from April through July, always after heavy rains (Femmer 1978; J. Briggler unpubl. data). Great Plains toads select rain-filled ditches, temporary pools, and flooded fields as breeding sites and prefer shallow water of 30 cm (11.8 in.) or less in depth (Femmer 1978). Calling is usually initiated at air temperatures greater than 14°C (57°F; Femmer 1978). Breeding takes place after a female has been mounted by the male (amplexed) in the water. Krupa (1988) studied this species in Oklahoma and found that males, during amplexus, used their hind legs to form a "basket" to capture and hold eggs as they were extruded by the female. A male held the eggs in this manner for up to three minutes, likely to make sure most of them were fertilized by his sperm. The eggs are laid in long, gelatinous strands in shallow water; often many females select the same general spot to lay their eggs. Female Great Plains toads in a central Oklahoma population were found to lay 1,342 to 45,054 eggs with an average of about 11,000 eggs (Krupa 1994). Females may have a double clutch within a single year (Krupa 1986). The eggs hatch in about a week or less. The tadpoles usually metamorphose to tiny toadlets (average 10 mm; 0.4 in.) in 17–49 days, depending on water temperature (Bragg and Smith 1943; Krupa 1994). Tadpoles in the warmer water of summer grow about twice as fast as those in the cooler water of spring. Femmer (1978) studied a population of toads in Howard County that had a tadpole development time between 18 and 20 days for eggs laid in late June.

The call of the Great Plains toad is a loud, metallic, chugging trill: "*chee-ga, chee-ga, chee-ga,*" lasting 20 to 50 seconds. The inflated vocal sac of calling males is sausage-shaped and extends forward and above the snout (see photo opposite).

Remarks: The Great Plains toad is listed as a species of conservation concern in Missouri because of its narrow range, destruction or modification of breeding wetlands, and limited knowledge on basic biology. Management of this species should focus on maintaining shallow, temporary wetlands for breeding and tadpole growth. Because this species occurs within agricultural areas of the Missouri River floodplain and has a voracious appetite for agricultural insect pests, they are considered economically important.

Fowler's Toad

Anaxyrus fowleri (Hinckley)

Adult Fowler's toad from Madison County.

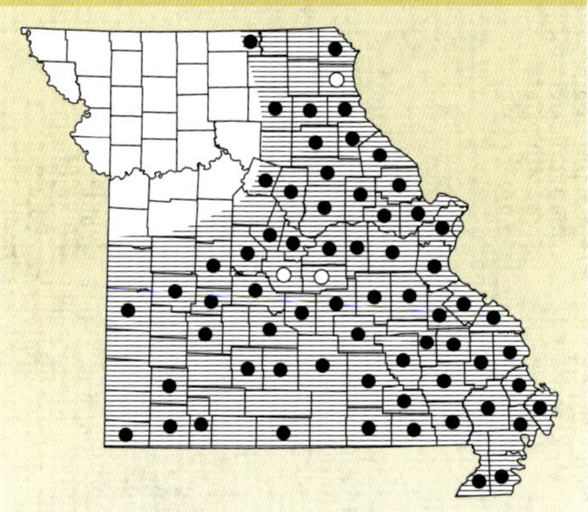

Distribution: Missouri: Northeastern, eastern, and southern Missouri. North America: Considered an eastern species and ranges from New Hampshire south to Georgia, northwestern Florida and into eastern Louisiana, and from Michigan and Illinois to southeastern Iowa, Missouri, and Arkansas to eastern Oklahoma (Powell et al. 2016).

Description: The Fowler's toad is a medium-sized toad that has a light-gray to green-gray background color with six paired, irregular dark-gray spots on the back. There are also dark markings on the sides and legs. Each dark spot on the back generally has three or more "warts." There is usually a light-tan or white line down the back. The belly is cream or white, and there is often a dark-gray spot on the chest. The parotoid gland is oblong and is connected to a rather shallow, bony cranial crest. Adult males are smaller than females, and during the breeding season, males have a dark-gray throat and dark, horny pads on the inside of each foreleg.

Adult Fowler's toads range in snout-to-vent length from 51 to 75 mm (2 to 3 in.) but have been known to reach 97 mm (3.8 in.; Daniel 2011; Powell et al. 2016).

Habits and Habitat: Fowler's toads occur in a wide variety of habitats, especially open areas with loose or sandy soils. This is the common toad of gravel and sand bars along

Fowler's toads are often seen along the gravel edges of streams, such as the Moreau River (above). Male Fowler's toad calling along a gravel bar on an Ozark stream in Carter County (left).

Missouri's many Ozark streams and rivers, as well as the sandy areas in the Mississippi Alluvial Basin. They are also seen in open wooded areas, urban and suburban yards, fields, and gardens. A study by Clark (1974) in Connecticut showed that Fowler's toads occupy home ranges that vary from 103 to 1,245 m^2 (1,109 to 13,401 ft^2).

They feed mostly at night on a variety of insects and earthworms.

Breeding: In southern Missouri, male Fowler's toads generally begin to call in late March into April, but in the rest of the state, they usually breed in May and early June. They will select flooded fields, ditches, swamps, marshes, and shallow, slow-water sections of streams, as well as ponds on occasions. Females are attracted to the breeding sites by a chorus of calling males. Females may lay over 8,000 eggs (Wright and Wright 1949). In Arkansas, clutch sizes range from 3,067 to 15,618 eggs per female with an average of 8,175 eggs (Trauth et al. 1990). The eggs are laid in long, paired strands in shallow water and hatch in less than one week (depending on water temperature). Tadpoles, which are black to dark brown, transform in 21 to 60 days (Mitchell and Anderson 1994; Wright and Wright 1949). Toadlets begin appearing by late June through mid-July. A study in northern Indiana showed that Fowler's toads take one to two years to reach maturity (Breden 1988).

The call of the male Fowler's toad is a short, medium-pitched, raspy, nasal "*w-a-a-a-a-h*," lasting one to three seconds.

Remarks: Fowler's toad was once classified as a subspecies of Woodhouse's toad (*Anaxyrus woodhousii*). This toad was elevated to a full species when genetic information and breeding call studies revealed that they are distinct from each other (Masta et al. 2002; Sullivan et al. 1996).

This species is presumed to hybridize with the American toad (*Anaxyrus americanus*) and Rocky Mountain toad (*A. woodhousii woodhousii*) in areas of overlap in Missouri (see species account for the American toad and Rocky Mountain toad).

Rocky Mountain Toad

Anaxyrus woodhousii woodhousii Girard

Adult Rocky Mountain toad from Callaway County.

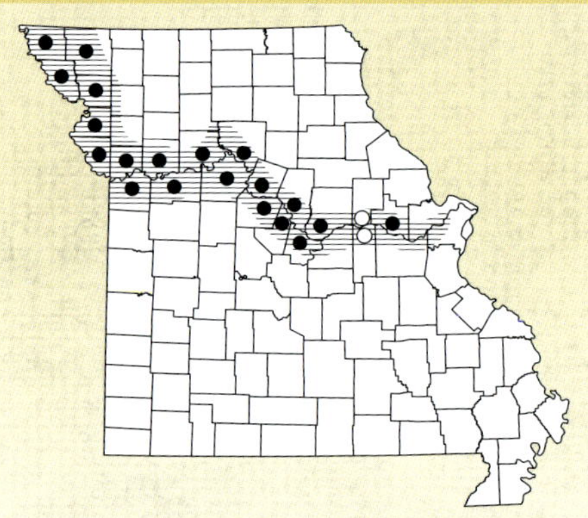

Distribution: Missouri: Primarily within the Missouri River floodplain from central to northwestern parts of the state. North America: Ranges from central and eastern Montana and most of North Dakota, south to Missouri and into Louisiana, and west to New Mexico, Utah, and Wyoming (Powell et al. 2016), with isolated populations in Idaho, Nevada, Oregon, and Washington (Dodd 2013a).

Description: The Rocky Mountain toad is Missouri's largest toad. This species is a medium-sized toad that has a number of irregular dark-brown or black spots on its back with one to six "warts" inside each spot. The general color varies from green, green gray, gray, tannish gray to brown. The belly is white and usually unspotted, although a single "breast" spot may be present on some individuals. A tan or white stripe is usually present down the back. The parotoid gland is oblong in shape; it is connected to a prominent bony cranial crest. Adult males are smaller than adult females. Males can be distinguished during the breeding season by a dark throat and dark, horny pads on the inside of each foreleg.

Adult Rocky Mountain toads range in snout-to-vent length from 64 to 100 mm (2.5 to 3.9 in.) but have been known to reach 127 mm (5 in.; Powell et al. 2016).

Habits and Habitat: This species prefers sandy lowlands, particularly river bottoms, and

JEFF BRIGGLER
JEFF BRIGGLER

Rocky Mountain toads breed in shallow, ephemeral pools along the Missouri River floodplain, such as this roadside ditch in Callaway County (above). Male Rocky Mountain toads call while perched on vegetation, such as this example from Callaway County (left).

open, dry areas adjacent to marshes, as well as around homes within urban and suburban areas. Rocky Mountain toads are the most abundant toad found in the sandy soils within the Missouri River floodplain. They can be observed in large numbers on roads after warm spring or summer rains. In May 2001, 135 Rocky Mountain toads were observed over a two-mile section of county road within the Missouri River floodplain in southern Callaway County (J. Briggler unpubl. data).

As with other toads, they hide in burrows by day, becoming active at night to hunt for a variety of insects and other invertebrate prey. A northern watersnake captured in northeastern Kansas was found to have eaten this toad (Smith and Powell 1993). Rocky Mountain toads also may be preyed upon by hog-nosed snakes, gartersnakes, and other species of watersnakes native to Missouri.

Breeding: Although individuals can become active in late March, this species will not begin to breed until late April or early May; breeding activity generally peaks in mid-May to early June and continues into July. In general, the Rocky Mountain toad breeds later and at warmer temperatures than the American toad. Groups of males congregate along the edges of river sloughs, shallow ditches, ponds, and flooded fields. They produce their distinct call until a female is attracted. Eggs are laid in long, paired strands during amplexus and are fertilized by the male. A female can lay a massive number of eggs. In Kansas, the average clutch size is 8,500 eggs (Collins 1993). However, Smith (1934) reported a clutch size as high as 25,644 eggs in Kansas, and Krupa (1995) reported a clutch of 28,500 eggs in Oklahoma. The eggs hatch in about a week or less; the black tadpoles begin to metamorphose into toadlets by late June or mid-July, which is generally 34 to 70 days depending on water temperature (Dodd 2013a). Recent metamorphs are about 10–15 mm (0.4–0.6 in.) in size and grow very rapidly; doubling body size within a month (Dodd 2013a). Rocky Mountain toads likely become mature during the first year after metamorphosis for males and the second year for females.

The male Rocky Mountain toad call is a short, nasal "*w-a-a-ah*," lasting one to two and one-half seconds, but lower in pitch than the Fowler's toad.

Remarks: The Rocky Mountain toad can hybridize with the American toad and Fowler's toad where their ranges overlap (see species account for the American toad and Fowler's toad). The Fowler's toad (*Anaxyrus fowleri*) was previously classified as a subspecies of the Woodhouse's toad, but the former has been elevated to a full species (Masta et al. 2002; Sullivan et al. 1996; see Fowler's toad account). With such full species changes (Masta et al. 2003), the Woodhouse's toad group (*Anaxyrus woodhousii*) has two recognized subspecies; the Southwestern Woodhouse's toad (*Anaxyrus woodhousii australis*) and the Rocky Mountain toad (*Anaxyrus woodhousii woodhousii*). The Rocky Mountain toad replaced the formerly recognized Woodhouse's toad common name (Johnson 2000). However, it is likely that additional taxonomic changes might be warranted between these two subspecies as more information becomes available.

Rocky Mountain toads are primarily found within the agriculture areas of the Missouri River floodplain. They can sometimes eat up to two-thirds of their own weight within a single night (Collins et al. 2010). Because of their appetite to consume large number of agricultural insect pests, they are considered economically important.

Family Hylidae

Cricket Frogs, Treefrogs, and Chorus Frogs

This family contains 50 genera and about 947 species inhabiting North and South America, Europe, extreme northern Africa, western Asia, and Australia (Pough et al. 2018). Hylids are most numerous in Central America and northern South America. The family seems to have originated in the tropics and spread north and south toward the temperate regions (Duellman and Trueb 1986). Most members of this family are diminutive, secretive creatures. Nine species of three genera (*Acris*-cricket frogs, one species; *Hyla*-treefrogs, three species; and *Pseudacris*-chorus frogs, five species) are native to Missouri.

Chorus frogs (*Pseudacris*) and cricket frogs (*Acris*) are only in North America. In this family, females are usually larger than males. Several of Missouri's treefrogs can change color; this characteristic seems to be associated with temperature, humidity, light, and even the activity and temperament of the frog. The majority of species have adhesive pads on their fingers and toes, which are helpful in climbing. Treefrogs are often around country homes where they may spend time on decks, patios, swimming pools, and bird houses (often when not being used by birds). At night, treefrogs are often near a porch light, catching insects attracted to the light.

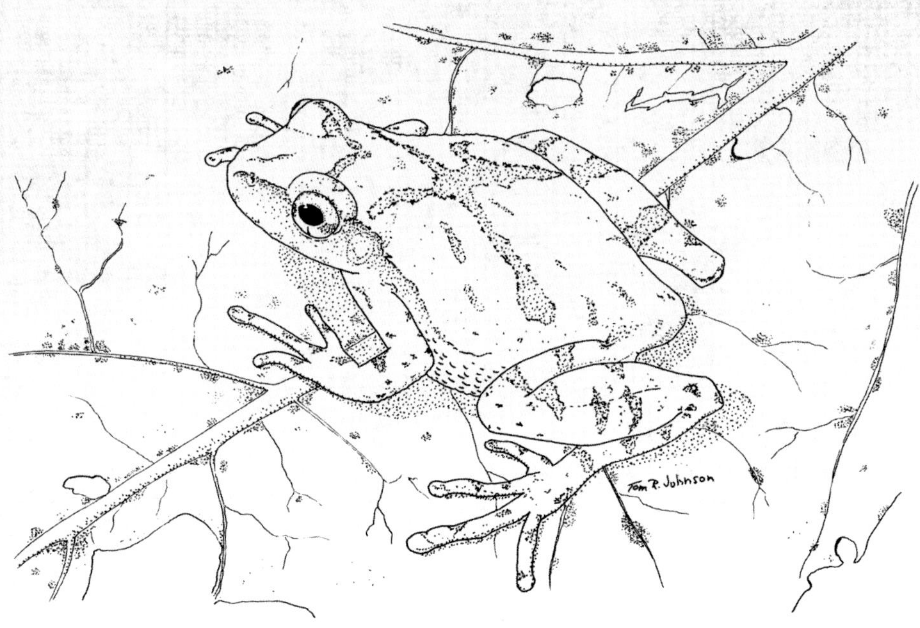

Spring Peeper

Blanchard's Cricket Frog

Acris blanchardi Harper

Adult Blanchard's cricket frog from Callaway County.

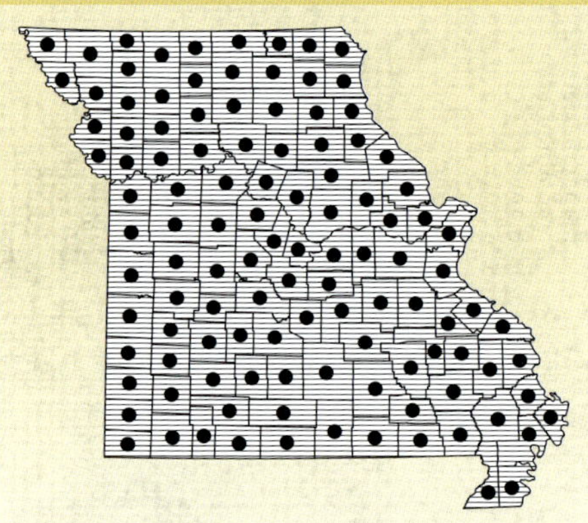

Distribution: Missouri: Statewide. North America: States north of the Ohio River and west of the Mississippi River from extreme eastern Arizona, and north to southeastern North Dakota (Powell et al. 2016). Surveys indicate that this species is gone or nearly gone from Minnesota, Wisconsin, northern Iowa, Illinois, Indiana, and Ohio.

Description: A small, nonclimbing, "warty" frog that exhibits a variety of colors. Three important color patterns are always present: a series of light and dark bars on the upper jaw, a dark triangle between the eyes, and an irregular black or brown stripe along the inside of each thigh. The general color is gray to tan, green tan, or brown. There is an irregular stripe down the back, which may be green, yellow, orange, or red. The belly is white. The feet are strongly webbed with small and poorly developed adhesive pads on the fingers and toes. The chin of males is spotted with gray and the throat may be yellow.

Adult Blanchard's cricket frogs range in snout-to-vent length from 16 to 38 mm (0.6 to 1.5 in.) but have been known to reach 41 mm (1.6 in.; Daniel 2011).

Habits and Habitat: This frog is usually active from late March to early November. During the spring and autumn, Blanchard's cricket frogs are active only during the day, but in

JEFF BRIGGLER

TOM R. JOHNSON

Calling male Blanchard's Cricket Frog from Moniteau County (above). A pair of Blanchard's cricket frogs in amplexus from Pulaski County. Male is on top (left).

warmer weather they are active both day and night. These alert little frogs will try to avoid capture by a series of quick, erratic hops. When approached, cricket frogs jump into the water but return quickly to shore. They prefer the open habitats, especially sandy, gravelly, or muddy edges of streams, ponds, and lakes. Cricket frogs are rarely found far from water but occasionally can be found on glades during the wet spring months. They are known to occasionally fake death by positioning themselves with their head down, eyes closed, and limbs tucked, and remaining motionless when threatened (McKnight et al. 2013). This tiny frog is known to overwinter on land by utilizing crayfish burrows, cracks in muddy banks, among tree roots, and in gravel slopes along the edges of ponds and streams (Dodd 2013a). A study on Rush Creek in Clay County during August and September found that cricket frogs preferred moist, muddy substrates that were close to shelter rocks and the water's edge (Smith et al. 2003).

The bulk of their food consists of a variety of small terrestrial insects (Johnson and Christiansen 1976). Adult Blanchard's cricket frogs have been observed being eaten by a fishing spider (*Dolomedes triton*) along the edge of a small stream in Pettis County (Johnson 2000).

Breeding: In Missouri, Blanchard's cricket frogs breed mainly between late April and July, but calling can begin as early as mid-March in the southern part of the state. Seale (1980) observed egg laying in Missouri from April to June. They select the shallow water in ponds and river backwaters where there is an abundance of aquatic plants. Warm air and water temperatures stimulate males to chorus. A study of cricket frogs in Indiana showed that a calling male may have several satellite or non-calling males within 5 to 30 cm (2 to 11.8 in.). These satellite males remained still and low to the ground but were seen amplexing with females that approached the calling male. It was presumed that the satellite males were less likely to become prey for bullfrogs and gartersnakes because they were less conspicuous (Perrill and Magier 1988). The satellite males that were successful in mating also saved a great deal of energy by not producing their breeding call. Amplexus, egg laying, and fertilization of the eggs take place in shallow water. A female may lay up to 400 eggs, either singly or in small packets of up to seven eggs, which are attached to submerged vegetation or resting on the pond bottom. The average number of eggs per female in Arkansas was 266 eggs (range 174–431 eggs; Trauth et al. 1990). In some populations females may lay two clutches of eggs or deposit partial clutches during the species' breeding season (Burkett 1969; Trauth et al. 1990). The eggs hatch in a few days, and the small tadpoles begin metamorphosis 5 to 10 weeks later (Burkett 1969), which is mainly in July and August (Seale 1980). In most cricket frog populations, their tadpoles have black-tipped tails that may be useful in avoiding capture by aquatic insects and other predators; the tail tip may distract a predator away from their head and body. This species has a short life span with the majority of individuals surviving no more than two years; however, some individuals may live three to five years (Beasley et al. 2005; Gray et al. 2005).

The call of the male Blanchard's cricket frog is a metallic "*gick*, *gick*, *gick*," somewhat like the sound of small pebbles being rapidly struck together. Although their breeding season may last until mid-July, the males can be heard calling day and night throughout the summer.

Remarks: Populations of this species have vanished or significantly declined from most of its northern range in the upper Midwest due to multiple factors (Beasley et al. 2005; Dodd 2013a). Although we have seen populations in Missouri decline during drought years, they have rebounded when rains returned in following years. Blanchard's cricket frogs were the most common species observed in surveys of 49 constructed wetlands in central and northern Missouri (Shulse 2011). Long-term monitoring from 2006–2017 for this species shows the population appears to be secure in the state. We do, however, need to continue to track Blanchard's cricket frogs in Missouri for signs of decline.

Cope's Gray Treefrog and Gray Treefrog

Hyla chrysoscelis Cope — *Hyla versicolor* LeConte Complex

JEFF BRIGGLER

Adult gray treefrog from Callaway County.

Description: Both species of gray treefrogs can be described together due to the extreme similarity of external characteristics. These frogs have "warty" or granular skin and prominent adhesive pads on their fingers and toes. Color varies from green to light green gray, gray, brown, or dark brown. Except for very light individuals, a few large, irregular dark, lichen-like blotches are usually present on the back. A large white or gray spot is always present below each eye. The belly

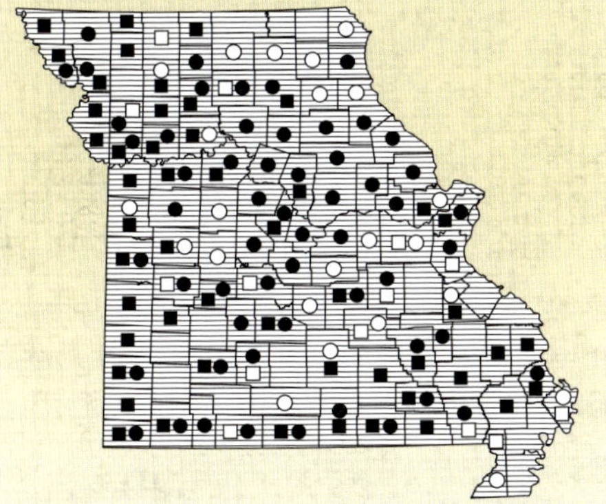

Distribution: Missouri: The gray treefrog (*H. versicolor*, circles) occurs mainly statewide with few to no reports in extreme southeastern and west-central Missouri along the Kansas border. The Cope's gray treefrog (*H. chrysoscelis*, squares) occurs mainly in southeastern, south-central, southwestern, western, and northwestern Missouri with few to no reports in northeastern Missouri. Both species have been found sympatrically in a number of counties. North America: Gray treefrogs range from southern Manitoba and Ontario, eastward to Maine, south to northern Florida, and west to Minnesota, eastern Nebraska, Kansas, and Oklahoma, and the eastern half of Texas (Powell et al. 2016).

JEFF BRIGGLER

JEFF BRIGGLER

A green colored gray treefrog from Callaway County (above). Calling male gray treefrog from Moniteau County (right).

is white to cream. The inside of the hind legs is yellow or orange yellow with gray or black mottling. Although both species of gray treefrogs in Missouri are morphologically identical, the Cope's gray treefrog tends to be slightly smaller and is more often green than the eastern gray treefrog (Dodd 2013a). During the breeding season, females usually appear heavier bodied, while males have dark throats. Newly transformed gray treefrog froglets are usually bright green.

Adult gray treefrogs range from 32 to 60 mm (1.3 to 2.4 in.) in snout-to-vent length but have been known to slightly exceed 62 mm (2.4 in.; Powell et al. 2016).

Habits and Habitat: These look-alike species of treefrogs appear to have the same general habits and habitat preference in Missouri, but *H. versicolor* is generally considered more of a forest species and *H. chrysoscelis* is generally more associated with grassland and oak savanna habitat. Gray treefrogs primarily live in small wood lots, in trees along prairie streams, in large tracks of mixed hardwood forest, and in bottomland forests along rivers and in swamps. Gray treefrogs spend much of their summer time in the canopy of deciduous forest occupying cracks, crevices, knotholes, bird holes, and other cavities (Mahan and Johnson 2007; Ritke

TOM R. JOHNSON

A pair of floating, gray treefrogs in amplexus (male on top) from Phelps County (above). A gray treefrog tadpole from a fishless pond in Warren County (left).

and Babb 1991). They are often seen with their heads sticking out of the opening and often return again and again to the same cavity (Pittman et al. 2008).

These frogs are frequently observed resting on large leaves or hiding in nooks and crannies of farm buildings, on porches or decks, within gutters, in bird houses, and climbing on windows at night. A bird study in Cole County that monitored nesting success of tree swallows (*Tachycineta bicolor*) in constructed bird nesting boxes surrounding a wetland showed that gray treefrogs frequently use these boxes during the spring and summer; one to four treefrogs per box may be observed with or without tree swallow use (J. Lundsted pers. comm.). In Missouri, gray treefrogs are normally active from late March through October.

During the breeding season, gray treefrogs may be seen on the ground at night en route to a breeding pond with the maximum recorded movement distance of 330 m (1,083 ft) from a breeding pond studied in Boone County (Johnson et al. 2007). Although the majority of the treefrog population, especially males, live within 60 m (197 ft) surrounding the breeding pond (Johnson and Semlitsch 2003), protection of forested habitat between 200 and 300 m (656 and 984 ft) surrounding the breeding pond is needed to ensure long-term stability of the population (Johnson et al. 2007).

Sonograms are "voice prints' that can be used to distinguish between species that are identical in appearance. The sonogram (top) of Hyla chrysoscelis *indicates 42 pulses per second, contrasted with the sonogram on bottom of* H. versicolor *with 19 pulses per second. The calls were recorded May 5, 1975, two miles north of Higginsville in Lafayette County. The two species were calling sympatrically at an air temperature of 22.2° C. The recordings were analyzed with a Kay model 6061A SonoGraph (narrow band).*

Green (1981) described how the toe pads of gray treefrogs enable them to climb or rest on vertical surfaces. He found that treefrog toe pads are "sticky" because of surface tension between a mucous layer produced by toe pad cells and the surface. A meniscus is produced all around the edge of the interface between each toe pad and the surface, causing surface tension that in turn causes a relatively strong bond.

Overwinter dormancy is mostly spent in the soil beneath the leaf litter that likely provides necessary moisture and temperature levels compared to arboreal tree cavities (Johnson et al. 2008). Gray treefrogs do not have to overwinter below the frost line because their liver produces a type of blood "antifreeze" called cryoprotectant that prevents freeze damage to their tissues (Costanzo et al. 1992; Storey and Storey 1987). The cryoprotectant in *H. chrysoscelis* is glucose compared to glycerol and glucose for *H. versicolor*. It allows these frogs to withstand temporary freezing events during the winter, but prolonged freezing may be fatal.

Gray treefrogs eat insects, spiders, and other invertebrates. A study in Boone County showed that 71 percent of the diet consisted of ants and beetles (Mahan and Johnson 2007). These frogs may, in turn, be preyed upon by bullfrogs, ribbonsnakes, gartersnakes, watersnakes, and wading birds (Hinshaw and Sullivan 1990). Their eggs and tadpoles can be food for fish, predacious aquatic insects, and salamander larvae. Fishing spiders, bullfrogs, and green frogs may eat gray treefrog froglets as they emerge from breeding ponds. Several common grackles (*Quiscalus quiscula*) were observed feeding on gray treefrog tadpoles in the shallows of a partially filled swimming pool in Jackson County (Smith and Bredehoft 2020).

Breeding: In Missouri, gray treefrogs mainly breed from early April to early July, although breeding into August occasionally occurs (J. Briggler pers. obs.). Males gather and begin calling at breeding sites when the night air temperature is around 16°C (60.8°F) or more. Preferred breeding sites include temporary wetlands, flooded field edges and ditches, and fishless, permanent ponds, such as sloughs, woodland ponds, and swamps. Vocalizing males may sit at the water's edge or station themselves on a plant, log, or branch above the water. Male Cope's gray treefrogs studied in western Tennessee were found to use up to three different calling sites during one night (Ritke et al. 1990). A study of the sympatric gray treefrogs in southern Missouri showed that *H. chrysoscelis* males called at ground level or very near the ground and had a slightly higher body temperature compared to *H. versicolor* males, which called from trees and bushes (Ptacek 1992). Females choose males based upon calling characteristics and calling site. The female initiates contact and moves with the male to the water to breed (Littlejohn 1958). A male will grasp a female with his forelegs when she comes near him. As they float in the water, the female will begin laying eggs. The male fertilizes the eggs while they are being laid. Females produce 900 to over 3,000 eggs in a clutch. Johnson and Semlitsch (2003) report an average clutch size of 1,018 in Missouri. The eggs are attached to floating vegetation in small, loose clumps of 6 to 45 eggs (Dundee and Rossman 1989). Each egg is light brown and a little over 1 mm (0.04 in.) in diameter. The protective jelly is weak and indistinct. Hatching takes place in about four or five days. Gray treefrog tadpoles have a high tail fin, which may be red or orange red, with large black blotches and a black border. In a Warren County study, gray treefrogs preferred more open canopy breeding sites that had warmer water and increased food, resulting in faster growth of tadpoles (Hocking and Semlitsch 2007; Hocking and Semlitsch 2008). The tadpoles transform into froglets in about one and a half to two months. The newly transformed gray treefrogs are 13–20 mm (0.5–0.8 in.) in length and are usually green. They spend most of the remainder of the summer in low vegetation near the breeding pond. Calls of the two gray treefrogs in Missouri are described in the remarks section of this account. Males can be heard calling away from breeding sites until early October.

Remarks: The gray treefrog (*Hyla versicolor*) and Cope's gray treefrog (*H. chrysoscelis*) can be distinguished in the field by comparing the calls of the males when heard at the same air temperature (Johnson 1966). In *H. versicolor*, the call is a musical, birdlike trill, which may vary from 17 to 35 pulses per second. The call of *H. chrysoscelis* can be described as a high-pitched buzzing trill, with 34 to 69 pulses per second. For proper identification of gray treefrog calls, especially when both species are not calling together, an analysis of a sound recording correlated with the treefrog calling temperature may be necessary. These two treefrogs also differ in the number of chromosomes: tetraploid in *H. versicolor* and diploid in *H. chrysoscelis*. In addition, the red blood cells of *H. versicolor* are larger than those of *H. chrysoscelis*. Natural hybridization between these two species has been reported but appears to be exceedingly rare (Gerhardt et al. 1994).

Green Treefrog

Hyla cinerea (Schneider)

JEFF BRIGGLER

Adult green treefrog from Butler County.

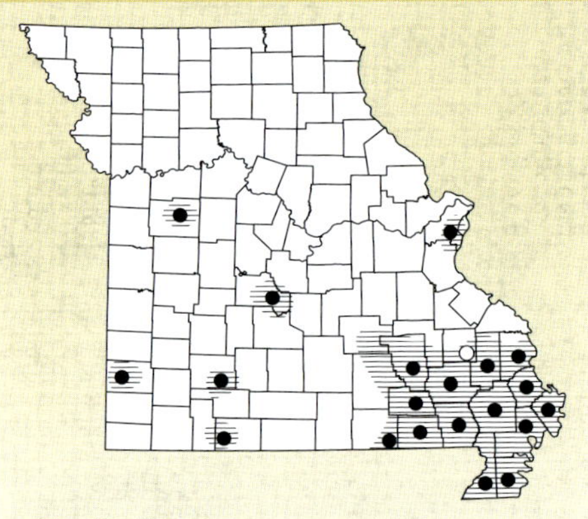

Distribution: Missouri: Mainly the southeastern corner of the state with several isolated populations. North America: Maryland and eastern Virginia, south through all of Florida, northwest to western edge of Kentucky and southern Illinois, west through southern Missouri and south through Arkansas, southeastern Oklahoma, and the eastern half of Texas (Powell et al. 2016).

Description: Green treefrogs are always green and can range from light to dark green. A white or pale-yellow stripe running from the upper lip and down the side is always present. In most of the green treefrogs in Missouri, this stripe stops halfway down the side, seldom reaching the groin. It may be outlined with a narrow black line on some individuals. Another white or pale-yellow stripe is present on the inside of the hind legs. The smooth back may have a few small gold spots. The belly is white or yellow. Males calling at night can be light green to yellow. Distinct, round adhesive pads are present on all digits. Adult males are smaller than females and have loose skin over their throats; the skin is inflated while they call.

Adult green treefrogs range in snout-to-vent length from 32 to 57 mm (1.3 to 2.2 in.) but have been known to reach 66 mm (2.6 in.; Powell et al. 2016).

Habits and Habitat: This species generally lives in vegetative areas near permanent

JEFF BRIGGLER

TOM R. JOHNSON

Green treefrogs live among the trees and breed in natural swamps, such as this example in Butler County (above). A pair of green treefrogs in amplexus (male is on top) from Bollinger County (left).

bodies of water, especially cattail marshes, cypress swamps, pond banks, or river sloughs and oxbows. They are also occasionally found around buildings, on porches or decks, within gutters or pipes, and other human structures. Green treefrogs often spend the day resting among green vegetation (e.g., long blades of cattails and grasses, among leaves of shrubs) where they are camouflaged by their green color. Small, grassy fields that border a forested wetland are an ideal place to find large numbers of foraging green treefrogs in Missouri. They mainly overwinter under logs, thick leaf litter, under tree bark, and within tree hole crevices (Fontenot 2011). To reduce water loss during dry conditions, they apply a mucous layer produced by a skin gland to their head and body by wiping with their limbs (Barbeau and Lillywhite 2005).

They are active on warm nights, climbing among vegetation in search of insects, especially flies, moths, crickets, grasshoppers, ants, and beetles.

Breeding: Male green treefrogs chorus from mid-April until early August, but egg laying probably occurs in late May into early July (Garton and Brandon 1975; J. Briggler unpubl. data). Males call from bushes, trees, and other plants along the water's edge, or while sitting on floating plants. They begin calling after dark. Amplexus will occur only when a female approaches and actually touches a male. Several non-calling, satellite males frequently are found close to a calling male. This position can result in successful mating of an approaching

A calling male green treefrog.

TOM R. JOHNSON

female by a satellite male (Gerhardt et al. 1987; Johnson 2000; Perrill et al. 1978; Perrill et al. 1982).

Clutch size is unknown in Missouri, but a female likely lays between 500 and 2,500 eggs, as reported in the neighboring states of Illinois and Arkansas (Garton and Brandon 1975; Trauth et al. 1990). A study of green treefrogs in eastern Georgia found that females can lay more than one clutch of eggs per season (Perrill and Daniel 1983). The eggs are fertilized by the male as they are released by the female and are laid near the water's surface on floating vegetation in small packets of 10–50 eggs. Hatching takes place in two to three days; transformation to froglets is generally five to seven weeks later between late June and early September. Treefrog froglets have been found to remain at or very close to their natal ponds or swamps until winter dormancy (Garton and Brandon 1975). Individuals can mature within one year (Garton and Brandon 1975) and are known to live up to six years (Snider and Bowler 1992).

The call of a male green treefrog is a series of regular nasal "*guank, guank, guank*" sounds with a ringing or metallic character. A chorus of green treefrogs resembles the sound of distant calling Canada geese.

Remarks: Populations of the green treefrog in Missouri represent the northwestern limit of this species' total range. This species naturally occurs in the Mississippi Lowlands of southeastern Missouri, but the species appears to be expanding its range north and west. It appears that these isolated records that occur in several counties are a result of accidental or intentional introductions, especially with several reports within the past decade. An introduced population of this species was established in Camden County at a private fish farm as early as the 1970s (D. Metter pers. comm.). Range expansion of the green treefrog within the southern upland habitat of the Ozark Highlands that border the Mississippi Lowlands is likely a natural expansion along river and stream drainages.

Even though swamp and marsh draining and channelization of streams in southeastern Missouri have destroyed a large part of their suitable habitat, this species is still quite common in Missouri and appears to be expanding its range northward. However, large areas of remaining swamps should be preserved so that this species can remain a part of the natural wildlife heritage of Missouri.

Spring Peeper

Pseudacris crucifer (Wied-Neuwied)

JEFF BRIGGLER

Adult spring peeper from Callaway County.

Description: The spring peeper is a small, slender frog with an X-shaped mark on its back. General color varies from pink to tan, light brown, or gray. The X-shaped mark may be very faint in light-colored individuals or prominent in darker frogs. There is a dark line running across the top of the head and between the eyes, and dark bars on the legs. The belly is a plain cream color. The tips of the fingers and toes all have distinct adhesive pads. Males are smaller than females; during the breeding season, they have darker throats than females.

The spring peeper ranges in snout-to-vent length from 19 to 32 mm (0.7 to 1.3 in.) but can reach up to 37 mm (1.5 in.; Powell et al. 2016).

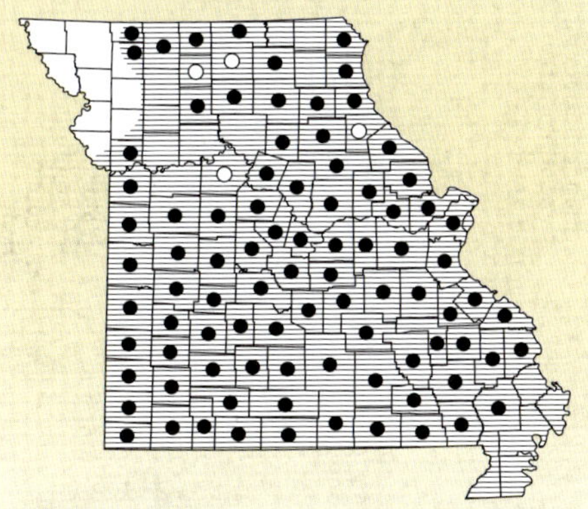

Distribution: Missouri: Throughout Missouri except in the extreme northwestern corner. North America: Throughout the eastern half of North America except for southern Florida (Powell et al. 2016).

Habits and Habitat: This is a forest species, living near ponds, streams, or swamps where there is thick undergrowth. Spring peeper abundance is positively associated with increased forest cover. The spring peeper usually remains hidden during the

JEFF BRIGGLER

Fishless, sinkhole pond used for breeding by the spring peeper from Shannon County.

day, becoming active at dusk; it may, however, become active during the day if heavy rains persist. This species can be active from late February through November in Missouri. As with a number of toads or frogs that overwinter in the soil, under leaves and logs, and beneath bark, spring peepers produce a kind of antifreeze in their blood. During autumn their liver releases glucose into their bloodstream, which protects their tissues from damage if they become frozen (Churchill and Storey 1996; Storey and Storey 1987).

They can be found on leaf litter and other surface debris during the active season, where they are foraging for a variety of small insects and spiders. Spring peeper tadpoles are eaten by fish, a variety of salamanders, and dragonfly larvae. Adults have been preyed upon by fish, birds, mammals, aquatic diving beetles, giant water bugs, and several species of snakes (see Dodd 2013a for review). On several occasions, northern watersnakes have been seen eating adult spring peepers in Missouri.

Breeding: The male spring peeper begins calling and breeding from late winter through spring (primarily late February to mid-May). In Missouri, migration of breeding adults entering ponds in Boone and Warren counties occurred between early March and late April (Hocking et al. 2008). Fishless, woodland ponds, temporary pools, water-filled ditches, or semipermanent, fishless swamps are favorite breeding sites, especially if brush, branches, and rooted plants are standing in the water. A study of a breeding population of spring peepers in eastern Maryland showed that both sexes are capable of breeding at two years of age. However, females may delay breeding for an additional year. Three- or four-year-old females were common in the study population (Lykens and Forester 1987). Males vocalize from sites at the edge of the water or from dead leaves or branches sticking out of the water. In a Missouri study, the distance between individual calling males averaged 50 cm (19.7 in.) and the range was 21–124 cm (8.3–48.8 in.; Gerhardt et al. 1989). While in amplexus, the female generally lays several hundred eggs that are fertilized by the male as they are laid. Average clutch size documented in Arkansas was 847 eggs (range 505–1,201; Trauth et al. 2004).

Calling male spring peeper from a fishless pond in Callaway County (above). Spring peeper eggs from a fishless pond from Warren County (left).

In late March 2003, an amplexus pair collected from Crawford County deposited 367 eggs in a bucket overnight (J. Briggler unpubl. data). The eggs are about 1 mm (0.04 in.) in size; they are laid singly and are attached to dead leaves, grasses, or sticks in shallow water. The eggs generally hatch in three or four days but may take up to 10 to 14 days depending on water temperature; tadpoles metamorphose in about three months. In Missouri, metamorph dispersal from ponds in Boone and Warren counties was primarily from late May to late July, with an estimated tadpole period between 84 and 138 days (Hocking et al. 2008). Individuals become sexual mature at the age of two.

The call of male spring peepers is a clear, high-pitched peep, with a slight rise at the end. The peeping call is repeated about once per second. This frog is one of the first species to begin calling in the spring. After the breeding season their call may be heard during the day or night from wooded areas, especially after a rain. Smith (1961) reported that males may be heard calling in the autumn in Illinois. Johnson (1975) heard several choruses of spring peepers calling on a rainy November night in eastern Missouri.

Remarks: This species was previously classified as a treefrog (*Hyla crucifer crucifer* Wied; Johnson 1987). Since then, laboratory analyses have shown that the spring peeper is actually a chorus frog (*Pseudacris*), and was placed in that genus (Cocroft 1994; Hedges 1986). At one time, two subspecies were generally recognized: southern spring peeper (*Pseudacris crucifer bartramiana*) and the northern spring peeper (*Pseudacris crucifer crucifer*), which occurred in Missouri. Genetic information no longer supports the recognition of the two subspecies (Moriarty and Cannatella, 2004).

Upland Chorus Frog

Pseudacris feriarum (Baird)

Adult upland chorus frog from Mississippi County.

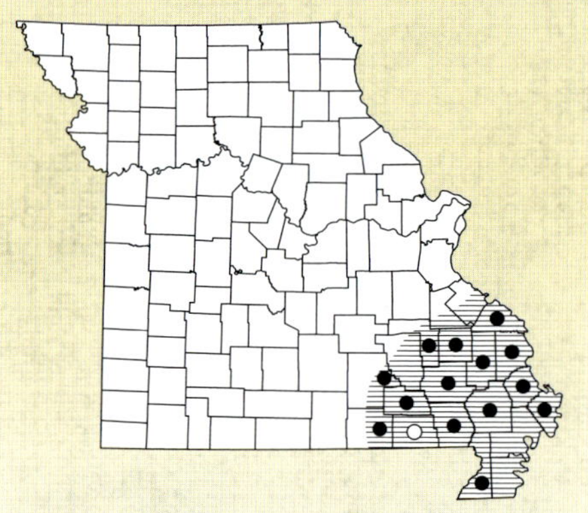

Distribution: Missouri: Southeastern Missouri consisting of the southeastern Ozark Highlands and Mississippi Alluvial Basin. North America: Southern Pennsylvania to a small part of northwestern Florida, westward through Mississippi and northward through extreme northeast Arkansas, southeast Missouri, and southern Illinois (Powell et al. 2016).

Description: The upland chorus frog is similar in appearance and natural history to the Cajun chorus frog (*Pseudacris fouquettei*) and the boreal chorus frog (*Pseudacris maculata*). The upland chorus frog is a small gray or tan frog with dark dorsal stripes that are narrow or broken into a series of dashes or spots. A few individuals may lack dorsal markings. There is a gray, irregular stripe that extends from the snout through the eye and down each side to the groin. A dark spot on the head between the eyes may be triangular. The belly is white. Adult females are larger than adult males; during the breeding season males have dark throats.

Adult upland chorus frogs range in snout-to-vent length from 19 to 35 mm (0.7 to 1.4 in.) but have been known to reach 39 mm (1.5 in.; Powell et al. 2016).

Habits and Habitat: This species tends to live in small patches of woods, forested swamps, and river bottomland forests in Missouri, especially those close to wetlands. They

JEFF BRIGGLER
JEFF BRIGGLER

Upland chorus frogs breed in temporary, fishless wetlands, such as this example in Stoddard County (above). Calling male upland chorus from Mississippi County (left).

typically remain close to the forested habitat but will cross open agricultural fields to seek breeding wetlands if necessary. Upland chorus frogs take shelter under leaf litter, under small logs or tree bark lying on the ground, or in small animal burrows during the day.

These frogs become active on damp nights when they search for prey (primarily insects and small spiders). An upland chorus frog was observed being eaten by a fishing spider (*Dolomedes triton*) in Virginia (Mitchell 1990). Outside the breeding season, upland chorus frogs are seldom seen.

Breeding: In Missouri, upland chorus frogs become active in February and breed in late winter or early spring, with the majority of the breeding occurring in March. Rainfall is necessary to stimulate breeding. Surveys conducted in six southeastern Missouri counties in 2019 documented peak breeding from March 12 to March 15. They are likely the first frogs to chorus in southeastern Missouri. Breeding takes place in temporary pools, flooded crop fields, ditches, and other shallow-water areas, especially those sites within or near forested habitat. However, they will move through large, open areas (e.g., grasslands and agricultural fields) to reach a breeding site if necessary. Males often sing in the water with their head and forelegs above the water, holding onto a blade of grass or plant stem. Females are attracted to the breeding pools, and amplexus may take place within an hour after sunset. Eggs are laid during amplexus and are fertilized by the male as they are laid. The eggs, which may number from 200 to over 500, are laid in small, elongated clusters of 40 to 60 eggs. The number of eggs per female from a North Carolina study ranged from 163 to 534 eggs (Alexander 1966). Females may lay two clutches per breeding season. The eggs are loosely attached to dead leaves, grasses, twigs, or plant stems in shallow water. Hatching generally takes place within one week, and transformation from tadpoles to froglets takes 6 to 10 weeks. Males are reproductively mature within one year whereas females often take up to two years (Alexander 1966).

The breeding call of male upland chorus frogs is similar to that of the cajun chorus frog and boreal chorus frog. All three species' calls sound like running a finger along the tines of a pocket comb. The upland chorus frog call is a clicking trill of "*crrreeeeek*" with a rising inflection as it ends. It is difficult to distinguish the upland chorus frog call from the boreal chorus frog call. The boreal chorus frog call is slightly raspier or more metallic in sound. Both the upland chorus frog and boreal chorus frog have a noticeably faster call rate and a shorter call length than the Cajun chorus frog (Elliott et al. 2009).

Remarks: Considerable taxonomic name changes have occurred within the trilling chorus frog complex of the genus, *Pseudacris* (Moriarty and Cannatella 2004; Lemmon et al. 2007). There are multiple species within this complex. Three very similar species occur in Missouri: upland chorus frog (*P. feriarum*), Cajun chorus frog (*P. fouquettei*), and boreal chorus frog (*P. maculata*). They can be distinguished by genetics, color patterns, slight differences in calls, and distribution. The distribution of the upland chorus frog and the Cajun chorus frog in Missouri is somewhat uncertain. The upland chorus frog appears to occur mostly within the Mississippi Lowlands of extreme southeastern Missouri, but they appear to overlap with the Cajun chorus frog as one moves west and possibly with the boreal chorus frog as one moves north in Missouri. It is possible that historic museum specimens identified as upland chorus frogs from the southeastern Ozark Highlands are actually Cajun chorus frogs and that upland chorus frogs are only found in the Mississippi Lowlands of extreme southeastern Missouri. Additional genetic studies are needed to better understand the boundaries of these three species, especially in potential areas of overlap.

Cajun Chorus Frog

Pseudacris fouquettei Lemmon, Lemmon, Collins and Cannatella

JEFF BRIGGLER

Adult female Cajun chorus frog from Stoddard County.

Description: This species is similar in appearance and natural history to the upland chorus frog (*Pseudacris feriarum*) and the boreal chorus frog (*Pseudacris maculata*). The Cajun chorus frog is a small tan to brown frog with dark dorsal stripes that are narrow or often broken into a series of dashes or spots. There is a gray, irregular stripe that extends from the snout through the eye and down each side to the groin. A dark spot on the head between the eyes may be triangular. The belly is white to cream in color. Adult females are larger than adult males; during the breeding season males have dark throats.

Adult Cajun chorus frogs range in snout-to-vent length from 19 to 25 mm (0.7 to 1 in.) but have been known to reach 30 mm (1.2 in.; Powell et al. 2016).

Habits and Habitat: Little is known about the basic life history of this newly described species in Missouri. The Cajun chorus frog primarily lives in a variety of habitats from forested areas to open grasslands and fields.

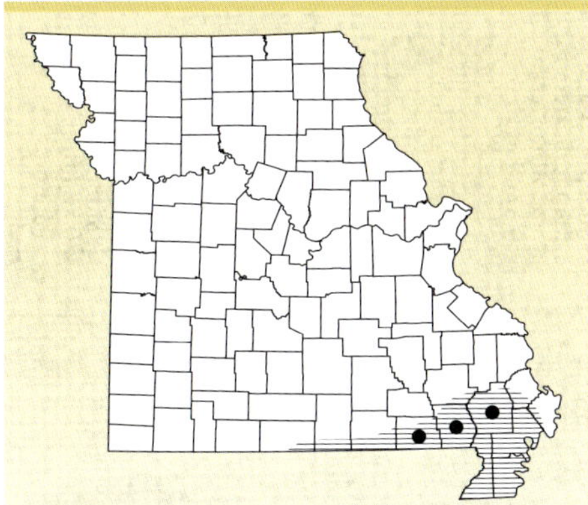

Distribution: Missouri: A few counties in southeastern Missouri. North America: Southeastern Missouri to eastern Oklahoma and southward to Gulf Coast of Texas, Louisiana, and Mississippi (Powell et al. 2016).

JEFF BRIGGLER

Cajun chorus frogs breed in ephemeral, fishless wetlands, such as this example in Ripley County.

They appear to be most common in bottomland forests, small patches of woods, and along the edges of swamps and marshes. Although commonly heard calling from breeding wetlands, they are rarely seen outside of the breeding season.

It is likely the Cajun chorus frog is similar to other chorus frogs in that it forages in moist leaf litter and feeds on a variety of small invertebrates, especially insects (Dodd 2013a).

Breeding: The reproductive biology of the Cajun chorus frog has not been studied extensively in Missouri. Cajun chorus frogs are late winter or early spring breeders. They have been heard calling in southeastern Missouri from February to early April with peak activity in March. This frog breeds in shallow, well-vegetated, temporary bodies of water (roadside ditches, flooded fields, water-filled depressions, swamps, and marshes) that are located within or near wooded areas. Amplexus, egg laying, and fertilization of the eggs take place in shallow water. A female may lay up to 1,500 eggs, in several loose, jelly clusters that are attached to twigs, sticks, and grasses. Female clutch size varies from 507 to 1,380 eggs in Arkansas (Trauth et al. 1990). A study in Texas showed that females may lay up to nine clusters of eggs within a single night (Livezey 1952). Each egg cluster averages about 43 eggs and ranges from 7 to 176 eggs. Hatching generally takes place in 4–11 days. The tadpoles generally live in the water for about 48 to 80 days depending on water temperature (Dodd 2013a). Emerging froglets usually remain within 5–10 m (16.4–32.8 ft) of the wetlands for several weeks (Livezey 1952).

The call of the male Cajun chorus frog is a clicking trill of "*crrreeeeeek*" that sounds similar to running a finger along the tines of a pocket comb. They have a much slower call rate and slightly longer call than both the upland chorus frog and boreal chorus frog.

Remarks: The Cajun chorus frog is a newly described species for Missouri and North America (Lemmon et al. 2008). Genetic data and differences in advertisement calls showed that this species was different from other trilling chorus frogs that once occurred in multiple states (Lemmon et al. 2007). Likely the Cajun chorus frog occurs in more locations in southeastern Missouri, but additional genetic data and calling surveys are needed to better understand the distribution of this new species (see species account for the upland chorus frog).

Illinois Chorus Frog

Pseudacris illinoensis Smith

JEFF BRIGGLER

Adult Illinois chorus frog from Dunklin County.

Description: The Illinois chorus frog is a medium-sized member of the treefrog family and our largest chorus frog. This species has a chubby appearance, with large, muscular, almost toad-like forelegs. The general ground color may vary from light tan to tannish gray. There is a distinct V-shaped mark between the eyes, a dark stripe from the snout to the shoulder, and usually a dark spot below the eye. A pair of large, dark, irregular V-shaped markings are present on the back behind the head. These markings are dark gray to brownish gray. The skin is rough or granular. The belly is white. Toe pads are small and round, and the webbing of the hind feet is poorly developed.

Adult Illinois chorus frogs range in snout-to-vent length from 25 to 41 mm (1 to 1.6 in.) but have been known to reach 48 mm (1.9 in.; Powell et al. 2016).

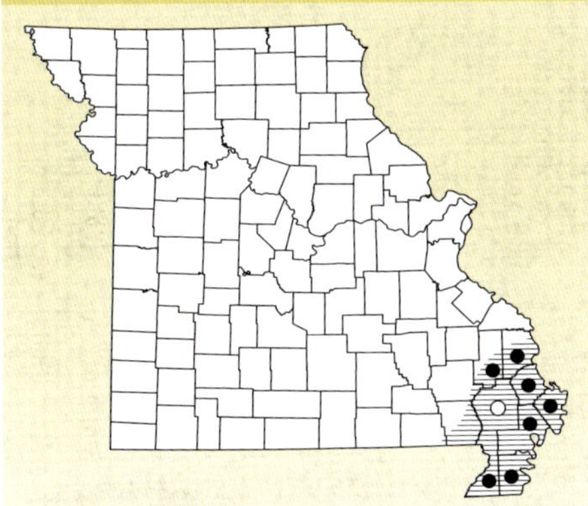

Distribution: Missouri: Restricted to the southeastern corner. North America: West central and southwestern Illinois, southeastern Missouri and one county in extreme northeastern Arkansas (Powell et al. 2016).

Habits and Habitat: In Missouri, this chorus frog prefers flat, sandy areas. Originally, Illinois chorus frogs were dwellers of the

Illinois chorus frogs breed in roadside ditches and flooded fields, such as this example in Scott County (above). Illinois chorus frog eggs in a temporary pool, Dunklin County (right).

natural sand prairies and sand savannas of southeastern Missouri, but this habitat has been nearly eliminated. However, Illinois chorus frogs continue to survive in soybean and cotton fields in the former sand prairie area. Unlike other fossorial species, such as members of the genera *Anaxyrus*, *Scaphiopus* and *Spea*, these frogs enter the soil head first using their strong forelegs and forefeet for digging (Brown et al. 1972). They are excellent burrowers in sandy habitat and are known to remain 15–20 cm (5.9–7.9 in.) below the surface for most of the year (Axtell and Haskell 1977), but during winter Illinois chorus frogs must burrow more than 25 cm (9.8 in.) to survive cold temperatures (Packard et al. 1998). Observations of burrowing behavior of Illinois chorus frogs in southwestern Illinois showed that adults' burrows were longer and deeper than those of newly transformed froglets. Also, the frogs tended to burrow in open areas away from any vegetation (Tucker et al. 1995). Their strong forefeet have good grasping ability; we have observed them hanging onto grasses as they sit in the water during the breeding season. Feeding likely occurs within the burrows underground, especially when aided by use of gasping forefeet to catch and hold prey (Brown 1978; McCallum and Trauth 2001).

This species is known to primarily eat dingy cutworms (a moth larvae), but spiders, beetles, flies, and ants are also eaten (Tucker 1997).

Breeding: The Illinois chorus frog breeds in late winter and early spring—usually late February to early April—depending on local weather conditions (J. Briggler unpubl. data). Breeding sites are flooded depressions in crop fields, roadside ditches, and other fishless, temporary pools. The males call while floating in the water or hanging onto grassy vegetation. A male will clasp a female about midbody; while they are floating in shallow water, the female

JEFF BRIGGLER

Illinois chorus frog tadpole from a roadside ditch in Mississippi County.

will begin laying eggs. In Arkansas studies, between 148 and 1,012 eggs with an average of 469 eggs per female are deposited (Butterfield 1988; Butterfield et al. 1989). The eggs are laid in masses with generally 8–100 eggs per mass (Brandon and Ballard 1998; Butterfield et al. 1989). Females can be observed using their forelimbs to grasp onto twigs or grass for support. As the female lays her eggs they are fertilized by the male. The eggs hatch in about a week. Illinois chorus frog tadpoles are large for a member of the chorus frog group, ranging up to 38 mm (1.5 in.) in total length. Their tadpoles are heavy bodied, with a high caudal fin containing many scattered, small markings. The tadpoles will take up to 60 days to transform into froglets. Most juveniles become reproductively mature within one year after metamorphosis, and adults generally live between two and five years (Tucker 1995; Tucker 2000). Further studies may show that this species could live longer due to their fossorial habits.

The breeding call of the male Illinois chorus frog is a clear, quick series of high-pitched birdlike whistles.

Remarks: This species was once recognized as a subspecies (*Pseudacris streckeri illinoensis*) of Strecker's chorus frog (*Pseudacris streckeri streckeri*), however, genetic data by Moriarty and Cannatella (2004) concluded that full species status was warranted.

Extensive agricultural practices in the sand prairie areas of the Missouri Bootheel have destroyed or modified nearly all natural ephemeral pools where this species formerly bred (Trauth et al. 2006). In addition, housing developments and other land uses have reduced or eliminated the sand prairie sites. Although this frog is still present in the highly cultivated areas of southeastern Missouri, there is concern whether these amphibians can withstand the continued destruction of their remaining habitat, especially wetland loss. Surveys over the years for this species have failed to locate breeding colonies in Pemiscot and the southern half of Dunklin counties, and it is likely the historic populations in these counties are extirpated. However, large numbers of frogs are still present in the core sand prairie regions of Scott, Mississippi, New Madrid, and the northern half of Dunklin counties. Long-term persistence of this unique, burrowing frog in southeastern Missouri is dependent on maintaining wetlands for them to reproduce. For these reasons, the Illinois chorus frog is listed as a species of conservation concern in Missouri.

Boreal Chorus Frog

Pseudacris maculata (Agassiz)

JEFF BRIGGLER

Adult boreal chorus frog from Nodaway County.

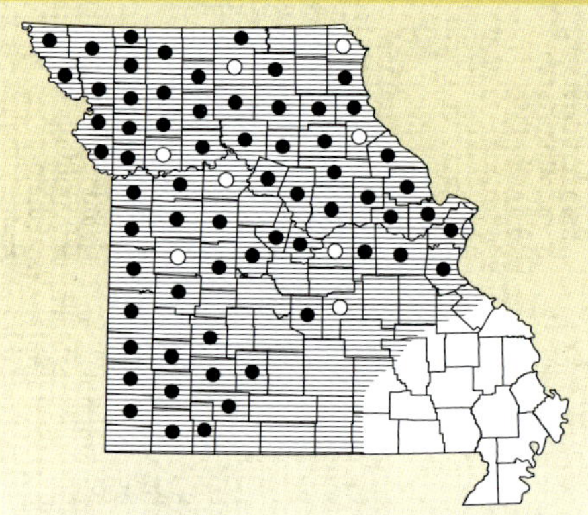

Distribution: Missouri: Found throughout the majority of the state, except in the southeastern quarter of Missouri. North America: Illinois west to Arizona, north through most of Canada with disjunct populations in southern Ontario and Quebec, and northern New York (Powell et al. 2016).

Description: The boreal chorus frog is a small gray or tan frog with dark stripes down the back. A broad stripe extends from the snout through the eye and down each side to the groin, which can be black, brown, or dark gray. The head, body, and legs may be gray, tan, or brown. The three dark stripes on the back may be broad and continuous from the head to the back of the legs or they may be broken into three rows of spots. There is usually a dark marking on the head between the eyes. The upper lip is white. Dark markings are gray or brown. The belly is white and there may be a few gray spots on the throat and chest. Adult females are larger than adult males; during the breeding season the males have dark throats.

Adult boreal chorus frogs range in snout-to-vent length from 19 to 35 mm (0.7 to 1.4 in.) but have been known to reach 39 mm (1.5 in.; Powell et al. 2016).

Habits and Habitat: This frog is most abundant in prairies but also may occur

JEFF BRIGGLER

JEFF BRIGGLER

Boreal chorus frogs breed in fishless, temporary wetlands throughout prairie and grassland areas, such as this example in Barton County (above). Calling male boreal chorus frog from Benton County (left).

on agricultural lands, in large river floodplains, along grassy edge of marshes, and in urban areas. After the breeding season, boreal chorus frogs take shelter in animal burrows, under boards, logs, or rocks, in clumps of grass, or in loose soil. Breeding sites are generally temporary bodies of water: flooded grain fields, ditches, woodland ponds, marshes, or river sloughs. They also will breed in fishless farm ponds with aquatic vegetation. Boreal chorus frogs have been reported to travel more than 600 m (1,969 ft) from a wetland, but most individuals move only between 20 and 50 m (66 and 164 ft) from their breeding wetland (Dodd 2013a; Spencer 1964). They do not burrow very deep and can tolerate freezing temperatures due to an increase in the level of glucose in their blood (Storey and Storey 1987). In Missouri, the boreal chorus frog is often the first frog to be heard in late winter, but it is rarely seen because it remains hidden in vegetation.

This species eats a variety of small invertebrates, especially beetles, flies, crickets, grasshoppers, ants, and spiders. Adult boreal chorus frogs are eaten mainly by a variety of birds, small mammals, and snakes. Tadpoles are eaten by aquatic invertebrates and fish. A study in northeastern Missouri showed that mosquitofish (*Gambusia affinis*) are a primary predator of the boreal chorus frog. Although mosquitofish are only 2.5 to 5 cm (1 to 2 in.) with tiny mouths, this aggressive fish greatly increases tadpole tail injuries (Shulse and Semlitsch 2014; Shulse et al. 2013).

Boreal chorus frog eggs from a flooded ditch in Callaway County.

Breeding: Boreal chorus frogs begin breeding activities in late February or early March, with peak activity during April and into May. They can be heard calling occasionally into July, especially in northern Missouri. Males have been heard in chorus at air temperatures as low as 1.7°C (35.1°F; Collins et al. 2010). Chorusing males were observed in a population in eastern Kansas and some calling males had one or more satellite (non-calling) males close to them (Roble 1985). Eggs are laid during amplexus; they are fertilized by the male as they are laid. The eggs are attached to submerged grasses and twigs just below the surface in loose, jelly clusters containing 15 to 190 eggs per cluster (Dodd 2013a). Five egg clusters observed in a temporary pool in late April in Linn County averaged 25 eggs (range 18–29; J. Briggler pers. obs.). A total of 300 to 1,500 eggs may be produced per female (Collins et al. 2010; LeClere 2013). Hatching takes place from a few days to nearly a week after being laid, depending on the water temperature. Transformation to froglets occurs six to eight weeks after hatching, and most individuals breed within one year. A study of boreal chorus frogs in several Great Plains states showed that there were no males older than three years of age in any of the populations (Platz and Lathrop 1993).

The call of the male boreal chorus frog is a vibrating or metallic clicking "*crrreeeeek*" with a rise in pitch at the end; it lasts about a second (Elliott et al. 2009). The sound is similar to running a fingernail over the small teeth of a pocket comb.

Remarks: This species was formerly recognized in Missouri as the western chorus frog (*Pseudacris triseriata*; Johnson 2000). Genetic information has shown that the boreal chorus frog occurs throughout most of Missouri, and the western chorus frog occurs east of the Mississippi River (Lemmon et al. 2007). Due to variation of markings and similar calls in *Pseudacris maculata*, *Pseudacris feriarum*, and *Pseudacris fouquettei*, it is difficult to distinguish these three species (see species account for the upland chorus frog). The boreal chorus frog is a common species throughout most of the state, especially the prairie regions. It has been heard calling in scattered areas within the central Ozark Highlands but is yet to be documented by voucher specimens. Additional genetic studies are needed to better understand this species' boundary in the southern and eastern part of its range in Missouri.

Family Microhylidae

Narrow-mouthed Toads

This large family of burrowing, secretive frogs has representatives in Asia, Malaysia, northern Australia, New Guinea, southern Africa, Madagascar, North and South America (Pough et al. 2018). The group probably originated in Asia. The family contains 63 genera with 557 species (Pough et al. 2018). In the United States this group is represented by two genera and four species; of these, two species of one genus (*Gastrophryne*) live in Missouri. Most members of this group are fossorial, spending most of their time in burrows or under rocks or logs. They tend to be small, plump, and squatty in appearance. All species in the United States have a characteristic fold of skin behind a small, narrow, pointed head. Although they may eat many types of small insects, these amphibians are mostly ant eaters.

Eastern Narrow-mouthed Toad

Eastern Narrow-mouthed Toad

Gastrophryne carolinensis (Holbrook)

JEFF BRIGGLER

Adult eastern narrow-mouthed toad from Callaway County.

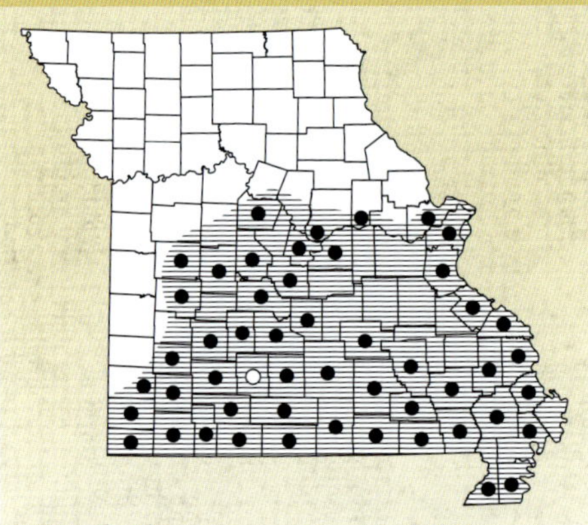

Distribution: Missouri: Southern half of Missouri. North America: Southern Maryland and eastern Virginia, westward to extreme southern Illinois, southern Missouri, eastern Oklahoma to the eastern half of Texas and eastward to all of Florida (Powell et al. 2016).

Description: The eastern narrow-mouthed toad is a small, plump amphibian with a distinct fold of skin behind a narrow, pointed head. General color is tan, brown, or reddish brown. The dorsal pattern forms a long, dark wedge with the narrow end at the head. This dark wedge is bordered by a wide lateral stripe of lighter color. There also is a dark stripe running from the snout to the side of the hind legs. Much of this pattern is obscured by the presence of numerous small, dark-brown, or black markings. The belly is mottled with dark gray. Characteristics of this burrowing species include a small, pointed head, a fold of skin across the back of the head, short legs, and the absence of both an external eardrum (tympanum) and webbing between the toes. Males can be distinguished from females by their deeply pigmented throat.

Adult eastern narrow-mouthed toads range in snout-to-vent length from 22 to 32 mm (0.9 to 1.3 in.) but have been known to reach 39 mm (1.5 in.; Daniel 2011; Toal and Collins 2003).

Eastern narrow-mouthed toads prefer to breed in temporary pools with a lot of aquatic vegetation, such as this example in Perry County (above). Eastern narrow-mouth toad eggs have one or more flat surfaces resembling a mosaic pattern (left).

Habits and Habitat: This toad lives in a variety of habitats from flooded swamps, open grasslands, and upland and bottomland hardwood woodlands in Missouri (Anderson 1954). This toad spends most of its time in loose, damp soil under rocks and logs. It prefers a habitat where shelter and moist soil are available, usually in the vicinity of ponds, swamps, and streams. In the Ozarks, however, individuals do occur under flat rocks on relatively dry glades. Once an eastern narrow-mouthed toad is uncovered, it tries to escape with a series of quick, short hops and a scramble into leaf litter or another nearby hiding place. Although they are mainly found on the ground beneath cover, an individual was found in a dead stump 2.4 m (7.9 ft) off the ground (Anderson 1954). They may create their own short burrow or utilize burrows created by other animals. An individual will often sit in a burrow with its head slightly visible at the surface awaiting prey.

The primary food of narrow-mouthed toads is ants, although they also eat termites, small beetles, and spiders. A study in Florida showed that ants, representing 43 different species, accounted for 95 percent of the eastern narrow-mouthed toad diet (Deyrup et al. 2013). They are often found near or in ant hills consuming numerous ants (Wood 1948). Narrow-mouthed toads produce a skin secretion that is noxious to many predators; the secretion also protects them from attacking ants during feeding (Garton and Mushinksy 1979).

Breeding: The eastern narrow-mouthed toad primarily breeds in large puddles, temporary pools, fishless flooded ditches, and fields, as well as flooded rice fields in southeastern Missouri. However, they will occasionally use permanent bodies of water, especially sites with thick vegetation and shallow edges (Briggler 1998). Choruses of males are not heard until late April and can take place during the summer months, especially in June or early July after heavy rains. The males usually hide under leaves, plants, or other debris at the water's edge or sit in thick vegetation within the water as they call. Once a male has attracted a female, he will clasp her with his forelegs and may continue to call during amplexus. Special glands on the belly of the male secrete a substance that causes the male and female to become firmly stuck together. It is possible that these special glands evolved in eastern narrow-mouthed toads due to their rounded bodies and the male's short arms, which makes amplexus difficult (Conaway and Metter 1967).

Females can lay over 1,000 eggs in several films on the surface of the water. The membrane envelope surrounding the egg is unique to *Gastrophryne* by being circular in shape with one or more flat surfaces that might resemble a mosaic pattern (Wright 1931). In Missouri, females have been reported to lay an average of 510 eggs with a range of 152 to 1,089 eggs (Anderson 1954). The eggs are fertilized by the male as they are laid. The eggs hatch in less than three days; the tadpoles are dorsoventrally flattened and dark brown to black. Their development is rapid, and they metamorphose 30 to 60 days later. Recent metamorphs in Missouri are 10 mm (0.4 in.) in size (Anderson 1954). Sexual maturity occurs when individuals are 21–23 mm (0.8–0.9 in.) in size, which is generally one to two years of age (Anderson 1954).

The call of a male eastern narrow-mouthed toad is a bleating, nasal "*baaaaa*"; it lasts one to four seconds, sounding like the call of a lamb. When one individual calls, it often causes other individuals to call simultaneously.

Remarks: In a few western Missouri counties, this species may occur in sympatry with the western narrow-mouthed toad, *Gastrophryne olivacea* (see the *G. olivacea* account for ways to distinguish the two species).

Due to the specialized diet of consuming ants of many species, including exotic and invasive species (Deyrup et al. 2013), narrow-mouthed toads not only might benefit from these new food sources, but could potentially be helpful in controlling invasive ants.

Western Narrow-mouthed Toad

Gastrophryne olivacea (Hallowell)

TOM R. JOHNSON

Adult western narrow-mouthed toad.

Description: The western narrow-mouthed toad can be identified by its plump body, short legs, uniform color, small pointed head, and a fold of skin across the back of the head. The general color is gray, tan, or olive tan. Small black spots may be scattered over the back and hind legs. There is no webbing between the toes. The belly is white. Adult females are larger than males; during breeding season the males have dark throats. This species can best be distinguished from the eastern narrow-mouthed toad by its lighter overall color and the absence of any prominent dorsal markings.

Adult western narrow-mouthed toads range in snout-to-vent length from 22 to 38 mm (0.9 to 1.5 in.) but have been known to reach 43 mm (1.7 in.; Powell et al. 2016).

Habits and Habitat: This species prefers grasslands, rocky and wooded hills, river floodplains, and areas along the edge of marshes. In Holt County, Bufalino and Easterla (1996) observed individuals on

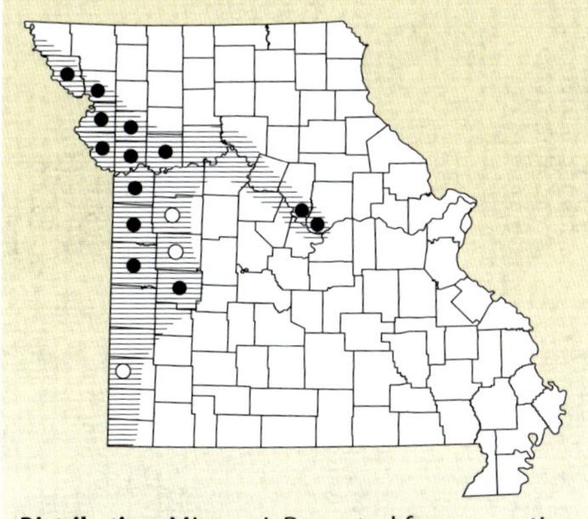

Distribution: Missouri: Reported from counties primarily in northwestern and western Missouri and from a few counties along the Missouri River. North America: Western Missouri to extreme southern Nebraska, south through the Great Plains states to southern New Mexico, most of Texas, and into Mexico (Powell et al. 2016).

JEFF BRIGGLER

Western narrow-mouthed toads breed in heavily vegetated wetlands, such as this example in Clay County.

southwest-facing slopes of the loess hills to the Missouri River floodplain. They are mainly active on the surface from April into October. These toads are very secretive and spend most of their time hiding in loose soil and under rocks, logs, or other objects; they have been known to take shelter in animal burrows, especially tarantula burrows (Blair 1936; Yearly 1979). In central Texas, McAllister and Tabor (1985) found eastern collared lizards in a burrow under a rock that was shared by a western narrow-mouthed toad. These tiny toads are known to move considerable distances. A study in Kansas reported that movements greater than 30 m (98.4 ft) appear to be quite common with this species and the greatest distance traveled by an individual from water was 610 m (2,001.3 ft; Fitch 1956a).

This species has evolved a toxic skin secretion that protects it from ant bites. It is known to sit on ant hills while eating ants, its preferred food. Although they are an ant specialist, they will also occasionally eat small beetles. Predators of the species are not well known, but Anderson (1942) reported a copperhead (*Agkistrodon contortrix*) eating a western narrow-mouthed toad in Jackson County.

Breeding: Usual breeding sites are ditches, temporary ponds, flooded fields, or pools in wooded areas. Warm, heavy rains will stimulate males to congregate at breeding sites and begin calling. Males call from thick vegetative cover along wetland margins or while floating among vegetation in shallow water. A male clasps a female just behind her forelegs. Special skin glands on the male's belly secrete a glue-like substance that helps hold them together during amplexus (Conaway and Metter 1967; Fitch 1956a). In Missouri, breeding primarily takes place from May to early July but can also occur later in the summer after warm, heavy rains. During amplexus the female generally lays from 500 to over 1,000 eggs as a film on the water surface (Freiburg 1951; Wright and Wright 1949). Similar to the eastern narrow-mouthed toad, the membrane envelope surrounding the eggs of this species can have a mosaic pattern. The male fertilizes the eggs as they are laid. The eggs hatch in two to three days, and the tadpoles metamorphose 20 to 30 days later (Collins et al. 2010). Sexual maturity is reached within one or two years with most males maturing earlier than females (Dodd 2013a).

The call of a male western narrow-mouthed toad is a high-pitched, short "*peep*" followed by a nasal buzz bleat, which lasts one to four seconds. The sound is similar to the buzz of a bee.

Remarks: Both western and eastern narrow-mouthed toads live in several western counties in Missouri, and sympatric populations occur in that area. However, the calls of both species are different enough to keep populations isolated from hybridization (Nelson 1972).

Because the western narrow-mouthed toad is quite secretive, limited to the eastern periphery of the species' geographical range, and little basic biology is known of this species in Missouri, this uncommon toad likely warrants additional attention.

Family Ranidae

True Frogs

The family Ranidae is the largest and most widespread family of frogs. It contains 365 species in 14 genera and probably had its origin in Africa (Pough et al. 2018). Representatives of this cosmopolitan family occur on every major land mass except New Zealand, Antarctica, most oceanic islands, the West Indies, and southern South America. The largest genus in the family in the New World (North and South America) is *Lithobates* (formerly *Rana*), with about 50 species. The *Rana* genus is considered restricted to the eastern hemisphere and western North America (Powell et al. 2016). In Missouri, the genus *Lithobates* is represented by eight species. Members of this family are the true frogs; they are typically medium- to large-sized, have long legs, smooth skin, and well-developed webbing between the hind toes. Another common characteristic is a glandular dorsolateral fold or ridge of skin along each side of the back. The American bullfrog (*Lithobates catesbeianus*), the largest member of the genus in the United States, does not have these ridges of skin. Natural food of true frogs consists of nearly any animal small enough to be swallowed, including worms, insects, spiders, crayfish, fish, other frogs, small turtles and snakes, birds, and small mammals. The tadpoles of most species of true frogs are herbivorous, but they occasionally may eat dead animal matter.

Pickerel Frog

Northern Crawfish Frog

Lithobates areolatus circulosus (Rice and Davis)

Adult northern crawfish frog from Pettis County.

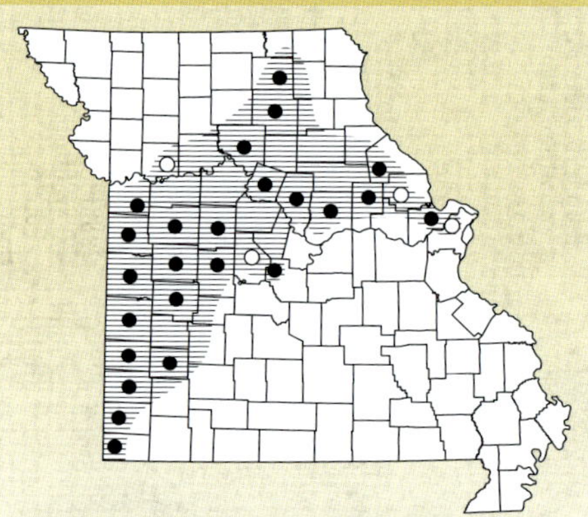

Distribution: Missouri: Rolling hills, prairies, and meadows of southwestern, extreme western, north central, and eastern Missouri. North America: Southeastern Iowa and northern Missouri, east to Indiana, south to central Mississippi, and northwest to eastern Oklahoma and Kansas. (Powell et al. 2016). This species is considered extirpated in south-central Iowa (J. Christiansen pers. comm.).

Description: This is a large frog with a light ground color and numerous, closely set dark spots. The head is disproportionately large in this species. A prominent dorsolateral fold extends from the eye to the thigh. Ground color may vary from light tan to light gray; the dark spots may be dark brown, gray, or nearly black and are often edged in white. A fine network pattern or spotting of dark pigment is usually present between the dark spots. The belly is white. Males can be distinguished from females by their smaller size, enlarged thumbs, and (during the breeding season) presence of a pair of saclike vocal pouches behind and below each eardrum (tympanum).

Adult northern crawfish frogs range in snout-to-vent length from 57 to 114 mm (2.2 to 4.5 in.) but have been known to reach 122 mm (4.8 in.; Powell et al. 2016).

Habits and Habitat: Northern crawfish frogs are active from March to October in Missouri. The species is restricted to native

JEFF BRIGGLER

TOM R. JOHNSON

Northern crawfish frogs live in grasslands and breed in temporary pools, such as this example in Macon County (above). A newly transformed northern crawfish frog (44 mm long; 1.7 in.) from a native prairie in Newton County (left).

prairie or former prairie areas in this state. Populations occur in or near low-lying hay fields, native grass pastures, prairies, and occasionally in river floodplains. This very secretive species spends most of its time hidden in burrows. Crawfish frogs primarily use crayfish burrows for retreats but may occasionally use burrows of other animals. A crayfish burrow being used by this species will have a noticeably flattened and denuded platform at the entrance. Because crayfish burrows can be about 1 m to more than 1.5 m (3–5 ft) deep and are often at or below groundwater level, the frogs are able to use them as winter retreats. In many circumstances, crawfish frogs use the same burrows from year to year (Heemeyer et al. 2012). Crawfish frogs become nocturnal (active at night) especially after warm, heavy summer rains. They are seldom encountered, however, except during their short breeding season.

They forage mainly at the entrance of their burrow while awaiting approaching prey. Food of this species consists of a variety of invertebrates, including insects, spiders, and small crawfish.

Breeding: In Missouri, northern crawfish frogs breed from March through mid-May with breeding occurring earlier in southern populations compared to northern populations. Males congregate in semipermanent pools and fishless ponds. Heavy rains and moderate temperatures seem to stimulate the males to migrate to breeding wetlands. Studies in Indiana showed crawfish frogs moving to a breeding wetland travel an average nearly 0.5 km (0.3 mi), with one frog traveling 1,020 m (3,346 ft) from its crayfish burrow (Heemeyer and Lannoo 2012; Heemeyer et al. 2012). Also, individuals migrating to and from the breeding wetland follow similar paths and utilize specific crayfish burrows along the way (Heemeyer and Lannoo 2012). Choruses in Missouri generally range from 5 to 10 males per breeding wetland and can be heard at distances up to 200 m (656 ft; Frese 2000; Frese and Sullivan 2000; J.

TOM R. JOHNSON

Calling male northern crawfish frog from Callaway County.

Briggler unpubl. data). Soon after the males start to call, the egg-laden females begin to arrive at the breeding site. A male will clasp a female behind her forelegs and while in amplexus in shallow water, the eggs are laid and fertilized. The large clumps of eggs are generally 125 to 150 mm (4.9 to 5.9 in.) across and are usually deposited in shallow water on a submerged plant stem or branch. Each female may lay between 2,000 and 7,000 eggs (Redmer 2000; Smith et al. 1948; Trauth et al. 1990). The eggs hatch in 3 to 10 days. Tadpoles grow rapidly and generally metamorphose and leave the wetland within 60 days, which is generally mid-May to early July. Newly metamorphosed froglets raised in captivity from eggs collected in the wild in Missouri were 23–26 mm (0.9–1 in.) long. Occasionally metamorphs are observed hopping on the prairie ground in July and August in western Missouri (J. Briggler unpubl. data). Maturity generally occurs in two to five years with males maturing earlier than females (Redmer 2000), with individuals living up to 10 years of age. We have observed that Missouri's crawfish frogs may not reproduce during dry, excessively cool springs.

The call of a male northern crawfish frog can be described as a deep, loud, snoring "*gwwaaa*," which can be heard from a considerable distance. As they call, their large, lateral vocal sacs vibrate the water surface, sending droplets of water in all directions. A number of males calling in chorus sounds like pigs during feeding time (Conant and Collins 1991). The males call while sitting in shallow water or floating in open water and have been reported to call while submerged (Barbour 1971; Frese and Sullivan 2000).

Remarks: Although nearly all native prairie regions of the state have undergone intensive cultivation, this species has persisted. Populations are more stable in the Osage Prairies of western and southwestern Missouri, but populations north of the Missouri River within the claypan prairie region are rapidly declining with only small, isolated populations remaining. These last remaining native prairies in Missouri are valuable for many reasons, but they are particularly valuable as habitat for northern crawfish frogs. Protecting, properly managing, and reestablishing prairies are vital to keeping this species of conservation concern on the landscape in Missouri. Because this species requires fishless ponds for breeding, the protection and addition of small, fishless ponds and pools in or near native prairies is highly recommended. Protecting the water table and reducing soil compaction from excessive use also are important. When soil becomes compacted or the water table becomes too low, burrowing crayfish can be eliminated, thus reducing the available burrows necessary for crawfish frog survival.

Plains Leopard Frog

Lithobates blairi (Mecham, Littlejohn, Oldham, Brown, and Brown)

JEFF BRIGGLER

Adult plains leopard frog from Livingston County.

Description: The plains leopard frog is a medium-sized frog with light-tan ground color and numerous rounded spots on the back. The head is wide and blunt, giving this species a stocky appearance. There is a distinct light line along the upper jaw. A white spot is almost always present on the tympanum. The dorsolateral folds are broken near the groin and a small posterior section of it is usually displaced further toward the midline. The numerous brown or green-brown spots on the back and legs are usually circular and uniform in shape. A dark spot is usually present on the snout. The belly is white with pale yellow near the groin and lower inner thighs.

This species can be distinguished from other leopard frogs and the pickerel frog by the narrow dorsolateral fold, which is broken and displaced dorsally, and by the absence of distinct white rings around each dorsal spot. See species accounts for the pickerel

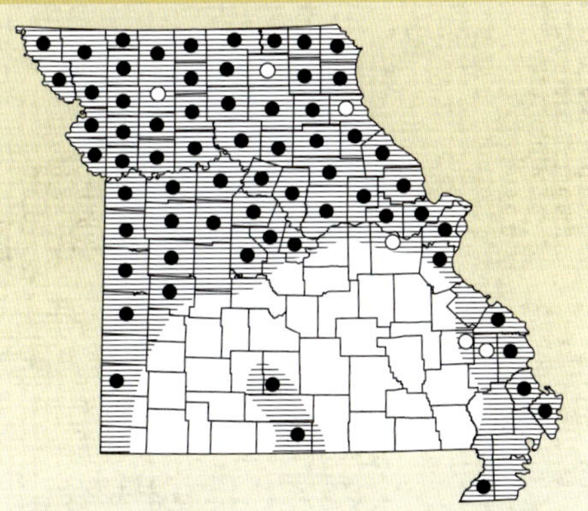

Distribution: Missouri: Throughout most of Missouri, including the Bootheel, but rarely present in the Ozarks. North America: The southern edge of South Dakota southward to eastern Mexico and northern Texas, and northeast to central Illinois. Isolated populations in northeastern corner of Arkansas, New Mexico and Arizona (Powell et al. 2016).

JEFF BRIGGLER

JEFF BRIGGLER

Plains leopard frogs prefer heavily vegetated wetlands for breeding, such as this marsh in Clark County (above). Plains leopard frog eggs from a water-filled ditch in Linn County (right).

frog (*Lithobates palustris*), northern leopard frog (*L. pipiens*) and southern leopard frog (*L. sphenocephalus*).

Adult plains leopard frogs range in snout-to-vent length from 51 to 95 mm (2 to 3.7 in.) but have been known to reach 114 mm (4.5 in.; Powell et al. 2016).

Habits and Habitat: The plains leopard frog is active from March to October. It lives mainly in the grasslands of the Great Plains states. In Missouri, this species occurs in the former prairie regions and associated river floodplains of northern, central, and western Missouri, as well as the sand prairie and river floodplains in the Bootheel. It uses a variety of aquatic habitats, including water-filled ditches, farm ponds, river sloughs, small streams, temporary pools, and marshes. During winter, these frogs retreat into the mud and dead leaves at the bottom of ponds and streams. In the summer, individuals are active in and around wet areas, but they will venture into grassy areas well away from water. This species can be quite abundant in favorable habitat. Hundreds have been seen on dirt roads adjacent to wetland marshes in Holt County during the night, presumably foraging for prey (J. Briggler pers. obs.).

TOM R. JOHNSON

Calling male plains leopard frog from Clark County.

The plains leopard frog eats a variety of small invertebrates, such as flies, beetles, spiders, worms, crickets, and grasshoppers. Leopard frogs are known to be eaten by snakes, including the red-sided gartersnake (*Thamnophis sirtalis parietalis*; Smith and Powell 1993).

Breeding: This species mainly breeds from mid-April to early June in Missouri, but individuals have been heard calling in mid-to-late March, especially in the southeastern part of the state. Males will gather at a marsh, pond, or temporary pool and begin calling after sunset. Females with eggs are attracted by the males' calls; amplexus takes place in the water. The eggs are fertilized by the males as they are laid. Egg masses are round or slightly oblong and each is surrounded by a thin coating of clear, protective jelly. The egg clusters are attached to submerged stems or branches in shallow water. Each female may produce 4,000 to 6,500 eggs. It takes two to three weeks for the eggs to hatch. Plains leopard frog tadpoles are grayish brown, around 51 mm (2 in.) long, and have only a few small, faint dark markings on the body and tail. Metamorphosis from tadpoles to froglets usually occurs during midsummer (about three months after egg laying), but the tadpoles that hatch late in the breeding season will overwinter in the breeding ponds and transform the following spring. Newly metamorphosed plains leopard frogs average 27 mm (1.1 in.) in length. In some years this species may breed during the autumn if there are a number of cool days and abundant rainfall.

The call of male plains leopard frogs can be described as a rapid series of guttural "*chuck-chuck-chuck*" sounds at a pulse rate of three per second (Mecham et al. 1973).

Remarks: The plains leopard frog occurs sympatrically with the southern leopard frog (*L. sphenocephalus*) in a number of Missouri counties and with northern leopard frogs (*L. pipiens*) in northwestern counties. Hybridization with these two species may occur, usually where habitats have been altered due to human activity. When a breeding habitat is changed, the environmental barriers that would normally isolate species during their breeding season are lost, and hybridization may occur (Sage 1992). See also the remarks section in the northern leopard frog species account.

American Bullfrog

Lithobates catesbeianus (Shaw)

Adult male American bullfrog from Callaway County.

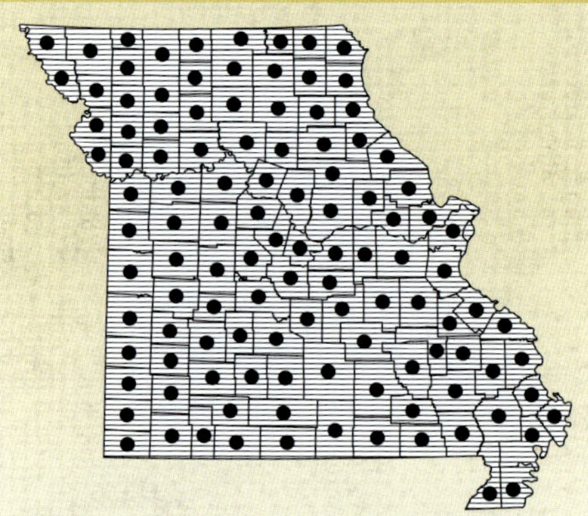

Distribution: Missouri: Statewide. North America: Historically, this large species ranged from New Brunswick and southeastern Quebec, southward to central Florida, westward to extreme southern Minnesota, into the Great Plains, and south into most of Texas. Bullfrogs have been intentionally introduced into many western states, as well as other countries (Powell et al. 2016).

Description: The American bullfrog is Missouri's largest frog. General color ranges from green to olive to brown. Some dark markings may be present on the back in the form of small brown spots or indistinct irregular blotches. The hind legs are marked with distinct, dark-brown bars. The belly is white, and the throat may have some gray mottling. The external eardrum is large and round. An adult male bullfrog has a tympanum much larger than its eye. Adult females have a tympanum the same size or smaller than its eye. Bullfrogs lack a dorsolateral fold. During the breeding season, adult males have bright yellow throats.

Adult American bullfrogs range in snout-to-vent length from 90 to 152 mm (3.5 to 6 in.) but have been known to reach 220 mm (8.7 in.; Powell et al. 2016).

Habits and Habitat: In Missouri, bullfrogs are active from mid-March to October (Moyle 1952; Willis et al. 1956). This is

JEFF BRIGGLER

TOM R. JOHNSON

American bullfrog tadpoles in a gravel stream in Ozark County (above). Freshly laid American bullfrog eggs from Linn County (left).

Missouri's most aquatic species of frog. Bullfrogs spend most of their time in or very near aquatic habitats such as lakes, ponds, rivers, large creeks, drainage ditches, sloughs, and permanent swamps or marshes. Because permanent water is required, this species may suffer a population decline during a drought year. A bullfrog is easily disturbed during the day and escapes by powerful jumps into the water. Young bullfrogs (less than two years of age) will often give out a short, high-pitched "*yelp*" when they jump into the water to escape an intruder. Bullfrogs can be approached at night with the aid of a flashlight. Both young and adult bullfrogs will move overland during heavy summer rains and can be seen crossing roads. In Missouri, Moyle (1952) and Willis et al. (1956) found that the majority of adult bullfrogs infrequently moved between wetlands, but overland movement up to 1.25 km (0.8 mi) could occur. However, considerable movement occurred by recently transformed individuals (Moyle 1952). This species burrows into the mud at the bottom of rivers or ponds to avoid winter temperatures. In Missouri, adults usually enter winter retreats in late October; young bullfrogs follow a week or two later (Moyle 1952; Willis et al. 1956). Young bullfrogs generally emerge from winter retreats during middle to late March and larger individuals appear about three weeks later (Moyle 1952; Willis et al. 1956).

A number of studies have been conducted on the bullfrog diet. The size or age of the bullfrog, time of year, and type of habitat are factors that influence the kinds of prey eaten. In a Missouri study, frogs living in farm ponds in the central part of the state ate the

TOM R. JOHNSON

Adult female American bullfrog from Pulaski County.

following foods (in order of frequency): insects, crayfish, amphibians, small mammals, fish, and birds (Korschgen and Moyle 1955). A slight difference was noted between the prey in an impoundment (large pond) population and bullfrogs living along rivers. Pond bullfrogs were found to eat dragonflies, spiders, ground beetles, crayfish, caterpillars, and moths, and to a lesser extent other invertebrates and a few vertebrates. Bullfrogs from rivers were reported to eat scarab beetles, ground beetles, caterpillars and moths, crayfish, cicadas, spiders, and dragonflies, as well as a variety of other invertebrates and a few vertebrates (Korschgen and Baskett 1963). Corse and Metter (1980) studied the diet of bullfrogs at a fish hatchery and found a wide variety of invertebrate and vertebrate prey with insects and small fish comprising the bulk of their diet. These reports show that this species will take advantage of the most abundant prey in a given habitat and does not have a preference for one particular prey species. Bullfrogs are opportunistic feeders, and generally speaking, their voracious appetite and size allow them to eat nearly any animal small enough to be captured and swallowed. Adult bullfrogs captured in Randolph County had eaten a prairie vole (*Microtus* spp.) and a young mink (*Mustela vison*; Beringer and Johnson 1995). An adult bullfrog was observed capturing and attempting to swallow a chestnut-sided warbler at the side of a small pond in Greene County (L. Rizzo pers. comm.). A medium-sized bullfrog was observed eating a hatchling western painted turtle (*Chrysemys picta*) in Boone County (J. Briggler pers. obs.). They are known to eat other frogs (Corse and Metter 1980). An adult bullfrog was observed eating a calling male gray treefrog in Maine (Hinshaw and Sullivan 1990). Bullfrogs also can be highly cannibalistic. Bullfrog tadpoles can be around 152 mm (6 in.) in length and often live in streams, ponds, and lakes where fish abound; it is presumed that they are toxic and too large for most fish. They may, however, be eaten by adult bullfrogs, wading birds, turtles, watersnakes, and gartersnakes. Bullfrog tadpoles were observed being eaten by giant water bugs and fishing spiders (*Dolomedes* spp.; Rogers 1996).

Breeding: In Missouri, bullfrogs begin to call in early April with large choruses being heard by mid-May into July (Moyle 1952; J. Briggler unpubl. data). Willis (1954) found breeding in farm ponds in central Missouri began in late May or early June, peaking during late June. Breeding has been documented in Missouri until the end of August, but a small amount may continue into September (Willis 1954). During the breeding season, males are territorial. Generally spaced every 2–6 m (6.6–19.7 ft) apart around a permanent wetland, they aggressively defend a calling station by mounting, pushing, kicking, bumping, or biting competing males.

In permanent wetlands (marshes, large ponds, or sloughs), males chorus on warm nights, usually between midnight and early morning (before sunrise). Larger, dominant males will call from open, shallow areas where the females lay their eggs. Such sites are protected from other males. A male will mount a gravid female and begin amplexus. As the female begins to lay her eggs they are fertilized by the male. The majority of egg laying in Missouri occurs from late May through July (Moyle 1952; Seale 1980), but eggs have been found in late August (Willis 1954).

The eggs are laid one egg deep on the surface as a large, wide floating mass. Each egg is quite small, around 1 mm (0.04 in.) in diameter. Female bullfrogs can lay over 20,000 eggs per clutch, but clutch size is normally between 10,000 and 12,000 eggs (Dodd 2013b). Larger and older females will produce significantly more eggs compared to smaller and younger individuals, and some females may produce two clutches of eggs during the summer (Emlen 1977). The eggs hatch in four or five days and the tadpoles, which feed on algae, grow quickly. Bullfrog tadpoles range from 51 to 152 mm (2 to 6 in.) in total length, depending on their stage of development. The tadpoles may be a plain olive color with a number of small spots on their back, sides, and tail, with a white or light-yellow belly, or boldly marked with black or dark-brown smudges and spots with a bright yellow belly. In Missouri, tadpoles usually metamorphose in 11 to 14 months (Willis et al. 1956). Transformation of tadpoles into froglets generally occurs from June through August in Missouri (Moyle 1952; Seale 1980; Willis 1954) with froglets' body length ranging from 30 to 60 mm (1.2 to 2.4 in.) with an average of 52 mm (2 in.; Moyle 1952). Bullfrogs may take an additional two or three years before adult size is reached. Female bullfrogs become sexually mature between 123 to 125 mm (4.8 to 4.9 in.) snout-to-vent length (Willis et al. 1956), while males attain sexual maturity at about 120 mm (4.7 in.) snout-to-vent length (Schroeder 1975). Bullfrogs are known to live slightly over seven years (Snider and Bowler 1992).

The call of a male bullfrog is a deep sonorous "*ger-a-a-rum*" or "*jug-o-rum*," which can carry a half a mile or more. In the summer, males call during the day and often right after sunset, but the most intense vocalization takes place between midnight and three or four in the morning during the breeding season. Bee and Gerhardt (2002) found that male bullfrogs can recognize voices of individuals and can tell familiar and unfamiliar males apart.

Remarks: A study, conducted in eastern North Carolina, discovered that bullfrog tadpoles, newly transformed froglets, and adult bullfrogs showed a resistance to the venom of copperheads and, to a lesser extent, to that of cottonmouths. The bullfrog tadpoles had the least resistance to the venom (Heatwole et al. 1999). Although bullfrogs are native to Missouri, this species has been introduced into many western states and countries around the world. In addition, bullfrogs appear to be resistant to lethal infectious diseases (i.e., amphibians chytrid fungus) but are often carriers of diseases infecting other amphibians (Daszak et al. 2004). Due to their voracious appetites and ability to spread infectious diseases, they are a threat to many other amphibian species and are considered an invasive species in many parts of their range. Bullfrogs (and green frogs) are classified as game animals in Missouri and are protected by a season and bag limit under the *Wildlife Code of Missouri*. The legs of these frogs are edible and considered by many to be a delicacy.

Green Frog

Lithobates clamitans (Latreille)

JEFF BRIGGLER

Adult green frog from Perry County.

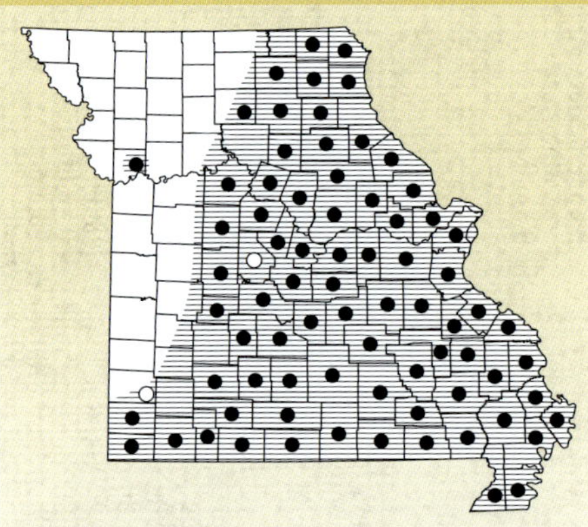

Distribution: Missouri: Occurs in about three-quarters of the state, southeast of a diagonal line from the southwestern to northeastern corners of the state. North America: Southern Canada and most of the eastern states south to central Florida and west to eastern Texas (Powell et al. 2016).

Description: The green frog is a medium-sized frog. General color varies from green to greenish tan to brown. The upper lip and head are usually green. Faint dark spots may be present on the back, and the legs usually have indistinct dark spots or bars. Sides of the belly often are marked with dark-gray vermiculation. The belly is white with some dusky markings. Adult males have a bright yellow throat. A distinct dorsolateral fold is present but extends only to midbody, not to the groin. The external eardrum (tympanum) is large and conspicuous, larger than the eye for males. During the breeding season, males have swollen thumbs and heavier forelegs. This species can be confused with the bullfrog (*Lithobates catesbeianus*) but is smaller and has a prominent dorsolateral fold, which bullfrogs lack.

Adult green frogs range in snout-to-vent length from 54 to 75 mm (2.1 to 3 in.) but have been known to reach 108 mm (4.3 in.; Powell et al. 2016).

JEFF BRIGGLER

JEFF BRIGGLER

Green frogs breed in permanent, woodland ponds, such as this example in Butler County (above). Male green frog along the edge of a wetland awaiting to call from Miller County (left).

Habits and Habitat: This species is mainly active between mid-March and mid-November in Missouri. We have noticed young green frogs active into early December during a mild autumn. Green frogs are rather solitary animals, especially in small stream habitats, where each deep pool may have a single adult in residence. In Missouri, this species is likely to live in creeks and streams, especially in the Ozarks. Other habitats include river sloughs, swamps, marshes, and ponds, and a small number of individuals are found in caves (Elliott 2007; Myer 1958b). They can also be found using golf course ponds and landscape water gardens around homes. When disturbed, a green frog will quickly jump into the water, often emitting a high-pitched squawk as it jumps (Smith 1961). A study in Boone County found that adult green frogs will change their behavior when bullfrogs are present by selecting habitat around the breeding site away from the more aggressive bullfrogs (Birchfield 2002). Overwintering sites mainly occur in aquatic habitats with many individuals preferring to hide under large rocks or other debris in flowing streams that do not freeze and provide plenty of oxygen. Green frogs in central Missouri moved from 550 to 900 m (1,804 to 2,953 ft) to reach overwintering sites. They were found to overwinter in deeper water along a creek, within partially submerged root tangles, and underneath large rocks along the edge of a creek (Birchfield 2002).

JEFF BRIGGLER

JEFF BRIGGLER

Green frog tadpole from Callaway County (above). Freshly laid green frog eggs from Osage County (right).

This species presumably eats a variety of insects, snails, slugs, spiders, and small crayfish (Dodd 2013b). Green frogs in southern Illinois were found to eat beetles, spiders, millipedes, snails, true bugs, and flies (Jenssen and Klimstra 1966).

Breeding: Although individual green frogs can be heard calling as early as late March in southern Missouri, breeding choruses occur from mid-April into mid-July with June being the peak month (J. Briggler unpubl. data). A study in central Missouri found that adult green frog migration into breeding ponds occurred from April through October with the peak from May to mid-June (Hocking et al. 2008). Any permanent standing water is used as a breeding site, including ponds, swamps, sloughs, beaver ponds, and slow-moving areas of rivers and streams. There is usually intense competition for choice calling areas, and the male with the best breeding site (one with an abundance of emergent plants) has a better chance of attracting gravid females (Wells 1977). Once a male has engaged a female in amplexus, the female will begin depositing her eggs in a wide, floating mass on the surface of the water. An amplexus pair was witnessed in late June in Osage County depositing eggs a few hours after dark (J. Briggler pers. obs.).

Each female can lay over 4,000 eggs (Wright and Wright 1949). Clutch sizes reported from studies in Arkansas ranged from 2,851 to 5,730 eggs (Trauth et al. 1990; Trauth et al. 2004). Females are capable of laying more than one clutch of eggs during the breeding season with considerably more eggs deposited during the first clutch (Wells 1976). The small, dark tadpoles begin hatching in three to five days, depending on the water temperature. Tadpoles range from 74 to 100 mm (2.9 to 3.9 in.) in total length and appear long and slender. Color is tan olive with extensive dark-gray mottling on the tail. Although some green frog tadpoles may transform the same summer as hatching, the majority will not metamorphose into froglets until the following summer. Newly transformed green frogs average 28 mm (1.1 in.) in body length. Hocking et al. (2008) captured over 3,000 metamorphs leaving woodland ponds in central Missouri from late May through October, with most individuals transforming between June and early September. Although green frogs have been documented to live for 7 to 10 years in captivity (Flower 1925), most individuals in the wild probably live five to six years (Martof 1956; Shirose and Brooks 1995).

The call of a male green frog can be described as an explosive "*bong*" that sounds like a loose banjo string. The sound may be emitted once or repeated three to four times.

Remarks: Two subspecies were once recognized for *Lithobates clamitans*: the green frog (*L. clamitans melanota* Rafinesque), and the bronze frog (*L. clamitans clamitans* Latreille). A recent genetic study did not support the two subspecies but concluded there are two distinct color phases (Austin and Zamudio 2008). The northern color phase is generally green to olive, while the southern color phase is more brown to bronze. The green color phase is found throughout the majority of the state except for southeastern Missouri, where the bronze phase typically occurs. Green frogs are classified as a game species in Missouri and can only be harvested for food or live bait during a defined season with a bag limit.

Pickerel Frog

Lithobates palustris (LeConte)

Adult pickerel frog from Perry County.

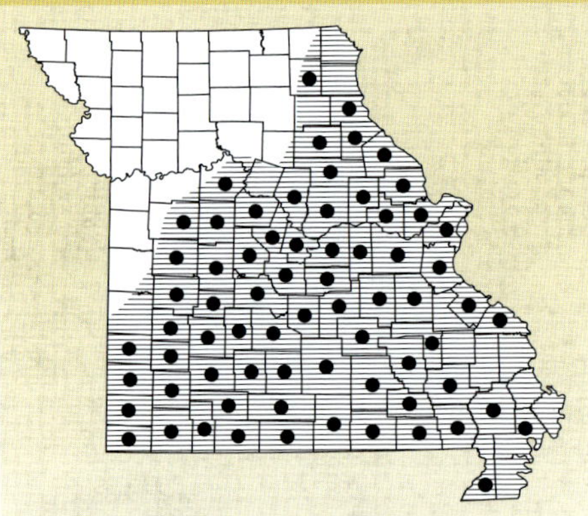

Distribution: Missouri: Common in southern and eastern Missouri. North America: New Brunswick, Nova Scotia, and extreme southern Quebec, southward to South Carolina, northern Georgia and Alabama, northward to Wisconsin, southeastern Minnesota, eastern Iowa, Missouri, to extreme northeastern and southeastern Oklahoma, eastern Texas, and most of Arkansas and Mississippi (Powell et al. 2016).

Description: A medium-sized frog with prominent dorsolateral folds and two parallel rows of squarish or rectangular spots running down the back between the folds. General color is gray, tan, or brown. A white line is present along the upper lip. The distinct dorsolateral fold extends down to the groin and may be white, cream, gray, yellow, or golden. Spots on the back are reddish brown, dark brown, or black. Dark bars on the hind legs are prominent. The underside of the hind legs and groin area are washed with bright yellow, orange yellow, or pink yellow. The belly is white. During the breeding season males can be distinguished from females by their enlarged thumbs with dark, thick pads. Adult males are generally smaller than adult females.

The pickerel frog may be confused with the three species of leopard frogs that occur in Missouri. This species can be distinguished from leopard frogs by the wide, unbroken dorsolateral fold, the two distinct rows of square or rectangular dorsal

JIM RATHERT
JEFF BRIGGLER

Shaded, damp and cool ravines are primary habitat of pickerel frogs in summer (above). Pickerel frogs overwinter in wet caves, such as this example in Franklin County (left).

spots and the yellow or orange yellow along the underside of the hind legs and groin area. See species accounts for the plains leopard frog (*L. blairi*) and southern leopard frog (*L. sphenocephalus*).

Adult pickerel frogs range in snout-to-vent length from 44 to 75 mm (1.7 to 3 in.) but have been known to reach 87 mm (3.4 in.; Powell et al. 2016).

Habits and Habitat: This species is generally associated with springs, cold streams, and cool, shaded woodland ponds. In Missouri, it occurs along Ozark streams under rocks at the water's edge. Pickerel frogs also live along streams in grassy areas, pastures, and near farm ponds, along shaded ravines in forests, in moist grassy fens, and under rocks on glades during the spring. In eastern Missouri along the Mississippi River, this species is associated with springs and creeks that flow from limestone bluffs. During the hot summer months in Callaway County, this frog is often seen in and around wet areas of yards, gardens, and water fountains (J. Briggler pers. obs.).

Pickerel frogs have a high association with caves, especially the entrance, throughout Missouri and their range (Elliott 2007; Fenolio et al. 2005; Prather and Briggler 2001; Resetarits 1986). A Boone County population of this species was intensely studied for several years; some of that information is provided below (Resetarits 1986; W. Resetarits pers.

comm.). Pickerel frogs use wet caves as a refuge from hot weather during late summer and from cold temperatures during winter. This is the only species of frog in Missouri that takes shelter in caves with regularity and in high numbers; a single cave may harbor hundreds of pickerel frogs. Resetarits (1986) conservatively estimated that 365 individuals overwintered in a single Boone County cave. The relatively constant temperature and moisture in a wet cave provide an ideal site for overwintering. The frogs may exit for short periods of time to feed, but between late September to early November, they will have moved to a deeper section of the cave where the temperature is more stable. Here they remain until spring (Resetarits 1986). During March or early April, the frogs leave the caves and locate a pond or slough for breeding. The frogs feed during the spring and summer; some begin to return to the cave in late July. Young-of-the-year pickerel frogs were found to have the greatest mortality during the winter. In some areas, this species overwinters in springs, in mud at the bottom of streams or ponds, and under large stones along the margins of small streams.

Pickerel frogs consume a variety of invertebrates, including ants, spiders, beetles, moth larvae, sowbugs, crickets, and earthworms (Gilhen 1984).

Breeding: Woodland ponds, sloughs, and even water-filled ditches are used as breeding sites. In Missouri, this species first begins to call in early March and continues until late May with peak chorusing from mid-April to early May. A study of breeding ponds in Boone County showed that adults migrated to the ponds between early March to early May (Hocking et al. 2008). Males usually chorus at one or two sites in a pond; over a dozen may call from a relatively small area (Wright and Wright 1949). Amplexus will take place as soon as a male is approached by a female. The female lays her eggs in shallow water in the form of a globular mass with an average diameter of 63 mm (2.5 in.); the mass is attached to a submerged stick or stem. The eggs are fertilized by the male as they are laid. Average clutch size in a Missouri population was 1,759 eggs, with a range from 704 to 2,896 eggs (Resetarits and Aldridge 1988). Larger females lay more eggs. The small tadpoles begin hatching in 10 days or more; depending on the water temperature, metamorphosis into froglets begins after three and one-half to four months. Pickerel frog tadpoles are generally 48 to 55 mm (1.9 to 2.2 in.) in total length, depending on their stage of development, and look similar to leopard frog tadpoles. In a Boone County population, most tadpoles completed metamorphosis from mid-July to mid-August with tadpoles spending from 125 to 157 days in the pond (Hocking et al. 2008). Pickerel frog tadpoles collected in late April in Pulaski County and kept at 21.1°C (70° F) transformed into froglets during mid-July (T. Johnson unpubl. data). Female pickerel frogs become sexually mature at about 59 mm (2.3 in.) snout-to-vent length, while males attain sexual maturity at about 45 mm (1.8 in.) snout-to-vent length (Resetarits and Aldridge 1988).

The call of male pickerel frogs is a low-pitched, descending snore lasting a couple of seconds. Although this frog calls from above water like most other frog and toad species, a number of authors have reported this frog vocalizing while under water (Morris 1944; Wright and Wright 1949; Barbour 1971).

Remarks: Skin secretions of pickerel frogs are known to be irritating or even toxic to small animals. If newly captured pickerel frogs are placed in the same container with other frog species, the latter will die within a short time due to these skin secretions (Morris 1944; Mulcare 1965). Most frog-eating snakes will not eat pickerel frogs. A study in west-central Kentucky showed that pickerel frogs that breed in ponds where fish are present are likely to be more toxic than frogs that breed in fishless ponds (Holomuzki 1995).

Even though pickerel frogs occupy somewhat specialized habitats of cool water springs, streams, and caves, they appear to be common throughout the karst, forested habitats of Missouri. This species is typically underrepresented in calling surveys but is commonly observed on land throughout much of its active season.

Northern Leopard Frog

Lithobates pipiens (Schreber)

Adult northern leopard frog from Mercer County.

Description: The northern leopard frog is a medium-sized frog with brown or green ground color and large, round, black dorsal spots. The head is wide with a short, blunt snout. There is a prominent white line along the upper lip. Tympanum color is rust brown; there is usually no light spot present in the center. The dorsolateral folds are wide, continuous to groin, and cream or fawn colored. Dark spots along the back and legs are reddish brown to black, prominently ringed with white. The large round spots on the sides are black and also are ringed with white or pale green. The snout usually has a large round black spot. The belly is white. The groin area and the underside of the hind legs are white or pale green.

This species can be distinguished from the southern leopard frog and the plains leopard frog (*L. sphenocephalus* and *L. blairi*) by the wider and continuous dorsolateral folds, and the distinct white rings around each dark spot (see southern and plains leopard frog accounts). During

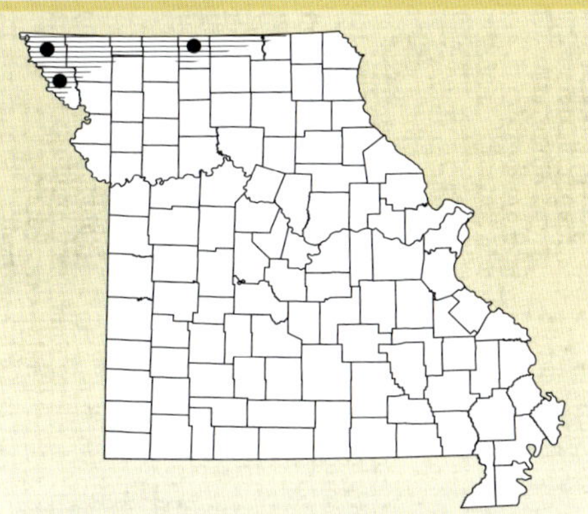

Distribution: Missouri: A few northern counties along the Iowa border. North America: Widespread in Canada, northeastern and north-central states. Ranges into the Pacific Northwest and into Arizona and New Mexico (Powell et al. 2016).

Northern leopard frogs live and breed in marshes, such as this example in Mercer County (above). Calling male northern leopard frog from Atchison County (right).

TOM R. JOHNSON

the breeding season, male northern leopard frogs can be distinguished from females by the presence of a vocal sac on either side of the head and enlarged thumbs.

Adult northern leopard frogs range in snout-to-vent length from 51 to 90 mm (2 to 3.5 in.) but have been known to reach 111 mm (4.4 in.; Powell et al. 2016).

Habits and Habitat: The northern leopard frog is active from March to October. This species lives in and near marshes, flooded ditches, floodplain pools and small ponds, and lakes in northern and eastern United States and Canada, especially grassland and open land habitats (Dodd 2013b). In Missouri, it can occur along the edge of small marshes and shallow drainage ditches. This species is known to disperse long distances from wetlands. In a Maine study, individual home ranges varied from 13 to 8,425 m^2 (140 to 90,686 ft^2) with an average of 1,096 m^2 (11,797 ft^2; Blomquist and Hunter 2009). A telemetry study in Iowa showed that northern leopard frogs can travel up to 1,217 m (3,993 ft) from breeding wetlands with many individuals traveling between 200–650 m (656–2,133 ft; Bartelt et al. 2001; LeClere 2013).

Like other leopard frog species, northern leopard frogs move into grassy areas during summer where they feed on insects. A variety of insects and spiders are included in the natural diet of this species. In autumn, leopard frogs move to permanent water where they retreat to the bottom or into the mud, remaining there throughout the winter.

Breeding: In Missouri, this species mainly breeds between late March and mid-April. Peak calling in the Mercer County population generally occurs during the first week of April. Males call, beginning at dusk, from small areas of open water in marshes or shallow ponds. They prefer clear-water wetlands (Dodd 2013b). Gravid females are attracted to the calling males; once amplexus occurs, each pair of frogs moves to shallow water where the eggs are laid. The eggs are fertilized during amplexus. Several globular masses of eggs are produced with a maximum of 6,000 eggs per female (Vogt 1981). In early April 2013, amplexus was observed and five freshly deposited egg masses were seen at the Mercer County population (J. Briggler unpubl. data). A study in Quebec, Canada, showed that females become sexually mature at two years of age. Also, reproducing females produced between 1,319 and 5,270 eggs (Gilbert et al. 1994). The egg masses are attached to submerged sticks, cattails, or grasses. Hatching takes place in 7–20 days, depending on water temperature (Dodd 2013b). Tadpoles are similar to those of southern leopard frogs and plains leopard frogs. Metamorphosis from tadpole to froglet occurs after two to two and one-half months of development (during June to early July). Juveniles were documented in a Michigan study to disperse up to 5 km (3.1 mi) from their natal wetland (Dole 1971). Northern leopard frogs are estimated to live four to five years (Leclair and Castanet 1987).

The breeding call of male northern leopard frogs is a deep rattling snore with occasional clucking grunts. The call has been compared to the sound made by rubbing a wet thumb slowly along the surface of an inflated balloon.

Remarks: This species was first discovered in the spring of 1985 in Atchison County (Johnson 1987), and another population was found later (1987) in Mercer County. This species may occur in a few additional northern Missouri counties with additional survey efforts. Northern leopard frogs are sometimes called "meadow" or "grass" frogs because they often move into pastures and lawns during the summer.

Males of this species usually have Mullerian ducts (vestigial oviducts) present along each kidney. Males of the other leopard frog species in Missouri do not have these vestigial organs.

This species and the plains leopard frog (*L. blairi*) occur sympatrically in the northwestern corner of Missouri. Studies of this population in northwestern Missouri and one in Mercer County have shown that the plains leopard frog appears to be hybridizing with northern leopard frogs and causing the decline of this rare species due to genetic swamping. However, the Mercer County *L. pipiens* populations appear to have more genetic vigor and show little hybridization with *L. blairi* (Sage 1992).

Although this species was probably never abundant in Missouri, the northern leopard frog is a species of conservation concern due to its limited range, loss of wetland marshes, and low population numbers. Preserving natural wetlands and surrounding grassland habitat are necessary for the long-term survival of this species and many other amphibians in northern Missouri.

Southern Leopard Frog

Lithobates sphenocephalus (Cope)

Adult southern leopard frog from Miller County.

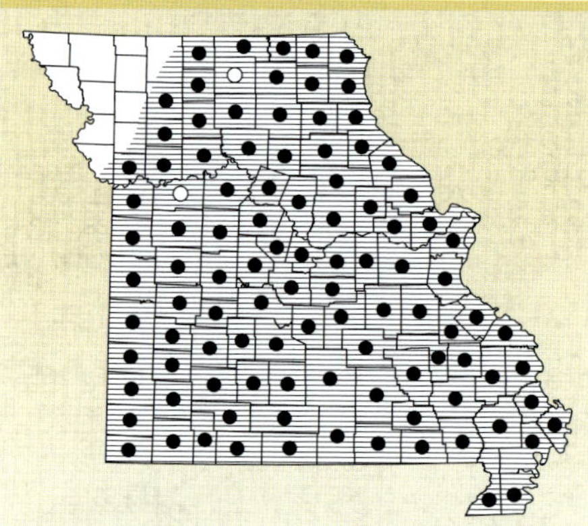

Distribution: Missouri: Throughout most of Missouri except for the northwestern corner of the state. North America: New Jersey south to Florida and west to southern Illinois, Missouri, eastern Kansas, Oklahoma, and Texas (Powell et al. 2016).

Description: The southern leopard frog is a medium-sized frog with a variable number of rounded or oblong dark spots on the back. The dorsolateral fold is narrow and distinctly raised, yellow or tan, and extends to the groin. The head appears long and the snout is pointed. General color is green, green brown, or light brown with some green on the dorsum. Dark markings on the hind legs appear as broken bars or elongated spots. There is usually no dark spot on the snout. A white line is present along the upper lip; the center of the tympanum usually has a distinct white spot. The belly is white. During the breeding season males can be distinguished from females by their smaller size, enlarged thumbs, and vocal sacs on either side of the head.

The southern leopard frog can be distinguished from the pickerel frog and other species of leopard frogs by the narrow and continuous dorsolateral fold, pointed snout, the absence of a snout spot (in most individuals), the lack of any yellow along

Southern leopard frogs breed in a wide variety of aquatic habitats but prefer those with abundant vegetation, such as this pond in Crawford County (above). A pair of southern leopard frogs in amplexus (male on top) from Linn County (left).

the groin area and inner thighs, and the absence of white rings around each dorsal spot. See species accounts for the plains leopard frog (*L. blairi*), pickerel frog (*L. palustris*), and northern leopard frog (*L. pipiens*). Specimens of dorsally unspotted southern leopard frogs have been found in New Jersey, Alabama, Kentucky, Mississippi, and southern Illinois (Redmer 1992). This lack of dorsal markings has thus far not been reported in Missouri's population.

Adult southern leopard frogs range in snout-to-vent length from 51 to 90 mm (2 to 3.5 in.) but have been known to reach 127 mm (5 in.; Powell et al. 2016).

Habits and Habitat: The southern leopard frog is normally active between late February and mid-October in Missouri. During the summer months, southern leopard frogs may venture far from water into pastures, meadows, or wooded areas where they search for insect prey. They have been known to travel up to 351 m (1,152 ft) within 24 hours in South Carolina (Graeter, 2005). When near an aquatic habitat, leopard frogs sit at the water's edge but quickly enter the water with a powerful jump if alarmed. This species uses a wide variety of aquatic habitats including creeks, rivers, sloughs, swamps, marshes, ponds, lakes, and flooded ditches. Due to a high rate of evaporative water loss, this species must remain near water or use burrows of mammals and crayfish to escape hot summer conditions. Recent metamorphs will actively burrow into the substrate or use existing burrows to avoid water loss during drought (Parris 1998).

Southern leopard frogs eat a variety of insects and other invertebrates (Crawford et al. 2009).

Calling male southern leopard frog from Callaway County.

JEFF BRIGGLER

Breeding: In Missouri, the southern leopard frog can have a prolonged breeding period occurring from late winter into autumn (Drake and Ousterhout 2011; Hocking et al. 2008; Johnson 2004; Seale 1980). Most calling occurs from early March into July with peak breeding from mid-March to mid-May. In a Boone County study, most adults migrated to breeding ponds in mid-March to early May (Hocking et al. 2008). In some years, southern leopard frogs bred during the autumn, and the tadpoles overwinter in the breeding wetland (Johnson 2004; Seale 1980; J. Briggler pers. obs.; T. Johnson pers. obs.). Ponds, sloughs, and flooded ditches are used as breeding sites. They prefer fishless ponds but will breed in ponds with certain small species of fish (Sexton and Phillips 1986). Males will call while floating in the water or, more often, will hide themselves among grasses or stems in shallow water. The eggs are laid and fertilized during amplexus. Several thousand eggs are normally laid in several clumps or masses, which are loosely attached to submerged sticks or stems. In Arkansas, the average clutch size is 2,959 eggs with a range from 1,700 to 5,537 (Trauth 1989). The eggs will take up to two weeks to hatch, depending on the water temperature. Southern leopard frog tadpoles range from 46 to 52 mm (1.8 to 2 in.) in total length and are olive gray with faint, gray, small markings on the body and tail. Due to the prolonged breeding season for this species and overwintering of tadpoles, metamorphosis can vary from mid-June through September (Hocking et al. 2008). Sexual maturity is probably reached within one to two years, with females taking slightly longer than males.

The breeding call of the male southern leopard frog can be described as a series of abrupt, chucklelike "*quacking*" sounds, repeated at a rate of 12 pulses per second (Mecham et al. 1973).

Remarks: This species occurs sympatrically with the plains leopard frog (*L. blairi*) in a number of Missouri counties; they are known to hybridize in some locations (Sage 1992; see species accounts for *L. blairi* and *L. pipiens*). The taxonomy of the southern leopard frog has been confusing for many years, and there is still some uncertainties related to the common name usage and subspecies distinction. Some authors recognize two subspecies: Florida leopard frog (*L. sphenocephalus sphenocephalus*) and Coastal Plains leopard frog (*L. s. utricularius*). Other authors do not recognize subspecies (Crother 2017; Powell et al. 2016). Although we chose to keep the name (common and scientific) southern leopard frog (*L. sphenocephalus*) in Missouri, it is likely that the subspecies named the Coastal Plains leopard frog (*L. s. utricularius*) occurs in Missouri.

Southern leopard frogs are widespread and quite common in Missouri. Long-term calling surveys in Missouri show that the number calling is highly influenced by weather conditions. During the 2012 drought in Missouri, very few southern leopard frogs were heard calling across the state, but they readily rebounded the following wet years. For such a common and widespread species, there is still considerable life history information unknown.

Wood Frog

Lithobates sylvaticus (LeConte)

TOM R. JOHNSON

Adult wood frog from Warren County.

Description: The wood frog is a medium-sized, tan amphibian with a distinct dark-brown "mask" on each side of the head extending from the snout to behind the tympanum. General color is pink tan, light brown, red brown, or dark brown. Scattered small, dark markings may be present on the dorsum, and the hind legs usually have brown or dark-brown longitudinal bars. There is a prominent white line along the upper lip. The dorsolateral fold is distinct and extends to the groin. The belly is white with scattered dusky markings. During the breeding season, adult male wood frogs can be distinguished from females by their smaller size, darker color, enlarged thumbs, and a vocal sac located just above each foreleg.

The only other frog native to Missouri that may be confused with wood frogs are young green frogs (*L. clamitans*), which may have a similar color but lack the distinctive dark-brown "mask" through each eye.

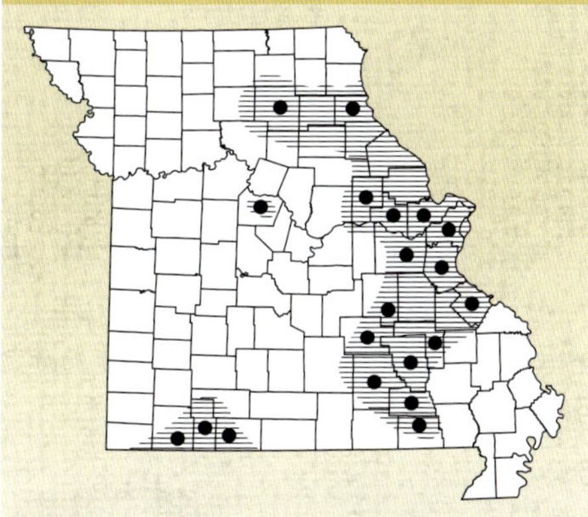

Distribution: Missouri: A few northcentral and eastern counties, and in the southeastern and southwestern Missouri Ozarks. North America: Alaska, western, central, and eastern Canada, northeastern states south to extreme northern Georgia, and Great Lakes states west to North Dakota. Isolated populations in Wyoming, Colorado, Missouri, Arkansas, and Alabama (Powell et al. 2016).

JEFF BRIGGLER

Wood frogs breed in temporal pools and fishless ponds, such as this example in St. Louis County (above). A pair of wood frogs in amplexus (male on top) from Warren County (right).

JEFF BRIGGLER

Adult wood frogs in Missouri range in snout-to-vent length from 35 to 70 mm (1.4 to 2.8 in.) but have been known to reach 83 mm (3.3 in.; Powell et al. 2016).

Habits and Habitat: This species is active between February and October. It is a secretive and solitary species and can be difficult to observe after its short breeding season. During the summer, wood frogs live along shady ravines, among dead leaves, or along north-facing rock outcroppings and bluffs. A study in Warren County showed the wood frog extensively uses forested, rocky ravines that are moist and cool (Rittenhouse and Semlitsch 2007a). Wood frogs travel an average of 110 m (361 ft) and a maximum of 394 m (1,293 ft) from the nearest breeding sites (Rittenhouse and Semlitsch 2007b). Also, movement within these habitats is mainly at night during or following rainfall. This species is also generally associated with cool, north-facing wooded hillsides, where there is ample shade and some moisture. Oak-hickory forest, often with sugar maple (*Acer saccharum*), is the preferred habitat. A wood frog among dead oak and maple leaves is nearly invisible. Rittenhouse and Semlitsch (2007a) found that most adult wood frogs completely buried themselves under the deciduous leaf litter during the spring and summer with an occasional eye or part of the body visible.

TOM R. JOHNSON

JEFF BRIGGLER

Wood frog tadpoles have a long, finely tapered tail fin (above). Freshly laid wood frog eggs from a fishless pond in Ripley County (left).

During summer drought conditions, they are often found within cave entrances in the Ozark Highlands (Prather and Briggler 2001).

This species overwinters on land under a deep layer of leaves or in the damp soil underbrush piles and logs. As with a number of North American anurans that overwinter close to the surface under cover objects, wood frogs protect themselves from freezing by producing a cryoprotectant (glucose) in their blood. This condition persists through their short breeding season (Storey and Storey 1987).

Wood frogs eat a variety of insects and other invertebrates. Several young wood frogs were captured during early June in Warren County and were observed for a few weeks. They ate small grasshoppers, leafhoppers, and small crickets (T. Johnson pers. obs.). As with other frogs in Missouri, wood frogs are preyed upon by snakes, raccoons, mink, and skunks. An adult wood frog was observed being eaten by an eastern yellow-bellied racer in a forest in Warren County in late May (J. Karel pers. comm.). Wood frog tadpoles prefer to eat algae and other plant material when available, but if such food sources become scarce they will shift their diet to eggs and larvae of amphibians, including wood frog eggs and larvae (Petranka et al. 1994, 1998).

Breeding: The wood frog is an explosive breeder during late winter and early spring. Small, fishless, woodland ponds and ephemeral pools are used for breeding. In Missouri, breeding takes place between early February and late March, depending on local weather conditions (Guttman et al. 1991; Hocking et al. 2008; J. Briggler unpubl. data).

Wood frogs require heavy, warm rains and an air temperature of at least 10°C (50°F) to stimulate breeding. Males move to a breeding pool as soon as the sun is down and normally vocalize until about midnight. During ideal weather conditions (warm and moist) wood frogs will call during the day. The spring peeper (*Pseudacris crucifer*) choruses and breeds along with wood frogs; its loud, high-pitched call can easily overpower the muted call of the wood frog. Gravid females begin arriving at the breeding site on the first night the males begin to call. By the second or third night, all the eggs are laid and few wood frogs remain at the pond or pool. The males normally call while floating in open water. Their paired vocal sacs expand and contract rapidly each time a call is produced. A gravid female will swim toward a calling male and amplexus will commence. While in amplexus the pair swims to a shallow area where there is an abundance of submerged branches or stems; the eggs may be attached to a branch or on the bottom. The eggs are fertilized by the male as they are laid. Most wood frog eggs observed were laid along the northern and northwestern section of each breeding pond, where the water temperature is usually warmest because of southern solar rays (J. Briggler pers. obs.). In Missouri, there can be 50 to 100 egg masses within a couple of square meters.

Wood frog egg masses measure up to 115 mm (4.5 in.) across and contain 500 to over 1,000 eggs. Egg masses averaged 867 eggs for a study in the northern Ozarks of Arkansas (Trauth et al. 1989). Individual eggs are encased in a large amount of clear jelly that causes them to be widely separated. Newly laid eggs are black with a small white dot on the bottom and average 2 mm (0.08 in.) in diameter. As egg masses age, they become greenish due to infiltration by symbiotic, green algae (*Oophilia amblystomatis*; Trauth et al. 2004). Wood frog eggs hatch within 10 days to two weeks. Development is rapid and metamorphosis takes place from May to mid-June. Hocking et al. (2008) captured 202 metamorphs exiting breeding ponds in early to mid-June in a Warren County population after the tadpoles spent from 97 to 104 days in the pond. Newly transformed wood frogs average 17 mm (0.7 in.) in body length; they are tannish gray and have the distinct, dark-brown "mask." Male wood frogs are able to breed for the first time at one to two years of age compared to two to three years of age for females (Bastien and Leclair 1992; Redmer 2002). Longevity appears to be relatively short for wood frogs with few females living beyond four years and few males beyond three years (Bastien and Leclair 1992).

The call of a male wood frog is a rapid, hoarse "*waaaduck*" sound that lasts about one second and may be quickly repeated three or four times; put simply, they sound like quacking ducks. The sound of a small chorus of this species carries only a short distance.

Remarks: Hurter (1911) first reported the presence of this species in Missouri. In 1973, the wood frog was classified as an endangered species in this state. During June 1978, a population was discovered in Warren County (L. Langbein pers. comm.) and reported to the Missouri Department of Conservation. Several populations were subsequently found in eastern and southwestern sections of the state and it was determined in 1982 that the species could be reclassified as rare. Wood frogs were successfully reintroduced into a fishless pond in St. Louis County in 1980 and have spread throughout the general area (Guttman et al. 1991). Today, the wood frog appears to be naturally expanding its range in Missouri but remains a species of conservation concern due to loss of temporary, fishless wetlands, conversion of forested habitat, and isolated populations. Management of this species should focus on maintaining and creating temporary wetlands needed for breeding and larval growth within large areas of intact forests. Preserving a 300 m (984 ft) buffer of mature forest around breeding wetlands within the broader forested landscape is important for maintaining healthy populations of wood frogs (Rittenhouse and Semlitsch 2007b). Connectivity of breeding wetlands with forested ravines is critical for wood frog movement and survival (Rittenhouse and Semlitsch 2007a; Rittenhouse and Semlitsch 2009).

Part II—REPTILES

(Class Reptilia)

Crocodilians, Turtles, Lizards, and Snakes

The first reptiles appeared on earth some 315 million years ago, evolving from salamander-like creatures. One important difference between amphibians and reptiles is the ability of reptiles to reproduce without having to return to an aquatic environment to lay eggs. Early reptiles were the first vertebrates to produce shelled eggs that protected their embryos from drying as they developed, thus giving these creatures the ability to live in dry habitats where amphibians could not survive. Scientists recognize about 10,700 species of reptiles on earth (Pough et al. 2018; Utez et al. 2020). These animals are placed in four major groups: Order Rhynchocephalia (tuatara), Order Squamata (lizards, amphisbaenians, and snakes), Order Crocodilia (crocodilians), and Order Testudines (turtles). Reptile species in Missouri are confined to turtles, lizards, and snakes. Most reptiles have scales (except for softshells, geckos, and one species of sea turtle) and claws (except for snakes and legless lizards). The eyes of semi-aquatic turtles, sea turtles, and crocodilians have an inner lid, known as a nictitating membrane, which protects their eyes when swimming or resting underwater. Like amphibians, most reptiles do not produce their own body heat by way of metabolizing food as do birds and mammals, which are endothermic. Instead, reptiles are ectothermic—they either remain the same temperature as their surroundings, bask in the sun or in shallow, warm water to increase their body temperature, or take shelter under rocks or in the soil to avoid temperatures too high or too low for their survival. Reptiles differ from the majority of amphibians when it comes to egg fertilization. Most amphibians fertilize their eggs after they are laid by the female (external fertilization). Reptiles use an intromittent organ (penis in turtles and crocodilians; hemipenis in lizards and snakes) to deliver the male's sperm inside the female (internal fertilization). There are about 395 species of reptiles native to the United States. Missouri has 72 native reptile species and two additional established nonnative species: 18 species of turtles, 11 species of native lizards, 2 species of nonnative lizards, and 43 species of snakes.

(Order Testudines)

Turtles

Turtles and tortoises represent the oldest living group of reptiles on earth. These generally hard-shelled reptiles are known from fossils as far back as the Triassic Period—over 200 million years ago—and have changed very little since they became established. All turtles and tortoises belong to the Order Testudines, with 341 living species in 14 genera (Pough et al. 2018). Most people are familiar with the general characteristics of turtles, and they are seldom confused with other animals. However, few people are aware of the natural history, habitat requirements, population status, or distribution of even the most common species of turtles. The turtles of Missouri can be divided into three groups: hard-shelled aquatic turtles, hard-shelled land turtles (box turtles), and softshells. The hard-shelled aquatic group has the largest number of species; it includes some of the smallest species as well as the largest freshwater turtle on earth. Box turtles of Missouri are represented by only two species.

Although many people call them tortoises, these reptiles are actually more closely related to basking turtles (sliders and painted turtles). Missouri has only two species of softshells. There are no tortoises native to Missouri. Turtle shells can be divided into two sections: the upper shell known as the carapace and the lower shell called the plastron. The shell of present-day turtles is composed of expanded ribs and bony plates covered by a layer of horny scales called scutes. Softshell turtles have reduced bony elements that are covered by tough skin instead of scutes. Turtles have other interesting features beside their shells. They lost their teeth through evolution and instead have a tough, horny beak. Also, a few species are known to vocalize while swimming, and hatchlings make sounds within their nests (Ferrara et al. 2013). All turtles lay their eggs on land. Females are particular about where they nest; they may travel long distances over land to locate a suitable site. Most turtles select well drained, sandy, or loose soil to deposit their eggs. The selected sites usually have some exposure to the sun. The eggs of softshells, and musk and mud turtles are hard-shelled, while snapping turtles, painted, map, and the rest of Missouri's turtles lay leathery shelled eggs. A total of 18 species with one additional subspecies representing four families of turtles are native to Missouri. Turtles have been around for a very long time, but their long heritage and stability may, in part, contribute to their decline. The swamps, marshes, and rivers to which turtles are adapted are all being rapidly altered by people. These changes in habitat include filling marshes and swamps, pollution, channelization, and elimination of basking and nesting areas. Biology of turtles worldwide is likely susceptible to climate change. Many of our native turtles do not have the ability to change or alter lifestyles in order to survive in a rapidly changing environment. Several of Missouri's turtles are on the state list of species of conservation concern. Indeed, these turtles are the most endangered reptiles in the state. And, as long as our wetlands, clean rivers, and other habitats diminish to make way for human "progress," we will continue to see more species decline in numbers. Persecution of turtles by needless (and unlawful) shooting is another reason turtle populations are declining. Some species are used as human food, which has resulted in overharvesting and declines in many turtle species worldwide. All our turtles play a role in the overall check and balance system we call nature. But now these long-lived, interesting animals need our help.

Western Painted Turtle

Family Chelydridae

Snapping Turtles

This family is composed of six species representing two genera (*Chelydra* and *Macrochelys*); it is restricted to North and South America (Pough et al. 2018). These species have large heads, powerful jaws, and small plastrons that restrict them from pulling in heads and limbs for complete protection. They are not great swimmers and prefer to walk on the bottom of freshwater wetlands. The snapping turtles (*Chelydra* spp.) range from Canada through the eastern and central United States, and south to Ecuador. Alligator snapping turtles (*Macrochelys* spp.) occur in southeastern and southern states. They are the largest freshwater turtles in the world.

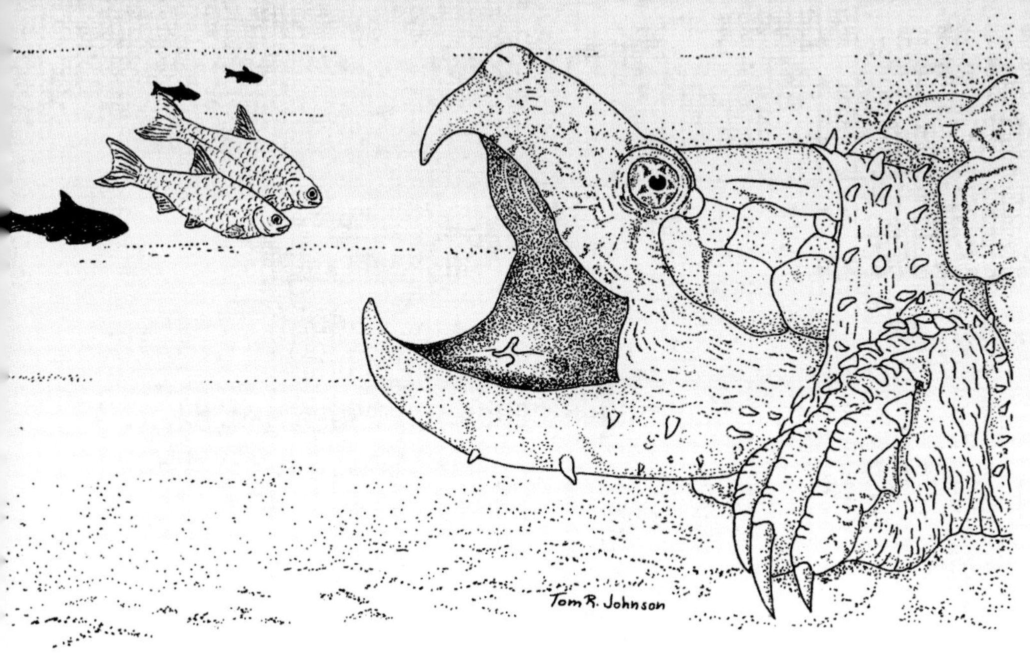

Alligator Snapping Turtle

Eastern Snapping Turtle

Chelydra serpentina (Linnaeus)

Adult eastern snapping turtle from Callaway County.

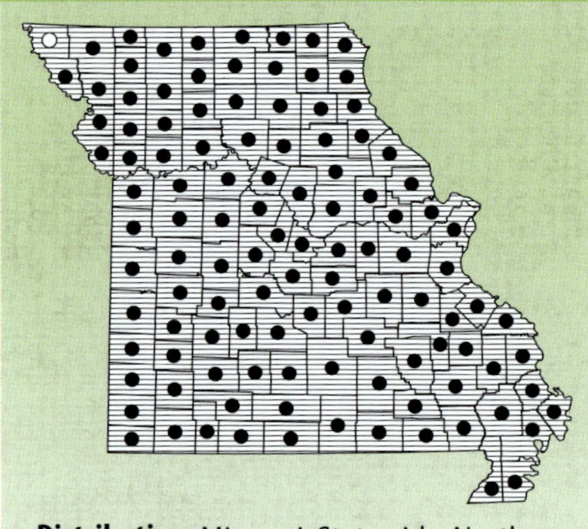

Distribution: Missouri: Statewide. North America: Nova Scotia and Ontario, south to Florida, west to eastern Montana, Wyoming, Colorado, New Mexico, and Texas (Powell et al. 2016).

Description: The eastern snapping turtle is a large aquatic turtle with a big pointed head, long thick tail, and small plastron. The carapace may be tan, brown, or nearly black; it often is covered with mud or algae. The head, tail, and limbs are brown. The head often is covered with numerous small black lines or spots. The plastron and underside of limbs are yellow white. The upper part of the tail has large, pointy scales in a saw-tooth row. The carapace has three rows of low keels that are prominent in young turtles but become less apparent with age. Adult males can be distinguished from adult females by their slightly smaller size. Also, the cloacal opening of males is further out from the edge of the plastron than it is in females. The young of this species have very rough shells and long tails.

Adult eastern snapping turtles range in carapace length from 203 to 360 mm (8 to 14.2 in.; Powell et al. 2016) but have been known to reach 494 mm (19.4 in.). Most adult snapping turtles weigh between 4.5

JEFF BRIGGLER

The plastron of a male eastern snapping turtle from Callaway County.

and 16 kg (9.9 and 35.3 lb). Record weight for a wild-caught individual is 34 kg (75 lb) and 39 kg (86 lb) for a captive-raised turtle (Powell et al. 2016).

Habits and Habitat: The eastern snapping turtle makes its home in a wide variety of aquatic habitats. It commonly occurs in farm ponds, ditches, marshes, swamps, sloughs, rivers, and reservoirs—anywhere there is permanent water. Snapping turtles prefer bodies of water with a mud bottom, submerged logs and snags, and especially abundant aquatic vegetation (Bodie et al. 2000; Lescher et al. 2013b). They tend to spend a lot of time hidden in the mud or vegetation in shallow water while awaiting prey to ambush or walking on the bottom searching for food (Collins et al. 2010; Trauth et al. 2004).

This turtle is primarily aquatic and seldom basks in the sun. As adult females travel overland during the egg-laying season, many are killed by vehicles. Turtles of either sex may travel overland if the pond in which they are living dries.

When taken out of water, snapping turtles will vigorously defend themselves; they will even become aggressive and lunge at any adversary. They are known for a defensive posture of dipping one side of the front shell while raising the opposite back side with a gaping mouth when encountered on land. Once in the water, however, they usually do not bite and will try to hide or escape rather than defend themselves.

Eastern snapping turtles are primarily active between March and November, with a peak in Missouri from May into July. They overwinter in water by burying themselves into the mud bottom, underneath logs and sticks, and within overhanging banks of ponds, lakes, swamps, marshes, or river backwaters (Ernst and Lovich 2009). Individuals are occasionally seen slowly moving on the bottom of pond under the ice during the winter in Missouri and other states.

TOM R. JOHNSON
JEFF BRIGGLER

A hatchling eastern snapping turtle from St. Louis County (above). The plastron of a hatchling eastern snapping turtle from Cole County (right).

Eastern snapping turtles eat a wide variety of animals and plant material. Natural food of this species includes insects, crayfish, fish, snails, earthworms, amphibians, snakes, small turtles, small mammals, birds, and aquatic vegetation (Ernst and Lovich 2009). Up to 36 percent of a snapping turtle's diet may consist of plant material (Alexander 1943). A large specimen captured in northeastern Illinois was found to have fed, almost exclusively, on duckweed (*Lemna* spp.; Budhabhatti and Moll 1990). Recent diet studies have shown the important role that turtles play in seed dispersal and germination of plants. A study in Christian County showed that eastern snapping turtles were important for seed dispersal of mulberries (*Morus* spp.), barnyard grass (*Echinochloa crus-galli*), and curly dock (*Rumex crispus*; Kimmons and Moll 2010). Carrion is often consumed, especially chicken livers and dead fish used to attract turtles to traps (Ernst and Lovich 2009).

Breeding: Courtship and mating likely take place any time between April and November, but most breeding activity occurs in late spring and early summer when water temperatures are warm. Mating takes place in the water. Courtship between a pair of snapping turtles usually involves the two turtles facing each other and waving their heads from side to side in opposite directions. Actual mating begins when a male mounts a female and grips her carapace with his claws. The male's tail moves beneath the female's tail so that their cloacal openings touch. The male then inserts his penis for sperm transmission. Like many species of turtles, the female snapping turtle can retain viable sperm for a number of years (Galbraith 1993). Sperm storage allows a single individual to colonize new locations, as well as allows a female to mate with multiple males. This will increase genetic diversity within a clutch laid by a single female (Galbraith 1993; Galbraith et al. 1993).

Mid-May into June is the usual period for egg laying. The female selects an area, often quite a distance from water, with deep sand or loose soil where she will dig out a bowl-shaped nest between 76.2 to 203.2 mm (3 to 8 in.) deep with her hind limbs (Ernst and Lovich 2009). If suitable nest sites are limited in an area, often elevated roadsides, railways, and levees are used. Clutch size has been reported from 4 to 109 eggs per female throughout their range with an average of 35 eggs (Ernst and Lovich 2009). Typical clutch size is 25–45 eggs with larger clutch sizes in the northern part of their range. A female was discovered depositing 49 eggs in early June in Lincoln County (Schuette 2005). The eggs are leathery shelled, cream colored, and about the same shape and size as a ping-pong ball. Hatching will occur between 55 and 125 days after the eggs are laid, depending on nest temperature and humidity (Ernst and Lovich 2009). Most eggs likely hatch between 75 and 95 days. Schuette (2005) reported an incubation period of 77 to 78 days for a clutch that hatched in late August in Lincoln County. Incubation temperature has been found to determine the sex of the hatchlings. In natural nests, the eggs on top are warmer than those eggs on the bottom, with a higher average temperature producing mostly females and a lower average temperature producing mostly males (Bull et al. 1982; Wilhoft et al. 1983). This is called temperature-dependent sex determination and occurs in other species of Missouri turtles. Predation on eggs within the nest is a major concern for snapping turtles and other turtle species. A study of eastern snapping turtles in Quebec, Canada, showed that up to 84 percent of nests were destroyed by a variety of mammalian predators such as raccoons, striped skunks, and mink (Robinson and Bider 1988). Hatchling eastern snapping turtles are 25 to 38 mm (1 to 1.5 in.) in carapace length, are dark brown to black, and have light spots along the edge of the upper shell and bottom shell. In Iowa, males reach sexual maturity in four or five years at a plastral length of 149 to 155 mm (5.9 to 6.1 in.); females mature at four to seven years at a plastral length of 123 to 175 mm (4.8 to 6.9 in.; Christiansen and Burken 1979). Eastern snapping turtles generally live up to 40 years of age (Moll 2001; Snider and Bowler 1992).

Remarks: The eastern snapping turtle is considered a game animal and is one of the few economically valuable reptile species in the state. Some people actively pursue this species for its meat, which is reported to make a fine stew and an excellent soup.

Over the years people have developed an intense dislike of this species caused by misinformation and a lack of understanding. Field studies have proven that this turtle will not harm game fish populations in natural bodies of water, and that a large part of its diet consists of aquatic vegetation, rough fish, crayfish, and carrion. Contrary to popular belief, snapping turtles do not cause substantial damage to waterfowl young under natural conditions (Lagler 1943). However, in artificial ponds where fish or waterfowl production is enhanced, this species may become a serious nuisance; control measures to reduce numbers of individuals may be required.

Alligator Snapping Turtle

Macrochelys temminckii (Harlan)

Adult alligator snapping turtle from Mississippi County.

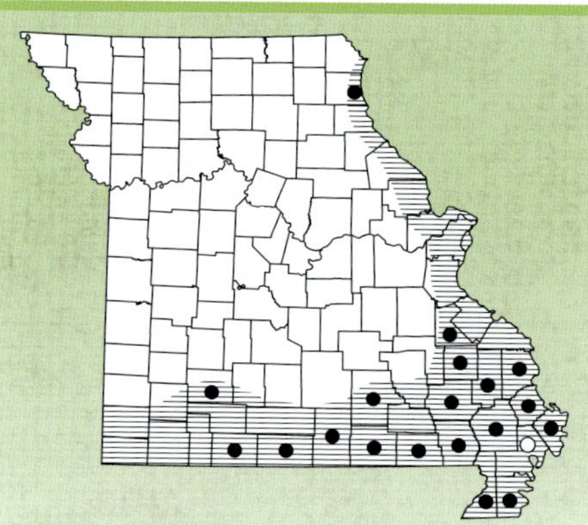

Distribution: Missouri: Mainly occurs in the large rivers, sloughs, and oxbow lakes of southern, southeastern, and eastern Missouri. North America: Alabama to eastern Texas, north to southern Kansas, across southern Missouri to historic isolated populations in Illinois and southern Indiana (Powell et al. 2016).

Description: The alligator snapping turtle is a huge aquatic species with a noticeably large head (compared to other species of turtles). The carapace has three prominent ridges—one along the centerline and one on either side. There is an extra row of scutes on each side of the carapace. The large head terminates in a sharp, strongly hooked beak. The tail is long and muscular with smooth round bumps. Skin on the head, neck, and forelimbs has a number of fleshy projections or tubercles. The plastron is reduced in size and affords little protection to the ventral area. Adults have dark-brown heads, limbs, and shells; skin on the neck and other areas may be yellowish brown. Males grow larger and have longer tails than females.

Adult alligator snapping turtles range in carapace length from 380 to 508 mm (15 to 20 in.) but have been known to reach 800 mm (31.5 in.; Powell et al. 2016). The record weight for this huge turtle is 113.9 kg (251.1 lb; Powell et al. 2016). A Dunklin County specimen, captured in 1994,

TOM R. JOHNSON

JEFF BRIGGLER

Alligator snapping turtles live in slow-moving rivers, such as this example in Butler County (above). Head of a juvenile alligator snapping turtle with a special worm-shaped appendage in its mouth used to attract prey (left).

weighed 58.1 kg (128 lb). A specimen weighing 50.3 kg (111 lb) was captured in 2019 in Cape Girardeau County (J. Briggler pers. obs.).

Habits and Habitat: The alligator snapping turtle is totally aquatic and seldom climbs out of the water to bask in the sun. Most specimens seen out of water are apparently females in search of an egg-laying site. The majority of their time is spent in deep water, hiding in root snags or among submerged logs (Lescher et al. 2013b). They spend daylight hours in hiding and become active at night. This species seldom attempts to swim; it normally moves about by slowly walking on the bottom of a body of water.

It appears the alligator snapping turtle cannot remain submerged for long periods of time, as some other aquatic turtle species can, especially in warm water (Ernst and Barbour 1972; Ernst and Lovich 2009). At a water temperature of 21°C (70°F), specimens were unable to remain underwater for longer than 50 minutes without surfacing to breathe. They typically extend their necks to reach slightly above the water surface to breathe. In addition, there is

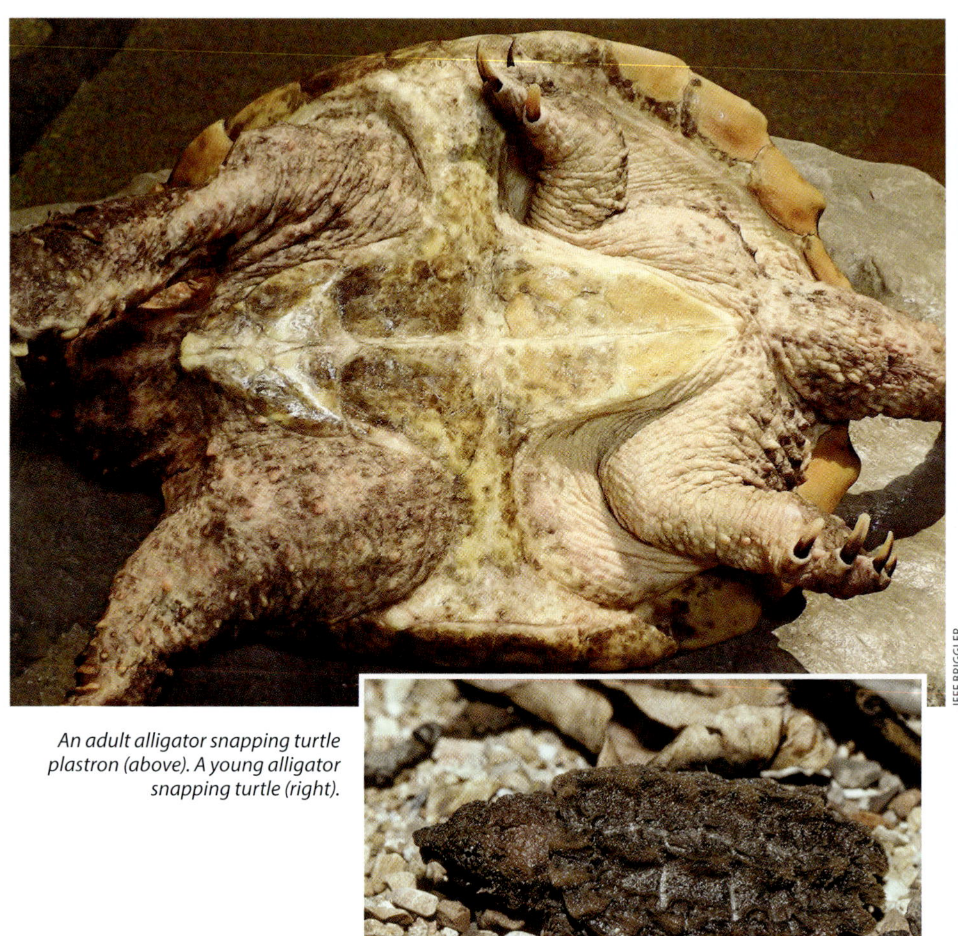

An adult alligator snapping turtle plastron (above). A young alligator snapping turtle (right).

some indication of pharyngeal respiration where the turtle moves water into and out of the mouth to increase oxygen consumption while underwater, but additional testing is needed to confirm this behavior (Allen and Neill 1950; Ernst and Lovich 2009).

This species is unique among North American turtles in having the ability to lure small fish to its mouth. Its tongue has a special appendage shaped like a stout worm; it can be moved at will by the turtle while it lies motionless on the bottom of a river or slough. Nearby fish are attracted to the wriggling "worm" and are captured and eaten when they venture too close.

This species' preferred habitat includes deep sloughs, oxbow lakes, and deep pools of large rivers. Other habitats include reservoirs and upland Ozark streams. Movements of this turtle species throughout the river channel can be extensive. One alligator snapping turtle in an Oklahoma river moved 27–30 km (16.8–18.6 miles) over three years (Wickham 1922). A telemetry study in southeastern Missouri showed considerable variation among individual turtles with a linear mean home range of 1,793 m (5,882.5 ft) within the river and ranged from 549 m (1,801.2 ft) to 3,322 m (10,899.0 ft; Shipman and Riedle 2008).

Although alligator snapping turtles will consume a variety of natural food, they feed mainly on fish but will capture and eat small turtles, snakes, small mammals, crayfish, mollusks, and even nuts and fruits (e.g., oak acorns, hickory and pecan nuts, persimmon fruits; Ernst and Lovich 2009). Young of the species presumably eat small fish, aquatic insects, crayfish, mussels, and snails. A study of alligator snapping turtles in northeastern Louisiana looked at the variety of fish eaten by wild-caught specimens. Sixty percent of the fish eaten were sunfish (bluegills and redear sunfish) and 19 percent were gizzard shad, which were the most common fish species in the turtle's wetland habitat (Harrel and Stringer 1997).

Breeding: Growth and reproduction biology of the alligator snapping turtles has not been well studied in Missouri. Studies in Louisiana found that individuals begin sexual maturation between 11 and 13 years of age, with males maturing earlier than females (Dobie 1971; Tucker and Sloan 1997). However, some individuals can take as long as 21 years to mature. The shell length of mature males is 38–42 cm (15.0–16.5 in.) compared to 33–37 cm (13.0–14.6 in.) in female alligator snapping turtles. Sperm were found in the *vasa deferentia* throughout the year in the Louisiana population, suggesting a prolonged breeding season in its southern range (Dobie 1971). Courtship and breeding take place in the water and mainly occur in the spring. Females emerge from the water presumably at night during late April through June to dig a nest and lay eggs. A study in Florida reported that the entire nesting event (digging chamber in soil, laying eggs, and covering nest with soil) took approximately four hours (Ewert 1976). It took this female turtle approximately 40 minutes to lay her 36 eggs. Clutch size across the species range is reported from 9 to 61 eggs with an average of 28 eggs (Ernst and Lovich 2009). Larger females generally produce more eggs than smaller females. The eggs are spherical, white, and have a rough, leathery shell. Egg incubation period is generally 80–114 days (Trauth et al. 2004). The egg diameter is 23 to 52 mm (0.9 to 2 in.; Ernst and Lovich 2009). Hatching probably takes place in late summer. Louisiana turtles laid one clutch of eggs per season, and there were indications that some adult females produce eggs only every other year (Dobie 1971). Newly hatched alligator snapping turtles normally have a brown or black shell and dark-gray skin with a long, slender tail; shell length averages 44 mm (1.7 in.; Allen and Neill 1950). A male captured as an adult lived for 70 years in captivity (Snider and Bowler 1992).

Remarks: The type locality for this species is Wolf River, Shelby County, Tennessee. The species name *temminckii* was named for Coenraad Jacob Temminck, a former director of the Leiden Museum, The Netherlands (Bour 1987).

The alligator snapping turtle is listed as a species of conservation concern. It is unlawful to capture, kill, or collect alligator snapping turtles in this state. Past overharvesting, water pollution, bycatch from fishing gear, and extensive habitat alteration are the main reasons for the decline of this species (Lescher et al. 2013a, 2013b). However, it appears that this species is expanding its range within the state, especially with the increased reports within reservoirs and upland Ozark streams in the southern part of the state. Raccoon predation of nests and drowning of individuals snagged on limb lines and trotlines are likely the primary threats this species faces today in Missouri.

Family Kinosternidae

American Mud and Musk Turtles

The Kinosternidae family is strictly New World (North and South America), comprising 25 species (Pough et al. 2018). Species within this family occur from Canada to Argentina and have been classified into five genera: *Staurotypus*, *Claudius*, *Cryptochelys*, *Sternotherus*, and *Kinosternon* (Pough et al. 2018). The last two, *Sternotherus* and *Kinosternon*, contain 11 species and are widely distributed in the United States. Both genera have representatives in Missouri comprising three species. This family is made up of rather small, dull-colored turtles. Indeed, one of the smallest turtles in the world, the eastern musk turtle (*Sternotherus odoratus*), is a member of the Kinosternidae. Eastern musk turtles have a very small plastron that affords little protection; only the front portion is movable. Mud turtles (genus *Kinosternon*), by comparison, have larger plastrons and the front and hind parts are both movable. All members of this family possess small scent glands along the seam between the carapace and plastron. When captured or roughly handled, musk and mud turtles produce a strong, unpleasant smell.

Yellow Mud Turtle

Yellow Mud Turtle

Kinosternon flavescens (Agassiz)

JEFF BRIGGLER

Adult yellow mud turtle from Barry County.

Description: The yellow mud turtle is a small, uniformly colored turtle of the Great Plains. Color of the chin and throat is normally yellow. Upper shell is somewhat flattened and usually olive brown. Populations in northeastern Missouri typically have darker shells. Scutes on the carapace are outlined in dark brown. Plastron is yellow brown; scutes normally have dark-brown margins. The limbs and upper parts of the head and neck are olive. The tail of the yellow mud turtle ends in a clawlike, horny tip. Adult males can be distinguished from females by concave plastrons and long, thick tails. An important characteristic that may be used to distinguish this species from other members of the family is the height of the ninth marginal scale on the carapace. This scale is much higher toward the middle of the upper shell than the eighth marginal. As with other members of the *Kinosternidae*, yellow mud turtles will give off an offensive, musky odor when captured.

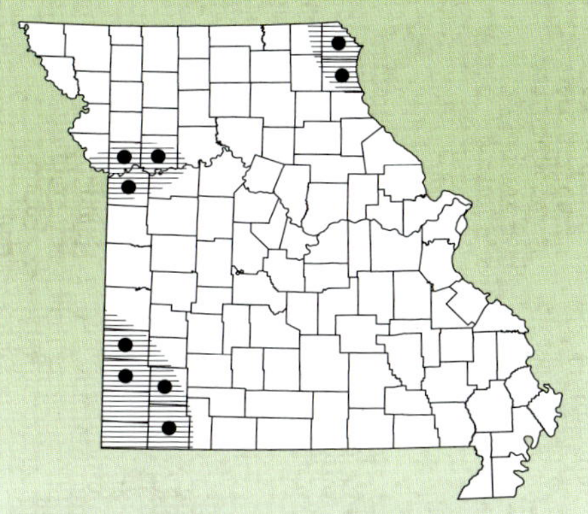

Distribution: Missouri: Restricted to a few counties in southwestern, west-central (Kansas City area) and northeastern. North America: Occurs in the Great Plains states, south to eastern New Mexico, most of Texas, and northern Mexico, with isolated populations in Missouri, Iowa, and Illinois (Powell et al. 2016).

JEFF BRIGGLER

Yellow mud turtles live in sand prairie habitat using wetlands for foraging and mating, such as this example in Clark County.

Adult yellow mud turtles range in carapace length from 100 to 125 mm (3.9 to 4.9 in.) but have been known to reach 168 mm (6.6 in.; Powell et al. 2016).

Habits and Habitat: Yellow mud turtles are one of the most studied reptiles in Missouri, especially populations in northeastern Missouri. Much of the natural history data for the yellow mud turtle in Missouri has been gleaned from these study publications (Kangas 2007; Kangas and Palmer 1982; Kangas et al. 1980; Kofron and Schreiber 1985). Yellow mud turtles are active between April and October. They live primarily in areas with open, sandy ridges close to wetlands in Missouri. A study in Illinois found that individuals require sandy ridges at least 4.5 m (14.8 ft) above the water level and up to 90 m (295.3 ft) from the water's edge (Tuma 2006). Individuals will begin to emerge from overwinter sites in early to mid-April to move into nearby wetlands (Kangas 2007; Kofron and Schreiber 1985). A variety of aquatic habitats, especially those with muddy or sandy bottoms, may be used by this species: marshes, rivers, sloughs, oxbow lakes, ponds, temporary pools, flooded fields, and water-filled ditches. Overland movement is common during the active period with most terrestrial activity taking place in early morning, evening, and at night. Two adult male yellow mud turtles were observed crossing a Barry County highway in the early morning in mid-to-late June. Most individuals are in wetlands for courtship, basking, and feeding from late April through early July (Kangas 2007; Kofron and Schreiber 1985). During the hot, dry summer months of July and August, individuals are buried in the sandy substrate at a depth of 25 cm (9.8 in.). Some turtles might emerge following summer estivation to move into wetlands for about a month (September into October) prior to settling in a location for the winter (Kangas 2007). Most individuals overwinter from October through March buried in the sandy soil with an average depth of 46 cm (18.1 in.) with a range of 34–61 cm (13.4–24 in.; Kangas 2007).

Yellow mud turtles eat a variety of aquatic animals, including aquatic and terrestrial insects, snails, crayfish, tadpoles, and dead fish. They occasionally eat aquatic plants. A study of the yellow mud turtle feeding habits in northeastern Missouri revealed that snails were the most frequently observed prey (Kofron and Schrieber 1985). Moll (1979) reported that captive yellow mud turtles ate night crawlers while they were buried in an aquarium filled with moist sand.

TOM R. JOHNSON

TOM R. JOHNSON

Color variation of the isolated population of the yellow mud turtle from northeast Missouri, Clark County (left). A yellow mud turtle plastron (right).

Breeding: Courtship and mating take place in shallow water between late April and mid-May. A male mud turtle will mount a receptive female and hold onto her upper shell with both fore- and hind limbs. During copulation the male may nudge, rub, or bite the head of the female (Lardie 1975). Eggs are laid in a hole dug by the female in well-drained sand or soil. Nesting sites usually have plenty of exposure to the sun. Six to eight white, elliptical, hard-shelled eggs are produced per female. Nests of the yellow mud turtle were observed in June and July during field studies in Clark County (Kangas and Palmer 1982). A study in northwestern Illinois showed that a female did not lay all her eggs at one time but waited a day or two and then laid again in the same general area (Tuma 1993). Such multiple nestings may reduce the number of eggs eaten by predators. Some females have been known to remain in the nest with the eggs, especially in dry years (Iverson 1990), but this has not been documented in Missouri. Hatching occurs from late August to mid-September; hatchlings are known to remain in the nest until the next spring. Hatchlings emerge in May in northeastern Missouri populations (Kangas 2007; K. Noel pers. comm.). Hatchlings average between 22 to 25 mm (0.9 to 1 in.) in carapace length. Sexual maturity takes six or more years and occurs at a carapace length between 8–12 cm (3.1–4.7 in.). Until recently, individuals were estimated to live about 30 or 40 years (Farrar 2001; Iverson 1991). Hedrick and Iverson (2017) documented that females are capable of living in the wild for at least 50 years and likely beyond 60.

Remarks: At one time, two subspecies of *Kinosternon flavescens* were recognized and both occurred in Missouri. The yellow mud turtle (*K. f. flavescens*) occurred in southwestern Missouri and the Kansas City area, while the Illinois mud turtle (*K. f. spooneri*) was found in extreme northeastern Missouri. Morphological and genetic information by Houseal et al. (1982) and Serb et al. (2001) proposed that the Illinois mud turtle is not a valid subspecies and indicated that it has a close affinity to the yellow mud turtle populations in the western part of their range. Although additional genetic studies are likely needed, current literature does not accept the validity of the Illinois mud turtle subspecies. Therefore, only the yellow mud turtle will be recognized in Missouri.

The yellow mud turtle is listed as endangered in Missouri because of its restricted range, limited habitat, and well-documented decline (Christiansen et al. 2012). Predation of eggs, as well as adults, by raccoons and other mammal predators is a continued threat to this species. Although considerable efforts are underway to ensure this species remains a part of Missouri's biodiversity, populations continue to decline and its long-term future in Missouri is uncertain.

Mississippi Mud Turtle

Kinosternon subrubrum hippocrepis Gray

Adult Mississippi mud turtle from Ripley County.

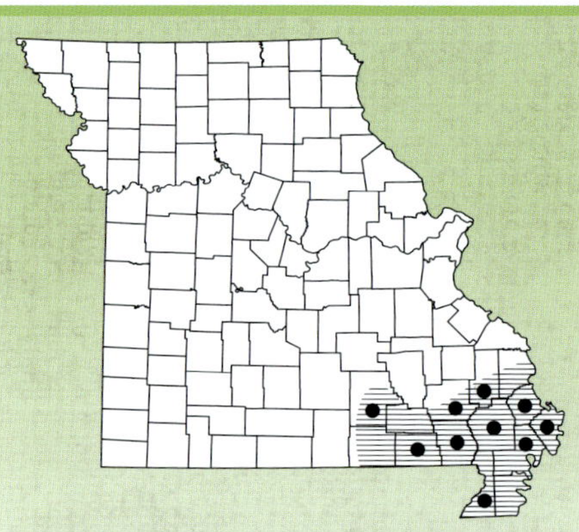

Distribution: Missouri: Mainly restricted to the counties of the Mississippi Alluvial Basin of southeastern Missouri. North America: Southern Illinois and southeastern Missouri, southwest to eastern Oklahoma and eastern Texas, south to southwestern Mississippi, and all of Louisiana (Powell et al. 2016).

Description: The Mississippi mud turtle is a small, dark turtle with yellow stripes along the side of the head and neck. The carapace is brown, dark brown, or nearly black (especially when the shell is wet). The plastron is yellow brown with brown extending along the margins of the scutes and along the center. Exposed fleshy parts are normally gray brown. On most individuals, there are usually two yellow stripes along the side of the head and neck; they may appear as a series of spots rather than stripes. Small, irregular yellow spots may cover parts of the head and neck on lighter-colored specimens. The ninth marginal scute on the carapace is not enlarged as in the yellow mud turtle. Adult males can be distinguished from females by concave plastrons and by longer, thicker tails terminating in a clawlike process.

Adult Mississippi mud turtles range in carapace length from 70 to 121 mm (2.8 to 4.8 in.) but have been known to reach 125 mm (4.9 in.; Powell et al. 2016).

Mississippi mud turtles live in natural swamps like this one in Bollinger County (above). The plastron of a Mississippi mud turtle from Butler County (left).

Habits and Habitat: This species lives in or near permanent and semipermanent water, including swamps, sloughs, oxbow lakes, ponds, ditches, and canals. It is most often observed in vegetated, shallow water with a soft bottom and seems to avoid flowing rivers. Although well equipped for an aquatic existence, this mud turtle spends as much time wandering about on land as it does in water. In Missouri, it is active from April to October. It is commonly seen crossing roads in southeastern Missouri during late spring into fall, especially following rains. In a study of this turtle in Oklahoma, Mahmoud (1969) observed two daily activity periods from June through August, between 4:00 a.m. and 9:00 a.m. and again between 4:40 p.m. and 10:00 p.m. Although this species will overwinter in the mud at the bottom of a wetland, many individuals move overland—sometimes considerable distances ranging from 55 to 600 m (180.4 to 1,968.5 ft)—to bury themselves under soil and leaves (Meshaka et al. 2017). They are known to burrow at depths of 10–15 cm (3.9–5.9 in.) to escape winter conditions in Oklahoma (Mahmoud 1969). In Missouri, the Mississippi

JEFF BRIGGLER

Mississippi mud turtles are often seen crossing roads in southeastern Missouri, such as this example from Ripley County.

mud turtle probably enters its overwintering retreat in late October. A study in Illinois found that this species spent a considerable amount of time in underground burrows (Skorepa and Ozment 1968). A three-day trapping survey of a 2 ha (4.9 ac) shallow wetland in Ripley County accounted for a density of nine Mississippi mud turtles per hectare (3.6 turtles per acre; J. Briggler unpubl. data).

The Mississippi mud turtle eats a wide variety of aquatic animals and plants. Mahmoud (1968) reported that insects, crayfish, mussels, and various amphibians were consumed by members of a population of *K. s. hippocrepis* studied in Oklahoma.

Breeding: Courtship and mating are presumed to occur from late April into June in Missouri. Anderson (1965) observed a pair copulating in shallow water in Stoddard County in mid-May. Females usually lay one to six eggs mainly from May into July. Mississippi mud turtles studied in Arkansas laid up to three clutches of eggs in one season (Iverson 1979). The eggs are normally laid in well-drained sandy soil. Anderson (1965) observed a female burying her eggs in early June in Pemiscott County. Upon careful digging of the nest, three eggs were found at 102 mm (4 in.) deep. The eggs are elliptical, pinkish white, and have a brittle, granular shell. Mississippi mud turtle eggs range in length from 22 to 30.5 mm (0.9 to 1.2 in.); incubation time ranges from 90 to 119 days (Iverson 1979). Females were found to reach sexual maturity between six and eight years of age at a carapace length of 80 to 85 mm (3.1 to 3.3 in.) while males mature at a carapace length of 76.6 to 95.4 mm (3 to 3.8 in.; Iverson 1979). A female that was collected as an adult lived 38 years in captivity (Pope 1939). It has been estimated that individuals can live 20 to 50 years in the wild (Frazer et al. 1991).

Remarks: Although the population of Mississippi mud turtles seems stable in Missouri, it is of utmost importance to preserve the natural habitats of this species, especially remaining cypress swamps, oxbow lakes, and sloughs surrounded by bottomland forest.

Eastern Musk Turtle

Sternotherus odoratus (Latreille *in* Sonnini and Latreille)

Adult eastern musk turtle from Crawford County.

Description: The eastern musk turtle is a very small, dark turtle with a smooth, domed upper shell and reduced lower shell. Carapace is dark gray brown to black. Plastron is much smaller than the carapace and the forward part is movable. It is usually yellow, brown, or grayish yellow. The fleshy parts are dark gray or black. There are normally two distinct yellow stripes along each side of the head and neck. Small projections of the skin called barbels are present on the chin and throat. Adult males can be distinguished from females by a longer, thicker tail ending in small, clawlike projection and by the presence of broad fleshy areas along the center of the plastron.

This is Missouri's smallest species of turtle. Adult eastern musk turtles range in carapace length from 51 to 115 mm (2 to 4.5 in.) but have been known to reach 150 mm (5.9 in.; Powell et al. 2016).

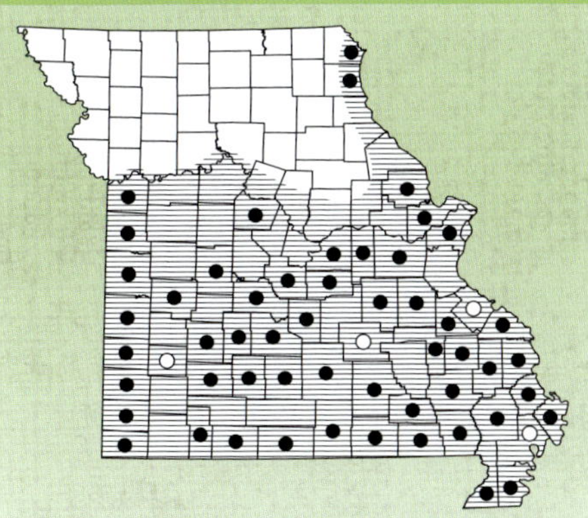

Distribution: Missouri: Statewide except for the northern and northcentral third of the state. North America: Southern Ontario, New England, west to central Wisconsin, south to Florida, west to eastern Kansas, Oklahoma to eastern and central Texas (Powell et al. 2016).

Habits and Habitat: The active season for eastern musk turtles probably lasts from

Adult male eastern musk turtle from Stone County (above). Eastern musk turtle plastron from Oregon County (right).

March to November. This species was studied by Mahmoud (1968, 1969) in Oklahoma and found to be most active from 4:00 a.m. to 10:20 a.m. and from 5:40 p.m. to 9:00 p.m. During daylight hours, this species prefers to remain buried in the mud or resting on the bottom (Ernst and Lovich 2009). This species prefers to crawl along the bottom in shallow water and seldom leaves the wetland. They rarely are seen basking but can occasionally be observed on logs, rocks, or small, horizontal tree trunks. Nickerson (2000) documented basking on logs and limbs of small woody saplings or bushes on the North Fork of the White River in Ozark County. In Missouri, this species is most abundant in slow-current sections of rivers and larger streams of the Ozarks, the swamps, sloughs, and small ditches of the Bootheel, and in a few rivers in the northeastern part of the state. They also can occur in reservoirs (Ford and Moll 2004). In a study in New Madrid County (Glorioso et al. 2010) musk turtles showed a preference for murky, shallow water. Eastern musk turtles overwinter in mud usually at a depth of 30 cm (11.8 in.) or so at the bottom of a wetland (Ernst and Lovich 2009). In preferred habitat, this species can be quite abundant. Glorioso et al. (2010) estimated a population size of 326 musk turtles in two wetlands in New

JEFF BRIGGLER

Eastern musk turtle hatchling from Butler County.

Madrid County. A study in Lake Springfield, Greene County, estimated a population size of 1,161 musk turtles with a density of 174 turtles per hectare (70.4 per acre; Ford 1999).

Eastern musk turtles are active bottom foragers searching for a wide variety of food. The food of this species includes aquatic insects, earthworms, crayfish, mollusks, small fish, tadpoles, algae, plants and their seed, and dead animals (Ernst and Lovich 2009; Ford and Moll 2004). The most frequently eaten items in a Missouri study were insects, mollusks, and plant seeds (Ford and Moll 2004). Anglers occasionally catch them on hook-and-line when using minnows, worms, or small crayfish for bait.

Breeding: Courtship and mating are known to occur in the spring and fall with the peak likely occurring between April and the end of June. In addition to the typical turtle nest construction of digging a hole and covering the eggs with soil, some musk turtles lay their eggs on the open ground or under debris (i.e., leaves, plants, rotting logs, or in sawdust piles; Ernst and Lovich 2009). Occasionally, several females will share the same nesting sites with many of the eggs intermingled (Cagle 1937; Edgren 1942). Eggs are likely laid in May through July in Missouri. Females lay about two to five eggs with larger females laying more eggs. The eggs are small (averaging 25.8 mm [1 in.] long), and white, elliptical in shape, with a thick brittle shell (Ernst and Lovich 2009). It takes between 65 and 86 days for the eggs to hatch with most individuals hatching in August and September. At hatching the young turtles average 23 mm (0.9 in.) in carapace length (Risley 1933). There is considerable variation in the age of maturity reported in the literature (Enrst and Lovich 2009). In nearby Oklahoma, males mature at four to seven years of age, while females mature at five to eight years of age (Mahmoud 1967). Although an eastern musk turtle lived slightly over 54 years in captivity (Snider and Bowler 1992), individuals in the wild probably live 20–30 years (Ernst and Lovich 2009).

Remarks: The common name for this species was formerly stinkpot. Musk glands located in the skin along their sides just below their carapace can give off a musky or foul-smelling odor when a specimen is captured, earning their name.

Family Emydidae

Pond and Box Turtles

The Emydidae family is one of the largest families of living turtles in the world. It comprises 12 genera, containing 52 species (Pough et al. 2018). In general, turtles in this family are small-to-medium sized and are adapted to a variety of habitats. This family includes a number of colorful species. Although the majority of species are aquatic, several kinds are either semi-aquatic or have taken to life on land (e.g., box turtles). Although native to Missouri, the red-eared slider has been introduced worldwide due to the pet trade and human food. This species is an invasive species in many countries throughout the world. In Missouri, this family is represented by seven genera with a total of 11 species and one additional subspecies.

Three-toed Box Turtle

Southern Painted Turtle

Chrysemys dorsalis Agassiz

JEFF BRIGGLER

Adult female southern painted turtle from Butler County.

Description: The southern painted turtle is a small aquatic turtle. The carapace is smooth, olive brown to almost black with a prominent yellow, orange, or red longitudinal stripe down the center. Outer edge of the carapace is yellow or orange. The plastron is plain yellow or tan; some individuals may have a faint brown blotch along the center. Exposed skin is dark brown or black and strongly patterned with yellow lines. Lines on neck and forelimbs can be orange or red. Adult males are smaller and have longer foreclaws and longer, thicker tails than females.

Adult southern painted turtles range in carapace length from 100 to 125 mm (3.9 to 4.9 in.) but have been known to reach 156 mm (6.1 in.; Powell et al. 2016).

Habits and Habitat: It is likely this species is active from late March into early November. This small, colorful turtle resides in the quiet water of shallow swamps, slow-moving streams, sloughs,

Distribution: Missouri: Mississippi Alluvial Basin in southeastern Missouri. North America: Southeastern Missouri and southern Illinois to western Alabama, Mississippi, Louisiana, Arkansas and to the northeastern Texas border area and southeastern Oklahoma border area (Powell et al. 2016).

Southern painted turtles live in bottomland forest wetlands, such as this example in Butler County (above). Southern painted turtle plastron from Wayne County (right).

JEFF BRIGGLER

JEFF BRIGGLER

oxbow lakes, and occasionally drainage ditches with aquatic vegetation and soft bottoms. They prefer wetlands embedded within bottomland forest consisting of tupelo, cypress, and oak tree species. They are often seen basking on logs, rocks, vegetation, and along banks. It is common to see many individuals basking on a single log, and they will readily take to the water with only slight disturbance. Southern painted turtles are occasionally seen crossing roads during the spring and early summer. Adults overwinter in the deeper parts of wetlands and usually are buried under the soft bottom mud. Southern painted turtles can be quite abundant. A study in New Madrid County estimated 103 southern painted turtles in two wetlands with most captures occurring in ditches that entered the wetlands (Glorioso et al. 2010).

Food consists mainly of aquatic insects, snails, crayfish, and plant material, with duckweed and algae readily consumed. Younger individuals consume more animal food in their diet compared to older adults.

Breeding: Little is known about the reproductive biology of this species in Missouri. Presumably courtship and mating occur in the water during the fall (Trauth et al. 2004).

JIM RATHERT

Southern painted turtles are often seen basking on logs.

Courtship behavior is similar to the western painted turtle. A female will depart the water to find a suitable egg-laying site that is usually a gentle, south-facing slope with loose soils. She will use her hind limbs to dig a flask-shaped nest that is 50 mm (2 in.) or slightly deeper. Two or more clutches from one to six eggs (average four) are laid within the nest and covered by the female (Moll 1973). Anderson (1965) reported finding a freshly laid clutch of southern painted turtle eggs in New Madrid County in mid-June that were 5.5 m (18 ft) from the water's edge and 3.7 m (12.1 ft) above the water level. Six elliptical, leathery-shelled eggs were present, with an average length of 29 mm (1.1 in.). The incubation period is from 65 to 80 days (Trauth et al. 2004). The eggs collected by Anderson (1965) were incubated at room temperature and later hatched on August 26–27 (an incubation period of 76–77 days). Hatchlings averaged 25 mm (1 in.) in carapace length (Anderson 1965). For eggs deposited in late July, hatchlings might overwinter in the nest chamber and emerge the following spring. Moll (1973) studied populations in Arkansas and Louisiana and found that males mature in two to three years compared to four years for females.

Remarks: Once considered a subspecies of the painted turtle (*C. picta*) complex, the southern painted turtle was elevated to full species status based upon genetic evidence (Jensen et al. 2015, Starkey et al. 2003). See remark section of the western painted turtle account for additional information.

The small size of the southern painted turtle is possibly due to competition with a number of sympatric species. The reduced size allows these turtles to use a niche that is usually not occupied by larger semi-aquatic species (Moll 1973). Although this species is considered common in Missouri, local populations have likely declined due to loss of wetland habitat and removal of bottomland forest.

Western Painted Turtle

Chrysemys picta bellii (Gray)

Adult western painted turtle from Callaway County.

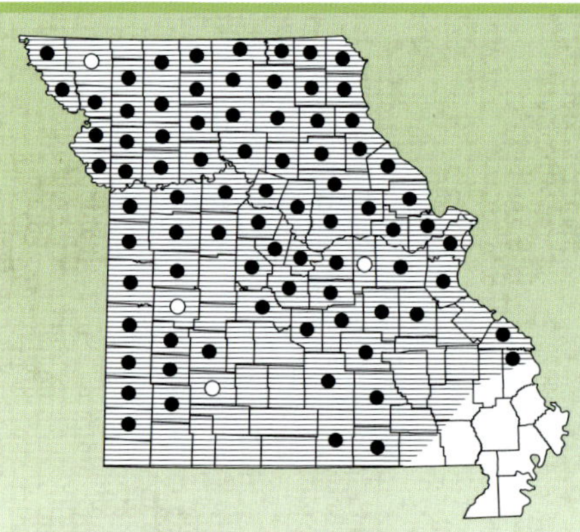

Distribution: Missouri: Statewide, excluding the state's extreme southeastern corner. North America: From south-central Canada to Wisconsin, south to southern Missouri, west to eastern Colorado, and north through eastern Montana. There are scattered populations in numerous western states (Powell et al. 2016).

Description: The western painted turtle is a small, brightly colored aquatic turtle with a smooth upper shell. The general color of the carapace is olive or olive brown to nearly black, usually with irregular yellow lines; marginal scutes may have one or more yellow bars and a red-orange outer edge. Exposed skin is dark brown or black and strongly patterned with yellow lines. Lines on neck and forelimbs can be orange or red. Plastron is yellow orange, bright orange, or red, with a prominent pattern of brown markings. Adult male western painted turtles have longer foreclaws and longer, thicker tails than females. Adult females are larger than males.

Adult western painted turtles range in carapace length from 115 to 200 mm (4.5 to 7.9 in.) but have been known to reach 254 mm (10 in.; Powell et al. 2016).

Habits and Habitat: This species is usually active from early April to October. A study in northeastern Missouri documented

Western painted turtle hatchlings emerging from their nest in Boone County (above). Western painted turtle plastron (left).

foraging activity mainly from April to late September (Kofron and Schreiber 1987). In Missouri, the western painted turtle occurs in slow-moving rivers, sloughs, oxbow lakes, ponds, drainage ditches, and marshes, especially where there is ample mud and abundant aquatic vegetation. They typically avoid waters with fast-flowing current. Painted turtles are active by day and sleep during the night in pools or ponds. Daytime activity consists of alternately basking and foraging in the water. These turtles require objects on which to climb and bask, such as partly submerged logs, rocks, mats of aquatic plants, or other objects. Although mainly aquatic, individuals are seen crossing roads, fields, and yards (LeClere 2013). During the cold winter months, adults overwinter in the deeper wetland bottom usually buried in the mud. The western painted turtle can be quite abundant, especially within the grassland areas of western and northern Missouri. During a study of a marsh in northeastern Missouri, Kofron and Schreiber (1987) captured more painted turtles than any other species of turtle.

Natural foods of this species include aquatic plants and algae, snails, crayfish, earthworms, leeches, insects, and fish. Western painted turtles will readily consume carrion. Young painted turtles have been reported to eat more animal matter than adults (Ernst and Lovich 2009).

Breeding: Courtship and mating mainly take place in shallow water from April to June but have been observed in the fall (Anderson 1965). A male will slowly follow a female, and once she starts swimming, the male will move in front of her and stroke her head and neck with his long claws. He will then swim a short distance and continue the stroking with his forelimbs. Once the female moves to the bottom of pond or marsh, mating will occur.

A female laden with eggs will leave the water and search for a suitable place to dig a nest and lay her eggs. A south-facing, gentle slope with loose dirt or sand and some low vegetation is an ideal location for nesting. In Missouri, the western painted turtle will produce eggs from mid-May through July. Eggs of this species are elliptical, white, and average 30 mm (1.2 in.) in length. Females will lay 4 to 23 eggs with an average of 12 (Ernst and Lovich 2009). A nest found in Lincoln County in May contained nine eggs (Smith and Powell 1993). Usually two or three clutches will be laid per season. It has been estimated that 30–50 percent of the female population do not reproduce every year (Ernst and Lovich 2009). Hatching normally takes place between 60 and 80 days. A female was observed depositing eggs in a yard in Boone County on June 15, 2009. Eight hatchlings emerged from the nest chamber on August 27 (J. Briggler pers. obs.). Newly hatched young may remain underground inside the nest until spring if the eggs were laid late in the summer (Vogt 1981). Hatchling painted turtles average 33 mm (1.3 in.) in carapace length. Maturity is usually obtained at two to four years of age for males and 6 to 10 years of age for females but does take longer in northern climates (Ernst and Lovich 2009). Painted turtles are known to live at least 30 to 40 years in the wild (Ernst and Lovich 2009).

Remarks: The taxonomic status of the painted turtle group is somewhat uncertain. Currently three subspecies of *C. picta* are recognized with the elevation of *C. dorsalis* as a full species (Jensen et al. 2015). Questions still remain on the validity of the three subspecies. There may be small populations of intergrades between midland painted turtle (*C. p. marginata*) in a few eastern Missouri counties (Ewert 1979). Also, painted turtles captured in the southeastern part of the Ozarks typically have a weak mid-dorsal stripe compared to *C. dorsalis* and less intense belly patterns compared to *C. p. bellii*. Additional studies are needed in these areas to verify these possible intergradations.

Surviving long periods of inactivity in cold, northern climates can be a challenge for many freshwater turtles. Painted turtles have unique adaptations to survive for an extended period of time within these cold environments. Not only do adult painted turtles significantly reduce their metabolic rates, but they also have a means to reduce deadly lactic acid buildup in their bodies during hibernation under water in oxygen-depleted environments. A large amount of lactic acid is transported into the turtle shell where it is buffered and stored (Jackson 2000). Hatchlings overwintering in the nest survive by supercooling and thereby resist actual freezing (Ernst and Lovich 2009). Such physiological adaptions allow adult turtles to survive oxygen-depleted aquatic environments and for hatchlings to survive freezing in terrestrial nests.

Western Chicken Turtle

Deirochelys reticularia miaria Schwartz

JEFF BRIGGLER

Adult western chicken turtle from Wayne County.

Description: The western chicken turtle is a small-to-medium-sized turtle with an oval shell and extremely long neck. The carapace may be light brown or olive, with faint, broad lines forming a netlike pattern. The western chicken turtle has a rather flattened appearance caused by the low, broad shape of the shell. The plastron is yellow with light-brown markings along scute seams. Its exposed skin is brown or black with numerous yellow or yellow-green stripes. Its forelimbs have a wide yellow stripe; its rump has distinct vertical yellow stripes. The underside of the head and neck of adults is plain yellow. Adult male western chicken turtles are smaller than females and have longer, thicker tails.

Adult western chicken turtles range in carapace length from 100 to 152 mm (3.9 to 6 in.) but have been known to reach 254 mm (10 in.; Powell et al. 2016).

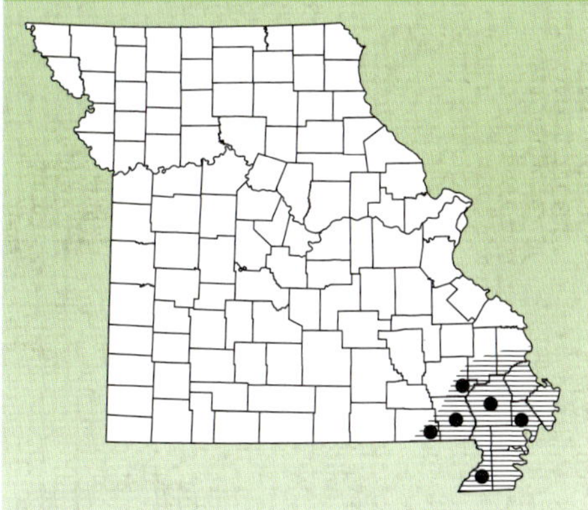

Distribution: Missouri: Mississippi Alluvial Basin in the extreme southeastern part of the state. North America: Southeastern Missouri south to western Mississippi, most of Louisiana, and west to southeastern Oklahoma and eastern Texas (Powell et al. 2016).

Habits and Habitat: In Missouri, its active season is probably mid-March into August.

Western chicken turtles forage and mate in natural cypress swamp habitat, such as this example in Butler County (above). Western chicken turtle plastron from Ripley County (right).

McKnight et al. (2015) found that western chicken turtles were only active for a brief period in the spring and early summer (mainly March to July) in Oklahoma and were buried underground for the remainder of the year. Western chicken turtles have been captured in Missouri wetlands from mid-April to mid-July (J. Briggler unpubl. data). Chicken turtles are semi-aquatic and spend nearly as much time wandering about on land as in the water. A study in Virginia showed that adult turtles take extensive overland excursions that allow them to locate other wetlands (Buhlmann 1995). Chicken turtles prefer shallow, still, or slow-moving aquatic habitats, including swamps, river sloughs, oxbow lakes, and ephemeral wetlands. This species will bask in the sun on partially submerged logs, often in association with other kinds of turtles. Considerable movement on land occurs during the nesting season, escaping hot, dry conditions by estivating beneath vegetation or leaf litter, or moving among wetlands. Unlike many other aquatic turtles, chicken turtles overwinter on land in open forests (Buhlmann 1995; Gibbons and Greene 1978). Many individuals overwinter 50–250 m (164.0–820.2 ft) from a wetland (Buhlmann and Gibbons 2001). Throughout the species range, population size is generally small with viable populations containing fewer than 40 adults (Buhlmann et al. 2008).

The western chicken turtle spends much time foraging in shallow water. Its long neck allows it to strike quickly to capture prey. It feeds upon a wide variety of invertebrates (e.g.,

JEFF BRIGGLER

Head of an adult western chicken turtle.

dragonfly and damselfly larvae, crayfish, and fishing spiders), tadpoles, and occasionally aquatic plants (Demuth and Buhlmann 1997; McKnight et al. 2015). A study in Oklahoma showed that crayfish and plants were most commonly consumed by western chicken turtles (McKnight et al. 2015). Missouri's population is presumed to have a similar diet.

Breeding: The reproductive biology of the Missouri population of the western chicken turtle has not been studied well, and until recently, most research has focused on the chicken turtle subspecies in southeastern United States (Buhlmann et al. 2008). Courtship and mating presumably take place in water during spring and early summer. In an Oklahoma study, the nesting season was May into July (McKnight et al. 2015). In Missouri, five female turtles were observed from June 9 through June 27 attempting to dig a hole, likely for egg laying, on a dirt road in Ripley County (J. Briggler unpubl. data). One of the females was x-rayed and contained 11 eggs. Females generally deposit multiple clutches (generally two) of 5 to 12 eggs each year (Trauth et al. 2004). Anderson (1965) reported 12 eggs produced in mid-July by a female collected in Butler County. This female contained developing egg follicles representing two additional clutches of 11 and 10 eggs. The eggs are laid in loose soil approximately 10.2 cm (4 in.) underground. The white eggs measure 36.7 to 39.9 mm long (1.4 to 1.6 in.; Anderson 1965). Incubation period is unknown for Missouri but generally has been reported to be 78–152 days for southeastern populations (Ernst and Lovich 2009). However, a study in central Arkansas found that most eggs hatch after a 68-day incubation period (Dinkelacker and Hilzinger 2014). Hatching probably takes place in the fall with potentially some hatchlings remaining in the nest through the winter; hatchlings have a carapace length of 28 to 32 mm (1.1 to 1.3 in.). Compared to most other turtles, chicken turtles apparently are not long-lived. It is likely few survive past 15 years of age (Gibbons 1987).

Remarks: The western chicken turtle is state endangered and extremely rare in Missouri. It was first reported in Missouri by Anderson (1957). Only a few specimens had been found since then until the spring of 1995, when a small population was discovered in Butler County (Buhlmann and Johnson 1995). This population in Butler County has declined dramatically, but a new population has been discovered in Ripley County. The chicken turtle requires wetlands and lowland forests to survive. This habitat has been greatly reduced in southeastern Missouri by the clearing of forests and draining of natural wetlands. Protection of the remaining wetlands and lowland forest, as well as the construction of wetlands and planting of trees, will ensure this animal remains a part of Missouri's biodiversity.

Blanding's Turtle

Emydoidea blandingii (Holbrook)

Adult Blanding's turtle from Clark County.

Distribution: Missouri: Extreme northeastern and northwestern corners of the state. Three specimens were captured in 1990 and 1991 in St. Charles County; however, no populations have been verified. North America: Southern Ontario and extreme southern Quebec, isolated populations in Nova Scotia, New England, and New York, around the Great Lakes states and into Iowa, with peripheral populations in Nebraska, South Dakota, and Missouri (Powell et al. 2016).

Description: The Blanding's turtle is a medium-sized turtle with an oval, moderately high-domed carapace and a long head and neck. The general color of the carapace is dark brown or black with many yellow spots or bars. The plastron has a hinge across the forward third of the shell; it is brown fading to yellow with large brown or black blotches on the outer portions of each scute. Its exposed skin is brown and yellow. The upper jaw may be covered with black pigment forming a "mustache." The underside of the head and neck is bright yellow. Adult male Blanding's turtles can be distinguished from adult females by longer, thicker tails that have the anal opening extending beyond the edge of the carapace and by a slightly concaved plastron.

Adult Blanding's turtles range in carapace length from 150 to 230 mm (5.9 to 9.1 in.) but have been known to reach 284 mm (11.2 in.; Powell et al. 2016).

Blanding's turtles live in natural marshes, such as this example in Clark County (above). Blanding's turtle plastron (left).

Habits and Habitat: This species is semi-aquatic and spends a considerable amount of time in shallow water along the edge of wetlands in prairie and grassland habitats. They frequent land, especially near wetlands. Blanding's turtles bask on logs or muskrat houses; their yellow throats are visible for a considerable distance. When captured, a specimen will withdraw into its shell and close the forward part of the plastron tightly against the carapace. In Missouri, this species is mainly active from late March through October (Kofron and Schreiber 1985). Habitats used by this species include natural marshes and sloughs, ponds, and drainage ditches. Abundant aquatic and emergent vegetation and a mud bottom are important habitat components. It overwinters in mud at the bottom of marshes or ponds. Kofron and Schreiber (1985) found two individuals buried under the mud in late October and mid-November in the shallow part of a marsh in northeast Missouri. Individuals radio-tracked in northwest Missouri entered overwintering sites by late October (Lehnhoff 2004).

Populations of this species were studied in both extreme northeastern and northwestern Missouri (Kangas 2007; Kofron and Schreiber 1985; Lehnhoff 2004); some of the information that follows is from these studies. The Missouri population was found to be more aquatic than reported in other states (Minton 2001; Vogt 1981). During the active season, it feeds in two phases: between early April and mid-July, and again from mid-August through mid-September. A radio-tracking study in northwest Missouri found that males moved greater distances and occupied larger home ranges (mean = 695 ha; 1,717.4 ac) than females (mean = 131 ha; 323.7 ac; Lehnhoff 2004).

The diet of Blanding's turtles in northeastern Missouri consisted of crayfish, aquatic larvae of insects, some terrestrial insects, fish, and frogs (Kofron and Schreiber 1985). A study of Michigan Blanding's turtles showed a similar diet (Lagler 1943). However, a study in northeastern Illinois found that Blanding's turtles in that area ate primarily snails and aquatic insects and that crayfish were not present in their wetland (Rowe 1992).

Breeding: Courtship and mating mainly take place in April through early June but have been observed in August in northwestern Missouri (Lehnhoff 2004). Vogt (1981) described breeding activity of Blanding's turtles observed in Wisconsin. Once a female is located in the water, a male will position himself behind the female and quickly mount her carapace. The male will stimulate the female by biting at her head and forelimbs. During mating, the male will expose his bright yellow throat in front of the female's head and wave it back and forth. The pair may remain together for several hours. Females will move overland to select a nesting site with sandy, well-drained soil and good exposure to the sun. Lehnhoff (2004) tracked three females to nesting sites in northwestern Missouri. These females began moving to nesting sites during the first week of June and stayed on land from 5 to 20 days. Two of the females nested in agricultural fields and the other female nested in a prairie with traveling distances ranging from 623 to 1,944 m (2,044 to 6,378 ft; Lehnhoff 2004).

The elliptical eggs are normally laid during June in Missouri. They are cream colored and pliable, with an average length and width of 38 by 25 mm (1.5 by 1 in.). Clutch size for the species ranges from 3 to 22 eggs (Ernst and Lovich 2009). Several nests have also been found in the northeastern Missouri population. Kangas (2007) found a nest with 13 eggs that hatched in mid-August. Two additional nests were found by radio-tracking females in early June (K. Noel pers. comm.). One female deposited 13 eggs in a nest that was 12–14 cm (4.7–5.5 in.) deep on a levee, and the eggs hatched in mid-August (K. Noel pers. comm.). The other nest was in a soybean field and contained 11 eggs at a depth of 10 cm (3.9 in.; K. Noel pers. comm.). Hatching takes place from mid-August into September; the newly emerged young average 31 mm (1.2 in.) in carapace length. Individuals that hatched in mid-August immediately burrowed at the base of a grass tussock (K. Noel pers. comm.). The tail of hatchlings is nearly as long as the length of the carapace. Sexual maturity occurs between 14 to 20 years of age. Blanding's turtles can live well over 40 years. An individual captured in Minnesota was determined to be at least 75 years old (Brecke and Moriarty 1989).

Remarks: Paul Anderson (1965) was the first person to discover this species in Missouri, locating a population in the northeastern part of the state. Another population was discovered in northwestern Missouri in 2000 (Durbian et al. 2001). The Blanding's turtle is listed as endangered in Missouri due to a small population and reduced natural wetland and grassland habitats. Small but consistent populations continue to exist in Missouri.

Northern Map Turtle

Graptemys geographica (LeSueur)

JEFF BRIGGLER

Adult female northern map turtle from Crawford County.

Description: The northern map turtle is a medium-sized, aquatic turtle with a low ridge or dorsal keel and strongly serrated marginal scales at the rear. Carapace may be brown or olive brown with a netlike pattern of fine, squiggly, yellow lines that give the shell the appearance of a road map. The plastron is light yellow; seams between scutes are dark brown. Coloration of head, limbs, and tail is dark brown to nearly black, with many narrow, greenish-yellow lines. A distinct small, yellow marking is located behind each eye that usually has a slight projection that is oriented toward the neck. Northern map turtles are strong swimmers; their limbs are fully webbed. Adult female northern map turtles are larger in size and have a much larger head than males. Males have a longer, thicker tail than females.

Adult males range in carapace length from 100 to 160 mm (3.9 to 6.3 in.). Adult females range from 170 to 292 mm in carapace length (6.7 to 11.5 in.; Powell et al. 2016; Vogt et al. 2018).

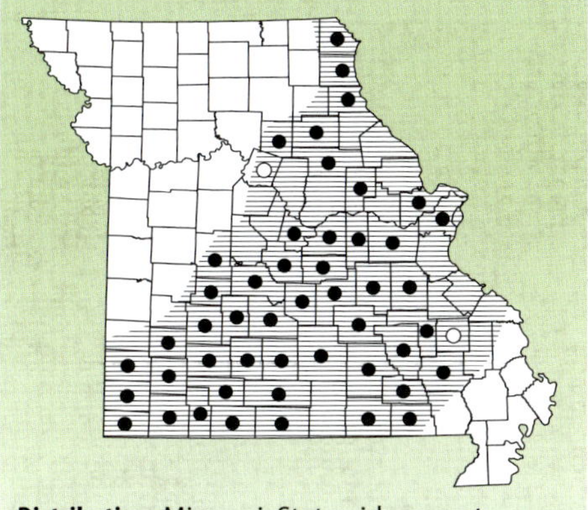

Distribution: Missouri: Statewide except for northwestern third and southeastern Missouri. North America: Southern Quebec and Ontario, west to central Minnesota, south along eastern Iowa, then southwest to southern Arkansas, east to northern Georgia, and then northeast to Quebec with numerous populations around this perimeter (Powell, et al. 2016).

JEFF BRIGGLER

Northern map turtles are commonly observed basking on logs on Ozark rivers and streams, such as this example in Ozark County (above). A northern map turtle nest with 12 eggs from Pulaski County (right).

TOM R. JOHNSON

Habits and Habitat: The active season for northern map turtles lasts from late March to October. Small-to-medium-sized rivers, reservoirs, sloughs, and oxbow lakes are primary habitats for northern map turtles in Missouri. They require many basking sites, a gravel or rocky bottom, and some aquatic plants. They are often seen basking on logs and rocks sticking out of the water during the active season but will take advantage of unusually warm, sunny days in winter to bask (Johnson 2000). Basking in the sun is common in a wide variety of species. In general, turtles bask not only to raise their core body temperature (thermoregulation), but basking also helps reduce the number of external parasites (such as leeches) and can reduce the amount of algae growing on their carapace (upper shell). In addition, warming in the sun increases their ability to digest and assimilate food, and direct sunlight allows their skin to produce vitamin D. Females will bask longer than males as this increases the speed of egg development (Vogt et al. 2018). Hatchlings and young adult northern map turtles will often bask away from adults on floating algae mats, small logs, and rocks. Adults may bask up to eight hours per day and can be seen basking in a group of 10 or more on logs or large rocks. They will share prime basking sites with other turtle species.

Several authors have reported that this species remains active in winter and has been seen swimming under the ice, with males being more active than females (Ernst and Lovich 2009; Vogt et al. 2018). In general, northern map turtles spend the winter on or near the

JEFF BRIGGLER
TOM R. JOHNSON

Northern map turtle hatchling from Crawford County (above). Northern map turtle plastron (left).

bottom in deep pools (3 to 6.1 m; 10 to 20 ft) and have been reported to congregate in groups of up to 100 individuals in a Virginia river (Graham and Graham 1992). They do not eat any prey during this time. The northern map turtle has been observed overwintering individually or in small groups (five or fewer) in rock crevices, between large rocks, inside large root wads, and under large, submerged logs along the bottom of several Ozark streams (J. Briggler pers. obs.).

Foraging for food takes place in early morning and late evening. Males, as compared to females, may utilize sections of rivers with more aquatic plants (Vogt et al. 2018). The diet of northern map turtles varies between the sexes; adult males eat snails, small mussels, and a variety of aquatic insects—especially caddisfly and mayfly larvae. The much larger adult females eat crayfish, snails, and a variety of freshwater mollusks and some insect larvae. Their larger head allows them to crush these prey, discard their exoskeleton or shell, and swallow the bodies. A study of northern map turtles in the Niangua River in Missouri showed that small snails were the most common prey followed by crayfish and then aquatic insects (White and Moll 1992). A Pennsylvania population of this species living in Lake Erie was reported to feed heavily on zebra mussels—an aquatic invasive species in North America (Lindeman 2006). Natural predators include raccoons, striped skunks, and red foxes that dig up and eat freshly laid turtle eggs. Newly hatched turtles may be eaten by crows, otters, and coyotes.

Breeding: Courtship and mating take place in the water, usually from late March through May. In several other states this species has been observed breeding in autumn (Ernst and Lovich 2009). Gravid females will leave the water and may move a considerable distance to locate a suitable site to lay their eggs. Nests may be dug in or along the edge of plowed fields, in patches of sand, or in clay banks. In Missouri, northern map turtles produce eggs from late May through early July. A study of this species in the Niangua River in southern Missouri found that females laid from 6 to 15 eggs per clutch (average 10 eggs) and may produce two clutches per season with a small number of females producing three clutches (White and Moll 1991). Vogt (1981) found that most females in Wisconsin lay just two clutches. A female was observed in mid-June laying a clutch of 12 eggs on a clay bank along the Big Piney River in Pulaski County (T. Johnson pers. obs.). Northern map turtle eggs are white, elliptical, and 32 to 35 mm (1.3 to 1.4 in.) long. Hatching typically takes place from August through September. A female northern map turtle was observed depositing eggs in early July in Randolph County. By September 8, hatching of these eggs (n=11) was completed, with an incubation period of 68 days (J. and H. Terrell pers. comm.). Eggs that were laid late in the season can overwinter in the nest and emerge the following spring (Gibbons 2013; Vogt 1981). Hatchlings, which are nearly round in shape, average 32 mm (1.3 in.) in carapace length. Generally, males become sexually mature around four years, and females are mature around 10 years throughout their range (Vogt et al. 2018).

Remarks: This is the most widespread map turtle species in the United States and is the most common turtle species observed when floating or snorkeling on Ozark streams. Until recently, this species was called the "common" map turtle.

Basking northern map turtles are easily disturbed by a passing canoe or boat and will quickly scramble into the water. The availability of prime basking sites may be as important to the survival of map turtles as the availability of prey. Pitt and Nickerson (2012) stated that the high use of many Ozark streams and rivers by canoes and inflated rubber tubes can negatively affect Missouri's northern map turtle population. Simply put, this traffic prevents turtles from basking and may, in turn, reduce survivorship.

Ouachita Map Turtle

Graptemys ouachitensis Cagle

JEFF BRIGGLER

Adult Ouachita map turtle from Jefferson County.

Description: The Ouachita map turtle is a medium-sized aquatic turtle with a dark, raised keel along the center of the carapace, forming a ridge. The rear of the upper shell is strongly serrated. The carapace color is olive green with usually one dark-brown or black smudge or blotch per scute. As with other map turtles, the carapace is covered with narrow, irregular yellow lines, which somewhat resemble a road map. The head, neck, limbs, and tail are dark brown or olive drab with numerous yellow lines or markings. Ouachita map turtles have a large, wide yellow marking behind each eye that is widest just behind the eye and becomes narrow on the top of the head. Neck stripes may or may not come in contact with the lower part of each eye. A large yellow spot is found under each eye and another directly below on the lower jaw. The eyes are cream to white and usually have a thin black horizontal line across the pupil. The plastron is flat and plain yellow with a faint gray streak along the seams of

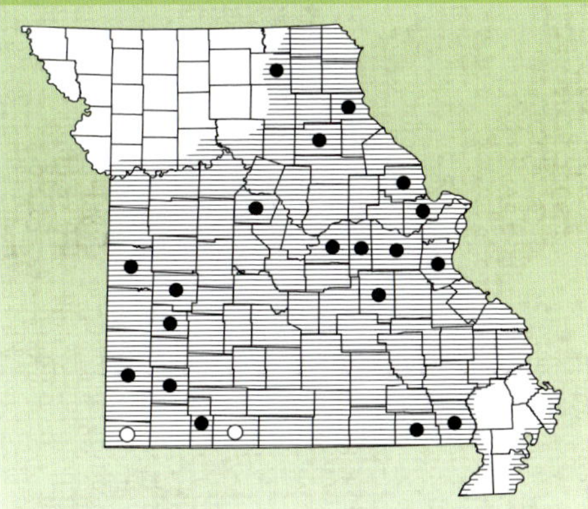

Distribution: Missouri: Southern and eastern Missouri and along the Mississippi and St. Francis rivers in the Bootheel. North America: Northeastern Louisiana and southeastern Arkansas and along the major tributaries of the Mississippi River from Minnesota to its mouth, with isolated populations from eastern Kansas to Ohio (Powell, et al. 2016).

JEFF BRIGGLER

JEFF BRIGGLER

Ouachita map turtle hatchling (above). Ouachita map turtle plastron from Osage County (right).

the scutes. These streaks are more pronounced on young turtles. As with other map turtles, female Ouachita map turtles are larger than males.

Adult males range in carapace length from 70 to 140 mm (2.8 to 5.5 in.). Adult females range up to 240 mm (9.4 in.) in carapace length (Powell et al. 2016).

Habits and Habitat: Generally, this species has habitat needs and life history similar to the northern false map turtle (*G. p. pseudogeographica*). It is considered a large river species but has been observed upstream in many tributaries and reservoirs throughout much of its range. It is sporadically captured throughout smaller Ozark rivers and streams. It is presumed to prefer the faster portions of rivers and streams (Trauth et al. 2004). Like other species of map turtles, the Ouachita map turtles can be observed basking on logs over the water. Rarely are they seen on land. They overwinter underwater, likely among rocks, tree debris, and undercut banks.

Ouachita map turtles have a more varied diet than other map turtle species (Ernst and Lovich 2009). In general, natural prey include aquatic insects, snails, crayfish, dead fish, and, occasionally, aquatic plants (Ernst and Lovich 2009). A study in west-central Illinois (Cass County) and extreme northwestern Tennessee reported on the diet of this species from still waters (lentic) and riverine (lotic) habitats (Moll 1976). Ouachita map turtles living in the Mississippi River with a sterile sand bottom were opportunistic feeders and ate mostly insects that fell on the river surface or were driven into the river by flooding or rain. Those that were living in a lake environment with a muddy bottom had a diet of mostly midge (fly) larvae. In a northern population of Ouachita map turtles, up to 32 percent of adults ate plant material (Vogt 1980).

Breeding: Information on the reproductive biology of this species in Missouri is lacking. They breed in the water from early April through May. There may be some occasional mating in late September into November (Vogt 1980). Courtship is similar to the false map turtle. A male will locate a female and swim directly in front of her and face her, extend his front legs, and vibrate his long claws against both sides of her head. If the female is receptive to mate, the male will swim above her, face forward, and copulation will occur. Eggs will be laid on land in open areas where there is loose soil or sand from May into July. From 6 to 15 eggs may be produced per female. Each female will lay two to three clutches per season (Ernst and Lovich 2009; Vogt 1980). Gravid females will usually dig their nests after sunset or just before sunrise in the morning. They have also been seen nesting on cloudy days or just after a heavy rain (Vogt 2018). Well-developed embryos within days of hatching and recent hatchlings have been known to make sounds within the nest (Geller and Casper 2019). Use of various sounds, especially clicking, might be a form of communication among siblings to begin hatching. Hatching will occur in late summer or early fall and rarely, if ever, do Ouachita map turtles overwinter in the nest (Ernst and Lovich 2009). Newly hatched young have a slightly oval-to-round carapace and are from 27 to 35 mm (1.1 to 1.4 in.) long. The first three central scutes (vertebrals) on the upper shell have a keel or ridge. The rear of the carapace is serrated. Coloration of the carapace of the hatchling is olive drab with thin yellow lines across each scute and a yellow, backward C-shaped mark on the pleural scutes. The plastron of hatchlings is light yellow with a complex dark-gray marking. Newly laid nests of Ouachita map turtles are preyed upon by raccoons, skunks, and foxes. Male Ouachita map turtles become sexually mature at two to three years of age. Females, on average, will be sexually mature between six and seven years old.

Remarks: This species was formerly classified as a subspecies of the false map turtle. Vogt (1993) elevated it to a full species after examining many specimens throughout their range. However, it hybridizes with *G. p. pseudogeographica* in central and eastern Missouri. Little is known about the ecology and population status of this species in Missouri. Capture rates are considerably lower in Missouri compared to some other states. In over 21 years of sampling in the middle Mississippi River, Ouachita map turtles only accounted for 0.1 percent of the captures of the turtle community (Braun and Phelps 2016). In another study in southeastern Missouri, only one Ouachita map turtle was captured out of 723 turtles of seven species in the floodplain backwaters of the Mississippi River (Wallace et al. 2007). Researchers snorkeling in Ozark streams can occasionally capture individual turtles (J. Briggler pers. obs.). Additional studies are needed to better understand the ecology, behavior, and status of this species in Missouri.

Northern False Map Turtle

Graptemys pseudogeographica pseudogeographica (Gray)

Adult male northern false map turtle.

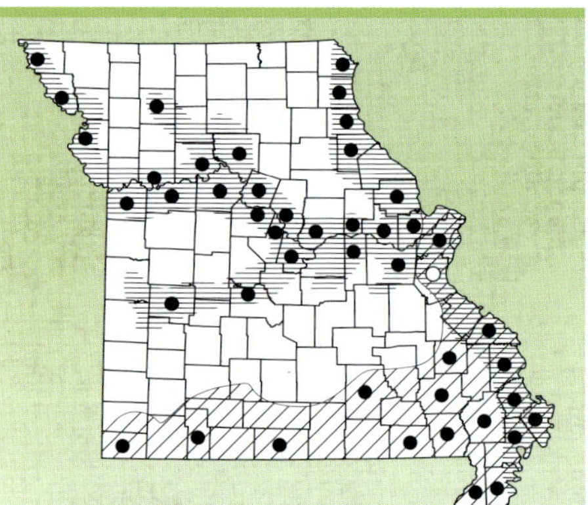

Description: The northern false map turtle is a medium-sized, aquatic turtle, with a prominent keel along the center of the upper shell and thin, yellow lines on the head, neck, and limbs. Its carapace is brown or olive, with light lines that follow the contour of each small, dark blotch on the marginal and costal scutes. The dorsal keel is usually high and forms a low ridge at the rear edge of each vertebral scute. The rear of the carapace is strongly serrated. The plastron is flat and plain yellow, but may have dark

Distribution: Missouri: Primarily found in the Missouri and Mississippi rivers and larger tributaries. The northern false map turtle occurs in the Missouri River in central and northwestern Missouri and the upper Mississippi River (horizontal lines), while the Mississippi map turtle subspecies occurs mainly in the lower Mississippi River, southern Ozark rivers, and southwestern Missouri (diagonal lines). North America: The northern false map turtle occurs mainly in the Missouri River in Missouri to North Dakota and Mississippi River from St. Louis north to Wisconsin and Minnesota. The Mississippi map turtle subspecies occurs mainly in the lower Mississippi River, lower Ohio River, and Arkansas River basin. Contiguous populations exist from eastern Mississippi and west to eastern Texas (Powell et al. 2016).

JIM RATHERT

TOM R. JOHNSON

Northern false map turtles are quite abundant in large rivers, such as the Missouri River (above). Northern false map turtle plastron (left).

lines or smudges along the scute seams, especially on young turtles. Its exposed skin is brown or olive with numerous narrow yellow lines. Eye coloration is usually white or light yellow; there may or may not be a dark, horizontal line crossing the pupil (see section on subspecies below). A yellow marking behind each eye extends upward then backward, forming a backward "L". However, these yellow markings are variable. Up to three thin, yellow neck lines reach each eye. There is a small round or oblong, yellow spot below each eye and at the forward tip of the chin. Adult males are smaller than females and have a longer, thicker tail and long fore claws.

Adult males range in carapace length from 90 to 150 mm (3.5 to 5.9 in.). Adult females range from 150 to 270 mm (5.9 to 10.6 in.) in carapace length (Powell et al. 2016).

Habits and Habitat: Northern false map turtles are active from late March to early October. They can be seen basking in the sun on logs or rocks projecting from the water. Feeding takes place in early morning. Overwinter dormancy takes place in the mud at the bottom of sloughs, lakes, and backwater areas of rivers. In the southern third of Missouri, these turtles may become semi-active during a mild winter. It is likely they do not eat any prey during winter months.

Adult Mississippi map turtle from Ripley County (above). Head of a northern false map turtle from the Missouri River (right).

Large rivers and their backwater are their preferred aquatic habitat, but northern false map turtles also occur in river sloughs, oxbow lakes, lakes, and reservoirs. Braun and Phelps (2016) found that this species prefers the main channel, especially around wing dikes, on the Mississippi River in southeastern Missouri. A muddy bottom, some aquatic vegetation, and numerous basking sites are important requirements. A four-year study of the habitat selection of this species was conducted along a section of the Missouri River, Cooper County, in central Missouri (Bodie and Semlitsch 2000a). The following information on habitat use is from this study. Major flooding along the Missouri River in 1993 and 1995 altered the floodplain and increased the availability of various wetlands to semi-aquatic turtles. These wetlands included flooded agricultural fields, flooded woodlands, and scour ponds—the formation of new and often permanent pools away from the main river channel. Female northern false map turtles

JEFF BRIGGLER

Mississippi map turtle plastron from Ripley County.

used the main river channel more often while males used the more permanent scour ponds. Hatchlings and young adults used shallow and temporary wetlands. These sites provided more available food and fewer predators. In addition, this and other studies found that unnatural winter lowering of water levels of floodplain wetlands will kill overwintering young northern false map turtles and other species. These water bodies were disconnected from the main channel and the dormant turtles were killed by freezing (Bodie and Semlitsch 2000a, 2000b; Christiansen and Bickham 1989). A study of Mississippi River turtles in southeastern Missouri found that false map turtles made up 82 percent of the turtles out of a total of six species that were captured in fyke and trammel nets (Barko et al. 2004).

Basking behavior of this species is similar to other map turtles. Much of the day may be spent basking on protruding logs or rocks, especially those objects that are along the shore. They are easily disturbed while basking and the passing of a boat will cause a quick dash into the water. If there is no additional disturbance, the turtles will swim back to and climb onto a basking object within 10 minutes or less (Lindeman 1998). A typical basking stance of this species is to stretch their head and neck, as well as all limbs, as far as they can and remain in this pose for minutes or hours. A study of northern false map turtles in western Wisconsin observed common grackles landing on basking map turtles to search for and eat leeches that were attached to the turtles' skin. The turtles paid little attention to this behavior and tolerated the pulling and pecking (Vogt 1979).

Northern false map turtles eat a variety of prey including snails, crayfish, aquatic insect larvae, dead fish, and aquatic vegetation (Ernst and Lovich 2009). False map turtles and other species of aquatic turtles have been observed feasting on the 13-year periodical cicadas (*Magicicada* spp.) as they drop into the waters of the Lake of the Ozarks (Powell and Powell 2011).

Breeding: Courtship and breeding take place in the water, usually in the spring. A male will use his long fore claws to stimulate a female by stroking her head while swimming in front of her. When the female sinks to the bottom the male will follow and mount her. Most breeding activity will occur in April through May. There have been reports of false map turtles breeding in the fall (Ernst and Lovich 2009). Soon after, breeding females begin swimming to the general area of nesting sites, such as islands with sandy beaches, the edge of agriculture fields, and open areas with loose soil and little vegetation. When ready to nest, they will move onto land and search for a suitable site. In stable aquatic habitats, this species is known to use the same nesting locations year after year (Vogt and Bull 1982). Nesting begins in early June and will last through July. A female uses her hind limbs to dig a flask-shaped hole from 89 mm (3.5 in.) to just over 152 mm (6 in.) in depth for egg laying. Northern false map turtles may lay from 8 to 22 eggs per clutch with an average of 14 eggs (Vogt 1980). Mississippi map turtles are reported to lay considerably fewer eggs (two to eight eggs per female). This species is reported to lay up to four clutches of eggs per season in Tennessee (Freedberg et al. 2005). In general, larger females lay more eggs. The white, elliptical, leathery eggs range from 32 to 37 mm (1.3 to 1.5 in.) in length (Vogt 1980). The eggs generally hatch in late summer or early autumn with an incubation period of about two months. Newly hatched northern false map turtles have a round to slightly oval shell and average 33 mm (1.3 in.) in carapace length. Males generally mature in four to six years while females take 8 to 12 years (Trauth et al. 2004). Individuals have been known to live between 30 and 35 years (Snider and Bowler 1992).

Subspecies: This turtle has two geographic races or subspecies in Missouri: the nominate race, the northern false map turtle, *Graptemys pseudogeographica pseudogeographica* (Gray), described above; and the Mississippi map turtle, *G. p. kohnii* (Baur). These two races are similar in appearance, and there is an area of intergradation of these subspecies in southeastern and eastern populations in Missouri. This makes it challenging to identify the map turtles in those areas (Vogt 1993). Important differences between the two subspecies is the shape of the yellow mark behind each eye. The northern false map turtle has a backward "L" yellow mark and narrow, yellow lines touching the posterior of each eye. Mississippi map turtles have a crescent-shaped line behind each eye; the yellow lines do not touch the posterior edge of each eye. Specimens from eastern and southeastern Missouri rivers tend to have eyes that lack a horizontal black line across the pupil. The Mississippi map turtle has a similar biology to that of the northern false map turtle.

Remarks: Although the false map turtle is the most common turtle species encountered in the Missouri and Mississippi rivers (Barko et al. 2004; Bodie et al. 2000; Braun and Phelps 2016), changes in riverine habitat, historical collecting, and drowning in fishing gear have likely affected this species. Habitat alteration and loss are important negative factors when trying to protect false map turtles in Missouri (Bodie 2001). Natural resource agencies need to restore and protect rivers by improving water quality and retaining natural flow. These efforts are critical when a species is having a difficult time maintaining a sustainable population. It has been shown that close to 90 percent of the nests of this species are destroyed by predators within 24 hours after its eggs are laid. Raccoons, river otters, skunks, and red foxes are the primary turtle nest predators. A fly species was discovered laying eggs in map turtle eggs right when the young turtles began hatching. The fly maggots ate the exposed egg yolk then entered the young turtles' belly and up to 36 percent of the hatchlings were killed before they could leave the nest (Vogt 1981).

Eastern River Cooter

Pseudemys concinna concinna (LeConte)

Adult eastern river cooter from Stone County.

Description: The eastern river cooter is a medium-sized, aquatic turtle with a broad shell that is usually oval. The carapace is olive brown, brown, or nearly black, with numerous yellow lines or markings; including backward facing C-shaped, yellow markings on the first, second, and third pleural scutes. The rear of the upper shell is weakly serrated. The head, limbs, and tail are dark olive to dark brown or nearly black with numerous yellow stripes. Scutes under the carapace and on the bridge are yellow with round or oval, dark-gray markings that often have a light center. These markings are centered along scute seams. The plastron is normally yellow; it may either be devoid of dark markings or have gray-brown markings along the scute seams, especially toward the anterior. Its exposed skin may be olive brown or black, with many yellow lines. A yellow, Y-shaped marking is usually present on either side of the head and neck. Adult males can be distinguished from adult females by their long foreclaws and a longer,

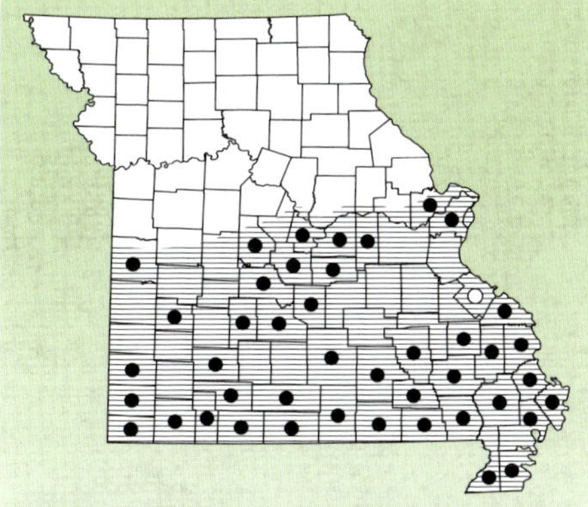

Distribution: Missouri: Southern half of the state. North America: Southeastern Kansas and southern Missouri east to Virginia, south through the Carolinas, Georgia, northern Florida, west to Texas and Oklahoma. There are small populations scattered in West Virginia, Kentucky, and Tennessee (Powell et al. 2016).

Eastern river cooters are often found in large well-vegetated reservoirs, such as Lake Springfield (above). Eastern river cooter plastron (right).

thicker tail. Also, adult females have a higher, more dome-shaped carapace than adult males. Old adults of both sexes can be melanistic (dark), causing yellow markings on their shell and skin to be faint or absent. Males are slightly smaller than females.

Adult eastern river cooters range in carapace length from 230 to 330 mm (9.1 to 13 in.) but have been known to reach 375 mm (14.8 in.; Powell et al. 2016).

Habits and Habitat: The ecology of the river cooter has not been studied well in Missouri. This species is presumably active between April and October (Ernst and Lovich 2009). This aquatic species lives in rivers and sloughs, as well as larger reservoirs, in southern Missouri, and is rarely seen traveling on land, except for females in search of egg-laying sites. River cooters spend a considerable amount of time basking on logs and will quickly slide into the water at the slightest disturbance. During the active season, they will forage for food during early morning and late afternoon. Winter dormancy takes place in mud at the bottom of rivers, river sloughs, or lakes. This species has been observed basking in the sun on warm days in winter. Studies in West Virginia and Illinois documented that radio-tracked cooters had home ranges varying from 1.2 to 5.3 ha (3 to 13.1 ac; Buhlmann and Vaughan 1991; Dreslik et al. 2003).

River cooters are predominantly vegetarian and consume a wide variety of aquatic plants. Turner (1995) examined 57 adult river cooters captured in a southwestern Missouri reservoir. This two-year study found that up to 90 percent of their diet consisted of various species of green algae. Large, floating mats of a mixture of algae were available to turtles in the study coves where the turtles were captured. The most abundant algae was the filamentous green algae, *Spirogyra*. In addition to algae, mulberries that have fallen into the water were readily eaten. Both males and females shared the same diet. Other studies have reported cooters eating mussels, crayfish, and insects (Ernst and Lovich 2009). Since algae and plant

TOM R. JOHNSON

Eastern river cooter hatchling from Morgan County showing "egg tooth."

materials are hard to digest, the river cooter has an extremely long intestine to increase digestion efficiency (Thomas et al. 1994). During the 13-year eruption of periodical cicadas (*Magicicada* species), Powell and Powell (2011) observed river cooters and other species of turtles feasting on cicadas as they dropped into the water of the Lake of the Ozarks.

Breeding: Courtship and mating take place during spring in the water. A male will swim above the female and occasionally stroke the head of the female with rapid vibrations of his long foreclaws. If copulation ensues, the pair will sink to the bottom and remain there until completion. Egg laying is presumed to take place mainly in late May through June in Missouri. A nest is dug by the female in loose soil or sand away from the aquatic habitat. Anderson (1965) documented a female depositing eggs about 15.2 m (50 ft) from the water's edge on an artificial sand beach at Lake Taneycomo on June 15. Gravid females were mainly captured from mid-April through June in southwestern Missouri (Turner 1995). Throughout their range, river cooters are known to deposit multiple clutches that contain 4–30 eggs (average 15 eggs) each year (Ernst and Lovich 2009). A study of river cooters captured in central Arkansas found that females produced two to three clutches per season, and clutch size was from 9 to 18 eggs (Iverson 2001). A study in southwestern Missouri found clutch size was 12–28 eggs and two to three clutches of eggs were laid per season (Turner 1995). Freshly laid eggs are elongated with a soft, pale-pink, leathery shell. River cooter eggs average about 39 mm (1.5 in.) in length. Their eggs mainly hatch during August and into September with an incubation period between 70 and 96 days (Ernst and Lovich 2009). Turner (1995) collected eggs from gravid females in Missouri and found that incubation time was 66 days when raised in captivity. Hatchlings may overwinter in the nest and emerge the following spring (Aresco 2004; Buhlmann and Vaughan 1991). River cooter hatchlings are round and from 27 to 39 mm (1.1 to 1.5 in.) in carapace length. The shell is green with yellow markings around the outer edge and numerous yellow lines crossing each scute. Their lower shell is light yellow and orange with brown or green markings along the scute seams. Nest predators include raccoons, skunks, and foxes. Newly hatched young may be eaten by large bass, great blue herons, and river otters. In Illinois, male river cooters begin breeding at about six years of age; females mature at 10-plus years (Dreslik 1997; Niemiller et al. 2013). Individuals in the wild likely live to 40 years of age or older (Dreslik et al. 1998).

Remarks: Little is known about the river cooter population in Missouri, but the species appears to be secure, especially within its preferred habitat. The river cooter and several other turtle species were annually harvested by licensed commercial fisherman from rivers in southeastern Missouri—even though they were not considered a commercial species in this state. They were illegally sold alive to regional fish/turtle dealers who shipped them to Asian markets, where they were sold for human consumption. Recent changes that stopped commercial harvest in Missouri will ensure this species and other turtle species remain common in Missouri.

Three-toed Box Turtle

Terrapene carolina triunguis (Agassiz)

Adult male three-toed box turtle from Oregon County.

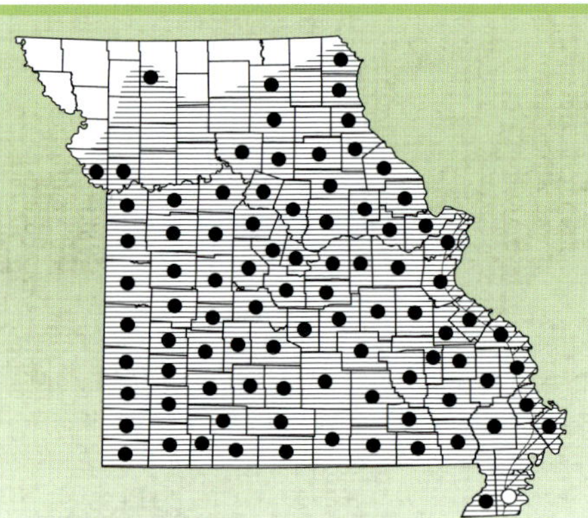

Distribution: Missouri: Statewide except for extreme northern and northwestern sections (horizontal lines). An area of intergradation with the eastern box turtle (*T. c. carolina*) along the eastern and southeastern edge of the state (diagonal lines). North America: Missouri, eastern Kansas, Oklahoma, and Texas, all of Arkansas, most of Louisiana and Mississippi and into southern Alabama and northwestern Florida (Powell et al. 2016).

Description: The three-toed box turtle is a small, terrestrial turtle with a high-domed shell and normally three toes on each hind limb. There is usually a ridge along the center of the carapace. The carapace may be olive or olive brown with faint yellow or orange lines radiating from the center of each large scute. The plastron has a distinct single hinge across the forward third of the shell; it allows the turtle to tightly close the lower shell against the inside edge of the upper shell. The plastron is plain yellow or yellow with brown smudges or lines that sometimes follow the scute seams. Its exposed skin is dark brown or black; scales on the head and forelimbs may be yellow or orange. The entire head and portions of forelimbs of males are often bright orange. Although most individuals have three toes on each hind limb, some may have four toes per hind limb. A study of three-toed box turtles and other subspecies showed that the number of toes on the hind limbs was much more variable than previously known.

JEFF BRIGGLER
JEFF BRIGGLER

Three-toed box turtle swimming across the Big Piney River in Texas County (above). Three-toed box turtle plastron from Dent County (left).

Therefore, within the species of *Terrapene carolina*, this characteristic is not a reliable way of telling the subspecies apart—especially if used as the only characteristic (Minx 1992). Most adult males can be distinguished from females by the greater amount of orange on the head and neck, thicker tail, slightly concaved plastron, and often red eyes. Adult females have a very small tail and most have yellow-brown eyes.

Adult three-toed box turtles range in carapace length from 115 to 125 mm (4.5 to 4.9 in.) but have been known to reach 179 mm (7 in.; Powell et al. 2016).

Habits and Habitat: The three-toed box turtle is mainly active from early April through late October and predominately inhabits open woodlands throughout Missouri, although grasslands are important during parts of the active season. This species was extensively studied by Schwartz and Schwartz (1974) and Schwartz et al. (1984) in central Missouri; much of the following information is based on their work. In central Missouri, this species generally becomes active soon after the last killing frost, between late March and late April. The turtles generally enter overwintering retreats shortly after the first killing frost of autumn, usually between mid-September and early November.

TOM R. JOHNSON

Three-toed box turtle hatchling from Cole County.

Daily activity begins with a period of feeding followed by basking in the sun in an open area. In warm weather, box turtles will crawl into a clump of dense grass or dead leaves and rest. Early evening activity usually consists of searching for a suitable retreat for the night. These retreats, called forms, are often under leaf litter or in a few inches of soil. The central Missouri population studied for 25 years showed a habitat preference for mature, oak-hickory forests with numerous openings and edge areas along brushy fields. The home range of adults varied from 2.2 to 10.6 ha (5.4 to 26.2 ac). Adults can find familiar habitat when displaced several kilometers. Schwartz (2000) found that some individuals stay within the same home range for up to 35 years.

Reagan (1974) studied a population in Arkansas and noticed a seasonal shift in habitat use. They are most often found in wooded areas near overwintering sites in late autumn and early spring, then move to grasslands during late spring, then back into wooded areas during the heat of summer, and finally back to grasslands in early autumn prior to moving to overwintering sites in wooded areas. A study in southwestern Missouri also showed a similar pattern in habitat use between grasslands and wooded areas (Sammartano 1994). The use of grasslands by box turtles coincided with mild temperatures and high moisture conditions. During hot, dry conditions in late summer, they are often found in temporary pools or in the water along the shallow margins of a pond. They will readily swim across rivers and lakes. Most individuals were observed swimming at the surface of the water, and some individuals were observed sitting on the bottom (McDowell et al. 2004; J. Briggler unpubl. data).

The three-toed box turtle's inability to dig deep enough into leaf litter and soil during cold weather is thought to cause a high incidence of winter mortality. However, a study of box turtles in southwestern Ohio showed that they overwinter in soil and under leaf litter at an average depth of only a few inches and never below 5 inches. Also, the turtles' body temperature was monitored during the winter; the specimens commonly reached a body temperature that was below freezing but saw no mortality due to freezing (Claussen et al. 1991). Another study in Ohio of eastern box turtles (*T. c. carolina*) found that box turtles produce a type of blood/

tissue antifreeze when their liver releases glycogen into the circulatory system. This allows the turtles to survive freezing during the winter and causes the heart and blood vessels to continue to function (Costanzo et al. 1993). Grobman (1990) studied emergence from overwintering sites in Missouri. Individuals begin to move closer to the surface at 7°C (45°F) and begin to emerge after five consecutive days of increasing temperatures. Once emerged, they remain near their overwintering site until warm, spring rains occur. A female box turtle in Callaway County was observed emerging from her overwintering site in a shallow depression of approximately 30.5 cm (1 ft) deep in early April (J. Briggler pers. obs.). This depression was filled with a large amount of leaf litter due to the surrounding forest canopy. She remained in this depression taking advantage of the sun for several weeks prior to moving to other locations for the spring and summer. She was observed back at this overwintering site in mid-October and remained at this site throughout the winter. During this time, she was observed sunning on several occasions up to November 10.

The three-toed box turtle can be abundant, especially in its preferred habitat of oak-hickory woodlands. One of the most extensive studies on the three-toed box turtle occurred in Cole County on a 22.2 ha site (54.9 ac; Kiester et al. 1982; Schwartz 2000; Schwartz and Schwartz 1974, 1991; Schwartz et al. 1984). During this 25-year study from 1965 to 1990, 1,743 turtles were observed. The densities ranged from 17.3 to 34.6 turtles per hectare (7 to 14 turtles per acre). Although very young turtles probably make up a sizable portion of a box turtle population, their secretiveness causes them to seem scarce. A study by O'Conner et al. (2015) in central Missouri estimated 1.85 turtles per hectare (0.7 turtles per acre) within a forested area compared to 4.14 turtles per hectare (1.7 per acres) within a mix of open field, woodland, and forest. A four-year study on 40 acres in Moniteau County captured 555 unique individuals via the use of dogs (34.3 turtles per hectare; 13.9 turtles per acre; D. Urich pers. comm.). Another study conducted within two small areas in Jefferson City, Cole County, found population estimates of 2.3 and 2.5 turtles per hectare (0.9 and 1 turtles per acre) within a highly fragmented, urban area (Riedle et al. 2017).

Food habits of the three-toed box turtle change as the turtle matures. Young turtles eat mostly insects and earthworms, but older turtles eat a proportionately larger amount of plant matter, including berries, mushrooms, and young shoots of various plants and grasses. Adults will occasionally consume earthworms, snails, and insects. A study in southwestern Missouri found the majority of their diet was dogwood berries in the fall and blackberries in the summer, although some insects were eaten (Sammartano 1994). Several adult three-toed box turtles in Cole County were observed eating cicadas in June 1998 during a large emergence of the periodical cicadas (*Magicicada septendecim*; J. Miller pers. comm.).

Breeding: Courtship and mating of the three-toed box turtle mainly occur from late April to late June or early July in Missouri. However, mating has been regularly observed in late summer into fall (Sammartano 1994; Schwartz and Schwartz 1974). In 2018 and 2019, four observations of mating in Callaway County were documented from late August to late September (J. Briggler pers. obs.). According to Ernst and Lovich (2009), a male courts a female by standing a few inches away from her, holding his head high and pulsating his throat. When the female moves closer to him, he begins to stimulate her by mounting her shell while he scratches with all four limbs. He may nip at the forward part of her shell, then move backward and rest the rear of his carapace on the ground, at which time mating takes place.

In Missouri, most egg laying takes place from mid-May to early July. A female will select an elevated, open patch of loose soil and dig a hole 7.6 to 10.2 cm (3 to 4 in.) deep with her hind limbs. This usually begins in late evening or at dusk, and the eggs are mostly laid at night. Generally, two to eight elongated white eggs will be laid by a female; the eggs are 24 to 40 mm (0.9 to 1.6 in.) in length (Ernst and Lovich 2009). Long-term studies in central

Adult eastern box turtle from Perry County (above). Eastern box turtle plastron from Perry County (right).

Missouri documented an average clutch size of about five eggs per female (Miller 2001). Multiple clutches may be laid each year. The baby turtles hatch in about three months; generally, 80–102 days (Trauth et al. 2004). Eggs laid late in the summer will hatch in the fall, and the hatchlings will remain within the nest until the following spring. A hatchling will have a carapace length of 30 to 33 mm (1.2 to 1.3 in.). Hatchling three-toed box turtles have a flatter carapace compared to adults and a distinct dorsal ridge. The carapace is brown gray, and there is a yellow spot on each large scute. Young box turtles are unable to close their plastron to protect themselves. Female box turtles have the ability to store viable sperm and produce fertilized eggs up to four years after mating. Sexual maturity is reached between 7 to 10 years of age in Indiana with males maturing earlier (Minton 2001).

Throughout their range, box turtles likely live between 50 to 80 years of age, but there are a few records of survival of over 100 years. The Schwartz and Schwartz (1991) study in Cole County estimated the oldest turtle to be 59 years old. This population was revisited in 1998 and 1999 (Miller 2001). Twelve turtles greater than 60 years of age were found, with one estimated to be between 65–74 years (Miller 2001). Schwartz (2000) estimated maximum longevity in the Missouri study to exceed 70 years.

Subspecies: The eastern box turtle (*T. c. carolina*) occurs east of the Mississippi River. In the summer of 1994, an adult eastern box turtle was captured crossing a gravel road in a wooded area in extreme eastern Pemiscot County, near the Mississippi River (Powell et al. 1994). This find represented the first record of this subspecies in an unincorporated area of the state and may represent a link to the Illinois population resulting from historic changes in the Mississippi River channel. Since then there have been additional confirmed reports of eastern box turtles in Missouri counties along the Mississippi River. It is assumed some eastern box turtles were brought to Missouri from Illinois by people. Others may have been left on the Missouri side of the Mississippi River as the river migrated across its historic floodplain years ago, or individuals may have swum the Mississippi River, thus colonizing eastern Missouri.

Remarks: The three-toed box turtle is familiar to most Missourians, and it is fitting to be named the official state reptile. Although this turtle is still quite common and widespread in Missouri, it does face many threats. Box turtle eggs and young are eaten by skunks, raccoons, and badgers. The primary causes of death in adults are diseases, extreme cold during the winter (with no snow cover for insulation), and human activities. Thousands of these reptiles are killed annually while crossing roads and highways, especially in May. We have received numerous verbal reports from Missourians finding 10 to 25 dead box turtles together in a relatively small area in a forest. During warm spells in the winter, individuals may emerge too early and are often killed by a quick decline in temperature (Neill 1948), or turtles die of a disease while in their overwintering burrows and are scavenged by mammals in midwinter. There is also a heavy demand for the box turtle overseas, especially for the European pet trade. This has resulted in the illegal commercial collecting of this species. Fortunately, Missouri Department of Conservation has protected the box turtle from being collected and sold as pets.

Many box turtles are captured by people on weekend outings and taken home to be kept as a pet. Such captive conditions are in direct conflict with the biological needs of these turtles; they may slowly starve to death or grow abnormal shells, claws, and mandibles if not maintained and cared for correctly. These reptiles are an interesting part of outdoor Missouri and should not be kept in captivity.

Ornate Box Turtle

Terrapene ornata (Agassiz)

Adult ornate box turtle from Clark County.

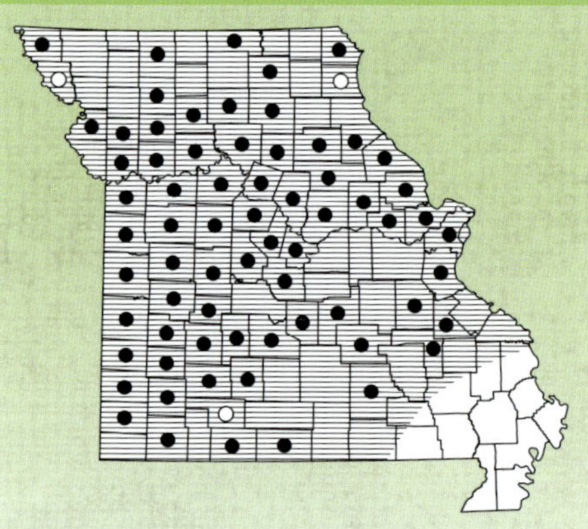

Distribution: Missouri: Statewide except for the southeastern corner. This species is more common in the northern and western parts of the state. North America: From southern North Dakota to southeast Louisiana and southern Texas; plus, a band from western Missouri to Illinois and eastern Iowa (Powell, et al. 2016).

Description: The ornate box turtle is a small, terrestrial turtle with a domed carapace and hinged plastron that allows it to close its lower shell. The carapace has a flattened appearance along the dorsal surface and normally has no ridge. The carapace is brown to nearly black with numerous yellow lines radiating from the center of each scute. A mid-dorsal yellow stripe is often present. The plastron is dark brown with bold light lines. Across the plastron, just anterior of the center, is a flexible "hinge" that allows the front and rear of the lower shell to press up against the upper shell, enabling the head, tail, and all four limbs to be protected from harm. Its exposed skin is gray brown, with faint yellow or orange-yellow spots or blotches; the chin and upper jaw are yellow or yellow green. There are normally four toes on each hind limb. Adult male ornate box turtles are slightly smaller than females and have a concaved area in the center of

Ornate box turtles prefer native prairies, such as this example in Clark County (above). Ornate box turtle plastron from Clark County (left).

their plastron, an enlarged and inward-facing first toe of the hind limbs, and red eyes. The eyes of females are yellow brown.

Adult ornate box turtles range in carapace length from 110 to 125 mm (4.3 to 4.9 in.) but have been known to reach 154 mm (6.1 in.; Powell et al. 2016).

Habits and Habitat: Missouri ornate box turtles are active between late March and mid-October. Three specimens in Chariton County were observed emerging from overwintering burrows in a prairie in early April (J. Briggler pers. obs.). In Kansas, the onset of warm, rainy weather will bring this species out of overwintering retreats (Legler 1960). Daily activity consists of basking for a short period soon after sunrise, followed by a period of foraging for food, a rest period during the heat of the day, and a short feeding session before resting at night. A study of this species in western Nebraska showed that they are active 20 percent of the day and inactive 80 percent (Converse et al. 2002). When inactive, ornate box turtles will shelter in clumps of grass, bury themselves in loose soil or sand, or enter burrows made by other animals. During the hot, dry days of mid-to-late summer, these turtles are generally inactive but will appear in numbers after a heavy rain. The study in western Nebraska observed ornate box turtles using the shallow water and mud of a prairie wetland during the hottest part of the summer (Converse et al. 2002). This species will escape the cold temperatures of winter by burrowing down below the frost line, which may be 45.7 cm (18 in.) or more; those that take shelter in wooded areas may not burrow as deeply. Studies of this species in southern Kansas showed that overwintering burrows were 22.9 to 73.7 cm (9 to

Ornate box turtle emerging from its winter retreat in early April from Chariton County.

JEFF BRIGGLER

29 in.) deep in open native prairie habitat (Metcalf and Metcalf 1970). A southwest Missouri study documented that ornate box turtles overwintered beneath the ground totally in open grassland habitat (Sammartano 1994). In their study area, overwintering ornate box turtles were observed to congregate in brushy areas, such as under blackberry bushes. During one particular Kansas winter (1975), 56 dead specimens were found aboveground with their head and forelimbs eaten. It was thought that many of the turtles had gone into overwinter dormancy while ill, and they died in their burrows and were subsequently dug up and eaten by coyotes or badgers once the flesh began to rot. Live and apparently healthy box turtles were found in the same area (Metcalf and Metcalf 1979, 1985).

Ornate box turtles are considered a prairie species, living where there is a mixture of prairie grasses and forbs. They will also live in open woodlands, rocky glades, and farm fields in former native prairie regions. Both the three-toed box turtle (*T. c. triunguis*) and the ornate box turtle have been found living in the same vicinity in a number of Missouri counties.

A two-year study of both box turtle species living in close proximity was conducted in southwestern Missouri (Sammartano 1994). Microhabitat selection was determined by each species' needs: feeding, basking, estivation, and overwintering. Not enough information was collected to determine nesting site selection. There was some overlap of food selected, but in general, ornate box turtles ate insects while three-toed box turtles ate vegetable matter. Three-toed box turtles spent the majority of their time in wooded areas, while ornate box turtles used the open prairie sites for the most part. Ornate box turtles were found in mammal burrows during August and September. In general, they spend periods of estivation in the soil, either in mammal burrows or in burrows they dug into damp soil in or near seasonally wet areas. This study showed there was little overlap between niche selection of the two box turtle species, allowing them to coexist with little competition.

An individual ornate box turtle may have a home range of about 2 ha (4.9 ac), and their density may be 0.4 per hectare (0.2 turtles per acre; Legler 1960). Another study found a density of 0.8 per hectare (0.3 turtles per acre) in east-central Kansas (Rose 1978). Adult males were found to have larger home ranges than females (Bernstein and Richtsmeier 2007).

Ornate box turtles are primarily insectivorous, with grasshoppers, crickets, beetles, and caterpillars comprising nearly 90 percent of their diet (Legler 1960). Earthworms are often eaten; especially in the spring (Anderson 1965). Berries, such as mulberries, wild strawberries, and dogwood fruit are a part of their diet (Sammartano 1994).

Breeding: Courtship and mating are most common in the spring. Breeding activity tapers off during the summer but may take place during early autumn. A courting male will nip at the edge of the upper shell of a female for a short time then mount her from behind and either clasp onto her carapace with his forelimbs or rest the rear of his shell on the ground and face skyward. The male's hind limbs grasp the female's plastron, her hind limbs wrap around his, and mating commences. The pair may remain in this position for up to two hours. Gravid females will locate an exposed area with loose soil or sand, dig a flask-shaped hole with their hind limbs, and deposit their eggs at the bottom. Nest burrows are from to 38 to 71 mm (1.5 to 2.8 in.) deep. Nesting will normally occur between mid-May and mid-June. Two to eight elongated, slightly brittle, white eggs may be laid per female. Often the female will empty her bladder just before or during egg laying. Their eggs are carefully covered with dirt or sand and left to incubate on their own. Most females will produce one clutch of eggs per season, but Collins (1993) reported that up to a third of adult females may lay a second clutch. Ornate box turtle eggs average 36 mm (1.4 in.) in length (Legler 1960). They hatch in two to three months and the hatchlings average 30 mm (1.2 in.) in carapace length. Newly hatched ornate box turtles have a round upper shell and are usually dark brown with medium-to-large yellow spots or irregular markings on some of the scutes, and a yellow dorsal stripe. Their plastron is yellow with an irregular brown central marking. The tail of a hatchling may be nearly half the length of the carapace. Hatchlings are unable to close their lower shell. It may take up to four years of growth before the plastron hinge becomes fully functional (Ernst and Lovich 2009). Freshly laid eggs of ornate box turtles are often eaten by raccoons, skunks, and badgers.

A study of ornate box turtles in Kansas noted that males are mature at a plastron length of 100 to 109 mm (3.9 to 4.3 in.) and eight to nine years of age (Legler 1960). Females were found to be mature at a plastron length of 110 to 119 mm (4.3 to 4.7 in.) and 10 to 11 years of age. A study in Texas reported that males mature at age seven and females at age eight (Blair 1976). The oldest specimen Blair (1976) found was 32 years old, and he discovered that the population had an almost complete turnover every 32 years. He concluded that it is improbable that ornate box turtles could live to 100 years of age. A Kansas study of 115 adults found that none lived longer than 27 years (Metcalf and Metcalf 1970). In general, most ornate box turtles likely live 30 years, with a few living up to 50 years.

Remarks: In recent years, two subspecies have been recognized for the ornate box turtle: the desert box turtle (*T. o. lutelola*), found in the southwestern U.S., and the Plains box turtle (*T. o. ornata*), found in Midwest and Great Plains states. Limited genetic information does not support the distinction of these two subspecies, and Missourians are most familiar with the ornate box turtle name. Thus, the authors kept the original common name, ornate box turtle, in this publication.

The future health of Missouri's population of ornate box turtles is of concern. Due to a lack of extensive movement of this species within a given area, habitat destruction and fragmentation can reduce the genetic viability of this species and other prairie animals (Bowen et al. 2004; Doroff and Keith 1990; Richtsmeier et al. 2008). The continued efforts to protect and increase the availability of native tallgrass prairie in Missouri is vital for this small reptile's survival. The periodic burning of native tallgrass prairies and savannas has become a normal management practice. However, if prescribed burning is scheduled during the active season of ornate box turtles, mortality will be certain (Frese 2003b). These reptiles cannot outrun a prairie fire. Such fires need to be scheduled in late fall and early winter to protect this species and other prairie-dwelling wildlife. Huge numbers of box turtles are killed on roads and highways each year in Missouri. Anderson (1965) found a local population decline of this species due to road mortality. For 29 years he kept track of the number of live ornate box turtles found along a stretch of highway during late May and early June. The species showed a noticeable reduction in numbers during that span of time. In addition, both species of Missouri's native box turtles need to be protected from illegal collecting for the worldwide pet trade.

Red-eared Slider

Trachemys scripta elegans (Wied-Neuwied)

JEFF BRIGGLER

Adult female red-eared slider from Butler County.

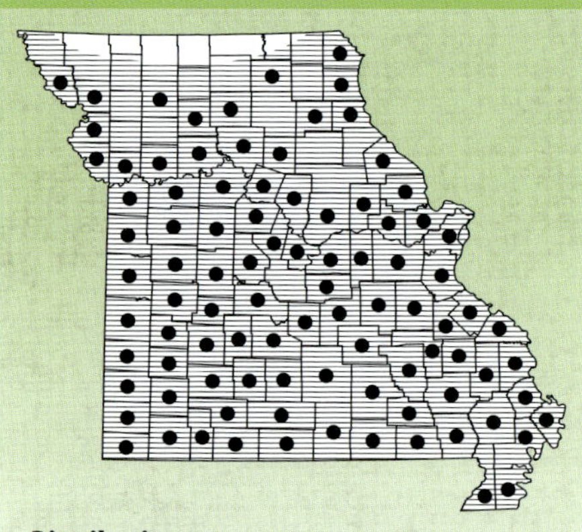

Distribution: Missouri: Statewide except for a few northern counties. North America: Southeastern Iowa, most of Illinois, western Indiana, scattered isolated populations in several states to New Jersey, most of Kentucky and Tennessee, Missouri, eastern, central, and southwestern Kansas, south to eastern New Mexico, through most of Oklahoma and Texas, all of Arkansas, Louisiana, Mississippi, and northwestern Alabama (Powell et al. 2016).

Description: The red-eared slider is a medium-sized, aquatic turtle with a patch of red on each side of its head. The carapace is olive brown with numerous black and yellow lines. The plastron is yellow, with each scute normally having a large dark-brown or black blotch. Its exposed skin is dark green with narrow black and yellow lines. A wide red or orange stripe is present on each side of the head behind the eye. Old individuals sometimes have an excess of black pigment that obscures most of the yellow stripes on the shell and skin and the red stripe behind the eye. This condition is known as melanism and is associated with old males. Adult male red-eared sliders have longer, thicker tails, longer foreclaws, and smaller shells than adult females.

Adult red-eared sliders range in carapace length from 125 to 203 mm (4.9 to 8 in.) but have been known to reach 302 mm (11.9 in.; Powell et al. 2016).

TOM R. JOHNSON

JEFF BRIGGLER

Melanistic male red-eared slider from Stone County (above). Red-eared slider plastron from Butler County (left).

Habits and Habitat: This species occurs in a wide variety of aquatic habitats, but prefers a mud bottom, plenty of aquatic plants, and abundant basking sites. Both natural waters (rivers, sloughs, and oxbow lakes) and man-made waters (ditches, ponds, and reservoirs) are used by this turtle. Red-eared sliders living in the Mississippi River prefer slow, muddy habitats (Braun and Phelps 2016). This turtle becomes active in March, when the air temperature reaches 10°C (50°F) or higher and remains active until mid-October through most of the state. A population of sliders studied in cooler climates in northeastern Missouri were active from late April to mid-September (Kofron and Schreiber 1987). Foraging takes place in early morning and late afternoon. On sunny days, basking in the sun on logs or other objects projecting from the water begins in midmorning and generally lasts until midafternoon. Sometimes a log is covered with many basking sliders. At night, red-eared sliders will sleep while resting on the bottom or floating on the surface. They are often seen moving about on land, especially males moving between ponds and females searching for nesting sites. Individuals overwinter underwater buried in the mud or under a cut bank. This species is usually the most abundant turtle observed in Missouri, especially in ponds and lakes. A study of a lake and borrow ditch in southeastern Missouri captured 785 sliders during one active season (Glorioso et al. 2010).

Hatchling red-eared slider from Washington County (above). Red-eared slider nest with 19 eggs from Clark County (right).

They eat a wide range of both aquatic plants and animals. Young individuals often consume more animal foods, but as they age, their diet shifts to more plant foods (Bouchard and Bjorndal 2006). Because sliders and several other species of turtles eat a variety of plant materials, they aid in the dispersal of plant seeds (Kimmons and Moll 2010). During periodical cicada eruptions, sliders and other turtles feed heavily on this available and abundant resource (Powell and Powell 2011).

Breeding: Courtship and mating take place between mid-March and mid-June. Several authors have stated that this species breeds in autumn also (Anderson 1965; Cagle 1950). The courtship ritual is similar to that of the western painted turtle. Females ready to lay their eggs leave the water and search for a suitable egg-laying site. Throughout their range, clutch sizes from 1 to 30 (average 10) eggs are laid per female between April and mid-July, with most eggs laid in May and June in Missouri. Females in Clark County have been observed depositing eggs in late May and early June, and in Cole County in mid-May (J. Briggler unpubl. data). One of the nests found in Clark County contained 19 eggs. Thomas (1993) found that annual reproductive potential was 22 to 27 eggs per female in Lake Taneycomo. Some females may lay more than one clutch during a season (Anderson 1965). The eggs are oval with a white, granular shell; they range from 25 to 40 mm (1 to 1.6 in.) in length (Ernst and Lovich 2009). Hatching usually takes place in late summer or early autumn.

Red-eared sliders are commonly observed basking on logs in ponds, lakes, sloughs, and rivers throughout Missouri.

The incubation period was 68–70 days for eggs hatched in a lab (Trauth et al. 2004). Young turtles usually remain in the nest and emerge the following spring. Newly hatched young from Clark County ranged from 28 to 39 mm (1.1 to 1.5 in.) with an average of 32 mm (1.3 in.; Kangas 2007). In a Missouri study of turtles in a cove of Lake Taneycomo, males generally matured at four to five years of age and females at seven to eight years (Thomas 1993). However, a slider population in a cold water cove took considerably longer to mature; seven to eight years for males and 13 to 14 years for females (Thomas 1993). Individuals have been reported to live up to 37 and 49 years of age in captivity (Ernst and Lovich 2009; Snider and Bowler 1992).

Remarks: Millions of baby red-eared sliders were sold as pets, most of them dying due to lack of proper care. The sale of these turtles was curtailed because of possible salmonella contamination that could be transmitted to humans by handling the turtles or the water in which they were kept. Red-eared slider hatchlings are still being produced by turtle "breeders" in several southern states. The majority of these young turtles are destined for the pet trade in Europe and Asia. Many are eventually released or escape to the wild and are causing a decline of native species, especially in southern Europe.

The yellow-bellied slider (*T. s. scripta*), which is native to southeastern states, is increasingly being found in Missouri, especially at high public use conservation lands. It appears that most of these animals were released pets. Efforts to remove this slider subspecies are being taken to reduce impacts to native turtles of Missouri.

Family Trionychidae

Softshells

Softshells occur in North America, Asia, and Africa. The family has 31 species in 13 genera (Pough et al. 2018). Only one genus, *Apalone*, with three species, occurs in the United States. External characteristics that separate turtles in this family from others include a round, flat appearance, carapace and plastron covered with skin, lack of any scutes, a long, tubular snout, and extensive webbing on both fore- and hind limbs. Missouri has two species of softshells. These turtles live primarily in rivers but also are known to occur in large reservoirs. To defend themselves, softshells compensate for the lack of a hard shell and scutes by having strong jaws and by being fast swimmers. Softshells also use their strong, sharp claws to defend themselves when picked up. Live softshells should be handled very carefully to avoid injury. These turtles are classified as a game species in Missouri. The Missouri Department of Conservation regulates a season and bag limit on them. These reptiles are economically valuable as a human food source worldwide; the meat of softshells is considered delicious.

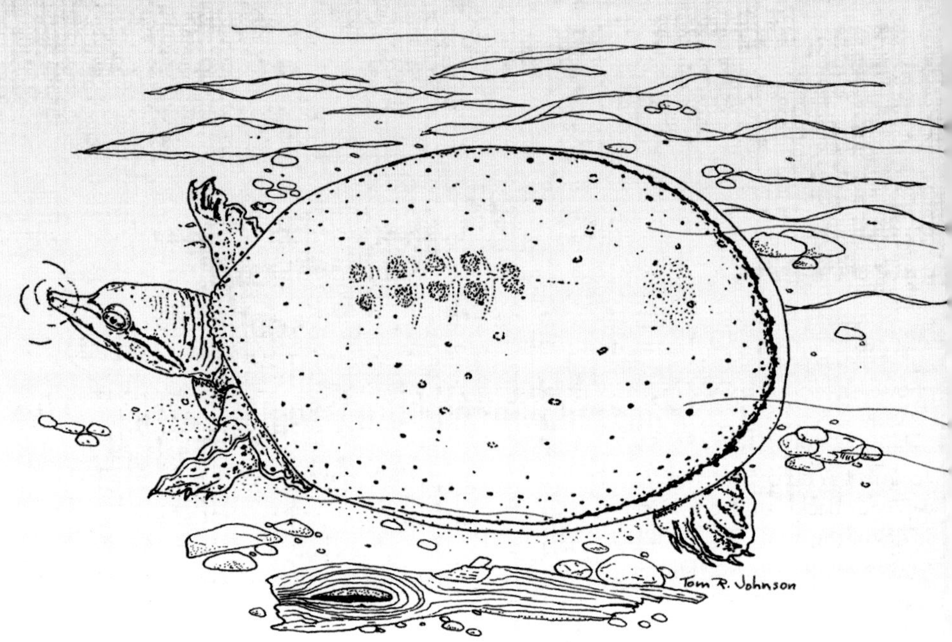

Eastern Spiny Softshell

Midland Smooth Softshell

Apalone mutica mutica (LeSueur)

Adult midland smooth softshell from Cole County.

Description: The midland smooth softshell is a rather plain-looking member of the softshell group. The front of the carapace lacks any small bumps or spines and the overall shell is quite smooth. The carapace may be olive gray or brown; males and young may have faint markings in the form of dots and dashes. Adult females have a mottled carapace with blotches of gray, olive, or brown. The plastron may be gray or cream colored and lacks any markings; however, the underlying bones are usually visible. Upper surfaces of neck and limbs are olive or gray; underside is white or cream colored. A light stripe bordered by black extends backward from each eye along the sides of the head. Adult males are smaller, have a longer, thicker tail, and longer foreclaws than females. Females have longer hind claws.

Adult female midland smooth softshells range in carapace length from 165 to 356 mm (6.5 to 14 in.); males range from 115 to 266 mm (4.5 to 10.5 in.; Powell et al. 2016).

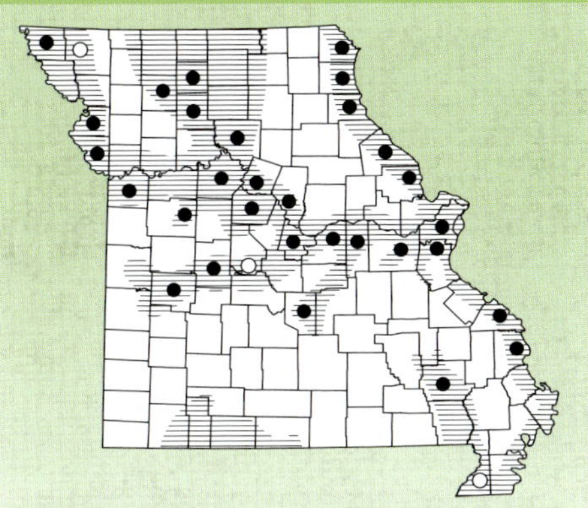

Distribution: Missouri: Nearly statewide; primarily in the Missouri and Mississippi rivers and their major tributaries, but uncommon in the Ozarks. North America: Western Pennsylvania, Ohio River drainage to southern Illinois, central and western Wisconsin, southeastern Minnesota, North and South Dakota, eastern Nebraska, Kansas, Arkansas and into Missouri, most of Oklahoma, and Louisiana, eastern New Mexico and Texas (Powell et al. 2016).

Midland smooth softshells use sandbars for basking and egg laying, such as the Missouri River (above). The flattened body of the midland smooth softshell (right).

Habits and Habitats: Midland smooth softshells live mainly in rivers and larger streams, especially the Missouri and Mississippi rivers, usually with moderate to fast current, and abundant sand or mud (Bodie et al. 2000). They are powerful swimmers. In the lower Mississippi River of southeast Missouri, they are most abundant along the border of the main channel and open side channels (Barko and Briggler 2006; Braun and Phelps 2016). A study in Kansas reported that most individuals use a 2–3 km (1.2–1.9 mi) section of river (Plummer and Shirer 1975). The midland smooth softshell is active from early April to mid-October in Missouri. It is more aquatic than other softshell species and spends considerable time in the water. When basking, it will usually do so on a gentle shoreline slope along a sandbar or mud flat within 1 m (3.3 ft) from the water's edge. The slightest disturbance will cause it to quickly dash into the water for protection.

Softshells are active during the daylight hours and search for food during the morning and late evening (Plummer and Shirer 1975). When resting they will bury themselves in mud or sand in shallow water with only the head and neck exposed. The turtle buries itself by diving head first into loose substrate using its powerful front and hind limbs to get deeper within the substrate. The turtle then stirs up the bottom substrate by a shuffling motion that allows the loose materials to settle over the turtle; concealing the shell with only its head slightly visible (Ernst and Lovich 2009). Air is taken in by protruding the tip of the snorkel-like snout out of the water. In this manner, the turtles are inconspicuous. These turtles also can remain submerged and carry on gaseous exchange from the water by pumping water in and out of the mouth and the cloaca. Softshells protect themselves from freezing weather by burying themselves in the mud of the river bottom.

This species is primarily carnivorous, feeding mainly on fish, crayfish, salamanders, tadpoles, frogs, snails, and aquatic insects (Ernst and Lovich 2009). They mainly sit and wait under loose substrate with only the head partially exposed to ambush prey by extending their long neck. Plummer and Farrar (1981) reported on the food habits of smooth softshells in eastern Kansas. Male and female turtles were found to differ significantly in types of food eaten. Up to 71 percent (by volume) of prey eaten by females consisted of aquatic insect larvae captured in deep water. Male softshells consumed terrestrial insects (67 percent by volume) that fall into the water along the shoreline. Softshells will also eat plant materials such as mulberries and cottonwood seeds when available (Trauth et al. 2004).

Breeding: Smooth softshells breed in April and May after emergence from overwintering sites. During the breeding season, a male will rapidly swim after a female with his neck extended. He will occasionally probe the underside of her shell with his head. If the female is receptive, she will become still and the male will mount her from above (Collins et al. 2010).

Egg laying takes place from late May into July. The number of eggs produced depends on the size of the female, and typically two clutches are laid per season. Later clutches within a season generally contain fewer eggs. Three to 26 eggs have been reported in a clutch, with an average clutch size of 18 eggs (Trauth et al. 2004). A female will select a sandbank, sandbar, or river island with exposure to the sun to dig her nest (Ernst and Barbour 1972). Most nests are located within 30 m (98.4 ft; range 4–90 m; 13.1–295.3 ft) of the water on higher ground (about 1–1.5 m; 3.3–4.9 ft) above the water level (Ernst and Lovich 2009). The female digs a cavity between 15 and 30 cm (5.9 and 11.8 in.) deep with her hind limbs, deposits her eggs, and covers them with soil. The spherical, white eggs generally hatch in August or September with an average incubation period of 68–70 days (Anderson 1965;, Trauth et al. 2004). Hatchlings average about 32 to 44 mm (1.3 to 1.7 in.) in carapace length. Young of this species are olive with faint distinct dots and short lines covering the carapace. Sexual maturity takes about six to nine years for females at a carapace length of 170 to 220 mm (6.7 to 8.7 in.). The smaller males mature at a carapace length of 110 to 126 mm (4.3 to 5 in.) in about four years (Ernst and Barbour 1972). Longevity of the species in the wild is unknown, but it is generally assumed to be more than 20 years.

Remarks: This species is somewhat of a habitat specialist (large rivers with sandbanks and sand islands for egg laying) and occurs at relatively low numbers in Missouri. In recent years, smooth softshells appear to be declining in a number of midwestern states. Channelization of rivers (altering the river flow and reducing egg laying sites), water pollution, increased predation of nests, and overharvesting has likely impacted this species (Tracy-Smith et al. 2012).

Both species of softshells are considered game species in Missouri. Recent studies have shown that unlimited harvest of softshells is not sustainable, and regulations were enacted in 2018 to eliminate the unlimited, commercial harvest of softshells in Missouri (Shaffer et al. 2017; Zimmer et al. 2014). There is a season and daily bag limit to allow a small number of softshells to be captured on hook-and-line by recreational anglers for local food, however. These changes will help to ensure that softshells remain common in Missouri rivers.

Although softshells may prey upon nearly any species of fish, there is no evidence to show that they harm fish populations in natural waters. Smooth softshells are less inclined to bite when captured compared to spiny softshells.

Eastern Spiny Softshell

Apalone spinifera spinifera (LeSueur)

JEFF BRIGGLER

Adult eastern spiny softshell from Butler County.

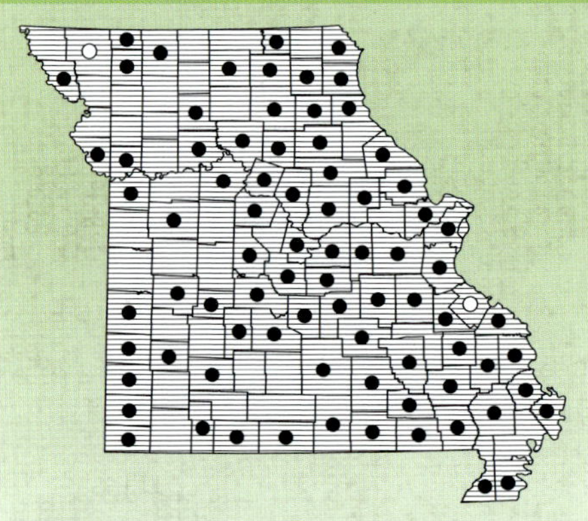

Distribution: Missouri: Statewide. North America: Southern Quebec, south to northwestern Florida and from eastern Minnesota and much of Wisconsin south to Gulf Coast states and west to much of Texas. There are five subspecies and isolated populations in several additional states (Powell et al. 2016).

Description: The eastern spiny softshell is a medium-to-large turtle with dark spots on fore- and hind limbs, a small ridge on each side of the snorkel-like snout, and small bumps or spines on the front of the carapace. The color of the carapace varies with sex and age. Young females and males of any age have an olive or gray-tan carapace with a black line along the margin and distinct small black dots and circles. The circular markings (known as ocelli) may have a dark center. Adult females have a dark-olive or tan carapace with brown and gray blotches. Spiny softshells have a plain, cream-colored plastron; underlying bones can be seen through the skin. Upper surface of the head, neck, and limbs are tan or olive with small brown or black markings; the throat is grayish white with small, dark-gray markings. A yellow stripe, bordered by dark brown, extends from the snout through each eye and along the sides of the head. Another light stripe runs from the angle of the jaw onto the neck. Adult males are smaller than

Female eastern spiny softshell digging a nest in early June on the shoreline of the Big Piney River in Pulaski County (above). Eastern spiny softshell egg from Cole County (left).

females; they also have dark spots and small bumps on the carapace and a large, thick tail with the anal opening well beyond the edge of the shell.

Adult female eastern spiny softshells range in carapace length from 180 to 540 mm (7.1 to 21.3 in.); males range from 125 to 310 mm (4.9 to 12.2 in.; Powell et al. 2016).

Habits and Habitat: The eastern spiny softshell resides primarily in large rivers and streams but also inhabits lakes, large ponds, and even roadside and irrigation ditches (Bodie et al. 2000). It is most common in tributaries and closed side channels of the Mississippi River (Barko and Briggler 2006; Braun and Phelps 2016). A wide variety of substrates, such as muddy, sandy, or gravel-covered bottoms are used. A study in Arkansas reported that most individuals generally move within a 2.5 km (1.6 mi) portion of stream (Plummer et al. 1997). The eastern spiny softshell is active between late March and October. Daily activity includes foraging for food in early morning, basking in the sun on logs or along the bank, and resting in shallow water with the shell covered by mud or sand. There is a short period of feeding during the afternoon and evening that may extend into the night. Most of the night is spent sleeping buried in the bottom substrate or among tree limbs or stumps (Ernst and Lovich 2009). Spiny softshells have a habit of floating near the surface in deep water but will quickly dive to safety with the slightest disturbance. To escape the cold temperatures of winter, this species will bury itself in 50.8 to 101.6 mm (2 to 4 in.) of mud or sand at the bottom of a river or lake starting in October or November (Ernst and Lovich 2009).

This species eats a variety of aquatic prey, including crayfish, insects, earthworms, snails, small mussels, fish, and tadpoles. Stomach analysis of turtles captured on the Osage River, St. Clair County, contained mainly crayfish and aquatic insects with an occasional fish (Anderson 1965).

Elevated sandbars along the river edge are preferred basking and egg laying sites, such as this example in Platte County.

JEFF BRIGGLER

Breeding: Courtship and mating occur in April and May. The courtship behavior is similar to that of the smooth softshell. Eggs are laid from late May through July. A gravid female will select a sand or gravel bar, or a sandy opening, usually near water to nest, but suitable nest sites can be a considerable distance from water (Daigle et al. 2002). A female was observed digging a nest in late June in fine gravel along the Niangua River, Camden County (Anderson 1965). A female digs a flask-shaped chamber between 10 and 25 cm (3.9 and 9.8 in.) deep with her hind limbs (Ernst and Lovich 2009). Three to 39 (average 18) round, white eggs may be laid, with larger females usually laying more eggs (Ernst and Lovich 2009). One, two, and potentially three clutches may be laid each season. Anderson (1965) reported 19 eggs taken from a female spiny softshell caught by anglers in early July in St. Clair County. Eggs of this specimen averaged about 28 mm (1.1 in.) in diameter. Hatching normally occurs from late August to October with an incubation period of 65–85 days (Ernst and Lovich 2009; Trauth et al. 2004). The young turtles are 30 to 40 mm (1.2 to 1.6 in.) in carapace length at hatching. Male spiny softshells become sexually mature at a carapace length of 125 to 140 mm (4.9 to 5.5 in.); females mature when their carapace reaches 250 to 280 mm (9.8 to 11.0 in.; Trauth et al. 2004). Maturity is about four years for males and eight to nine years for females (Trauth et al. 2004). A female spiny softshell was reported to live 25 years in captivity (Snider and Bowler 1992).

Remarks: Once two subspecies of *Apalone spinifera* were recognized in Missouri: the nominate subspecies, the eastern spiny softshell (*Apalone s. spinifera* LeSueur), and the western spiny softshell (*A. s. hartwegi* Conant and Goin). Molecular data showed these two subspecies are the same, and therefore *A. s. hartwegi* was placed into the *A. s. spinifera* group (McGaugh et al. 2008).

A comparative study of the forage habits of smooth and spiny softshells was undertaken in Iowa (Williams and Christiansen 1981). It was found that spiny softshells are less specialized in their habitat requirements than smooth softshells. Spiny softshells searched for food on the bottom, while smooth softshells located food in the water column.

Spiny softshells are a game species in Missouri with a season and daily limit. Recent regulation changes to reduce harvest will ensure that softshells will remain common throughout Missouri. Unlike the smooth softshell, spiny softshells are defensive and will try to bite when captured.

(Order Squamata)

Lizards and Snakes

(Suborder Sauria)

Lizards

Because lizards and snakes have a number of similar characteristics, these two types of reptiles are grouped into one order (Squamata) and placed in separate suborders (Sauria: lizards; Serpentes: snakes). About 60 percent of the reptile species residing on earth are lizards (over 6,500 species) and snakes (over 3,700 species; Uetz et al. 2020). Lizards live on every continent except Antarctica (Pough et al. 2018). Lizards are grouped into 41 families (Pough et al. 2018). The general concept of lizards as a group is an animal with four limbs, a long tail, eyelids, ear openings, scales, claws, and the capability of quick movement. Although this description fits most of Missouri's lizards, it fails to describe many that occur elsewhere. There are many tropical lizards, known as geckos, that have no eyelids—the eyes are instead protected by a clear, nonmovable eye cap. Some lizards, such as the horned lizard (genus *Phrynosoma*), have very short tails. Some species of skinks have prehensile tails. Several types of lizards have no limbs, such as slender glass lizards (genus *Ophisaurus*). The lower jaw of both lizards and snakes is attached to the upper jaw by a special bone, known as the quadrate bone, which allows the jaw to open wide when prey is being swallowed. In lizards, however, the front of the lower jaw is fused, unlike in snakes, and cannot expand to swallow large prey.

Many lizards also have evolved special tail vertebrae that enable the tail to be easily broken off if grabbed by a predator. This is an effective means of self-defense. There are special muscles along the tail that constrict at the break and prevent any blood loss. A new tail will eventually replace the broken section but will not have the same color or scale pattern as the original tail. The regenerated tail is supported by a cartilaginous rod rather than vertebrae. Most of Missouri's lizards can lose their tail to escape a predator, except eastern collared lizards (*Crotaphytus collaris*) and Texas horned lizards (*Phrynosoma cornutum*).

There are only three species of lizards in the world that are venomous: the Gila monster (*Heloderma suspectum*) occurs in the southwestern United States and Sonora, Mexico; the Mexican beaded lizard (*H. horridum*) lives in western Mexico and northern Central America; and the Komodo dragon (*Varanus komodoensis*) is found on several Indonesian islands. The Komodo dragon is the largest living lizard, reaching total length of 3 m (9.8 feet). Missouri has 11 native species of lizards with two additional subspecies. These are classified into five separate families. In addition, two nonnative (introduced) lizards from two additional families have become established in Missouri. Missouri's lizards are nonvenomous and beneficial to people because they eat a variety of insects.

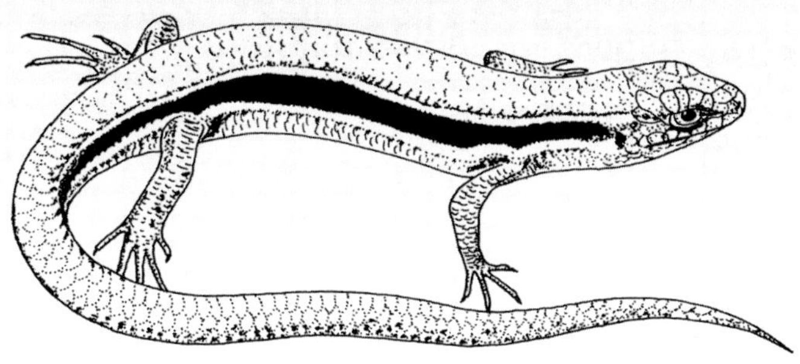

Southern Coal Skink

Family Crotaphytidae

Collared and Leopard Lizards

Family Crotaphytidae includes 12 recognized species consisting of two genera (Pough et al. 2018). This group is restricted to North America from eastern Missouri to the west coast and south into northern Mexico (Pough et al. 2018). These lizards have large heads, robust bodies, and the ability to run on their hind legs. They require a dry, open habitat. In Missouri, this includes rocky outcrops (glades) in open forest and sparsely vegetated fields. Other characteristics of this family include teeth located on the inner surface of the lower jaw and a form of communication employing head bobbing and body pushups. Males use these gestures to defend small territories or to court a female. Lizards in the family defend themselves by biting and scratching. Missouri has one species (eastern collared lizard) in the family Crotaphytidae.

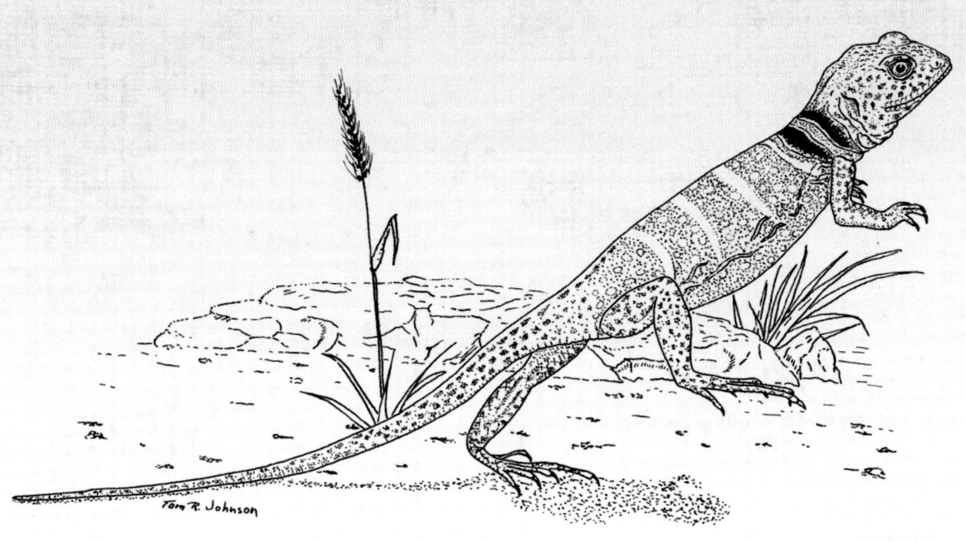

Eastern Collared Lizard

Eastern Collared Lizard

Crotaphytus collaris (Say *in* James)

Adult male eastern collared lizard from Wright County.

Description: The eastern collared lizard is a colorful, long-tailed lizard with a large head. The color is most conspicuous on males during the breeding season (May and June); general ground color is tan, yellow, green, or blue green. There are usually a number of small light spots scattered over the upper body and limbs and dark bands across the dorsum. Both males and females have two dark-brown or black irregular lines across the neck producing a "collar." Females are yellow tan or light brown with faint light spots. Females also have black dots along each side of the cloaca that are absent in males. Females heavy with eggs have red spots or bars on the sides of the body and neck (Ivanyi 2009). Newly hatched young have dark bands and yellowish crossbars.

Adult eastern collared lizards range in total length from 203 to 356 mm (8 to 14 in.; Powell et al. 2016).

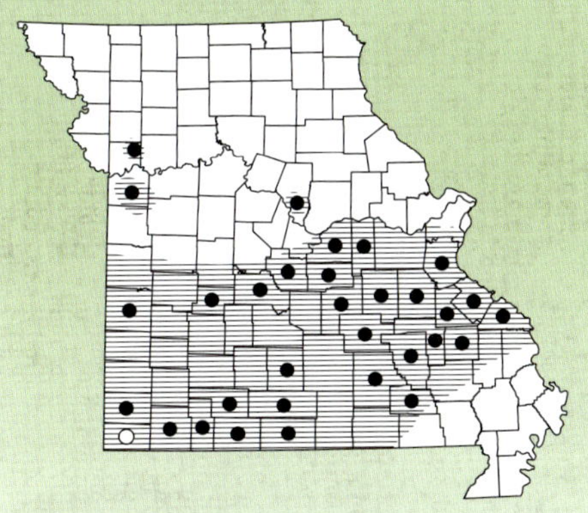

Distribution: Missouri: Throughout most of the Missouri Ozarks and on glades of the St. Francois Mountains. Relict populations occur south of the Missouri River within the Kansas City area. Populations occurring north of the Missouri River appear to be introduced. North America: Missouri south through northern and western Arkansas, west to Utah and Arizona, and south into Mexico (Powell et al. 2016).

Habits and Habitat: In Missouri the eastern collared lizard lives among rocks on dry,

Eastern collared lizards live among the rocks on glades in the Ozarks, such as this example in Carter County (above). Adult female eastern collared lizard from Douglas County (right).

JEFF BRIGGLER

JEFF BRIGGLER

open, south- or southwest-facing limestone, sandstone, and granite glades. They can also be found in human-altered landscapes such as rock quarries and rocky riprap along reservoir dams. Adult males prefer to observe the world from a perched location on a large boulder.

As with all native lizards in Missouri, the eastern collared lizard is active by day, especially when the weather is sunny and warm. The preferred air temperature for this species is between 73° and 93°F. Seasonal activity usually lasts from April to September for adults, with a peak from May to late July, but young lizards may remain active through October (Angert et al. 2002; J. Briggler unpubl. data). A great deal of time is spent basking in the sun on exposed rocks, but these lizards are quick to take shelter under large rocks or in rock crevices when approached. If caught in an open area, a collared lizard can run very fast to escape; it will often run on its hind limbs with the forward part of the body held upright. Each lizard defends a home territory by chasing away other collared lizards when a territory is violated. The bright colors of males are used to ward off other males from invading an individual's home range. A study in central Oklahoma found that young males use several types of avoidance behavior when traveling into or through a dominant male's home range. They try to remain hidden, move as little as possible, and refrain from display head-bobbing

JEFF BRIGGLER

A hatchling eastern collared lizard from Dent County.

or pushups. If spotted by the dominant male, young males retreat immediately. It was found that the young males who are successful living in or near a dominant male's home range will have a much better chance of taking over the site if the male leaves or is removed by a predator (Baird et al. 1996). Long-term population studies of the collared lizard in Shannon County documented that most glades can generally support 15–25 animals, but some population estimates for glades can exceed 50 animals with favorable conditions (Templeton et al. 2011).

Collared lizards overwinter in burrows at depths of 20.3 to 30.5 cm (8 to 12 in.) under large rocks with a small tunnel leading to the edge of the rock (Collins et al. 2010). Two adults were found together in Dent County in early September in a torpid state under a mostly embedded large rock (J. Briggler pers. obs.).

Collared lizards eat a variety of insects such as grasshoppers, beetles, and moths. They also eat spiders, small snakes, and other lizards (Best and Pfaffenberger 1987; McAllister 1985). This species is preyed upon by snakes, hawks, and, in southwestern Missouri, roadrunners. They will escape predators by running fast, seeking shelter, and even swimming at the water's surface (J. Briggler pers. obs.).

Breeding: Courtship and mating mainly take place from mid-May to early June. An adult male will court a female by showing off his brightly colored throat and body as he prances around her. This is normally followed by the male walking around and over the female, bobbing his head up and down as he moves. If the female is willing to mate, the male will mount her back, grasp a few folds of her neck skin in his mouth, and move the rear part of his body beneath her tail, at which time copulation takes place. A study in Oklahoma by Baird

(2004) showed that reproductive coloration of females increases courtship behavior by males. A courtship and mating event was observed in late June on a glade in Reynolds County. A female eastern collared lizard was seen approaching a male sitting in the sun on a large bolder. When the female was close to the male, he grasped her by the neck and twice moved over her back and copulation began. At one point the female moved on top of the male, slid her head under his belly, and moved beneath him, at which time they mated again. Within a four-minute period the two lizards copulated four times (Salveter and Overby 1998). Four to 24 creamy white, leathery eggs are laid by a female. However, the number of eggs seldom exceeds 12 (Anderson 1965; Fitch 1956b). The average clutch size is seven to eight eggs. Some females are known to deposit a second clutch within a single season (Anderson 1965). Egg laying usually takes place within 20 days after breeding; eggs are laid in a burrow that has been excavated beneath a large rock (Fitch 1956b). On two separate occasions in late June, a clutch of five eggs were found under a large rock with an excavated tunnel: one clutch in Maries County and another clutch in Douglas County (J. Briggler unpubl. data). Once the eggs are laid, the female plugs the nest's entrance with dirt to better protect the eggs from predators, allowing the eggs to develop and hatch on their own. Hatching takes place within two to three months (usually late August into September), and in Missouri, hatchlings average 38 mm (1.5 in.) in snout-to-vent length (Sexton et al. 1992). Hatchlings grow rapidly and can double in size prior to winter. Reproductive maturity is reached at 77 mm (3 in.) snout-to-vent length (Sexton et al. 1992). With plentiful food and resulting rapid growth, individuals may reproduce within their first or second year of life. The lifespan of this species rarely exceeds five years.

Remarks: A popular common name for this lizard is mountain boomer, implying this lizard makes a booming sound. This name probably originated in southwestern United States where settlers may have seen the lizard on rocks while hearing the barking call of a local frog species. In reality, the collared lizard is voiceless.

This species of conservation concern was once quite rare in the state due to lack of management of glade habitat throughout Missouri. Glades became overgrown with cedars due to lack of periodic fires, and many populations of collared lizards were greatly reduced or disappeared because of unsuitable environmental conditions. A study of this lizard in Missouri showed that collared lizards living in marginal habitats (overgrown glades) had a slower growth rate and smaller body size and found that young lizards had reduced activity. Also, these lizards showed a shorter seasonal activity, and female reproductive activity was shorter compared to populations in states to the west or south. In general, the lizard's glade habitat has been degraded by a lack of fire; the occasional natural fire prevented trees from overtaking the open areas required by this species. This, in turn, caused an overall lowering of mean temperatures during the lizard's active period and reduced prey availability (insects and lizards; Sexton et al. 1992). Wildlife managers and foresters worked to improve collared lizards' glade habitats throughout the Missouri Ozarks in the 1980s and 1990s. Habitat improvements resulted in population increases in existing locations and the successful introduction of lizards throughout many other locations in Missouri (Templeton et al. 2011; J. Briggler unpubl. data). Today, the eastern collared lizard appears to be quite common and secure in Missouri. Maintaining the open nature of glade habitat, minimizing human disturbance of rocks, and reducing the illegal collections of animals will help keep this species common in Missouri

Family Phrynosomatidae

Spiny and Sand Lizards

This family consists of nine genera with 148 species ranging from southern Canada, most of the United States, and through Central America (Pough et al. 2018). Most of these lizards prefer dry, open habitats with rocky outcrops or loose soils. Similar to the family Crotaphytidae, this family has teeth located on the inner surface of the lower jaw and communicates by head bobbing and body pushups. These gestures are used by males to defend their territories and court females. Horned lizards are represented by 17 species (Powell et al. 2016). These are small lizards with flat, stout bodies and short tails that prefer loose, sandy soils. Most are known for their ability to squirt blood from the eyes when stressed, and some species are distasteful to predators due to chemical properties from their ant prey. Spiny lizards, in the genus *Sceloporus*, consist of nearly 100 species (Pough et al. 2018). Most perch on rocky outcrops, large rotting stumps and logs, fences, and walls. Most males have blue patches on their belly and throat. Missouri has two species in the family Phrynosomatidae (Texas horned lizard and prairie lizard).

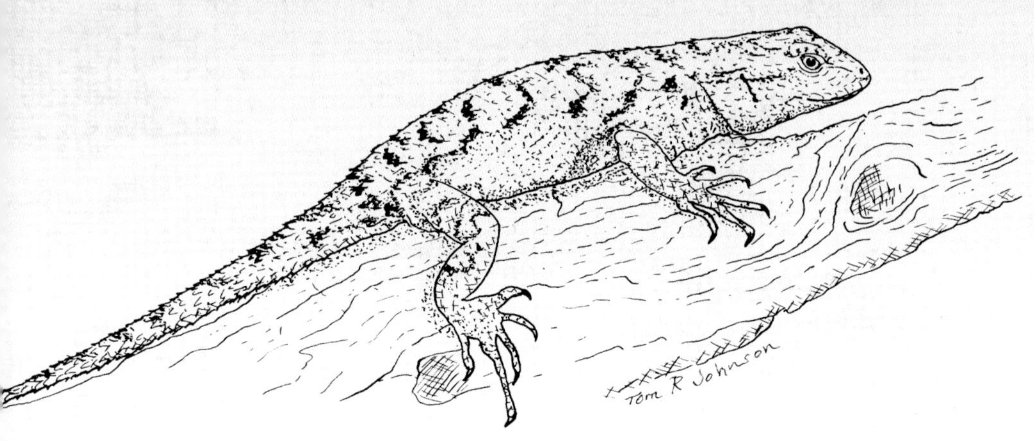

Prairie Lizard

Texas Horned Lizard

Phrynosoma cornutum (Harlan)

TOM R. JOHNSON

Adult Texas horned lizard from Vernon County.

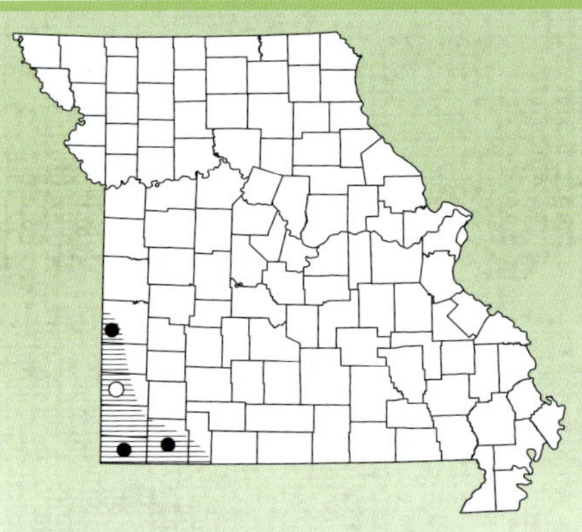

Distribution: Missouri: Southwestern corner of the state. North America: Northern Kansas, southeastern Colorado, southwestern Missouri to northern Louisiana, most of Oklahoma and Texas, southern New Mexico and southeastern Arizona, and into northern Mexico (Hodges 2009).

Description: The Texas horned lizard is a stocky, short-tailed lizard with several large "horns" protruding from the back of the head. Its general color is tan, grayish brown, or reddish brown. There are two large, dark-brown spots behind the head and a series of brown markings on the back. A white or yellow line extends down the center of the back. Scales on the limbs, sides, and tail are large and pointed; the head is heavily armored with large scales, some modified to form "horns." The belly is white with several small gray spots.

Adult Texas horned lizards range in total length from 64 to 100 mm (2.5 to 3.9 in.) but have been known to reach 181 mm (7.1 in.; Powell et al. 2016).

Habits and Habitat: Texas horned lizards prefer open, dry areas with loose or sandy soil and an abundance of rocks. This species is active by day as long as the sun is out and the temperature is high. The active season is from April to early October. When inactive,

Juvenile Texas horned lizard, likely released, from Jackson County (above). Head of Texas horned lizard (left).

the Texas horned lizard seeks shelter just below the surface of the soil. Horned lizards are occasionally seen along the edges of dirt or gravel roads. They are best observed during sunny mornings when they bask in open areas. Although this species will obtain water by licking dew from vegetation, it has evolved the ability to collect rainwater along its back. During a rain, it will arch its back upward, flatten its body, lower its head, and drink the water as it collects along its lips. Between the scales on their back are specialized, microscopic channels that help direct rainwater to their downward-pointing head (Sherbrooke 1990). A telemetry study in Texas found that individuals used from 291 to 14,690 m^2 (348.0 to 17,569.1 yd^2) of area with considerable overlap among individuals (Fair and Henke 1999).

Horned lizards eat primarily ants, especially harvester ants (*Pogonomyrmex* spp.) but occasionally eat other insects (beetles and grasshoppers) and spiders (Hodges 2009). Because of its specialized diet and need for high temperatures, this species is very difficult to keep in captivity.

Breeding: Courtship and breeding of the Texas horned lizard have not been observed in Missouri. Breeding presumably takes place soon after the active season begins (likely May and June). Females produce an annual clutch of 13 to 49 eggs (Degenhardt et al. 1996) with an average of between 23 to 25 eggs (Collins et al. 2010; Hodges 2009). The female digs a nest in loose soil or under large rocks in which the eggs are deposited. The eggs will hatch within one to two months (Collins et al. 2010). Hatchlings are 29 to 32 mm (1.1 to 1.3 in.; Conant and Collins 1991) in total length. They have been known to live up to 10 years of age in captivity (Baur 1986).

Remarks: Unlike many other species of lizards, the Texas horned lizard is not capable of breaking off its tail when grabbed by a predator. They rely on their cryptic coloration, ability to burrow under the soil, quick sprinting and freezing behavior, and an unusual habit of squirting blood to avoid and escape predators. They can squirt a few drops of blood from the corners of the eyes when captured. Basking lizards have been observed emitting a small amount of blood from their eyes when the lizard's head becomes warmer than the rest of the body (Collins et al. 2010). A study of this behavior found that canid predators (dogs, coyotes, and foxes) could stimulate captive lizards to squirt blood from their eyes in 80 percent of the trials (Middendorf and Sherbrooke 1992).

The Texas horned lizard is listed as a species of conservation concern in Missouri due to extremely low numbers and little suitable habitat. This lizard has been very popular in the pet trade for many decades and that has resulted in over collecting and declines throughout their range. Today, most southwestern states no longer allow this species to be collected. There are only seven documented reports of this lizard in the state, with the most recent capture of a juvenile being found in July 2013, in Jackson County. A number of Missourians have brought live specimens from Texas or Oklahoma while returning from vacation trips and released them near their homes in several parts of this state (Johnson 2000). It remains to be seen if natural populations occurred in Missouri or if reported individuals are released pets primarily found in urban areas.

Prairie Lizard

Sceloporus consobrinus Baird and Girard

JEFF BRIGGLER

Adult female prairie lizard from Callaway County.

Description: The prairie lizard is a small, gray-to-brown, rough-scaled species. General color may be tan, gray, brown, or red brown. Male and female prairie lizards differ in color. Males are either dark gray or brown on the dorsum with little or no pattern. The belly is strongly marked with deep, iridescent blue bordered with black. The throat also is blue. Females have distinct wavy lines crossing the dorsum, some orange or red at the base of the tail, and a white belly with faint dark spots. There may be some pale blue along the sides of the belly.

Adult prairie lizards range in total length from 90 to 191 mm (3.5 to 7.5 in.; Powell et al. 2016).

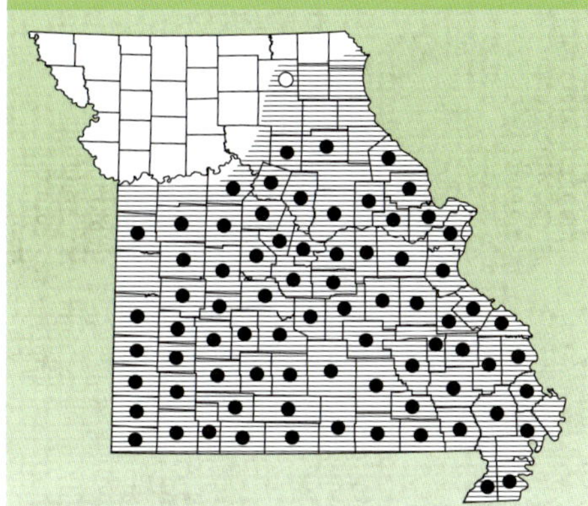

Distribution: Missouri: Occurs throughout the southern half of the state and into northeastern Missouri in the Mississippi River Hills. North America: Mississippi west through Texas into New Mexico, north to extreme southern South Dakota, southeast to Mississippi River in eastern Missouri (Powell et al. 2016).

Habits and Habitat: In Missouri, prairie lizards live in open woodlands or along the edges of woods and fields and on rocky glades. Tree stumps, downed trees, rock piles, and brush piles are often a part of this lizard's territory. They are commonly observed around homes where they utilize split rail fences, firewood

Male prairie lizard showing ventral coloration (above). Prairie lizards are often seen climbing trees within woodlands and along forest edges, such as this example in Callaway County (right).

piles, old lumber piles, decorative railroad ties, and rock walls. This species becomes active on sunny days, usually basking in the sun or foraging for insects between 8:00 a.m. and 1:00 p.m. It becomes inactive during the heat of the day, then active again during the late afternoon and evening. The seasonal activity of this species is normally from late March to mid-October (Marion 1970). Males defend a small territory that may include two or three adult females. When encountered, this lizard will run along a dead, fallen tree, escape into a rock crevice, or quickly climb a tree—always remaining on the opposite side from the intruder. Prairie lizards retreat into underground burrows, rotting tree roots and stumps, under large rocks, and within rock piles in the winter (Trauth et al. 2004).

This species eats a wide variety of insects but also eat spiders and other invertebrates.

Breeding: The reproductive season extends from March into early August. Courtship and mating begin in the spring soon after the adults emerge from winter retreats and will continue into summer. Males are territorial and display bright blue throat and belly colorations while bobbing their head and performing pushups. This behavior wards off other males, especially during the competitive mating season (Lahti and Leaché 2009). Mating occurs when the female allows the male to approach her. Egg laying generally occurs from mid-April into late July with females producing 4 to 17 eggs each.

The reproductive biology of this species has been studied in eastern Missouri (Marion 1970). The study showed that females can produce eggs when less than one year old, but only one clutch of eggs is laid the first season. Two-year-old females produce two clutches per season and are the major contributors to the population. There was no indication that any females laid a third clutch of eggs in the eastern Missouri population. The first clutch is laid during late May or early June and another is produced in July.

A gravid female will locate a suitable nesting site, such as a small patch of loose dirt or old sawdust pile, and dig a flask-shaped hole. She will deposit her eggs within the hole, cover

Prairie lizard hatchling from eggs in Cole County (above). Prairie lizard nest containing 15 eggs from Cole County (left).

them with dirt, and abandon the nest. Incubation of the eggs is generally six to eight weeks (Trauth et al. 2004), and hatchlings begin to emerge from nest in mid-summer through early autumn. In late May 1980, a large female was observed in Cole County digging a hole in an open area (Johnson 2000). Upon visiting the nest site the next morning, 15 eggs were observed in the nest. The eggs were deposited in five layers containing three eggs each, and each layer of eggs was covered with a small amount of dirt. The nest was 83 mm (3.3 in.) deep. The eggs were between 11.5 and 13 mm (0.45 and 0.51 in.) in length and were collected and incubated at an average temperature of 27°C (80°F). Hatching took place 52 days later in mid-July. Total length of hatchlings was between 46 and 49 mm (1.8 and 1.9 in.). Another partial clutch containing three eggs was collected in early July 2013 in Cole County (J. Briggler unpubl. data). These eggs hatched in late July. Smith and Powell (1993) documented 13 eggs in a Lincoln County nest that was 55 to 75 mm (2.2 to 3 in.) deep in sandy soil. The eggs hatched in mid-August. Sexton and Marion (1974) made observations on egg retention in this species. They reported that when the second clutch of eggs is delayed by being retained in the female, the incubation time is shortened; this allows the young to hatch sooner and have time to grow before the first frost. According to Marion (1970) few fence lizards live longer than three years under natural conditions.

Remarks: The prairie lizard was formerly named the northern fence lizard (*Sceloporus undulatus hyacinthinus*). Prairie lizards are common throughout Missouri, especially within open woodlands and along woodland edges. They are often the most common lizard observed by people around homes and buildings. A long-term study conducted by the Missouri Department of Conservation in the Ozarks showed an increase in number of prairie lizards immediately following tree removal that resulted in opening of the forest (Rota et al. 2017). Maintaining and restoring open woodlands, as well as reducing feral cats, will help keep this species common in Missouri.

Family Scincidae

Skinks

This is a large group of lizards with species on nearly every continent and many oceanic islands. There are over 1,580 species worldwide, representing about 115 genera (Pough et al. 2018). The smallest species may be a mere 76 mm (3 in.) long; the largest (in Australia) may reach a length of over 610 mm (24 in.). Most species have smooth, overlapping scales and live either on or under the ground. Five genera with about 15 species occur in the United States (Powell et al. 2018). A few nonnative species have been introduced into southern Florida. Missouri skinks are represented by two genera with a total of six species and one additional subspecies. All our skinks can quickly break off their tails if grasped by a predator. A new tail will regenerate but is usually dull gray brown.

Female Common Five-lined Skink

Southern Coal Skink

Plestiodon anthracinus pluvialis (Cope)

JEFF BRIGGLER

Adult southern coal skink from Iron County.

Description: The southern coal skink is a small, shiny, brown-tan lizard with broad, dark lateral stripes. Its general color is tan, brown, or olive brown. The broad lateral stripe can be brown or black; it is bordered by a thin light line above and below and ranges from two to four scales wide. There are no light stripes on the head. The chin normally has a single postmental scale, but occasionally some Missouri individuals were observed to have two postmental scales (Smith 1946; M.A. Nickerson pers. comm.). During the breeding season, adult males can be distinguished from adult females by the presence of dark orange on the sides of the head. Hatchling southern coal skinks are black with faint dorsolateral lines and a blue-gray tail.

Adult southern coal skinks range in total length from 125 to 178 mm (4.9 to 7 in.; Powell et al. 2016).

Habits and Habitat: The active season of southern coal skinks is mid-March to early

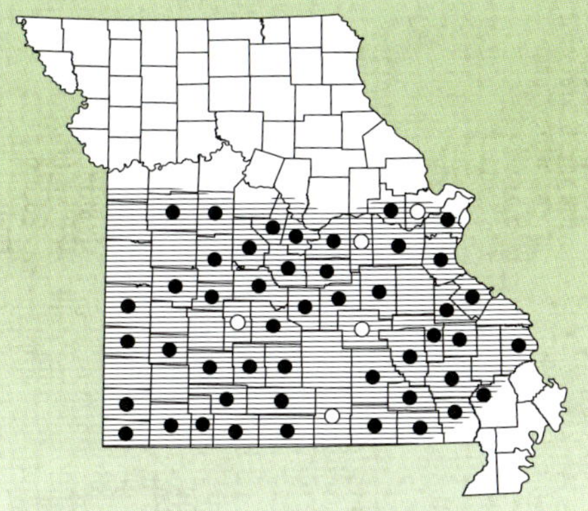

Distribution: Missouri: Restricted to the southern half of the state, except for the Bootheel. North America: Eastern Kansas, southern Missouri, all but the eastern edge of Arkansas, eastern and southern Oklahoma, eastern edge of Texas, northern and northwestern Louisiana, with scattered populations in Mississippi, Alabama, Georgia, and the panhandle of Florida (Powell et al. 2016).

Southern coal skinks live on rocky, open glades, such as this example in Shannon County (above). Young southern coal skink from Miller County (right).

JEFF BRIGGLER

TOM R. JOHNSON

October. They are most easily found in April. This shy lizard becomes active midmorning on sunny days but will quickly take shelter in dead leaves or under rocks or logs when approached. Coal skinks have been observed to run into small streams to avoid capture. This species lives in forests and woodlands from along moist stream bottomlands to drier rocky hillsides. They prefer open, damp, rock-strewn woods where they take shelter under rocks, logs, leaf litter, or bark. They are commonly found on open, rocky glades along hillsides during the spring. They also inhabit rock quarries and rocky roadcuts. During the winter, they likely retreat underground into rocky crevices and animal burrows and beneath large rocks.

This species eats a variety of small insects and spiders. Anderson (1965) found that young southern coal skinks readily ate live termites. The tail of a skink can easily be broken off as protection from a predator. Natural predators are snakes, large lizards, birds such as shrikes and hawks, and mammals such as badgers and skunks. Domestic cats are known to capture and eat small lizards.

Breeding: Courtship and mating occur from April through May, and egg laying takes place during May and June. A female will lay between 4 to 10 eggs, presumably beneath rocks and logs. The eggs average 12.5 by 7.5 mm (0.5 by 0.3 in.) in length and width (Anderson 1965). The eggs are guarded by the female until they hatch around 28 to 35 days later. Anderson (1965) reported a female from Camden County that laid eight eggs in late May. In another report, a clutch of eight eggs were laid in captivity from a female collected from Oregon County (Smith and Smith 1952). Hatchlings are about 48 to 51 mm (1.9 to 2 in.) in length (Powell et al. 2016). Individuals likely reach sexual maturity in two years.

Remarks: Although this species is seldom seen compared to other woodland skinks in Missouri, the species appears to be secure. Maintaining glades and adjacent rocky wooded hillsides will ensure this species remains common throughout the Ozarks of Missouri.

Common Five-lined Skink

Plestiodon fasciatus (Linnaeus)

JEFF BRIGGLER

Adult male common five-lined skink from Callaway County.

Description: The common five-lined skink has shiny scales and a dark ground color with light stripes. Its color varies with age and sex. Adult females are normally brown with a dark-brown lateral stripe, five tan stripes, and a blue or blue-gray tail. Adult males are uniform olive or tan with a faint, dark, lateral stripe and a few light stripes. During the breeding season, the male's head is bright red orange. Adult males are slightly larger than adult females. Hatchlings and young adults are black with five yellow stripes from head to base of tail and a bright blue tail. Common five-lined skinks have seven upper labial scales along the upper lip where the fifth labial is the first to contact the orbit of the eye. Also, two postlabial scales are present on the head.

Adult common five-lined skinks range in total length from 125 to 222 mm (4.9 to 8.7 in.; Powell et al. 2016).

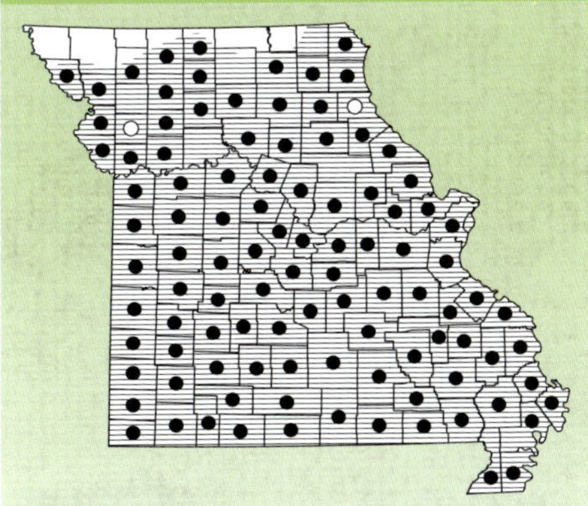

Distribution: Missouri: Occurs statewide, except for a few counties in extreme northern Missouri. North America: Eastern New York, southern Quebec, lower Michigan, eastern and central Wisconsin, southern Minnesota, northeastern Iowa, southern and central Illinois, Missouri, eastern Kansas, Oklahoma, and Texas to northern Florida (Powell et al., 2016).

Habits and Habitat: This species resides in open woods, along wooded bluffs and on

Common five-lined skinks are often found in open, woodlands, such as this example in Phelps County (above). Female common five-lined skink attending a clutch of 13 eggs (three eggs out of view) from Callaway County (right).

rocky, south-facing hillsides. It requires shelters in the form of rocks, downed logs, stumps, and standing dead trees. This skink will climb trees, especially when threatened or searching for insects. This species will often live around farm buildings, rock gardens, rock piles, patios, and firewood piles. In Missouri, the common five-lined skink is mostly active from March to October with a peak from April into early June. As with the majority of skink species, the common five-lined skink will allow its tail to be broken off when grabbed by a predator as a form of defense. Very little blood is lost when this happens. A new tail will eventually grow back, but it will be shorter than the original and a drab gray color.

Common five-lined skinks eat a variety of insects, spiders, snails, and smaller lizards. Natural predators may include hawks, opossums, armadillos, skunks, moles, shrews, snakes, and larger lizards. A prairie kingsnake captured in Cass County had eaten a common five-lined skink (Smith and Powell 1993). Domestic cats are known to extensively prey on lizards and small snakes.

Breeding: Courtship and mating likely occur from April into June. During the breeding season, males become aggressive, and the orange-red head seems to help in sex recognition. Courtship behavior has not been documented in Missouri. During mating, the male will grasp the skin behind a female's head. In a genetic study of several midwestern populations of common five-lined skinks, it was found that most females will breed with more than one male. In addition, females will retain sperm from one breeding season to the next and up to 65 percent of egg clutches may be fertilized by more than one male (Bateson et al. 2011).

TOM R. JOHNSON

A young common five-lined skink from Osage County.

Eggs may be laid in leaf litter or under rotten logs, tree stumps, or flat rocks. In Missouri, egg laying takes place in early May and June. Four to 14 eggs may be deposited (Fitch 1985); the female remains with the eggs until they hatch. Eggs hatch within 27 to 45 days, usually in mid-July into August. Numerous nests have been documented in Missouri. Anderson (1965) reports four clutches ranging from 6 to 12 eggs laid from mid-May to early June. Two nests were found in Greene County in late June 2015; one clutch contained 12 eggs, and the second nest had 11 eggs (Edmond 2015). A Lincoln County female was attending a nest of eight eggs under a limestone rock in mid-July (Smith and Powell 1993). Another nest with a female from Lincoln County was found in a railroad tie in early July. This nest contained nine eggs and hatched in mid-July. A final nest with a female and six eggs was discovered under a corkboard in Franklin County in early July. These eggs hatched in late July (Smith and Powell 1993). Five clutches of captive common five-lined skinks were studied by Groves (1982); these clutches were collected from Virginia and North Carolina. Each of the clutches had eggs that spoiled (addled) and every one of these spoiled eggs was eaten by the attending female. During the incubation period, each female frequently licked and reoriented their eggs. If an addled egg was found, the female would bite one end of that egg, and the contents were consumed by licking and jaw movements. After the egg was empty and the shell collapsed, the shell was grasped and eaten. It is presumed that the removal of spoiled eggs from the nest protected the rest of the clutch from being contaminated by fungi or discovered and eaten by a predator (Groves 1982). The shiny black hatchlings with five stripes along the back, and a bright blue tail average 56 to 60 mm (2.2 to 2.4 in.; Smith and Powell 1993). The bright blue tail color of juvenile skinks is believed to protect young lizards from attack by aggressive adult males. This allows the young to remain in the same area where adult males are active (Clark and Hall 1970).

Remarks: This species and the broad-headed skink (*Plestiodon laticeps*) can be easily confused; see the description section in the broad-headed skink species account. The common five-lined skink is found throughout the forested areas of the state. It is one of the most abundant lizards in Missouri and often seen by people.

Broad-headed Skink

Plestiodon laticeps (Schneider)

Adult male broad-headed skink from Callaway County.

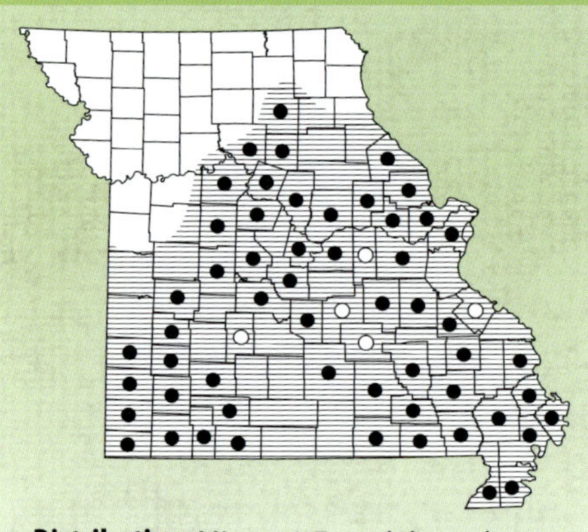

Distribution: Missouri: Found throughout the southern two-thirds of the state. North America: Mainly in southeastern United States from southeastern Pennsylvania south to northern half of Florida, west to eastern edge of Kansas, and south to eastern Texas (Powell et. al 2016).

Description: The broad-headed skink is a large, shiny, smooth-scaled lizard that lives in forest openings. Its color varies considerably between sexes and age classes. Adult males are normally olive brown with few or no stripes along the head and sides. During the breeding season, males develop a swollen head, which has given rise to the species' common name. At this time, their head has an orange-red color; lacking in adult females. The general color of females is tan to brown, with light- and dark-brown or black stripes down the back and sides. The wide, dark stripe down each side is usually the most prominent. Hatchling broad-headed skinks are jet black with five narrow, yellow lines along the back and sides and have a bright blue tail. The common five-lined skink (*Plestiodon fasciatus*) can be confused with the larger broad-headed skink. The former has two postlabial scales on the head that are absent or small on broad-headed skinks. In addition, broad-headed skinks have eight or nine upper

JEFF BRIGGLER

TOM R. JOHNSON

Male broad-headed skink basking on a tree in Lincoln County (above). Female broad-headed skink with a clutch of 21 eggs from Cole County (left).

labial scales where the sixth labial is the first to contact the orbit of the eye. The common five-lined skink has seven upper labials where the fifth labial is the first to contact the orbit of the eye.

Adult broad-headed skinks range in total length from 165 to 324 mm (6.5 to 12.8 in.; Powell et al. 2016).

Habits and Habitat: As with all Missouri lizards, the broad-headed skink is active during the day mainly between April and October. Most of their time is spent in or near trees. They are the most arboreal lizard species in the state. In general, this species resides in forest openings. They will take shelter in tree cavities, often many feet above the ground. This species can also be seen on old stumps, large logs, or around dilapidated farm buildings. A study in Kansas found broad-headed skinks prefer open patches of mature forest with fallen logs of 2 m (6.6 ft) or greater in length (Hullinger 2018). Many broad-headed skinks have been seen or captured within the city limits of Jefferson City; usually basking in the sun on firewood piles, large tree trunks, old railroad ties, rock gardens, and wooden decks.

Its diet includes insects, spiders, land snails, small lizards and their eggs, and snakes, as well as baby mice. Although broad-headed skinks are considered a strict carnivore, an individual was observed eating a ripe strawberry in Florida (Jackson 2012).

Breeding: In a study of broad-headed skinks collected in Alabama and South Carolina, it was found that the male's enlarged, orange-red head allows females to identify it as an appropriate candidate for courtship and mating (Cooper and Vitt 1988). Courtship and mating presumably take place in late April through early June. Courtship was observed in late May in Montgomery County. The male pursued the female on the ground. When their bodies were in contact the male stroked the back of the female's head with his left forelimb. The female did not move and mating occurred and lasted about five minutes (Schuette et al. 1992). Females will begin brooding a clutch of eggs around 30 days after mating. Nests are usually located in a rotting stump or log, within tree cavities, under the bark of a rotting log, or in a pile of rotting lumber. Hullinger (2018) discovered three nests under bark on rotting logs of oak trees in Kansas. A study in South Carolina found that brooding females will move their entire clutch to another nest site if the original nest is flooded by heavy rain or severely damaged by another animal (Vitt and Cooper 1985). Anderson (1965) and Smith (1946) stated that clutches with more than 10 eggs may indicate a communal nest (eggs deposited by more than one female). Johnson (1987, 2000) speculated that large clutches of broad-headed skinks could indicate a communal nest produced by two or more females (also known as "egg dumping"). A study of this species in South Carolina found that the larger general body size of female broad-headed skinks can allow them to produce more eggs compared to other species of skinks, and no communal nesting was seen (Vitt and Cooper 1989). The number of eggs laid by broad-headed skinks generally ranges from 6 to 25 eggs (Mount 1975; Trauth et al. 2004). A female will remain with her egg clutch until they hatch in July and August. Several nests of the broad-headed skink have been documented in Missouri. Two clutches of broad-headed skinks were found in Bollinger County: one clutch contained 14 eggs and the other 17 eggs (Schuette 1980). A gravid female from Lincoln County laid 18 eggs under a pile of wood in mid-June and hatching was completed by mid-July, with an incubation period of 35 days (Schuette 2005). Another nest from Lincoln County was found in an earthen bank in early July (Smith and Powell 1993). Nine of the 15 eggs hatched in early August. Additionally, two broad-headed skink nests were found in Cole County; both contained 22 eggs. One of the clutches was retained and incubated. These eggs averaged 12 mm wide by 15 mm in length (0.5 by 0.6 in.) and hatched in early August (Johnson 2000). Hatchlings may measure from 57 to 86 mm (2.2 to 3.4 in.; Powell et al. 2016). Individuals reach sexual maturity in about two years and likely live between seven and eight years (Niemiller et al. 2013; Snider and Bowler 1992).

Remarks: As with all skink species, broad-headed skinks can lose part of their tail for distraction and protection when grabbed by a predator. This harmless lizard will certainly try to bite a person's hand when captured, but the bite seldom breaks the skin. The bite is a little painful, but not dangerous. An adult male may be called a "scorpion" by some people who mistakenly believe it to be venomous.

Great Plains Skink

Plestiodon obsoletus Baird and Girard

JEFF BRIGGLER

Adult Great Plains skink from Jackson County.

Description: The Great Plains skink is the largest skink species in the United States. They are light tan to gray in background color with the legs, back, sides, and tail scales edged in brown to black. These dark markings can combine to form irregular, longitudinal, narrow strips on the back, sides, and tail. There can be a wash of yellow and a few faint orange spots along the sides of the head and just above the belly. Belly color is a plain light gray. Scales along the sides of their body are in oblique rows—no other Missouri lizard has such a scale pattern. Adult males and females look very much alike, but males have a slightly swollen head during the breeding season. The hatchlings are black with bluish tails and a series of white or orange spots on the anterior sides of the body onto the head.

Adult Great Plains skinks range in total length from 165 to 349 mm (6.5 to 13.7 in.; Powell et al. 2016).

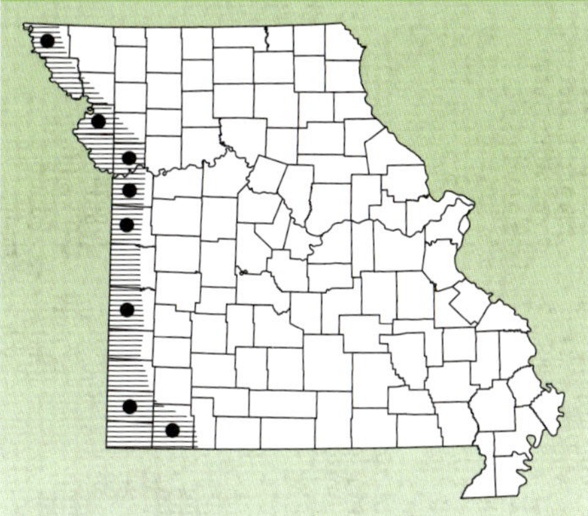

Distribution: Missouri: A few counties along the western edge of the state. North America: Southern and southwestern Nebraska, Kansas, western edge of Missouri, northwestern corner of Arkansas to southwestern Colorado, most of New Mexico, central and southeastern Arizona, east to most of Texas, and south to northern Mexico (Powell et al. 2016).

JEFF BRIGGLER

DAREN RIEDLE

Great Plains skinks are often found under rocks or within rocky crevices in grassland habitats, such as this location in Jackson County (above). Young Great Plains skink with broken tail (right).

Habits and Habitat: This species is active from March into October in Missouri. Individuals have been observed during every month from March 22 to October 13, with peaks from April to early July (Anderson 1965; J. Briggler unpubl. data). This species has been extensively studied in eastern Kansas; much of the information that follows was gleaned from those published works (Collins et al. 2010; Fitch 1955; Hall and Fitch 1971; Hall 1972). Great Plains skinks are excellent burrowers. They spend a great deal of time in burrows under rocks, burrows in loose soil, or within rocky crevices. In these locations, they are safe from a variety of predators, can find insect prey, and can be warmed by sun-heated shelter rocks. They will also venture into grassy openings to search for prey. This usually occurs at an air temperature above 21°C (70°F) between 10:00 a.m. and 4:00 p.m. Great Plains skinks are quick and use their speed to escape predators. They will readily detach their tails to distract a predator if needed. If captured, they will readily bite to defend themselves. The ideal natural habitat for this species is reported to be native prairie with low hills, sparse vegetation, and numerous rocks embedded in the soil. However, Easterla (1972) captured one individual in a live mouse trap in late April on a western slope of a loess mound that did not contain rocks in

TOM R. JOHNSON

Head and upper body of the Great Plains skink.

Atchison County. In Missouri, Great Plains skinks have been found beneath rocks, on rocky roadcuts, and around buildings with landscaping rocks. The winter is spent below the frost line in burrows, under rocks, or deep within rock crevices. In preferred habitat, their home range was found to average 15.2 m (49.9 ft) in diameter (Collins et al. 2010).

This species is known to eat a variety of insects, including crickets, grasshoppers, and beetles, as well as spiders. Young lizards of other species are occasionally eaten. Land snails are also eaten and young Great Plains skinks were found to eat smaller snails (Fitch 1955; Hall 1972). Natural predators are snakes, hawks, shrikes, and mammals—such as badgers and skunks.

Breeding: No study of the breeding biology of this species has been undertaken in Missouri. In eastern Kansas, courtship and breeding take place mainly during May (Collins et al. 2010). A male was observed to pursue a female and touch her with flicks of his tongue. The male then grasped the female behind her head at which time copulation took place. Eggs are usually laid in burrows excavated by the female under a flat rock in June or July and number from 5 to 32 (Fitch 1955; Hall and Fitch 1971). Larger females were reported to produce a higher number of eggs, and some females do not breed each year (Fitch 1955). Each female will guard her clutch of eggs until the eggs hatch (from one to two months later). Newly hatched young are about 64 mm (2.5 in.) in total length (Powell et al. 2016) and have shiny, black bodies and blue tails. Sexual maturity is reached in about three years. In captivity individuals may live eight or more years (Caron and Swann 2009).

Remarks: The Great Plains skink is a peripheral species in Missouri and may never have been common. There continue to be a few scattered sightings for this species in western Missouri, with the most recent individual being found on the floor of a shop on May 16, 2019. Reports in Kansas show this reptile is quite common and occurs in numerous counties that border Missouri (Collins, et al. 2010). The Missouri Department of Conservation considers the Great Plains skink a species of conservation concern due to few records and lack of rocky habitat throughout the grassland areas of the state.

Northern Prairie Skink

Plestiodon septentrionalis septentrionalis Baird

Adult male northern prairie skink from Harrison County.

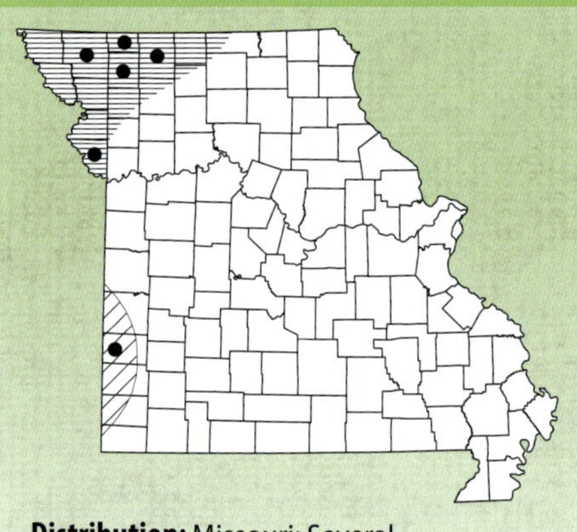

Distribution: Missouri: Several northwestern counties (horizontal lines). The southern prairie skink is known from Barton County, close to the Kansas border (diagonal lines). North America: From southcentral Manitoba, Canada, and Minnesota, northwestern Wisconsin and the eastern Dakotas, south to southcentral Texas (Powell et al. 2016).

Description: The northern prairie skink has many stripes and a long tail. A narrow, light line along the dorsum has a prominent dark stripe on each side that extends onto the tail. There are additional thin dark stripes along the sides; ground color is tan to olive brown. Prairie skinks have two postmental scales. During the breeding season males have reddish orange on their head, which is not found on females. Young are darker in color than adults, and they have a bright blue tail.

Adult northern prairie skinks range in total length from 133 to 224 mm (5.2 to 8.8 in.; Powell et al. 2016).

Habits and Habitat: Prairie skinks are mainly active from April to September, with peak activity in April into June. Easterla and Meadows (1993) captured numerous individuals in late May and mid-June in Harrison County. Their preferred habitat is native prairies and open grasslands with flat rocks or other shelter material near streams, marshes, lakes, and ponds. They spend

JEFF BRIGGLER

JEFF BRIGGLER

Rocky slopes along small prairie streams is typical habitat of the southern prairie skink, such as this example in Barton County (above). Adult female southern prairie skink from Barton County (left).

much time under rocks or in grass thatch with short periods of sunning during the morning or early afternoon. Easterla and Sleep (1999) captured several individuals under boards, under piles of hay, and under bark at the base of a dead tree in Worth and Gentry counties in mid-June and early August. Prairie skinks will quickly run under rocks, into thick vegetation, or into water to escape a predator (Collins et al. 2010; LeClere 2013). In early May 2017, an adult southern prairie skink in Barton County escaped capture by running into a nearby stream and remaining motionless on the stream bottom until it was disturbed (J. Briggler pers. obs.). Prairie skinks are known to overwinter in burrows excavated by themselves or other small animals (LeClere 2013).

This species eats a variety of insects (e.g., grasshoppers, crickets, beetles, caterpillars), spiders, snails, and young lizards. Although unusual, there is a record of a prairie skink eating a discarded bait minnow on the bank of a man-made lake in southeastern Nebraska (Somma 1991). Natural predators include snakes, hawks, badgers, skunks, and birds. Three adult prairie skinks were found in Harrison County in late May impaled on a barb wire fence by loggerhead shrikes (*Lanius ludovicianus*).

Breeding: The reproductive biology of the prairie skink has not been studied in Missouri. Courtship and egg laying likely take place in May and June. Courtship is presumed to be similar to that of other skinks. Females lay eggs in moist, shallow burrows under rocks, logs, or boards. Five to 18 eggs may be laid during mid-to-late June (Collins et al. 2010). In Iowa, three nests were found in late June under a rock, a piece of tin, and a piece of concrete (Frese 2003a). Females remain with the eggs until they hatch. A study in Nebraska found that brooding females attempt to keep their egg clutch properly hydrated by moving eggs or soil with their snout within their burrow system (Somma 1985). In addition, brooding

Adult female northern prairie skink from Harrison County (above). Adult male southern prairie skink from Barton County (right).

females aggressively protect their eggs from predators and intruders. Eggs hatch in one to two months (July or August). Hatchlings have a bright blue tail and an average total length of 51 mm (2 in.; Powell et al. 2016). Young reach sexual maturity in two years (Collins et al. 2010).

Subspecies: Two prairie skinks live in Missouri: the northern prairie skink (*Plestiodon septentrionalis septentrionalis*, Baird), described above, and a subspecies, the southern prairie skink *Plestiodon septentrionalis obtusirostris* (Bocourt). The southern prairie skink is slightly smaller than the northern subspecies, with a total length of 125 to 201 mm (4.9 to 7.9 in.; Powell et al. 2016). This subspecies has fewer or fainter stripes on the back and sides. The widest lateral line is only two scales wide. The southern prairie skink was first reported in Missouri in mid-April 1992, in Barton County (Toal and Reiserer 1992). To date, this is still the only known location for the southern prairie skink in Missouri. Natural history of the southern prairie skink is similar to that of the northern prairie skink.

Remarks: Recently these two subspecies have been recognized as full species in some references (Collins et al. 2010; Powell et al. 2016). Fuerst and Austin (2004) presented genetic evidence separating *P. s. septentrionalis* and *P. s. obtusirostris*; however, Crother (2017) decided against the elevation of these subspecies due to inadequate geographic sampling of these taxa. Until additional genetic studies over a wider range of the species is completed, we choose to continue subspecies designation.

Until 1992, only one record existed for the northern prairie skink in Missouri from 1949 (Anderson 1965). Easterla and Meadows (1993) found numerous individuals in a remnant native prairie in Harrison County, but records still remain scarce in the state. Both prairie skinks subspecies are species of conservation concern in Missouri due to their rarity in the state.

Little Brown Skink

Scincella lateralis (Say *in* James)

JEFF BRIGGLER

Adult little brown skink from Ozark County.

Description: The little brown skink is Missouri's smallest species of lizard. The little brown skink has a brown or gray-brown color with a wide, dark-brown or black dorsolateral stripe. This stripe extends from the snout to the middle of the tail. Small dark flecks are usually present on the back and sides. The belly is light yellow, white, or gray. The little brown skink is the only species of lizard in Missouri that has a clear scale on each lower eyelid, allowing this ground-dwelling lizard to see when its eyelids are closed. Thus, its eyes are protected from damage as it quickly moves through thick leaf litter. Adult females are larger than adult males.

Adult little brown skinks range in total length from 75 to 146 mm (3 to 5.7 in.; Powell et al. 2016).

Habits and Habitat: This species is normally active from March to November. It is a woodland lizard, seldom venturing into open areas unless there is an abundance

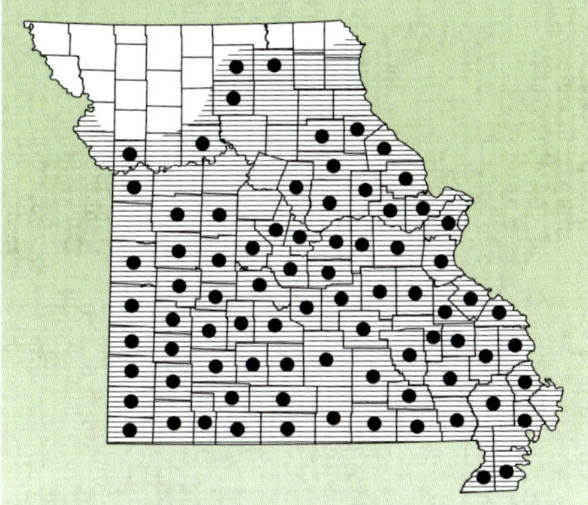

Distribution: Missouri: Southern, eastern, and northeastern Missouri. North America: New Jersey south to most of Florida, west through western West Virginia, in southern Ohio, Indiana, and Illinois, most of Missouri, and eastern Kansas, most of Oklahoma and the eastern two-thirds of Texas (Powell et al. 2016).

JEFF BRIGGLER

TOM R. JOHNSON

Little brown skinks live among the leaf litter and downed logs within forested habitat, such as this example in Butler County (above). Little brown skink eggs from Taney County (right).

of natural shelter, such as leaf litter, short grass, or low plant growth with scattered rocks or logs. On warm, sunny days, little brown skinks can be seen or heard as they quickly move through leaf litter on the forest floor. Little brown skinks escape predators by a rapid, lateral, snakelike movement. They are also known to enter shallow water of a stream or puddle if no other shelter is handy. This species seldom climbs onto rocks, bushes, or trees. On cool days during the active season this species will take shelter in leaf litter or under flat rocks or logs. Their tail easily breaks off when being captured.

A study in northern Florida showed that male little brown skinks had no set home range and moved over a much larger area than females, which tended to have a fixed home range (Brooks 1967). An ecological study of forest-dwelling reptiles in south-central Ozarks of Missouri found that this species was the most common reptile in their study sites (Rota et al. 2017). Sampling conducted between 1992 and 2014 captured 13,287 little brown skinks.

This species eats a variety of small insects, spiders, and small earthworms (Collins et al. 2010). Because they are small and always on the ground, little brown skinks fall prey to a variety of predators: snakes, other species of lizards, large wolf spiders, birds, and mammals such as shrews, skunks, and armadillos. An eastern yellow-bellied racer captured in Camden County had eaten a little brown skink (Smith and Powell 1993). Eastern bluebirds were observed feeding little brown skinks to their nestlings in North Carolina (Beane and Trail 1991).

TOM R. JOHNSON

Hatchling little brown skink from Taney County.

Breeding: Courtship and mating are presumed to occur in the spring or early summer. Fitch and Greene (1965) and Fitch (1970) reported that females normally produce two clutches of eggs each season, with two to seven eggs (average about three or four) per clutch. Their eggs are often retained inside the female for a considerable period, and the embryos are well developed when the eggs are laid. This shortens the incubation period, which may subsequently last only 22 days. Eggs are laid in rotten logs or stumps, under rocks, or in leaf litter. Unlike most other skinks, little brown skinks do not stay with the eggs after laying them. Anderson (1965) reported a female from Greene County that laid four eggs in late May and a second clutch of three eggs in late June. The white eggs were 8.1 to 9.2 mm long (0.3 to 0.4 in.). The first clutch hatched in early August, after 63 days, and newborn young were 43 to 48 mm (1.7 to 1.9 in.) in total length. A female captured in Lincoln County laid three eggs in mid-July, and two of the three eggs hatched in early September after 48 days. Both of the hatchlings measured 44 mm (1.7 in.) in total length (Schuette 1992). Another female collected from Callaway County deposited four eggs in early July, and two hatched later that month after 22 days (Schuette 2005). Sexual maturity typically is reached within one year and most individuals live only about three years.

Remarks: The former common name for this species was "ground skink." The type locality for *Scincella lateralis* is the banks of the Mississippi River below Cape Girardeau (Say *in* James, 1822 "1823"). Relatives of this species are found in western Asia. This species is the most abundant forest-dwelling lizard in the state, and its population is considered secure.

Family Teiidae

Racerunners, Whiptails, and Their Relatives

This is a family of lizards containing 146 species in 10 genera (Pough et al. 2018). The family is restricted to the Western Hemisphere and includes species in a surprising variety of sizes and forms. The smallest species is 76 to 102 mm (3 to 4 in.) long, and the largest—the caiman lizard of South America—can reach 1,220 mm (4 ft) in length. Several West Indian teiids have been introduced into Florida. The majority of species range throughout Central America and the northern half of South America. Only one genus, *Aspidoscelis* (formerly *Cnemidophorus*), is native to the United States; it contains species commonly called racerunners and whiptails. These are medium-sized lizards with strong hind limbs and long, thin tails. Scales on the limbs and body are small granules; belly scales are large, quadrangular in shape, and placed in uniform, transverse rows. Scales on the head are large symmetrical plates. There are about 10 species of *Aspidoscelis* lizards native to the United States; the majority reside in western states. Several members of *Aspidoscelis* group are all females and reproduce through parthenogenesis. Missouri has one species with an additional subspecies consisting of males and females.

Six-lined Racerunner

Prairie Racerunner

Aspidoscelis sexlineata viridis (Lowe)

JEFF BRIGGLER

Adult male prairie racerunner from Shannon County.

Description: The prairie racerunner is a long, slender, fast-moving lizard that lives in dry, open areas. The ground color is a dark brown or black. There are normally seven longitudinal stripes extending from the head along the back and sides onto the tail. The stripes are yellow, white, gray, or pale blue. The tail is gray or brown and rough to the touch. The head and forward part of the body are tinged with blue or green, particularly on males. The belly is gray or blue gray on males and salmon pink to creamy white on females. Adult male prairie racerunners have a broader head than females, and females have a heavier body.

Adult prairie racerunners range in total length from 152 to 267 mm (6 to 10.5 in.; Powell et al. 2016).

Habits and Habitat: This fast, alert lizard is normally active on warm, sunny days between 8:00 a.m. and 3:00 p.m. (Fitch 1958a). Its active season may last from mid-April to mid-September. On cool or cloudy

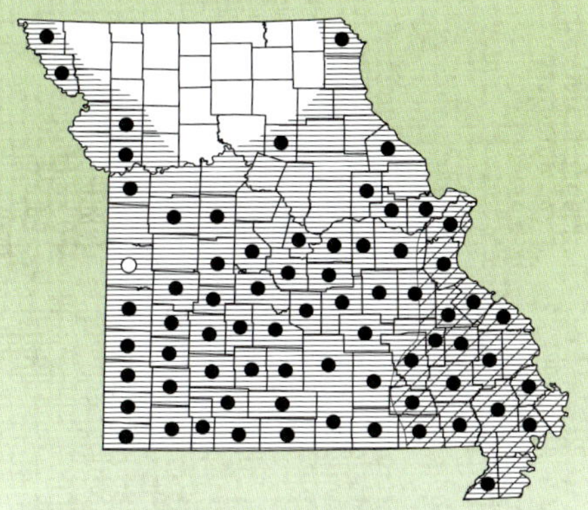

Distribution: Missouri: Throughout most of the southern half of Missouri and along the Mississippi and Missouri rivers in northern Missouri. There is the likelihood of intergradation with the eastern geographic race, the eastern six-lined racerunner, in a few counties along the eastern and southeastern edge of the Ozarks (diagonal lines). North America: Virginia south to most of Florida, from western Wisconsin south to the Gulf States, west to extreme southern South Dakota, and south through eastern Colorado and New Mexico, and into most of Texas (Powell et al. 2016).

Prairie racerunners are commonly found on open, rocky glades throughout the Ozarks, such as this example in Shannon County (above). Adult female prairie racerunner from Miller County (right).

days racerunners take shelter in burrows they have dug or in the burrows of other animals. They also burrow under objects such as flat rocks or boards. Extreme heat also causes this species to seek shade or cooler temperatures in burrows underground. Racerunners are ground dwellers and are not known to climb trees or bushes. Ayres (1973), however, observed a racerunner climb a bush while in pursuit of an insect. Their tails break off when grasped by a predator, which might allow them to escape. Racerunners will not hesitate to enter creeks or pools of shallow water to escape being captured. Racerunners have been observed fleeing into water while avoiding capture in South Carolina and west-central Arkansas (Johnson 2000; Trauth et al. 1996). These lizards burrow into loose soil on south- or southwest-facing slopes to escape the harsh weather of winter. Juvenile racerunners enter overwintering burrows after adults and emerge earlier in the spring (Trauth et al. 2004).

These lizards prefer open areas with loose soil or sand and sparse vegetation, especially in fields, grasslands, along river floodplains, and rocky, south-facing hillsides throughout the state. They can be quite abundant in open, rocky, south-facing hillsides, such as glades, in the Ozarks. They are also encountered along the edge of sand and gravel roads, railroad tracks, and abandoned rock quarries (Easterla 1972; J. Briggler pers. obs.). In a long-term study (18 years) of amphibians and reptiles that inhabit a 1.5 ha (3.7 ac) glade and open woodland habitat in Shannon County, prairie racerunners were found to be the most abundant species (up to 40 individuals were found during a one-day visit; J. Briggler unpubl. data). In a Kansas study, its home range was confined to about 0.08 to 0.10 ha (0.20 to 0.25 ac; Fitch 1958a).

Racerunners eat insects, spiders, scorpions, and other invertebrates. A study of this species in northwestern Illinois sand prairie habitat reported that 52 percent of the prey it consumed were grasshoppers and crickets (Warner 2000). In an Oklahoma study, young racerunners were found to eat leafhoppers and jumping spiders (Paulissen 1987). Natural predators include other lizards, such as eastern collared lizards (when they share the same habitat), snakes, birds, such as shrikes and roadrunners, and mammals, including skunks, badgers, and armadillos.

Breeding: Courtship and mating take place between late April and June. Breeding activity was observed in St. Clair County in late April (Anderson 1965). A male courts a female by displaying his colorful throat and chest. If the female is receptive, the male will grasp the loose skin on the back of her neck, mount her, and mate. When copulation has ended, the male may follow the female with his tail elevated and his cloaca dragging the ground (Anderson 1965; Collins et al. 2010). Females two or more years of age will normally lay two clutches of eggs per season. Trauth et al. (2004) reported from one to eight eggs may be laid per female. A burrow is dug in loose soil by a female who will open a chamber where her eggs are laid. The egg clutch is then abandoned. Racerunner eggs average 16 by 9.1 mm (0.6 by 0.4 in.; Anderson 1965). A two-year study of nesting six-lined racerunners in North Carolina observed their use of old sawdust piles for egg deposition. Sixty-four nests were found within eight sawdust piles with an egg count from one to five eggs (Brown 1956). Anderson (1965) also found seven freshly laid eggs in a sawdust pile in St. Clair County in early July that hatched in mid-September. Young hatch after an incubation period of two months (August and September). The young average 90 mm (3.5 in.) in total length (Anderson 1965) and sport a pale blue tail and vivid yellow body stripes. Young grow quickly and can reach maturity in one to two years and may live about four or five years (LeClere 2013).

Subspecies: The eastern six-lined racerunner, *Aspidoscelis sexlineata sexlineata* (Linnaeus) is a subspecies of the prairie racerunner; it has six stripes instead of seven. The natural history of this subspecies is similar to that of the prairie racerunner. The distribution of these two subspecies in eastern and southeastern Missouri needs study. There appears to be a broad area of intergradation in southeastern and eastern Missouri with the eastern six-lined racerunner.

Remarks: Racerunners are sometimes called "field streakers" or "sandlappers" due to their quick speed and association with sandy habitat (Conant and Collins 1991). With reported speeds of 28.9 kph (18 mph; Vogt 1981), this species can be extremely difficult to capture by hand during the active season. Maintaining the open nature of rocky glades throughout the Ozarks will help keep this species common in Missouri.

Family Anguidae

Glass and Alligator Lizards, and Their Relatives

This is a wide-ranging family of lizards with a great diversity of forms. A number of species are long, slender, and legless, causing them to be easily confused with snakes. There are species that have a reduced number of toes or only two limbs; there also are some species that are more "typical" lizards with four normal limbs. The Anguidae family contains about 130 species representing 14 genera (Pough et al. 2018). Members of this family occur in North and Central America, Europe, Asia, and northern Africa (Pough et al. 2018). Four genera, with a total of 14 species, occur in the United States: *Anniella*, legless lizards, with five species; *Gerrhonotus*, the Texas alligator lizards, with one species; *Elgaria*, western alligator lizards, with four species; and *Ophisaurus*, the glass lizards, with four species.

Members of the family Anguidae are distinguished by the presence of bony plates (osteoderms) in each scale and by a prominent deep groove located along each side of the body. The plates reinforce the scales but reduce flexibility. It is thought that the groove provides flexibility of the body for movement, breathing, and swallowing prey, and also for egg development in females. The groove is lined with tiny scales (granules).

Missouri has one species of glass lizard (genus *Ophisaurus*) native to the state. Glass lizards (also called glass "snakes") have very long, fragile tails that break off easily if grasped by a predator or struck with an object. Glass lizards will try to avoid a predator by a speedy retreat through grass. If captured they will try to escape by biting and will break off their tail as a last resort.

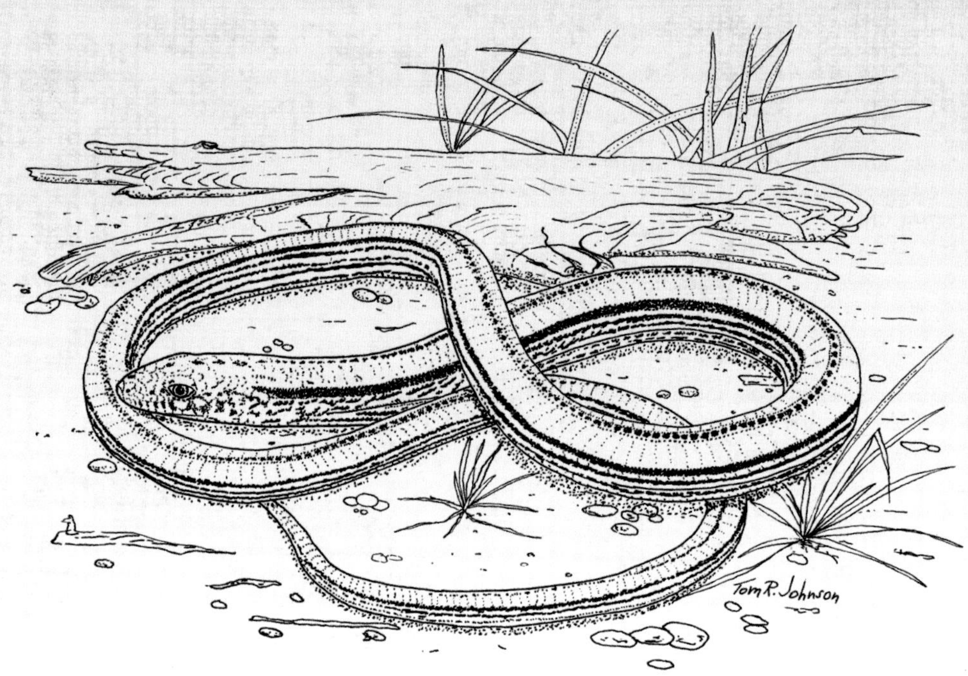

Western Slender Glass Lizard

Western Slender Glass Lizard

Ophisaurus attenuatus attenuatus Cope

Adult western slender glass lizard from Pettis County.

Description: The western slender glass lizard is a snake-like lizard that is long, slender, and legless. Its ground color is normally gray, tan, or brown, with black stripes on the back and sides. Narrow, dark stripes located below the lateral groove are prominent on juveniles and subadults but become faded once adulthood is reached. The dorsal stripe normally changes into a series of dark spots or dark crossbands as the lizard matures. The belly is white and the underside of the tail normally has dark stripes. Although glass lizards superficially resemble snakes, they do have some notable differences. The eyes are protected by movable eyelids; there is an ear opening on either side of the head; and a lateral groove runs down the entire length of the body. Snakes have none of these characteristics. This species has a long tail that is normally two-thirds of the total length of the reptile (unless the tail has been broken off and is being regenerated). There are no visible differences between males and females in Missouri individuals.

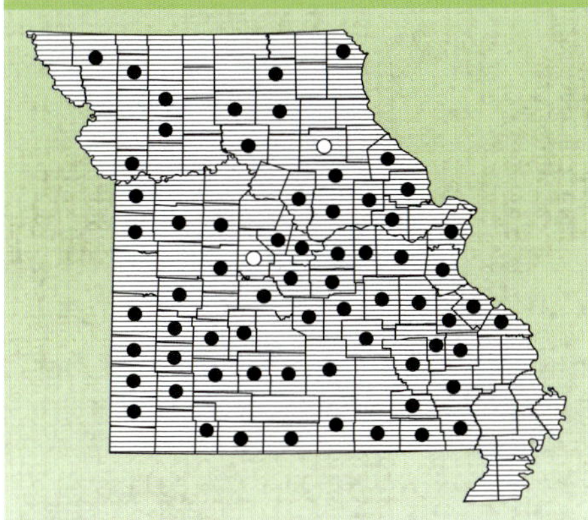

Distribution: Missouri: Presumed to occur mostly statewide, except for the southeastern portions of the state. Most common in counties with former prairie and savanna habitat. North America: Northwestern Indiana through Illinois, southeastern Iowa, Missouri, eastern half of Kansas, most of Oklahoma, eastern half of Texas, most of Arkansas, and the western half of Louisiana. There is an isolated population in southwestern Wisconsin (Powell et al. 2016).

TOM R. JOHNSON

Head of western slender glass lizard.

Adult western slender glass lizards range in total length from 560 to 1,181 mm (22 to 46.5 in.; Powell et al. 2016).

Habits and Habitat: This species is active on the surface mainly from April to October. Trauth (1984) studied this species in Arkansas and reported two periods of activity: April and May (coinciding with breeding) and October and November. Slender glass lizards are active during the day as long as air temperatures range from 10°C to 32°C (50°F to 90°F). Although this lizard often takes shelter in clumps of grass or small mammal burrows, it also will burrow into loose soil. Because of its rapid movement and its coloration that blends so well with the vegetation, a glass lizard sighted in tall grass can quickly escape. If captured, a glass lizard will frantically thrash about and the tail can quickly detach from the body. They seldom bite when captured and often spin their bodies to escape.

In Missouri, the western glass lizard occurs on prairies, pastures, in open woods, on dry, rocky hillsides, glades, and fens. They are most commonly found in the prairie and grasslands areas of the state. Anderson (1965) reports specimens being plowed up by farmers working in grain fields. Glass lizards are also seen on moist, vegetative fens throughout much of the Missouri Ozarks. In Kansas, Fitch (1989) reported a home range size of about 0.4 ha (1 ac).

Glass lizards consume a variety of insects and other invertebrates, including land snails and spiders; they also eat other lizards, frogs, small snakes, eggs of ground-nesting birds, and occasionally young mice (Collins et al. 2010; LeClere 2013). This species of lizard may be preyed upon by snakes, hawks, and badgers.

Breeding: The reproductive biology of this species has not been studied in Missouri but has been studied in nearby Kansas and Arkansas. This species probably mates from late April into early June; a female will produce 5 to 17 eggs during June and early July (Fitch 1985). Each female will produce one clutch of eggs per season. The clutch of eggs will be attended by the female until they hatch in 50 to 60 days. Eggs may be laid in a rotten log, under a rock, or under other cover. Hatchlings average 185 mm (7.3 in.) in total length. Young take three to four years to reach adulthood (Collins et al. 2010; Trauth 1984). This species of lizard has been documented to live at least nine years in the wild (Fitch 1989).

Remarks: Glass lizards are so snake-like in appearance they are often mistaken for a snake. Local names for this lizard include glass snake and joint snake. When handled, the tail will break into many pieces to distract a predator. There is no truth to the belief that this or any other lizard can rejoin itself once its tail is detached. This species can be difficult to detect and several records were unfortunately obtained in Missouri due to being killed by prescribed burns in April and May.

(Suborder Serpentes)

Snakes

Snakes are long, slender reptiles covered with scales; they are limbless and lack ear openings and eyelids. The eyes of snakes are protected by a clear, nonmovable scale. Although snakes are limbless, they can move about with ease. They have an extremely flexible backbone composed of 200 to 400 vertebrae. Each vertebra has a pair of ribs that is attached to muscles. Snakes move forward by a side-to-side movement or in a straight line using muscles to move their skin forward in waves. The internal organs of snakes are elongated so they fit into the tubular body cavity. Paired internal organs are offset so that they fit into the elongated body cavity. Most snakes have an elongated right lung and no left lung. A snake's heart is normally located about one-third of the length from the head.

Snakes shed their outer layer of skin periodically as they grow. During the active season, young snakes usually shed once every four or five weeks; adults may shed once every six to eight weeks. With rattlesnakes, a new segment is added at the base of the rattle at each shedding. In Missouri, rattlesnakes may shed two to four times per year. As the rattle becomes longer, the old segments weaken and may break off, so it is not possible to determine the age of a rattlesnake by counting the number of segments.

The snake's long, forked, extendable tongue often is thought to be dangerous; in fact, it is completely harmless. Snakes use their tongue to pick up odors from their environment, which are transferred to special sense organs in the roof of the mouth. Snakes use their tongue to identify and find prey or locate a mate during the breeding season. The two sections of their forked tongue allow snakes to detect differential odor levels. That is, if the right section brings in more odor, that would indicate a prey animal slightly to the right of a snake's head. The food habits of Missouri's snakes are as varied as the types of snakes. Rough greensnakes, for example, eat insects, insect larvae, and spiders. Watersnakes, on the other hand, eat fish, frogs, tadpoles, and crayfish. The western ratsnake and bullsnake eat rodents, small birds, and bird eggs. Kingsnakes also eat rodents and are important predators of lizards and other snakes, including venomous species. Snakes must swallow their food whole; some can engulf animals three times the diameter of their head. The lower jaws of snakes are loosely joined to the skull, and the upper jaws are movable. A snake usually grasps its prey (whether fish, frog, or mouse) by the head and engulfs it by advancing first one side of the jaw and then the other. Snakes' teeth help them swallow prey as well. The teeth are sharp and curve toward the rear of the mouth. They hold the prey and prevent its escape. Some snakes, such as watersnakes and gartersnakes, eat their prey alive, while venomous snakes usually inject venom into the animal and swallow it after it is dead. Several Missouri snakes, such as ratsnakes, kingsnakes, milksnakes, and bullsnakes, kill their prey by constriction. A constrictor grasps prey in its mouth and immediately wraps several tight coils around it. Thus, prevented from breathing, the prey animal dies. The prey is then swallowed.

In Missouri, snakes normally breed in the spring, soon after they emerge from winter dormancy. A few species, however, breed in autumn. About half of Missouri's snake species lay eggs (oviparous); the rest give birth to fully developed young (ovoviviparous). Some egg-laying species are the western ratsnake, bullsnake, kingsnakes, racers, wormsnake, ring-necked snake, and greensnakes. The size of the egg depends on the species; the number of eggs produced depends on the size of the female. In general, the larger the female, the more eggs she can produce. Snake eggs are elongated and have a tough, leathery shell. Females select rotten logs, stumps, or leaf litter in which to deposit their eggs. As young snakes develop within the egg, a small "egg tooth" grows on the tip of the snout; this is used to slit the shell when hatching. Afterwards, it is shed. Snakes usually hatch in late summer or early autumn. Snakes that retain their young until they are completely developed are watersnakes, gartersnakes, brownsnakes, copperheads, cottonmouths, and rattlesnakes. From an evolutionary perspective, this form of reproduction is slightly advanced over egg laying.

Each young snake is protected inside the female in a thin, sac-like membrane. The membrane also contains yolk for nourishment. Some of the young snakes break through the membrane while inside the female and emerge from her in a tight coil; others break through after being born. Snakes that develop inside the female are normally born in mid- to late summer.

Herpetologists currently recognize approximately 3,700 species of snakes throughout the world (Uetz et al. 2020) in approximately 25 families (Pough et al. 2018). Snakes in Missouri are represented by two families: Colubridae and Viperidae. Missouri, with its wide variety of natural habitats—prairies, Ozark hills and valleys, swamps, and marshes—has a total of 43 species of snakes with an additional four subspecies. Most of our snake species (86 percent) are nonvenomous. Although many may bite in self-defense, their bite usually produces nothing more than shallow scratches. There are six species of venomous snakes in Missouri. The smallest snake native to Missouri is the flat-headed snake (*Tantilla gracilis*), which averages about 200 mm (7.9 in.) in total length. The largest is the bullsnake (*Pituophis catenifer sayi*), which can reach a total length of 2,667 mm (105 in.; Powell et al. 2016).

SNAKE MYTHS

Although most people in the United States have had some biology courses during their school years, it is amazing how many myths about snakes persist. Perhaps because snakes are so different from us, there has been a tendency for people to believe fantastic stories about these reptiles. The following snake myths are still prevalent in Missouri. There is no biological evidence to support any of these beliefs, but when it comes to snakes, some people have difficulty separating truth from fiction.

Myth: Snakes are slimy.
Truth: Like all reptiles, snakes have tough, dry skin that protects them from the harsh life as a land-dwelling animal. Salamanders and frogs have moist or slimy skin, not snakes.

Myth: A snake will swallow young snakes to protect them.
Truth: No species of snake has this ability. Any snake swallowed by another will quickly die inside the stomach due to a lack of oxygen and strong digestive juices.

Myth: Snakes move in pairs.
Truth: All snakes are predators and compete for food. Snakes may be together in the general vicinity of prime habitats, but this is due to convenience, not because they live in pairs. During the breeding season, however, several male snakes will follow the scent trail of a female, which would give the impression that snakes move in pairs.

Myth: Hoop snakes.
Truth: People may claim to have seen a snake grab its tail in its mouth and roll along the ground like a hoop, but this has never been documented—nor is it possible for any snake to do such a thing.

Myth: Snakes cannot bite while underwater.
Truth: Snakes can and do bite underwater—that's how they capture aquatic prey (fish, tadpoles, or salamanders). Watersnakes and the venomous cottonmouth can defend themselves while underwater.

Myth: Snakes can steal milk from cows.
Truth: This falsehood probably originated in areas where kingsnakes and ratsnakes enter barns and farm sheds to search for mice. Seeing a snake in a milking shed at about the time a family cow begins to slow down in milk production could have caused a farmer to blame the snake for taking milk. Milk from a cow is not a natural food of any reptile. It is likely that

snakes do not have the proper enzymes to digest milk. Also, the many small, sharp teeth in a snake's mouth would hinder their ability to "steal" milk from a cow.

Myth: Western ratsnakes, also known as black snakes, breed with copperheads or rattlesnakes.
Truth: Under no circumstance will a black snake mate with a venomous species. This would be like expecting a chicken to breed with a hawk.

Myth: A snake must be venomous if it vibrates its tail.
Truth: Although rattlesnakes and other venomous snake do vibrate their tails, many types of harmless snakes (e.g., ratsnakes, kingsnakes, racers, bullsnakes, etc.) can and will vibrate their tail when alarmed or threatened.

The biology and natural history of Missouri snakes are both interesting and enjoyable to learn. Snake myths and misunderstandings make colorful stories but should not be confused with scientific facts.

Bullsnake

Eastern Copperhead

Family Colubridae

Typical Snakes

This snake family contains at least 1,800 species in approximately 255 genera (Pough et al. 2018), more than all other snake families worldwide and nearly 50 percent of all snake species. Most of the species in this family are small-to-medium in length. Members of the Colubridae family usually have large scales on the top of the head, and the scales along the back are either smooth (e.g., kingsnakes, racers, or ring-necked snakes), or have a ridge or keel, which makes them appear rough (e.g., watersnakes, gartersnakes, and hog-nosed snakes). Although some species of colubrids found in the world are venomous to humans, none of the members of this family that occur in Missouri are dangerous to people.

There are seven subfamilies represented in the family Colubridae (Pough et al. 2018). Although some authorities have elevated these subfamilies to family level, we chose to follow classification outlined in Pough et al. (2018). Species of snakes in Missouri are found in three subfamilies: Colubrinae, Dipsadinae, and Natricinae.

The subfamily Colubrinae contains the harmless egg-laying snakes. It includes the most diverse species worldwide (726) and represents many North American snake species (Pough et al. 2018). This group contains 101 genera (Pough et al. 2018), and Missouri has eight genera (*Cemophora*, *Coluber*, *Lampropeltis*, *Opheodrys*, *Pantherophis*, *Pituophis*, *Sonora*, and *Tantilla*) with 16 species.

The subfamily Dipsadinae is comprised of the reared-fanged or swivel-jaw snakes. This group is represented by 96 genera and 752 species, with most occurring in central and South America (Pough et al. 2018). There are five genera exclusively found in North America, and Missouri has four genera (*Carphophis*, *Diadophis*, *Farancia*, and *Heterodon*) with six species.

The subfamily Natricinae contains the harmless, live-bearing snakes with 36 genera and 225 species worldwide (Pough et al. 2018). This group is characterized by the North American semiaquatic snakes. In Missouri, this group has eight genera (*Clonophis*, *Haldea*, *Nerodia*, *Regina*, *Storeria*, *Thamnophis*, *Tropidoclonion*, and *Virginia*) with 15 species.

There are 37 species of snakes native to Missouri, with an additional four subspecies that are members of the Colubridae family; they are grouped into 20 different genera.

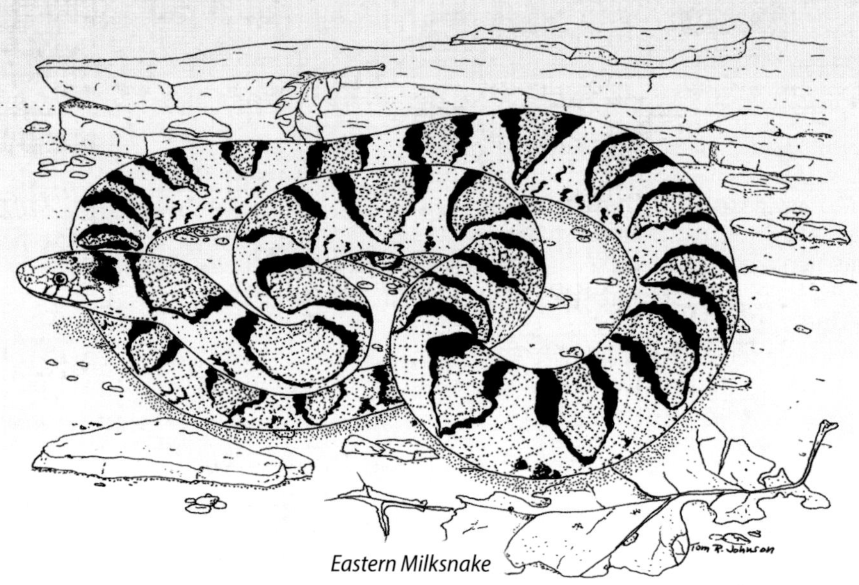

Eastern Milksnake

Western Wormsnake

Carphophis vermis (Kennicott)

Adult western wormsnake from Shannon County.

Description: The western wormsnake is a small two-toned snake that lives in wooded areas or rocky hillsides. Its dorsal color is purple brown to black; the ventral is unmarked with a salmon-pink color extending up the sides. The head is flattened to aid in burrowing. The tail terminates in a harmless spike that also aids in maneuvering through soil. The western wormsnake has smooth scales and a divided anal plate. Males have longer tails than females, and they normally have keels or ridges on the dorsal scales above the anal plate. Adult males are smaller than females. Wormsnakes have a pair of enlarged teeth at the rear of their mouth which are presumed to help with swallowing earthworms and other prey.

Adult western wormsnakes range in total length from 190 to 280 mm (7.5 to 11 in.) but have been known to reach 391 mm (15.4 in.; Powell et al. 2016).

Habits and Habitat: This small species is secretive and seldom seen. Western

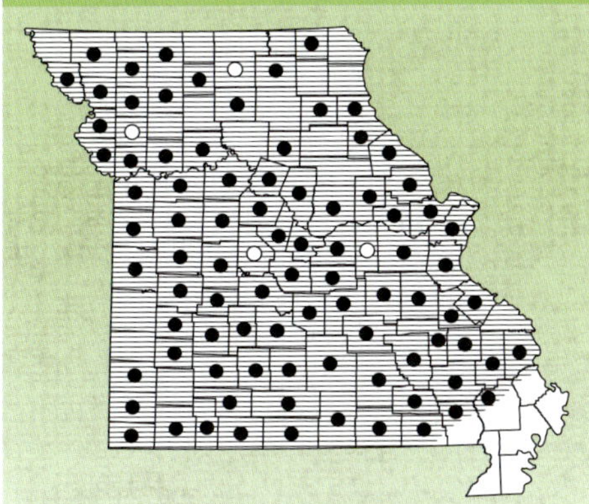

Distribution: Missouri: Statewide except for the southeastern counties. North America: Southern Iowa, southeastern Nebraska, most of Missouri, eastern Kansas and Oklahoma, northeastern corner of Texas, western half of Arkansas and the northeastern and northwestern corners of Louisiana. There is an isolated population in extreme southwestern Wisconsin (Powell et al. 2016).

TOM R. JOHNSON

TOM R. JOHNSON

A western wormsnake hatchling (above). Western wormsnake eggs from Taney County (right).

wormsnakes in Missouri are active between March and October. However, they become inactive during hot, dry summers and burrow down into the soil to estivate. These small snakes are not known to bask in the sun on the surface. Instead, western wormsnakes burrow under rocks exposed to the sun and become warm when the rock is heated by the morning sun. Individuals mainly overwinter in small rodent burrows, in crevices among large rocks, or under rotting logs and stumps. A few western wormsnakes were found overwintering in a cave in Jefferson County (Drda 1968).

The western wormsnake is considered a forest or forest edge species. They are most often encountered on rocky, forested hillsides and on open, rocky glades; especially in the spring and fall. They require a slightly moist micro habitat and meet this need by living in thick leaf litter and under rocks or rotting logs on the forest floor. Specimens have been found in deep leaf litter that had accumulated in shaded, steep-sided ravines in a forest in Warren County. Old trash embedded in the soil can harbor western wormsnakes. In the right habitat, a number of these snakes may be found in a relatively small area. Up to 25 specimens were found during a few hours of search effort by Anderson (1965).

The natural prey of this species consists of earthworms, slugs, and some slender, soft-bodied insect larvae such as beetle grubs (Ernst and Ernst 2003). Western wormsnakes were reported to eat young prairie ring-necked snakes (Clark 1970; Hurter 1911). Natural predators of this species include kingsnakes and milksnakes, and predatory mammals such as moles and shrews. Opossums have been reported to prey on western wormsnakes (Reynolds 1945). The study took place in central Missouri and documented that up to 2.5 percent of

JEFF BRIGGLER

When threatened, western wormsnakes coil tightly to protect their heads and display the bright colored belly, such as this example from Callaway County.

the total volume of 68 opossum stomachs examined was that of western wormsnake remains. The opossums were captured during the spring. However, this mammal is known to consume dead wildlife on roads and highways, thus, the snake remains could have been road carrion–not snakes that were captured alive and eaten by opossums.

Breeding: Western wormsnakes likely breed during the spring, but some could breed in the fall. Adult males store sperm in the *vasa deferentia* throughout the year, but the highest concentration of sperm was available in the fall (Aldridge and Metter 1973). Egg laying takes place from June to early July. The eggs are laid in a burrow under a rock or inside a rotting stump or log. Three eggs in the process of hatching were found while a research team was digging a trench in a Warren County forest. The eggs were about 200 mm (7.9 in.) under the surface (Altnether 2003). A female will produce from one to six eggs (Anderson 1965). A study in eastern Kansas showed that larger females will produce more eggs—up to 12 per female, and it is likely that only one clutch is produced per season (Clark 1970). The eggs are white, elongated and become somewhat clear as the embryos develop. A freshly laid clutch of western wormsnake eggs can be variable in length: 20 eggs varied from 14 to 29 mm (0.6 to 1.1 in.) in total length (Anderson 1965). Hatching normally takes place from the middle of August to the middle of September and the young are from 89 to 102 mm (3.5 to 4 in.) in total length (Anderson 1965). Newly hatched western wormsnakes have a purplish-black dorsum and the ventral is orange red to pink in color (Ernst and Ernst 2003). Males are sexually mature at age two; females are mature at three years (Fitch 2003). A long-term study in Kansas found that females may live up to 10 years (Fitch 1999).

Remarks: This fossorial, harmless species is not known to bite people. When captured it is difficult to hold onto individuals because of the species' smooth scales. A specimen will attempt to escape by trying to work its head between the handler's fingers and will press its harmless tail tip against the handler's skin.

Northern Scarletsnake

Cemophora coccinea copei Jan

JEFF BRIGGLER

Adult northern scarletsnake from Scott County.

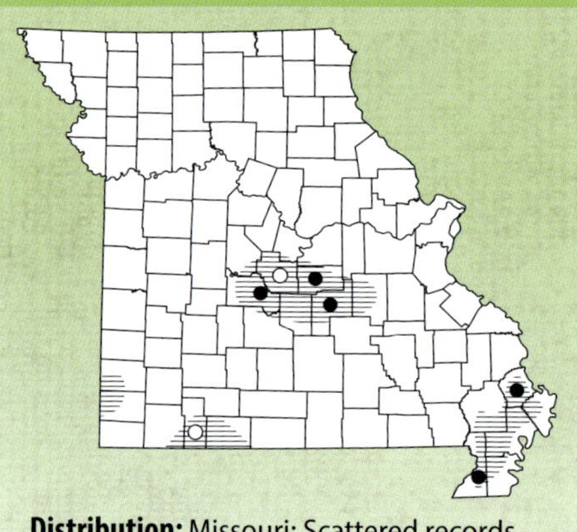

Distribution: Missouri: Scattered records in central, southwest, and southeastern parts of the state. North America: Southern New Jersey south to northern Florida, and west to eastern Oklahoma and Texas. Numerous disjunct populations are known throughout the northern part of their range (Powell et al. 2016).

Description: The northern scarletsnake is a small-to-medium-sized, multicolored snake with a pointed head. This snake has wide orange or red incomplete bands bordered by narrow black bands over a white or light-yellow ground color. There may be some black pigment present in the red bands on large individuals. The snout is pointed and generally red or orange. An important characteristic of this species is a spotless white or cream-colored belly. Dorsal scales are smooth and the anal plate is not divided. Male scarletsnakes have a longer tail and more dorsal bands than females.

This uncommon species can be confused with the more common eastern milksnake (*Lampropeltis triangulum*) that has bold black markings on a white belly.

Adult northern scarletsnakes range in total length from 360 to 510 mm (14.2 to 20.1 in.) but have been known to reach 828 mm (32.6 in.; Powell et al. 2016).

NOPPADOL PAOTHONG

JEFF BRIGGLER

Although northern scarletsnakes can be found in a variety of habitats, they are most commonly observed in the sand prairies and savannas in southeastern Missouri (above). Ventral coloration of an adult northern scarletsnake (left).

Habits and Habitat: The scarletsnake is a secretive species that spends much time underground, except on warm nights or after heavy summer rains. In Missouri, habitat preference varies from mixed hardwood-pine woodlands to sand prairie grasslands and savannas. The few specimens taken in central and southwestern Missouri were found on wooded, rocky hillsides. More recently, numerous individuals have been found in the sandy and loamy soils in southeastern Missouri. They can be found under flat rocks, logs, leaves, or other objects. This snake can quickly burrow into the sandy substrate by thrusting its head and neck back and forth until concealed beneath the loose soil. The active season is probably April into October. Individuals have been found from late April to mid-August with most individuals found in July in Missouri (J. Briggler unpubl. data). Scarletsnakes are very mild mannered and rarely if ever bite when handled.

Scarletsnakes predominantly eat the eggs of turtles, lizards, and other snakes; they may swallow the eggs whole or break the eggshell with their teeth and then swallow the contents. The rear maxillary teeth are large and sharp and aid in slitting egg shells (Trauth 1993). This species may occasionally eat lizards, small snakes, and mice. The prey is killed by constriction. Although a colorful snake, its specialized diet of reptile eggs and its inclination to remain hidden makes this snake a poor pet.

JEFF BRIGGLER

Pointed snout of the northern scarletsnake.

Breeding: Very little is known about the reproductive biology of this species range-wide, and no information is available for Missouri. Mating is presumed to occur in the spring, and the eggs are probably laid during June and July. Females deposit white, elongated eggs in underground burrows, under rocks, or under plant material (Ernst and Ernst 2003). A clutch of eggs was found in Georgia buried in pine straw (Trauth 1982). The reported clutch size of the scarletsnake contains two to nine eggs. In Arkansas, an average clutch size was 4.3 eggs based upon ovarian counts (Trauth et al. 1994). Most eggs are presumed to hatch in September, and hatchlings average 147 mm (5.8 in.) in total length (Ernst and Ernst 2003). Reproductive maturity is probably reached at two to three years of age.

Remarks: The northern scarletsnake is a species of conservation concern in Missouri. Because so few specimens of the northern scarletsnake have been reported in Missouri, it is difficult to determine the status of the species. The scattered records found throughout most of the Ozarks are old reports prior to the 1970s. However, there have been numerous individuals found since 2010 in the sand prairie and savanna areas of southeastern Missouri. In South Carolina, Nelson and Gibbons (1972) documented that scarletsnakes were more common than previously reported; because of the species' secretive nature, they concluded that it is an infrequently encountered species. Likely, more individuals would be discovered in Missouri with intensive sampling.

Kirtland's Snake

Clonophis kirtlandii (Kennicott)

JEFF BRIGGLER

Adult Kirtland's snake from Clark County.

Description: The Kirtland's snake is a small-to-medium-sized snake with numerous black or dark-brown blotches. The dorsal color is red brown to gray brown with four alternating, round spots extending along the back. The head is black or brown with a cream-to-yellow-colored chin and throat. An important characteristic of this snake is a pink-to-red-colored belly with dark stippling along each side. Dorsal scales are keeled, and the anal plate is divided.

Adult Kirtland's snakes range in total length from 360 to 457 mm (14.2 to 18 in.) but have been known to reach 662 mm (26.1 in.; Powell et al. 2016).

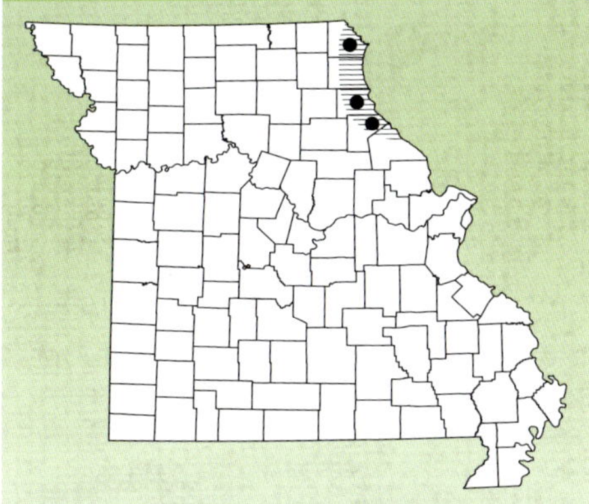

Distribution: Missouri: Restricted to a few counties along the Mississippi River in the northeast corner. North American: Populations ranging from western Pennsylvania, Ohio, Indiana, Illinois, northeastern Missouri, extreme southeastern Wisconsin, southern Michigan, and a few disjunct populations in northwestern Kentucky and Tennessee (Powell et al. 2016).

Habits and Habitat: The Kirtland's snake is fossorial (meaning it spends most its time underground). It mainly uses crayfish burrows in grassland habitat that are damp and typically adjacent to a river, creek, or wetland, but it has been reported in wooded areas (Anton et al. 2003; Capparella et al. 2012; Minton 2001). Most individuals

Kirtland's snakes live in damp, grassland habitats with the best remaining habitat occurring along roadsides in northeastern Missouri (above). Head of a Kirtland's snake from Clark County (right).

in Missouri have been found in relict, wet grassland habitat along roadsides. Some of the highest population numbers are found in vacant lots and trash dumps in heavily populated urban areas in Ohio, Indiana, and Kentucky (Minton 2001; Smith 1961). Kirtland's snakes are most active in the spring, especially late April thru early June, and in autumn (October) prior to overwintering (Conant 1943; Minton 2001). Individuals are rarely seen during the day and often hide beneath rocks, boards, or other artificial objects, such as roadside trash, but during the dry summer months and cold winter months they retreat underground. In the winter, this species has been known to use crayfish burrows in Missouri. This species likely has a small home range based upon repeated observations of individuals being found under the same or nearby cover objects (Wilsmann and Sellers 1988).

The diet of Kirtland's snake consists primarily of earthworms and slugs, but it will occasionally feed on crayfish, leeches, small fish, and soft-bodied insects.

Breeding: The reproductive biology of this species has not been studied in Missouri. Courtship and mating occur mainly in the spring (Minton 2001; Powell and Parmerlee

JEFF BRIGGLER

Colorful ventrum of the Kirtland's snake from Clark County.

1991; Smith 1961) but have been reported in the autumn (Anton et al. 2003). Females give birth to 4–15 young with an average of eight between late July and late September (Conant 1943; Tucker 1976). Newborn young are generally 100 to 175 mm (3.9 to 6.9 in.) in length. The young have indistinct dorsal blotches, a darker body, and a deeper red belly than adults. Reproductive maturity is reached by the age of two (Ernst and Ernst 2003).

Remarks: Kirtland's snakes have a limited distribution and are only known to occur in nine Midwestern states, including Missouri. This species is endangered, threatened, or rare throughout its range (Ernst and Ernst 2003). In Missouri, this species is considered a species of conservation concern that is extremely rare with only four known locations in northeast Missouri. Until 2006, this snake was considered likely extirpated from the state due to only a single specimen collected in 1964 (Shulse 2006). A few new records have occurred since 2006, mainly due to specimens being found dead on roads. This species has likely declined in Missouri due to the loss of native prairies and associated wetlands. The best remaining habitat occurs along highway right-of-ways and streams. If an individual is observed, please photograph and report it to the Missouri Department of Conservation.

Eastern Yellow-bellied Racer

Coluber constrictor flaviventris Say

JEFF BRIGGLER

Adult eastern yellow-bellied racer from Madison County.

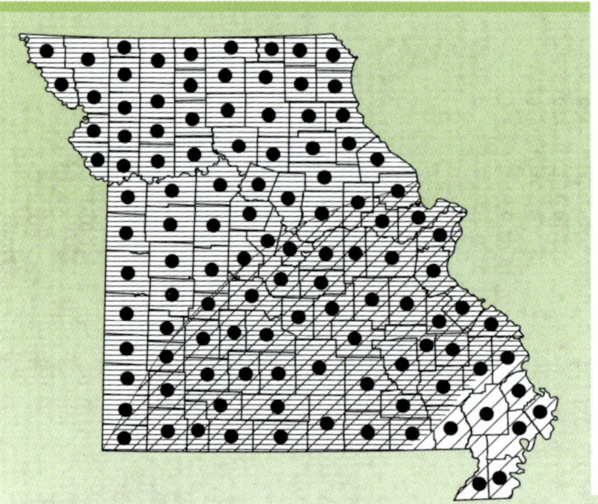

Description: The eastern yellow-bellied racer is a medium-to-large, smooth-scaled snake with variable dorsal color. Adult eastern yellow-bellied racers (three years of age or older) can be tan, brown, olive, blue, gray, or nearly black. The belly is yellow, ivory, or light blue gray. Racers have a slender body, proportionately long tail, smooth body scales, and a divided anal scale. Hatchlings and young racers are strongly patterned with closely spaced gray or brown mid-dorsal blotches and smaller, alternating spots on

Distribution: Missouri: Occurs nearly statewide (horizontal lines); it is replaced by the southern black racer in the southeastern part of the state (diagonal lines). North America: Eastern yellow-bellied racers are found throughout most of the Great Plains states from North Dakota and Montana south into southeastern Texas. The southern black racer ranges from southern Indiana and Illinois, southwest into southeast Missouri, most of Arkansas, and into eastern Oklahoma and extreme northeastern Texas. In addition, it is found in the southern Carolinas south to Florida and west to eastern Louisiana (Powell et al. 2016).

JEFF BRIGGLER

Racers vary in color throughout much of Missouri and may represent intergrades of several geographic races (subspecies), such as this specimen from Callaway County.

the sides over a tan ground color. The belly of young racers is normally cream colored with some dark-gray speckling. The juvenile pattern fades with age, and by the third season most or all dorsal spots disappear.

Adult eastern yellow-bellied racers range in total length from 580 to 1,270 mm (22.8 to 50.0 in.; Collins et al. 2010) but have been known to reach 1,905 mm (75.0 in.; Powell et al. 2016).

Habits and Habitat: The eastern yellow-bellied racer is active from March to November with a peak from late April to early August. Fitch (1963a) conducted an extensive study of this species in eastern Kansas and reported that racers are active at air temperatures between 15.6°C to 32.2°C (60°F to 90°F) but seem to prefer a temperature range of 25.6°C to 26.7°C (78°F to 80°F). Racers are diurnal and can be observed searching for food or basking on sunny days. This fast-moving snake depends on speed and agility to escape a predator or overtake prey. When alarmed, racers will vibrate their tails. Racers usually struggle violently, bite viciously, and discharge musk and waste matter from their vents when captured. Eastern yellow-bellied racers live in a wide variety of habitats, including native prairies, grasslands, pastures, old fields, glades, brushy areas, forest edges, and agriculture fields.

During spring and fall, racers are often found on rocky, wooded, south-facing hillsides. This species will overwinter in such habitat (beneath rocks, within rock crevices, and within rotting logs and stumps) or will select animal burrows (mammals and crayfish) in open habitat as a winter retreat. Racers (along with western ratsnakes) were reported to use a small sandstone cave as a hibernaculum. They were observed using a thermal gradient inside the cave to select optimum overwintering conditions (Sexton and Hunt 1980). Drda (1968) found a few racers overwintering in a Jefferson County cave. They will also use human structures such as stone walls, building foundations, abandoned wells, and cisterns. Many racers, along with several other species of snakes (western ratsnakes, eastern gartersnakes, and brownsnakes), were found overwintering in rock-lined old wells in several counties in Missouri (J. Briggler unpubl. data; R. Daniel pers. comm.). Although not territorial, racers

Hatchling eastern yellow-bellied racer from Cole County (above). Eastern yellow-bellied racer eggs from Cole County (right).

JEFF BRIGGLER

TOM R. JOHNSON

have been found to have a home range averaging about 10.1 to 12.1 ha (25 to 30 ac) in size in studies in Kansas (Fitch 1963a; Klug et al. 2011).

Eastern yellow-bellied racers eat a variety of animals, including many small mammals, especially rodents, as well as lizards, small snakes, frogs, birds, and insects (Fitch 1963a). A specimen captured in Camden County had eaten a ground skink (Smith and Powell 1993). Racers collected in Holt County had eaten grasshoppers, rodents, and small snakes (R. Seigel pers. comm.). Racers are not constrictors; they quickly seize prey with their mouth and swallow it alive. Larger prey items often are pinned to the ground and chewed upon until death prior to swallowing.

Breeding: Courtship and mating occur soon after racers emerge from overwintering retreats, usually during early April and throughout the spring. Anderson (1965) documented mating in Jackson County in mid-to-late April. Adult males locate females by scent. Once a receptive female is found, the male will court her by moving alongside and on top of her, while his body ripples spasmodically. If the female moves the male follows, remaining in contact and continuing the courtship. Once the female becomes passive, the male moves his tail under hers and copulation occurs. Egg laying takes place from mid-June to late July. Generally, 8 to 21 eggs are laid per female, depending on the size of the female. The eggs may be laid under logs or large rocks, or in rotten stumps, sand, compost piles, sawdust piles, or abandoned mammal burrows. A clutch of nine eggs was found under a large flat rock in Cole County in early August with well-developed embryos. Racer eggs are cream colored with a rough texture, somewhat like sandpaper; they average 29.7 by 17.2 mm (1.2 by 0.7 in.; Fitch 1963a). A specimen captured in Lincoln County laid 16 eggs in early June; the eggs averaged 30.4 by 24.4 mm (1.2 by 1.0 in.; Smith and Powell 1993). Another much larger clutch of 29 eggs was found in an abandoned mammal burrow in the corner of a Greene County vegetable garden (Edmond 2011). These eggs averaged 29 mm (1.1 in.) in length. An unusual egg deposition site was reported in Barton County: 23 eggs were discovered in mid-June in a 50.8 mm (2 in.) diameter drainage pipe (Hoisington and Durbin 2019). Racer eggs generally hatch in about two months in August and September. Depending upon temperature and moisture level, incubation time can range from 40 to 97 days (Ernst and Ernst 2003). Anderson (1965) reported an incubation period for a clutch of eggs in Missouri to be 65 to 66 days. Eggs collected by Edmond (2011) in Greene County hatched in late July with a minimum incubation of 42 days, but the date of egg laying was unknown. Hatchling eastern yellow-bellied racers from Lincoln County averaged 253.9 mm (10 in.; Smith and Powell 1993), while hatchlings from Greene County averaged 256.3 mm (10.1 in.; Edmond 2011). Reproductive maturity is reached in two to three years. Individuals are known to live about 10 years, but it is likely they can live considerably longer (Fitch 1999).

Subspecies: The southern black racer (*Coluber constrictor priapus* Dunn and Wood) is known to occur in the Mississippi lowlands of southeastern Missouri. Adult southern black racers are normally uniform dark gray to blue black and have a prominent white chin and throat. Throughout much of the Ozark Highlands, there appears to be more of a southern black racer influence based upon coloration (R. Krager pers. comm.), but it is unclear if southern black racers are more widely distributed in Missouri or if this coloration represents intergradation of the southern black racer and eastern yellow-bellied racer. In addition, racers captured in the St. Louis area have a blue body coloration more similar to the blue racer (*Coluber constrictor foxii*) found east of the Mississippi River (R. Krager, pers. comm.). It remains to be seen if these colorations are indicative of individual variation or if the distribution of these subspecies is changing. Additional genetic research is likely warranted to better understand the distribution and taxonomic status of the racer complex.

Little is known about the natural history of the southern black racer in Missouri. They have been observed along swamps near limestone bluffs, soybean fields, and brushy areas near drainage ditches (Anderson 1965). They appear to be most abundant within the remaining sand prairie and sand savanna habitat of southeastern Missouri. This subspecies is known to eat frogs, lizards, small snakes, and rodents (Anderson 1965; R. Daniel pers. comm.).

Remarks: The local name for racers in Missouri is "blue racer." The name racer is fitting due to their incredible speed as they are often seen fleeing quickly across a road or into their surroundings. Racers are also good climbers and can be found basking in small trees or bushes. They are highly visual and often seen raising their head or "periscoping" in grassland habitat to view the world around them.

Eastern Coachwhip

Coluber flagellum flagellum Shaw

TOM R. JOHNSON

Adult eastern coachwhip from Maries County.

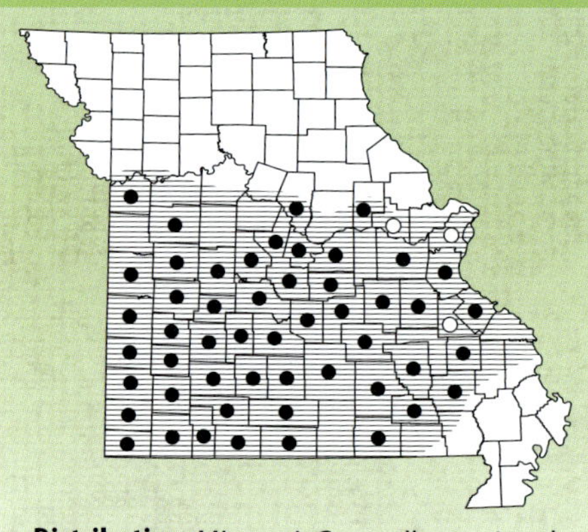

Distribution: Missouri: Generally restricted to the Missouri Ozarks and the Ozark border areas. North America: Southeastern Kansas through the Missouri Ozarks, south through Arkansas, the western half of Louisiana, extreme eastern Texas, the eastern half of Oklahoma, and into southeastern states to the southern tip of Florida (Powell et al. 2016).

Description: The eastern coachwhip is a large, slender, fast-moving snake with dark color anteriorly and light color posteriorly. Its color varies, but generally the anterior two-thirds is dark brown or black; the posterior one-third of the body is tan or red brown. Belly may be brown, tan, light yellow, or pink. Individuals from southwestern Missouri often are completely black or dark brown. Occasionally some individuals will have broad, pink bands along the back. Juvenile eastern coachwhips are marked with numerous dark-brown crossbands over a tan ground color anteriorly, which fades to an overall tan color on the posterior. The dark markings disappear with age. Dorsal scales are smooth; the anal plate is divided.

Adult eastern coachwhips range in total length from 1,067 to 1,520 mm (42 to 59.8 in.) but have been known to reach 2,590 mm (102 in.; Powell et al. 2016).

Habits and Habitat: This species is active during sunny days from April through

TOM R. JOHNSON

TOM R. JOHNSON

Eastern coachwhips occasionally have pink bands on their bodies and tails (above). Ventral coloration of an eastern coachwhip from Maries County (left).

early November. In Missouri, the eastern coachwhip lives on dry, rocky glades, or brushy or wooded hillsides. It also occurs along the edge of prairies where there is ample brush and other shelter. They are primarily active during the daytime hours. When approached, a coachwhip will normally escape with an explosive burst of speed to retreat under a nearby rock or up a tree. If an individual is cornered, it will maneuver into a defensive coil, vibrate its tail, fight savagely and bite to defend itself. On cool days coachwhips hide under flat rocks or in small mammal burrows. They are known to occasionally raise the upper portion of their body to get a better view of the surrounding area. This species overwinters in rock crevices, under large rocks, or in mammal burrows. Home range size can be quite large and variable among individuals. A study in eastern Texas found an average home range of 70.4 ha (174.0 ac), which varies from 16.1 to 268.4 ha (39.8 to 663.2 ac; Johnson et al. 2007). Another study in Georgia found an average home range of 102.9 ha (254.3 ac), which varied from 59.1 to 132.6 ha (146.0 to 327.7 ac; Howze and Smith 2015).

This species eats grasshoppers, cicadas, rodents, small birds, lizards, and other species of snakes (Ernst and Ernst 2003). Coachwhips will climb trees to eat bird eggs and nestlings. A study in Arkansas found lizards to be the predominate food item for the coachwhip (Trauth and McAllister 1995). The coachwhip is not a constrictor. They actively search for prey. Once

TOM R. JOHNSON

Hatchling eastern coachwhip from Maries County.

prey is located, they quickly capture the prey with their mouth and immediately swallow it. Young coachwhips consume insects and small lizards.

Breeding: Little is known about the courtship and mating of this species. Mating is likely to occur in April or May soon after emerging from winter retreats (Ernst and Ernst 2003; Perkins and Lentz 1934). However, a pair of coachwhips were observed mating along the edge of a paved road in early June in Cedar County where the male repeatedly bit the neck of the female and quickly intertwined his tail with jerking motions (T. and M. Green pers. comm.). Mating may occur in the fall (Trauth et al. 2004). Four to 24 eggs (average of 11) are laid per female (Ernst and Ernst 2003). Eggs are laid during late June through July in loose soil, animal burrows, rotting logs, or decaying leaf litter (Collins et al. 2010; Vermersch and Kuntz 1986). The eggs are white, have a granular texture, and have an average size of 48 by 26 mm (1.9 by 1 in.; Anderson 1965). A female kept in captivity in St. Clair County laid 12 eggs in early July (Anderson 1965). Hatching takes place in late August into September with an average incubation period of 64 days (Ernst and Ernst 2003). Newly hatched eastern coachwhips average 256 mm (10.1 in.) in length. Age and size of sexual maturity is unknown. An individual, captured as an adult of unknown age, lived over 18 years in captivity (Snider and Bowler 1992).

Remarks: The eastern coachwhip is considered Missouri's fastest snake, reaching a speed of up to 5.8 kph (3.6 mph; Mosauer 1932). Although fast, it cannot move as fast as a person can run. A young eastern coachwhip captured in Barton County was observed feigning death just after being released at the site of capture (Pflanz and Powell 1990).

Prairie Ring-necked Snake

Diadophis punctatus arnyi Kennicott

JEFF BRIGGLER

Adult prairie ring-necked snake from Barry County.

Description: The prairie ring-necked snake is a small, dark snake with a yellow or orange ring around the neck that is one or two scale rows wide and usually has a posterior edge of black. The dorsal color is normally shiny dark gray but may be gray brown. The top of the head is usually darker than the body. Coloration of the belly is yellow with numerous irregularly placed, small black spots; the belly changes to a bright orange or red along the underside of the tail. Body scales are smooth; anal scale is divided. Young ring-necked snakes have a darker dorsal color, which is often shiny black. Ring-necked snake adult males are smaller than females.

Adult prairie ring-necked snakes range in total length from 254 to 380 mm (10 to 15 in.) but have been known to reach 706 mm (27.8 in.; Powell et al. 2016).

Habits and Habitat: This is a common but secretive snake that takes shelter under rocks, logs, bark slabs, or other debris. Ring-necked

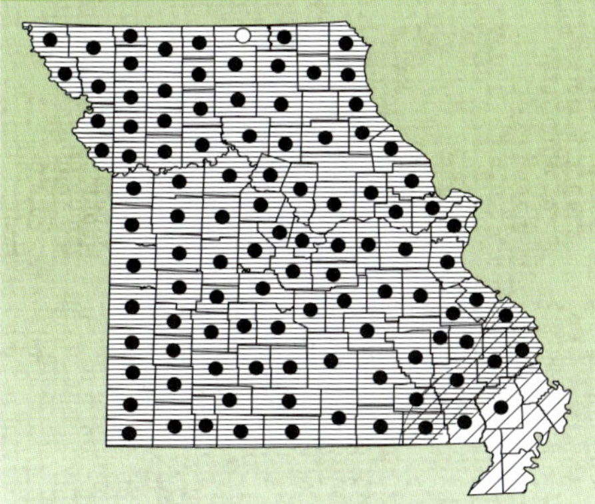

Distribution: Missouri: Nearly statewide (horizontal lines) but is replaced by the Mississippi ring-necked snake in the southeastern corner of the state (diagonal lines). North America: The prairie ring-necked snake ranges through most of the central and southern Great Plains states and into Missouri and Arkansas. The Mississippi ring-necked snake occurs from southeastern Missouri south to the Gulf of Mexico and into eastern Texas (Powell et al. 2016).

Colorful ventrum of a prairie ring-necked snake with sharply pointed tail from Shannon County (above). Prairie ring-necked snake eggs from Taney County (right).

snakes are active between late March and early November. Optimal body temperature is maintained by resting under sun-warmed rocks or other objects on the ground; thus, they do not openly bask in the sun. A study in eastern Kansas found that an optimum feeding temperature was between 25°C and 35°C (77°F and 95°F; Henderson 1970).

Prairie ring-necked snakes are the most common snakes in some locations in Missouri. They occur in a variety of habitats: native prairies, pastures, open woods, along the edge of woodlands, and on dry, rocky glades. They can be found in towns and small cities with empty lots and unkept yards that are littered with rocks, slabs of old concrete, old tin, or boards used by snakes as hiding places (Riedle 2014). This study, conducted in Cole County, found that a local population was partially sustained by two-year-old ring-necked snakes moving into a different population. The ring-necked snake can be found with several other small snake species using the same flat rock, piece of tin, board, or other shelter objects. In spring, one flat rock in the preferred habitat may harbor 10 or more ring-necked snakes. A rock 61 by 43 cm (24 by 17 in.) was lifted in early May to reveal 17 ring-necked snakes (Hurter 1911). During the hot, dry months of July, August, and early September, ring-neck snakes will burrow into the soil to locate cooler temperatures and moisture. They will overwinter in burrows of other animals or move underground into deep crevices in bedrock or into the rotting roots of old, large stumps. Wilkinson (1962) collected many ring-necked snakes in talus slides on south-facing slopes during the winter in Boone County. Predators include other snakes, such as kingsnakes, milksnakes and racers, collared lizards, shrikes, hawks, shrews, and domestic cats. Opossums were reported to eat ring-necked snakes in a study conducted in central Missouri; but it is unsure if the opossums were locating ring-necked snakes in leaf litter or were eating road-killed individuals (Reynolds 1945). Newly hatched ring-necked snakes have, several times, been observed in the webs of spiders (*Theridion* spp., cobweb weavers) in an old, rock-walled basement of an Osage County farmhouse (G. and I. Emich pers. comm.). This same spider-eating-snake scenario was reported for the northern ring-necked snake (*D. p. edwardsii*; Groves and Groves 1978; Klemens 1993). A ring-necked snake was found in a black widow spider web in Virginia (Ernst and Ernst 2003).

Natural prey is primarily earthworms, but other prey animals include slugs, insect larvae, small salamanders, and small frogs, such as immature chorus frogs. A pair of enlarged teeth at the back of the mouth help ring-necked snakes swallow their prey. As with many species of snakes, ring-necked snakes locate their prey by scent. It is likely earthworms make up nearly all of the prairie ring-necked snake's diet in Missouri. An earthworm diet allows this species to live in close proximity to people in towns and small cities (Riedle 2014).

Breeding: This species will breed soon after it emerges from overwintering in late March or early April. However, in a Boone County study, breeding mainly occurred in the autumn (Wilkinson 1962). Courtship and breeding have not been studied in Missouri, but a breeding pair was observed in the grass on a Perry County glade in early May (J. Briggler, pers. obs.). The majority of adult females reproduce annually (Fitch 2003). More than one clutch may be laid per season. It has been found that the eggs show advanced development before being laid (Ernst and Ernst 2003). In Missouri, eggs are laid in June through early July (Anderson 1965; Wilkinson 1962). On average, a female will lay about three or four eggs; larger females will lay more eggs, with a maximum of 10 (Fitch 1975). Fitch (1975) reported that ring-necked snake eggs are "somewhat curved and sausage-shaped." Several authors have reported communal nesting by ring-necked snakes (Blanchard 1942; Gilhen 1970). The eggs are white to yellow, have thin, leathery shells, and are 23 to 28 mm (0.9 to 1.1 in.) in length. A female will deposit her eggs in abandoned small mammal burrows, under flat rocks or piles of tree bark, or in old sawdust piles—as long as there is ample moisture. The eggs will hatch during late August through early September. Prairie ring-necked snake hatchlings average 111 mm (4.4 in.) in total length. Males reach breeding age at two years; females mature at three years (Fitch 1975). In a Kansas study, Fitch (1975) found a small number of individuals in a population that may reach 15 or more years of age.

Subspecies: In all, ring-necked snakes of the *Diadophis punctatus* lineage contain 12 subspecies or geographic races (Ernst and Ernst 2003). The prairie ring-necked snake, described above, ranges through all of Missouri except for the southeastern corner, where the Mississippi ring-necked snake (*Diadophis punctatus stictogenys* Cope) replaces it. There is a potential area of intergradation between the two races just north and west of the Bootheel that is in need of study (see species range map). A field collection of 13 specimens, reported to be intergrades between the two subspecies, were found in Ste. Genevieve County (Anderson 1965). This southern geographic race is reported to be slightly smaller than the prairie ring-necked snake. Its coloration is similar, but the yellow neck ring of the Mississippi ring-necked snake is usually narrower and can be "broken" or interrupted dorsally. Its belly is yellow with small black spots that are arranged in two or three distinct, longitudinal rows. This ring-necked snake has a similar natural history to the prairie ring-necked snake covered above.

Remarks: When first uncovered, a specimen may tightly coil its tail and expose its brightly colored underside to draw attention away from its head. This small snake is not known to bite a person but when first handled, will discharge a pungent, unpleasant musk from glands at the base of its tail which is, simultaneously, mixed with fecal matter from its cloaca. This offensive behavior is found in several snake species, including some venomous snakes.

Western Mudsnake

Farancia abacura reinwardtii Schlegel

JEFF BRIGGLER

Adult western mudsnake from Ripley County.

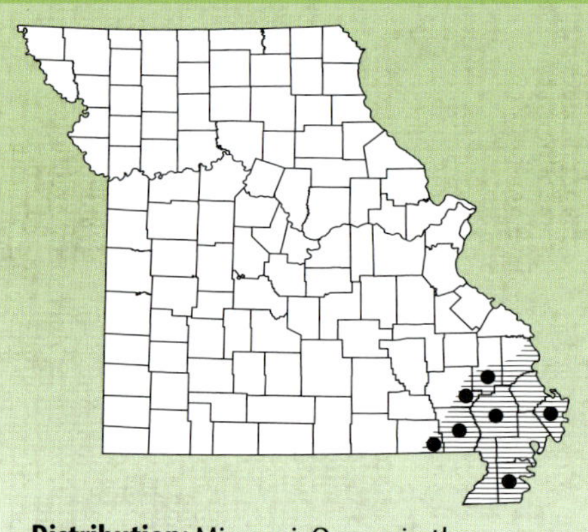

Distribution: Missouri: Occurs in the natural swamps of the Mississippi Alluvial Basin of southeastern Missouri. North America: Extreme southern Indiana and Illinois, and south to the Gulf Coast states of eastern Alabama, Mississippi, Louisiana, and eastern Texas (Powell et al. 2016).

Description: The western mudsnake is a medium-sized, smooth, glossy black snake with a red- and black-checkered belly. Dorsal color is black or blue black and extends onto the belly to form irregular bands. These bands of black are separated by the red belly color extending onto the sides. The head and body appear slightly flattened. The dorsal scales are smooth; anal plate is divided. The tip of the tail normally terminates in a sharp but harmless spine. Young western mudsnakes are the same color as adults but have more of a sharp tail tip compared to the blunt tail tip in adults.

Adult western mudsnakes range in total length from 1,020 to 1,370 mm (40.2 to 53.9 in.) but have been known to reach 2070 mm (81.5 in.; Powell et al. 2016).

Habits and Habitat: This semi-aquatic snake is active from late March into early October, with the majority of activity in Missouri occurring from April through June. Mudsnakes are highly secretive. They

Western mudsnakes live in shallow swamps with an abundance of downed logs (above). A sharp scale at the end of the tail of an adult western mudsnake (left).

are seldom active during the day, preferring to take shelter under logs or in animal burrows. They are swamp dwellers and prefer shallow areas where there are numerous rotten or water-soaked logs. They can be found in flooded, hardwood forest bottomlands, marshes, swamps, margins of streams, oxbow lakes, moist soil units, and occasionally in drainage ditches (Schepis 2013; J. Briggler pers. obs.).

Mudsnakes are not known to bite to defend themselves, but when freshly captured they will attempt to prick one's skin with the tip of the tail. This behavior may be startling, but the tail tip cannot break the skin. A female mudsnake captured in southern Mississippi feigned death, much like a hog-nosed snake (Doody et al. 1996). Also, when threatened, it will tuck its head beneath its coils and curl the tail to show the red belly coloration.

A telemetry study in the northern limits of the western mudsnake range in Wayne County documented that the greatest distances moved by individuals occurred mostly in June and July (Schepis 2013). Males tended to move greater distances in the spring and non-gravid females moved more in mid to late summer. Its home range was extremely variable. The average home range for males was 11.3 ha (27.9 ac), and for non-gravid females it was 21.9 ha (54.1 ac). The home range of a single gravid female was 0.025 ha (0.06 ac; Schepis 2013). Mudsnakes may overwinter in decaying stumps, cavities at the base of trees, and burrows along banks overhanging water. Snakes tracked by Schepis (2013) typically overwintered 4 to 6 m (13.1 to 19.7 ft) from the edge of natural wetlands and ditches, mainly using sites that had an eastern, southern, or southeastern exposure.

The western mudsnake has a specialized food preference: the aquatic amphiuma or lesser siren. Adult mudsnakes use the sharp tail tip to prod its prey, causing it to uncoil for easier swallowing. They occasionally eat other species of salamanders, tadpoles, and frogs.

JEFF BRIGGLER

Colorful ventrum of an adult western mudsnake from Ripley County.

Breeding: Little is known about the courtship and mating of this species in Missouri and range wide. Western mudsnakes presumably breed in the spring, probably April through June. A pair was observed mating in Stoddard County during an afternoon in mid-June (M. Nickerson pers. comm.). Mating has been observed in the water and on land (Langford and Borden 2004; Meade 1937). Langford and Borden (2004) observed courtship in mid-June in Alabama. The pair was intertwined in a loose ball with the male biting the female's neck. Eggs are presumed to be laid in animal burrows, rotten logs, or cavities at the base of trees. Nesting is likely to occur from July to early September. The leathery, creamy-white eggs average 36 by 25.4 mm (1.4 by 1 in.; Meade 1937). Eleven to 50 eggs are laid; female western mudsnakes remain with their clutch of eggs until they hatch (Cagle 1942; Meade 1940; Mount 1975; Riemer 1957). Usually, the female will lie in a loose coil around the eggs. The eggs hatch in August to early October. A clutch of 26 eggs with an unknown egg-laying date was found in Butler County, and six hatched in mid-August with an average total length of the hatchlings from 243 and 250 mm (9.6 to 9.9 in.; Anderson 1965). Schepis (2013) documented a female entering a nesting cavity at the base of a tree in early July. Later, 20 eggs were observed with the attending female. The female generally remained with the eggs throughout incubation until all eggs hatched in early October. However, she did make a few movements of a very short distance only to return to the nest cavity. Newly hatched mudsnakes were found to remain near their nest site and not move into swamps until the following spring when small amphiumas and young lesser sirens were more abundant (Semlitsch et al. 1988). In an Arkansas study, females required at least 2.5 years to become reproductively mature (Robinette and Trauth 1992). Maturity of males is unknown, but it is likely to occur sooner than females. A wild-caught adult female lived over 18 years in captivity (Snider and Bowler 1992).

Remarks: Two myths associated with this harmless snake have persisted for years. One myth involves the sharp scale at the end of the tail, which is erroneously believed to be a stinger that can deliver a deadly venom. This has given rise to the local name stinging snake. The second myth claims that this swamp dweller can grasp its tail in its mouth and roll across the ground with great speed. There is no truth to either of these myths.

The western mudsnake is listed as a species of conservation concern in Missouri due to loss of wetland and bottomland forest habitats and limited distribution in southeastern Missouri. Because this species requires a natural swamp habitat, it is important that Missouri's last remaining native cypress swamps and surrounding bottomland, hardwood forest be preserved.

Dusty Hog-nosed Snake

Heterodon gloydi Edgren

JIM RATHERT

Adult dusty hog-nosed snake from Scott County.

Description: The dusty hog-nosed snake is a stout-bodied, small-to-medium-sized, brown to brown-gray snake with longitudinal rows of dark-brown blotches and an upturned snout. The dorsal pattern of blotches consists of a row along the midline and two rows of brown spots on each side. The number of dark, dorsal midline blotches is fewer than 32 in males and fewer than 37 in females. There is usually a dark-brown diagonal line through the eye to the angle of the jaw and an elongated dark-brown blotch running along the side of the head onto the neck. The rostral scale is upturned and elongated, appearing shovel-like. The belly is wide, mostly colored black from the neck to the underside of the tail; it is edged in yellow blotches on some individuals. Scales on the body are keeled, and the anal plate is divided.

Adult dusty hog-nosed snakes range in total length from 380 to 635 mm (15 to 25 in.) but have been known to reach 914 mm (36 in.; Powell et al. 2016).

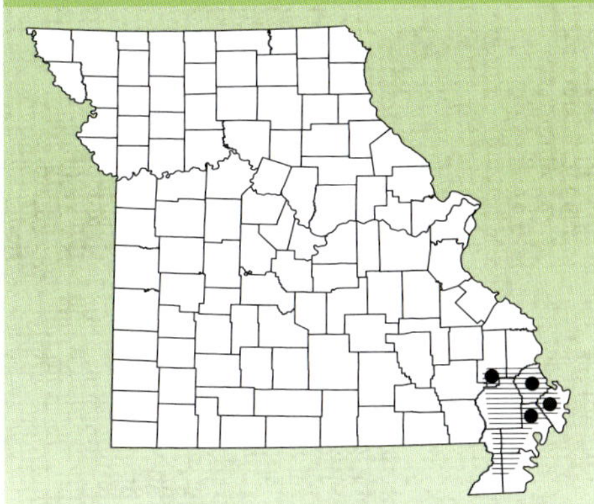

Distribution: Missouri: Restricted to a few counties in southeastern Missouri. North America: The dusty hog-nosed snake has disjunct, isolated populations in southwestern Illinois, southeastern Missouri, southeastern Kansas, and eastern parts of Oklahoma and Texas (Powell et al. 2016).

Dusty hog-nosed snakes live in sand prairies and savannas, such as this example in Scott County (above). Head of a dusty hog-nosed snake from Scott County (right).

Habits and Habitat: Little is known about the natural history of this species in Missouri. This species is active from mid-April to mid-October with most observations being reported in May, June, September, and October in Missouri. It is mainly active during daylight hours, except in the hot summer months in Missouri. In the summer, they are likely active at dusk and dawn when temperatures are more suitable. At night, hog-nosed snakes spend their time in temporary burrows created by using the sharply upturned snout, similar to a shovel, to dig in the loose, sandy soils (Edgren 1955; Platt 1969). The well-drained, sandy prairies and savannas in southeastern Missouri are the preferred habitat. They are most commonly encountered in the few remaining natural sand prairies and savannas but occasionally can be found on the edges of agriculture fields and urban areas where loose, sandy soil exists. Overwintering sites are presumed to be in burrows below the frost line dug by mammals or by the snake itself.

When threatened, this harmless species may react in a variety of ways. An individual may try to escape by crawling toward shelter; it may coil and try to hide its head; or it may hiss loudly, spread its head and neck, and strike at the intruder. If the intruder is persistent, the snake will writhe about, open its mouth, extend the tongue, regurgitate any freshly eaten food, roll on its back, and "play dead." This death-feigning behavior appears to be less frequent than the eastern hog-nosed snake. If left alone, the snake will eventually right itself and move away.

The dusty hog-nosed snake find prey by sight and odor. It uses the upturned snout to dig animals out of their burrows in the sandy soil it prefers. Prey consists of amphibians (especially frogs and toads), lizards, birds and their eggs, eggs of many reptiles, and small rodents (Ernst and Ernst 2003).

JEFF BRIGGLER

Belly and underside of tail of a dusty hog-nosed snake from Scott County.

Breeding: No information exists on the reproductive biology of this species in Missouri, and very limited information is available throughout its entire range. Based upon information from more studied *Heterodon* species, courtship and mating presumably occur soon after emergence from overwintering in mid-April into May and occasionally in the late summer into fall. Females likely search for nesting sites in June and July. Likely, 2 to 23 eggs are laid beneath the sandy soils. The eggs are positioned more in a line instead of a cluster due to the tunnel burrowing habits, similar to the plains hog-nosed snake. A female collected in mid-June in Texas contained 16 eggs (Anderson 1965). The elliptical, white-to-cream-colored eggs with smooth leathery shells average 31 by 17 mm (1.2 by 0.7 in.; Anderson 1965). Eggs likely hatch from August into September. Females presumably reach sexual maturity within two to three years, while males mature in one to two years.

Remarks: The western hog-nosed snake complex (*Heterodon nasicus*) formerly consisted of three subspecies: dusty hog-nosed snake (*H. n. gloydi*), Mexican hog-nosed snake (*H. n. kennerlyi*), and plains hog-nosed snake (*H. n. nasicus*). All three subspecies were elevated to species status. Two of these new species occur in Missouri: the dusty hog-nosed snake (*H. gloydi*) in southeastern Missouri and the plains hog-nosed snake (*H. nasicus*) in northwestern Missouri. These two species can be told apart by where they are found and by the number of dark, dorsal blotches (see individual species description accounts for details).

Hog-nosed snakes have a pair of enlarged teeth at the rear of the mouth on the upper jaw; it is presumed the teeth are used for holding and deflating toads that are full of air, and to subdue prey by injecting a toxic saliva by prolonged chewing (Kapus 1964; Kroll 1976; McKinstry 1978). Although hog-nosed snakes rarely bite people, a bite can be painful and result in slight bleeding, discoloration, and swelling at the site of the bite (Averill-Murray 2006; Weinstein and Keyler 2009). Bites do not pose significant danger to humans (Weinstein and Keyler 2009), but care must be taken when handling hog-nosed snakes, especially individuals conditioned to being hand-fed mice in captivity.

At one time the dusty hog-nosed snake was considered extirpated in Missouri because the last documented specimen was found in 1961 (Anderson 1965; Johnson 2000), but an individual was captured in 2004 from Scott County (Briggler 2004). Since this discovery, numerous individuals have been documented from Scott and New Madrid counties. Due to restricted distribution, collection pressure, and loss and alteration of sand prairie and savanna habitat in southeastern Missouri, the dusty hog-nosed snake is listed as a species of conservation concern. Conservation of this species requires an understanding of the natural history of this species in Missouri, which, unfortunately, is lacking.

Plains Hog-nosed Snake

Heterodon nasicus Baird and Girard

TOM R. JOHNSON

Adult plains hog-nosed snake.

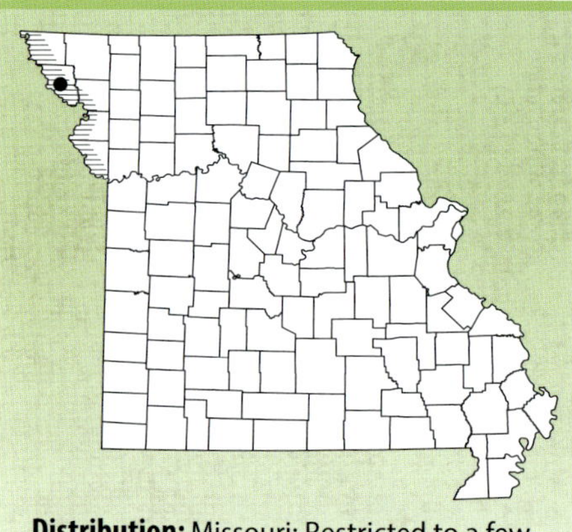

Distribution: Missouri: Restricted to a few counties in northwestern Missouri. North America: Southern Saskatchewan and Manitoba Canada south through the Great Plains states to central and eastern New Mexico and northwestern Texas. Disjunct populations occur in Illinois, Missouri, Iowa and Minnesota (Powell et al. 2016).

Description: The plains hog-nosed snake is similar to the dusty hog-nosed snake in appearance. This species is a small-to-medium-sized, gray-tan snake with a stout body. There is a row of dark-brown blotches down the back, two rows of smaller dark spots along the sides, and an upturned snout. The number of dark, dorsal blotches is more than 35 in males and more than 40 in females for the plains hog-nosed snake. There is usually a dark-brown diagonal line through the eye to the angle of the jaw and an elongated dark-brown blotch running along the side of the head and onto the neck. The rostral scale is upturned and elongated, appearing shovel-like. This species' belly is wide with jet-black color that extends from the neck onto the underside of the tail; it is edged in yellow blotches on some individuals. The scales on the body are keeled, and the anal plate is divided.

Adult plains hog-nosed snakes range in total length from 380 to 635 mm (15.0 to 25.0 in.) but have been known to reach 918 mm (36.1 in.; Powell et al. 2016).

DAVID STONNER

TOM R. JOHNSON

Plains hog-nosed snakes prefer loose or sandy soils of the loess hills and associated floodplains, such as this example in Atchison County (above). Ventral coloration of a plains hog-nosed snake (left).

Habits and Habitat: Little is known about the natural history of this species, with only three confirmed specimens from northwest Missouri. The plains hog-nosed snake resides in dry, sandy or loose soil in prairie and savanna habitats. In Missouri, it has been found where the steep loess hills meet the floodplain of the Missouri River. Platt (1969) studied this snake in east-central Kansas and found that it is mainly active during the day from late April to mid-October. The plains hog-nosed snake takes shelter in burrows it digs in sandy or loose soils, as well as mammal burrows (Degenhardt et al. 1996; Smith 1961). Such shelters are used at night during the active season and during the winter.

Similar to the other hog-nosed snake species, this species shows a wide variety of bluffing tactics when threatened; even faking its own death (see dusty hog-nosed snake account for more detail).

Movement of the species between points of capture may be restricted to a home range, but the size of the area is variable and dependent on the availability of food and shelter (Platt 1969). Platt (1969) found mean movement at two studies in Kansas to be 93–255 m (305.1–836.6 ft) for females and 79–207 m (259.2–679.1 ft) for males.

The plains hog-nosed snake locates its prey by sight and smell. By use of the upturned, shovel-like snout it can dig prey out of a burrow. Although hog-nosed snakes are known for their toad eating habits, they also consume many other amphibians, birds and their eggs, small rodents, and reptile eggs (Ernst and Ernst 2003). Turtle eggs are commonly consumed by the plains hog-nosed snake throughout its range (Kolbe et al. 1999; Platt 1969). It is particularly adapted to eating toads due to its large mouth, flexible jaws, and pair of enlarged teeth at the rear of the mouth on the upper jaw. These teeth allow for maneuvering and presumably deflating toads, as well as injecting a toxic saliva (Kroll 1976).

Breeding: Mating normally takes place in the spring, but some will mate in autumn. The eggs are laid in a shallow burrow in loose or sandy soil. From 4 to 23 eggs are laid in June or July, and average 32.5 by 17.9 mm (1.3 by 0.7 in.; Collins et al. 2010). The white or cream-colored eggs are elliptical with smooth leathery shells. Clutch size averages about 10 or 11 eggs. Female plains hog-nosed snakes produce eggs on a biennial cycle (Platt 1969). The eggs hatch in August or September with an incubation period of 50–60 days (Collins et al. 2010). Hatchlings are from 140 to 200 mm (5.5 to 7.9 in.; Conant 1975) in total length; they are the same color as adults but often brighter. Sexual maturity is attained in one to two years in males and two to three years in females (Ernst and Ernst 2003). An individual, which was caught as an adult, lived almost 20 years in captivity (Snider and Bowler 1992).

Remarks: With the elevation of the plains hog-nosed snake to a species (see remarks section in dusty hog-nosed snake account), the common name western hog-nosed snake is no longer necessary and has been replaced by the plains hog-nosed snake. The plains hog-nosed snake is considered to no longer exist in Missouri since the last documented specimen was from 1961 (Anderson 1965; Johnson 2000). There have been more recent anecdotal reports of this species in the Kansas City area, but these reports lack documentation (e.g. specimens or photographs).

The plains hog-nosed snake is a species of conservation concern due to restricted distribution and preference of a specialized habitat (sandy, loess hills) in northwest Missouri. Although this species has not been seen for many years in the state, it is important to keep a watchful eye for this secretive snake.

Eastern Hog-nosed Snake

Heterodon platirhinos Latreille

Adult eastern hog-nosed snake from Dent County.

Description: The eastern hog-nosed snake is a medium-sized snake with a heavy body and a pronounced upturned snout. Compared to other snakes of similar size, the hog-nosed snake has large eyes and a short tail. Its color is highly variable. Its ground color is gray, tan, yellow, brown, olive, or orange. Individuals can either have a series of brown dorsal blotches (20 to 30) with a smaller light marking between them or can be dull colored and lack dorsal markings, except near the head. With heavily marked specimens, there are several additional dark markings on the head: a V-shaped marking behind the eyes, as well as a dark bar across the head between the eyes, and a diagonal dark bar from each eye to the corners of the mouth. Individuals may be jet black in some parts of the species' range (Powell et al. 2016). The belly is gray, yellow, or pink, mottled with gray or greenish gray. The underside of the tail is normally lighter than the belly. Even on individuals with no dorsal markings, there is always a pair of large,

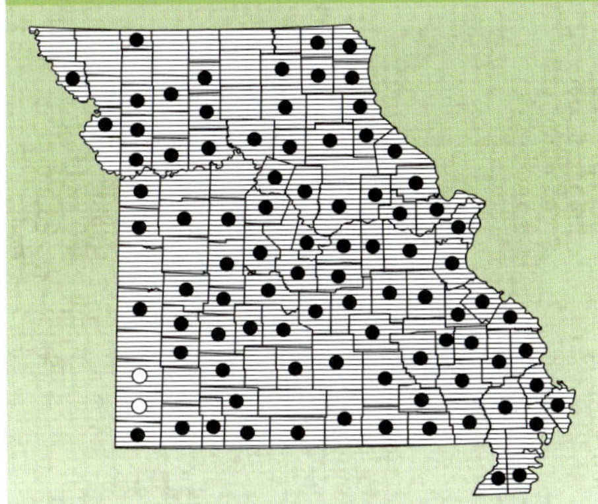

Distribution: Missouri: Presumed to occur statewide. North America: The entire eastern half of the United States from southern New England through Florida, west to southeastern South Dakota, and south to the eastern half of Texas (Powell et al. 2016).

Adult eastern hog-nosed snake from Cole County (above). Adult eastern hog-nosed snake in a defensive posture from Ozark County (right).

dark-brown or black blotches behind the head. Newly hatched eastern hog-nosed snakes are more colorful than adults with numerous brown, black, tan, yellow, or orange blotches that can form bands toward the tail. Dorsal scales are keeled, and the anal plate is divided. Males are smaller than females; the tail of males is longer than females.

Adult eastern hog-nosed snakes range in total length from 510 to 840 mm (20.1 to 33.1 in.) but have been known to reach 1,268 mm (49.9 in.; Powell et al. 2016).

Habits and Habitat: This species is active during the day. It takes shelter in the burrows of small mammals or burrows into loose soil or sand and will on occasion hide under a flat rock or various man-made objects (e.g., board or old pieces of tin) during the active season. Overwinter dormancy lasts from early November to April. A young of the year was observed in a sunny spot along the sandy floodplain of the Jacks Fork in early November in Shannon County (T. Crabill pers. comm.). Eastern hog-nosed snakes overwinter in abandoned small mammal burrows, primarily those made by moles (Plummer 2002), burrows dug by themselves, or within old logs and stumps (Ernst and Ernst 2003). Observations of eastern hog-nosed snakes have shown that this species has a poorly defined home range and makes long-distance movements that often vary considerably from year to year (Collins 1993; Plummer and Mills 1996, 2000). Home ranges are generally quite large with an average of 50.2 ha (124 ac) in a northern Arkansas study (Plummer and Mills 2000).

JEFF BRIGGLER

TOM R. JOHNSON

An eastern hog-nosed snake playing dead from Ozark County (above). An eastern hog-nosed snake eating an American toad (left).

Eastern hog-nosed snakes live in areas of sandy or loose soil in open woods, old fields, open river floodplains, and on rocky, wooded hillsides. This snake will most often be found when it is crossing a country road, woodland trail, or moving across a cultivated farm field. Eastern hog-nosed snakes eat toads. Their prominent, upturned snout, found on few other species in Missouri, is used to dig and prod loose or sandy soil in a deliberate effort to locate toads. The search for prey usually takes place in the morning (Ernst and Ernst 2003).

Close examination of the open mouth of a newly captured eastern hog-nosed snake will reveal the presence of small, black objects. They are only a few millimeters (about 0.1 in.) in size and there may be from one to six or more of them. These are, in fact, parasitic flatworms. They are known as flukes (Class Trematoda, *Ochetosomatid aniarum* and *O. digeneans* to name just two). These internal parasites are adult stages of native amphibian parasitic flukes, which live as tiny larvae in the bladder and lungs of toads and frogs and their tadpoles. When a hog-nosed snake eats an infected toad or frog, the fluke's larvae (metacercaria) are eventually released into the snake's digestive system. The life history of these common amphibian internal parasites is quite complex, but hog-nosed snakes, racers, gartersnakes, and watersnakes (*Heterdon*, *Coluber*, *Thamnophis*, and *Nerodia* spp. respectively) are hosts to them as part of the parasite's life cycle (Dyer 1999). At least four different animal hosts may be part of a fluke's life cycle.

Local or regional names such as "spreadhead," "hissing viper," and "puff adder" are used for this harmless species because of their elaborate means of defending themselves. When threatened, an eastern hog-nosed snake will flatten its head and neck, hiss loudly, and even strike; but this is an act and the mouth is closed. If this action does not drive away an intruder or the snake is further provoked, the hog-nosed snake will go into convulsions,

Hatchling eastern hog-nosed snake from Cole County (above). Hatchling eastern hog-nosed snake from Cole County (right).

thrash around, open its mouth, let the tongue hang out, regurgitate any prey animal from its stomach, release feces from its cloaca, and roll over on its back, and play dead. This dead-looking snake will give itself away as a fake death if it is gently rolled over onto its belly; it will immediately roll over onto its back again. If the intruder moves away, and the hog-nosed snake is left alone, it will eventually slowly roll over onto its belly, cautiously look about, flick its tongue to make sure it is safe, and move away to safety. It has been documented that newly hatched eastern hog-nosed snakes can use both forms of protective display (Raun 1962). Munyer (1967) saw an eastern hog-nosed snake feign death while in the water; it appeared bloated and dead with the belly up and the mouth closed. Sexton (1979) remarked on the importance of the hog-nosed snake turning its head toward their body while exposing the coiled tail as a means of protecting its head. Because of their hissing and defensive behavior, hog-nosed snakes are often misidentified as dangerous and killed because of unwarranted fear.

Hog-nosed snakes are the primary predator of toads in North America. This includes our true toads of the genus *Anaxyrus* and spadefoots, *Scaphiopus* and *Spea*. Toad skin secretions are toxic and few predators eat them. But species of snakes in the genus *Heterodon* (hog-nosed snakes) are immune to toad toxins (bufotoxin); allowing them to eat toads with no ill effects (Tyler et al. 2001). However, they are also known to eat a variety of frogs, terrestrial salamanders, lizards and their eggs, ground-nesting bird eggs, and small mammals, such as mice and shrews (Ernst and Ernst 2003; LeClere 2013). Predators include hawks, owls, kingsnakes, racers, and coachwhip snakes. Their eggs are likely eaten by raccoons, skunks, and badgers.

Breeding: The eastern hog-nosed snake mates in April and May with another smaller peak occurring in September and October (Plummer and Mills 1996). Their courtship has not been observed in Missouri. Males locate females during the breeding season by following scent trails (Plummer and Mills 1996). Often females will mate with more than one male during the breeding season (Plummer and Mills 1996). Normally only one clutch of eggs is produced per summer; the eggs are laid in June or July. Platt (1969) studied the reproductive biology of this species in east-central Kansas, and his data are included in this account. Four to 61 eggs are laid per female, with an average of 22. The eggs are usually laid in a shallow burrow in sand or loose soil. Hog-nosed snakes have also been known to lay their eggs in old sawdust piles, beneath rocks, and probably in small mammal burrows (Ernst and Ernst 2003). A captive female from Cole County laid 24 eggs in late June (Johnson 2000). The white eggs averaged 35 by 23 mm (1.4 by 0.9 in.) in size when freshly laid. Their eggs are reported to incubate for 50 to 65 days. Hatching takes place in August or September. Their brightly colored young have an average total length of 234 mm (9.2 in.) at hatching. Eastern hog-nosed snakes become mature at about two years of age.

Remarks: All black or melanistic specimens of eastern hog-nosed snakes have been reported in several states, including Alabama, Georgia, Indiana, Kentucky, Ohio, and North Carolina (Barbour 1971; Conant 1951; Minton 1972; Mount 1975; Palmer and Braswell 1995). Hurter (1911) reported dark-brown to black specimens from four Missouri counties (Stone, Howell, Jackson, and Crawford). His description of the specimens follows: "The color above is dark brown to black, beneath dirty white to yellowish. The lower side of the tail is always yellow. Sometimes the faint markings on the back of the eastern hog-nosed snake can be perceived." He also stated that he had never seen melanistic young eastern hog-nosed snakes in this state. Prodigy of melanistic adults are dark gray at hatching and darken as they develop to become a satin-black color as adults (Ernst and Ernst 2003). In Roger Conant's first edition of the Peterson field guide of reptiles and amphibians (1958), he stated that jet-black or plain gray eastern hog-nosed snakes can be common in some locations. Another all black specimen was photographed south of Eureka in St. Louis County (R. Krager pers. comm.).

Hog-nosed snakes have enlarged teeth on the posterior maxillary bones at the back of their mouth, which are thought to assist in swallowing their prey. In addition, this species has a gland (called Duvernoy's gland) that produces a mild venom. These are in association with their salivary glands, and their secretions will enter a prey animal as it is being swallowed. This toxin helps subdue the prey for ease of swallowing. A study of hog-nosed snake saliva indicated it is somewhat toxic to amphibians but not to laboratory mice (McAlister 1963). Anderson (1965) conducted an experiment to determine if the saliva of hog-nosed snakes was toxic to humans. He concluded that some individuals may develop a burning or tingling sensation with a little swelling of the hand from their saliva; but these symptoms were short-lived. It should be emphasized that hog-nosed snakes are not aggressive, rarely if ever bite people—even when freshly captured—and are not considered a threat to humans.

Prairie Kingsnake

Lamprepeltis calligaster (Harlan)

Adult prairie kingsnake from Chariton County.

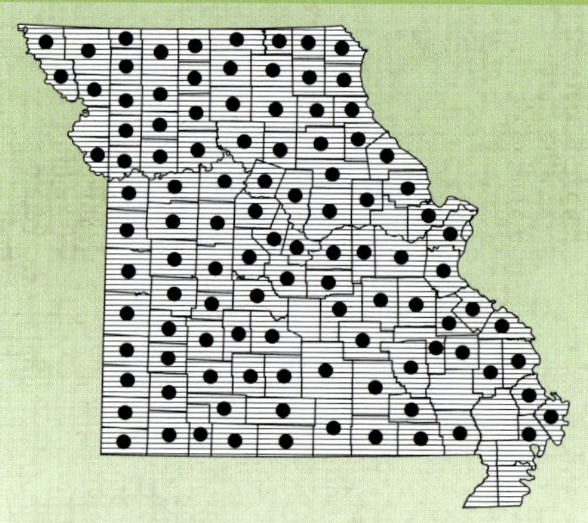

Distribution: Missouri: Presumed to occur statewide but less common in the southeastern corner of the state. North America: Southeastern Nebraska and southern Iowa east and south to the southern half of Illinois, western Indiana, Kentucky, Tennessee, and northern Mississippi, south to the western half of Louisiana and the eastern half of Texas, north to most of Oklahoma and west-central Kansas (Powell et. al. 2016).

Description: The prairie kingsnake is a medium-sized, tan or gray snake with numerous brown blotches. Up to 60 brown or red-brown, black-edged blotches occur along the back with two rows of smaller blotches along the sides. In many individuals, the dorsal markings are more like saddles or bands than rounded blotches. Older individuals often have a darkened ground color; this is usually observed on individuals from the southern half of Missouri. These darkened individuals will often have their faded large blotches fused with the darkened ground color causing a striped appearance. The top of the head usually has a backward-pointing, arrowhead-shaped marking, and a thin dark marking across the head between the eyes and down to the corners of the mouth. Scales along the upper and lower jaws as well as the chin are normally white. The belly is yellow with rectangular brown markings. Newly hatched young are lighter and more colorful than adults. The dorsal scales are smooth and the anal plate is single.

JEFF BRIGGLER

TOM R. JOHNSON

Hatchling prairie kingsnake from Cole County (above). Prairie kingsnake hatchlings from Andrew County (left).

Adult prairie kingsnakes range in total length from 760 to 1,067 mm (29.9 to 42.0 in.) but have been known to reach 1,430 mm (56.3 in.; Powell et al. 2016).

Habits and Habitat: The active season of this species is normally April to early November. Adult prairie kingsnakes have been seen above ground on sunny days during a warm winter. This species has been found active in November, December, and February in Illinois (Smith 1961). In the spring and autumn, prairie kingsnakes are active in the morning and early evening, but they become nocturnal during the summer. As with most kingsnakes, this species is rather secretive; it takes shelter under objects such as logs, rocks, and man-made materials (weathered boards and old pieces of tin), or in small mammal burrows. This kingsnake will use overwintering dens in rocky, wooded, south-facing areas, road embankments, and animal burrows. Prairie kingsnakes commonly use crayfish burrows located in drier, upland prairies in northern Missouri (J. Briggler pers. obs.). However, a field study of this species in west-central Illinois found that overwintering sites were mostly mammal burrows that were not shared with other individuals or snake species. In addition, most snakes moved to their selected overwinter sites starting in mid-September (Richardson et al. 2006). This study confirmed the consensus that kingsnake species are secretive and remain hidden; prairie kingsnakes fitted with transmitters remained underground at a minimum of 73 percent of the time tracked.

The name prairie kingsnake is somewhat misleading because this common harmless species lives not only in native prairie habitats but also along the edge of crop fields, hayfields, fallow farm fields, or the edge of open wood lots, on rocky, wooded hillsides, and near farm buildings. Prairie kingsnakes have been found in empty lots on the edge of small towns in northern and central Missouri.

When alarmed, a prairie kingsnake will vibrate its tail; if captured, it may try to bite to defend itself. However, their bite causes nothing more than shallow scratches. When first captured, a specimen will usually release musk from glands at the base of its tail.

Food consists of small mammals (e.g., mice, voles, and shrews), lizards, and occasionally small snakes. This species kills its prey by constriction. As with other kingsnake species, the prairie kingsnake is immune to the venom of the copperhead, cottonmouth, and rattlesnake (Weinstein et al. 1992). They have been found to prey on the eggs or young of ground-nesting birds (Ernst and Ernst 2003). Several studies in surrounding states have shown that small mammals comprise 60 to over 79 percent of the snakes total diet (Fitch 1982; Klimstra 1959; Trauth and McAllister 1995). A specimen captured in Cass County had eaten a five-lined skink, and another specimen captured in Jackson County had eaten a prairie vole (*Microtus o. ochrogaster*; Smith and Powell 1993). Predators of this species and their eggs include hawks and owls, kingsnakes, raccoons, opossums, skunks, badgers, and domestic cats.

Breeding: This species mates in early spring soon after emerging from winter dormancy (late March-early April) into early June. A pair of adults were found copulating along a grassy roadside ditch in Callaway County in late March during a warm spring (J. Briggler pers. obs.). Males locate females by following their scent. Copulation ensues soon after a short courtship and the pair are together for one hour or less. Males have been observed in "combat dance" behavior on several occasions, and this behavior is thought to play a role in one male becoming an area's dominant breeder (Fitch 1978; Ernst and Ernst 2003).

Prairie kingsnakes in Missouri lay their eggs mainly during June and July. A female may lay from 3 to 21 eggs, with an average clutch size of 10 eggs (Ernst and Ernst 2003; Fitch 1978). The eggs are laid under rocks, and in mammal burrows, old logs and stumps, old sawdust piles, and plowed fields (Anderson 1965; Ernst and Ernst 2003); the white, smooth eggs are elongated and adhere to each other. Usually, only one clutch is produced, and not all females breed during their active season. The eggs hatch between late July and mid-September (Fitch 1978). A female collected in DeKalb County laid 12 eggs in late June (Anderson 1965). The eggs had an average measurement of 47 by 21 mm (1.9 by 0.8 in.). These eggs were incubated at 23.9°C (75°F) and hatched in early September with an incubation period from 72 to 74 days. Another female collected from Audrain County laid eight eggs in late June. These eggs were incubated at 28°C (82°F) and hatched in mid-August with an incubation period of 48 days (Johnson 2000). The newly hatched young ranged from 222 to 294 mm (8.7 to 11.6 in.) in total length (Anderson 1965; Johnson 2000). Prairie kingsnakes have been found to reach adulthood at age two to three with males maturing earlier (Fitch 1978). A wild-caught adult lived a little over 23 years in captivity (Snider and Bowler 1992).

Remarks: The prairie kingsnake was described as a species from specimens collected near Saint Louis in 1827 (Schmidt 1953). Once three subspecies were recognized as part of the *Lampropeltis calligaster* group with *L. c. calligaster* in Missouri (Johnson 2000). McKelvy and Burbrink (2017) elevated all subspecies to full species distinction. Another common name for this species is the yellow-bellied kingsnake.

A prairie kingsnake lying dead on the road unfortunately is a common sight in many parts of the state. Prairie kingsnakes are often misidentified as copperheads and killed because of unwarranted fear. The dorsal markings of this species are round or saddle shaped; dorsal markings of the copperhead are hourglass shaped. Because prairie kingsnakes consume many rodents, they provide a valuable check on rodent populations.

Speckled Kingsnake

Lampropeltis holbrooki Stejneger

JEFF BRIGGLER

Adult speckled kingsnake from Miller County.

Description: The speckled kingsnake is a medium-to-large, shiny, black snake covered with small yellow spots. Ground color is dark brown or black. Each dorsal scale normally has one white or light-yellow spot that causes the snake to appear speckled. Scales along the top of the head also have these yellow spots. Occasionally the light spots form crossbars along the back, giving it a chain-like pattern. This is apparent on hatchlings and young speckled kingsnakes but will change to an overall speckled appearance with age. Adult speckled kingsnakes found in western Missouri may have deep yellow spots that cover over 50 percent of each scale. The upper and lower labial scales are yellow, with bold black or dark-brown bars along their edges. Belly coloration is light yellow with a series of irregular dark-gray to black, half-moon to rectangular markings that become larger and more numerous near the tail. Terminal scale of the tail is a sharp spike. The dorsal

Distribution: Missouri: Presumed to occur statewide. North America: A few isolated pockets in Nebraska, southern edge of Iowa, south to all of Missouri, Arkansas, and Louisiana, most of Texas except along the border with Mexico, north through most of Oklahoma and Kansas (Powell et al. 2016).

JEFF BRIGGLER
TOM R. JOHNSON

Hatchling speckled kingsnake from Shannon County (above). Ventral coloration of a speckled kingsnake from Cole County (right).

scales are smooth, and the anal plate is single. Male speckled kingsnakes are larger than females and have a longer tail.

Adult speckled kingsnakes range in total length from 900 to 1,220 mm (35.4 to 48 in.) but have been known to reach 1,829 mm (72 in.; Powell et al. 2016).

Habits and Habitat: This species is mainly active between April and October, with most surface activity occurring in April into June. Speckled kingsnakes are active during the day in spring, early summer, and autumn. They become active at night in summer to avoid hot temperatures. This species is secretive. It takes shelter under rocks, under logs, in rotten stumps, and in small mammal burrows, especially mole burrows. This species spends its winter dormancy underground in small animal burrows, rock crevices, hollow logs and stumps, sawdust piles, and foundations of abandoned buildings (Ernst and Ernst 2003). Although rare, a speckled kingsnake was found overwintering in a cave in Jefferson County (Drda 1968). Individuals have been observed using crayfish burrows on drier, upland grassland sites in northern Missouri for both summer and winter retreats (J. Briggler pers. obs.).

This handsome snake occurs in a wide variety of habitats: prairies and along prairie streams, brushy areas, edges of forest and old farm fields, rocky, wooded hillsides, and along the edges of swamps and marshes. In Missouri, it is often found on rocky, wooded hillsides with some openings.

Speckled kingsnakes kill their prey by constriction and consume mice and shrews, bird eggs, small birds, lizards, and snakes—including venomous species. This snake, as with all kingsnakes, is immune to the venom of the Missouri's various native pit vipers (Weinstein et al. 1992). Prey animals and their eggs are located by scent. In some areas, the diet of speckled kingsnakes is primarily reptiles and their eggs. They can be cannibalistic and are known to eat their own young (Ernst and Ernst 2003). A speckled kingsnake was observed eating freshly laid map turtle eggs on the bank of a river in extreme southern Mississippi (Brauman and Florillo 1995). Natural predators include hawks and owls, kingsnakes and racers, and mammals, such as raccoons and skunks, which will eat their eggs or hatchlings. Domestic cats likely kill young kingsnakes.

Breeding: Courtship and mating occur in April through early June. There have been reports of a pair of male speckled kingsnakes involved in a "combat dance" during their breeding season to establish breeding dominance (Ernst and Ernst 2003). A male will locate a female by following her scent trail. Courtship involves a male examining a female with numerous tongue flicks, jerking movements of his body along hers, and grasping of the female's neck in his mouth, at which time copulation will occur. They remain joined for about one hour. There are reports of females mating with more than one male, and a female will usually lay more than one clutch of eggs per season (Ernst and Ernst 2003). Larger females will produce more eggs than smaller individuals. A female may lay her eggs in a rotten log, stump, old sawdust pile, or old small mammal burrow in June or July. The eggs, which are white to cream colored, adhere to each other and average 38 mm (1.5 in.) in length. Throughout their range, up to 29 eggs may be produced with an average of 10 eggs in a nest (Ernst and Ernst 2003). Clutch sizes reported from surrounding states range from 2 to 23 eggs per female (Collins et al. 2010; Trauth et al. 1994). Boyer and Heinze (1934) captured a Jefferson County female that laid 14 eggs in early July. A St. Clair County female laid 14 eggs in early July, and these eggs hatched in early September with an incubation period of 69 days (Anderson 1965). Hatching generally takes place around 60 days after being laid. This usually happens in late summer or early fall (mostly August into September). Hatchlings average 271 mm (10.7 in.) in total length (Ernst and Ernst 2003). The young are brighter in color than adults and have distinct light crossbands along the back. Sexual maturity is reached at the age of two or three (Zweifel 1980). Captive specimens have lived slightly over 14 years (Snider and Bowler 1992), but compared to many other kingsnake species, the speckled kingsnake likely can survive over 20 years.

Remarks: Until recently there were five subspecies of kingsnakes in the *Lampropeltis getula* group (combined and called "common kingsnakes"), which were elevated to full species by Pyron and Burbrink (2009). This group includes the speckled kingsnake, *L. g. holbrooki*, now classified as *L. holbrooki*. Likely this species intergrades with or is replaced by the eastern black kingsnake (*L. nigra*) in southeastern Missouri (see remarks section in the eastern black kingsnake account).

Speckled kingsnakes vibrate their tail when alarmed and can produce a short, loud hiss. Freshly collected specimens may try to bite and will smear a foul-smelling musk onto your hand from glands at the base of the tail; but they quickly calm down and can be easily handled. This snake is known locally as the "salt-and-pepper" snake.

Eastern Black Kingsnake

Lampropeltis nigra (Yarrow)

An adult eastern black kingsnake from Scott County.

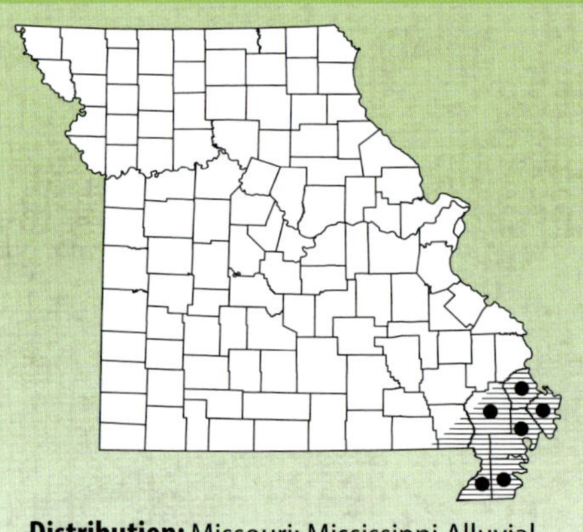

Distribution: Missouri: Mississippi Alluvial Basin of southeastern Missouri. North America: Mostly east of the Mississippi River from southern Illinois, Indiana, and Ohio, south to the gulf shore of western Alabama, all of Mississippi, and eastern Louisiana. Populations occur west of the Mississippi River in Louisiana, northeast Arkansas, and southeast Missouri (Powell et al. 2016).

Description: The eastern black kingsnake is similar in appearance to the speckled kingsnake. A medium-to-large snake that is mostly black with small yellow to cream speckles or with faint, white or cream chainlike markings along the sides. A series of yellow or white dotted crossbands are present along the back from the neck to the tip of the tail, especially in young. These crossbands usually fade with age. The top of the head is mostly black. The belly is cream to pale yellow, with black, checkered markings that are mostly concentrated along the middle of the belly. The dorsal scales are smooth, and the anal plate is single.

Adult eastern black kingsnakes range in total length from 900 to 1,140 mm (35.4 to 44.9 in.) but have been known to reach 1,480 mm (58.3 in.; Powell et al. 2016).

Habits and Habitats: Little is known about the natural history of this species in Missouri. Eastern black kingsnakes are likely active

JEFF BRIGGLER

JEFF BRIGGLER

Head of an adult eastern black kingsnake from Scott County (above). Ventral coloration of an eastern black kingsnake from Scott County (left).

from late March through October. Documented specimens have occurred in Missouri from late March to mid-July (R. Daniel pers. comm.). They are mostly diurnal during the warmer months of the year, especially in spring, early summer, and again in autumn. In nearby states, this species occurs in a wide variety of habitats from open lands to forest, including dry, rocky hillsides, open woodlands, overgrown fields, and areas with loose soil for burrowing (Boundy and Carr 2017; Minton 2001; Niemiller et al. 2013). Also known to occur in urban areas, especially around old buildings. An individual captured in late May in Scott County was found among riprap of a fishing lake spillway (Daniel and Edmond 2014). Some individuals may be aggressive (coiling or S-shape loops of the body, vibrating tail, and biting) when threatened, while others are calm. During the winter months, individuals retreat underground into animal burrows, rock crevices, hollow logs and stumps, and old abandoned buildings (Ernst and Ernst 2003).

Eastern black kingsnakes prey on a variety of animals, especially reptiles and their eggs. They are known to feed on numerous species of snakes, including venomous species, and turtle eggs. They also eat small mammals, lizards, birds and bird eggs (Ernst and Ernst 2003). If prey is not swallowed immediately, it will be killed by constriction. Individuals will grab larger prey by the head and quickly constrict them (Niemiller et al. 2013).

Breeding: No information exists on the reproductive biology of this species in Missouri. Courtship and mating begin when individuals emerge from overwintering sites and continue into the June. Minton (2001) found individuals mating in early to mid-May in Indiana. About 3 to 30 eggs are likely laid in animal burrows, rotting logs or stumps, and sawdust piles in June or July (Boundy and Carr 2017; Niemiller et al. 2013). Clutch sizes of 3, 12, and 15 eggs have been observed in Indiana from late June into July (Minton 2001). The white, elongated eggs generally adhere to each other in small clusters. Hatching likely occurs from August into September in Missouri. In Illinois, a clutch of eggs hatched in late August (Smith 1961). The incubation period is between 56 and 81 days (Minton 2001; Niemiller et al. 2013). Reproductive maturity is likely reached at two to four years of age, with males maturing earlier than females.

Remarks: The eastern black kingsnake was formerly considered a subspecies of the common kingsnake group (*L. getula*) and recently elevated to full species status (Pyron and Burbrink 2009). Although the eastern black kingsnake was recently documented from two counties in Missouri (Daniel and Edmond 2014; Edmond and Daniel 2014), this species apparently has been in the state since 1932 (Anderson 1965; Burt 1935). These authors acknowledged the eastern black kingsnake in southeastern Missouri based upon color and pattern, but assumed it was an intergrade with the Tennessee population. More recent genetic data, along with coloration and pattern, confirmed the presence of the eastern black kingsnake in southeastern Missouri (Daniel and Edmond 2014). This species likely intergrades with the speckled kingsnake along the boundary between the Ozark Highlands and Mississippi Alluvial Basin. Additional studies are needed to better understand its distribution and natural history in the state.

Eastern Milksnake

Lampropeltis triangulum (Lacépède)

JEFF BRIGGLER

Adult eastern milksnake from Callaway County.

Description: The eastern milksnake is a brightly colored, medium-sized snake with smooth scales. The ground color is white to pale yellow with a series of 20 to 30 red, orange-red, red-brown, or light-brown dorsal blotches that are bordered by black. Markings on the neck just before the head and on the head are variable; often the blotch on the neck merges with a marking on the head forming a "Y" or "U" shape. There is usually a narrow marking between the eyes on top of the head and another narrow marking from each eye to the corners of the mouth. They have a white or light-gray snout. Eastern milksnakes from a few northeastern Missouri counties may have light-brown blotches (which are bordered in black). There may be a row of smaller red, brown-to-black markings along the sides between the larger dorsal blotches; however, these may be missing or very small on specimens from western or southwestern Missouri. The top of the head may be red or orange, and the belly is usually white

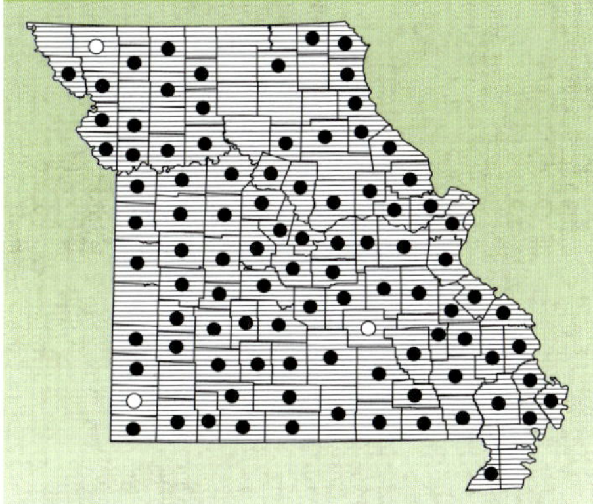

Distribution: Missouri: Presumably statewide. North America: Southeastern Ontario and southern Quebec, Canada, southern tip of Maine, New England states to western Virginia, western North Carolina to northern tips of Georgia and Alabama, northern and western Mississippi, west to northern and western Louisiana, northern and eastern Arkansas, Missouri, and eastern, central, and western edge of Iowa, southeastern Minnesota to southern half of Wisconsin (Powell et al. 2016).

Eastern milksnakes are often found on open, rocky glades, such as this example from Camden County (above). Ventral pattern of an eastern milksnake from Cole County (right).

and boldly marked with black squares and rectangles. An adult eastern milksnake collected during mid-April in southwestern Benton County had a dark-gray belly with some white to light gray from the chin to the underside of the neck (R. Krager pers. comm.). Dorsal scales are smooth, and the anal plate is single.

Adult eastern milksnakes range in total length from 610 to 900 mm (24 to 35.4 in.) but have been known to reach 1,321 mm (52 in.; Powell et al. 2016).

Habits and Habitat: Eastern milksnakes are active from April until late October. They are secretive and seldom seen in the open. They generally do not bask in the sun but instead coil up under a large or flat rock and are warmed by their shelter rock as it is warmed by the sun (Fitch and Fleet 1970). This secretive snake requires slightly damp places to hide and will usually take shelter alongside or under deeply embedded rocks or logs (Anderson 1965). During hot weather in July, August, and September, this species will move underground in mammal burrows or under large rocks. Winter dormancy is spent in rodent burrows, crevices along rocky hillsides, foundations of old buildings, and old wells and cisterns. It is common for milksnakes to overwinter with other snake species, including kingsnakes, ratsnakes, racers, and copperheads. When first encountered, a milksnake will try to escape into thick grass, dead leaves, or under the nearest rock. If grabbed it will strike at a hand and give off

JEFF BRIGGLER

Hatchling eastern milksnake from Osage County.

a short hiss as it strikes, it will vibrate its tail very rapidly, smear feces on the hand holding it, and at the same time release a smelly musk from glands at the base of its tail. Even after a specimen has settled down while being held, it will suddenly and deliberately open its mouth and just start chewing on a finger or loose skin between the fingers. Their bite is harmless; this is not a venomous species.

Milksnakes are active during the day in the spring and autumn, but become nocturnal during the summer (LeClere 2003). This species was found to have an average home range of 7.1 ha. (17.5 ac) in a 10-year study of marked specimens in eastern Kansas (Fitch 1999). Eastern milksnakes were observed frequently reusing the same shelter rock during their active season and from year to year (Fitch and Fleet 1970; Fitch 1999).

This species resides under rocks on open, sparsely wooded, south-facing, rocky hillsides, especially glades, as well as along the edge of forests with rocks and downed trees. Milksnakes have been found under a pile of boards, rusting sheets of tin, or other discarded junk laying on or embedded in the ground. An eastern milksnake was observed under a pile of old corrugated tin along the edge of a forest in central Clark County (T. Johnson pers. obs.).

Eastern milksnakes prey on lizards, small snakes and their eggs, mice, voles, and shrews. This species will also eat the young of its own kind, as well as young venomous snakes (Ernst and Ernst 2003). Milksnakes in other states have also been found to eat bird eggs and the young of ground-nesting birds (Ernst and Ernst 2003). All prey are killed by constriction except small reptiles, bird eggs, or newborn rodents, which are eaten as is. A Shannon County specimen was found to have eaten eight skink eggs (R. Daniel pers. comm.). Hawks and owls may be able to prey on milksnakes, but the snake's secretive nature may allow them to be safe from aerial predators. Milksnake eggs and young are likely eaten by kingsnakes, other milksnakes, scarletsnakes, and racers, as well as raccoons, skunks, and badgers.

Breeding: Courtship and mating occur in the spring, likely from early April through May; and there could be some mating in the autumn (Fitch 1999; Ernst and Ernst 2003). Because of their secretive nature, few copulating pairs of milksnakes have been observed in the wild, however, this species has been successfully bred in captivity. A female's scent is picked up by the male's tongue flicking, and he will follow the trail until she is encountered. He will then move over her body, cover her with several loops of his body, and make several jerking movements as he advances his head toward hers. During this time, he may also rub her body and neck with his chin. Once the female is still he will grasp the skin of her neck area in his mouth and copulation will quickly follow. They will remain together for one-half to a little over an hour (Ernst and Ernst 2003).

Egg laying will occur from 30 to 45 days later. Female eastern milksnakes will lay their eggs in late May through early July. They will lay up to 24 eggs, with an average of 7.6 eggs per female reported range wide (Ernst and Ernst 2003). Numerous observations of egg laying were made by R. Krager (pers. comm.) from females collected in Missouri and held in captivity. Seventeen females laid between 1 to 13 eggs, with an average of five eggs per clutch (R. Krager pers. comm.). In general, larger female milksnakes will lay more eggs per clutch. Their white, elongated eggs average 38.2 mm (1.5 in.) in length and adhere to each other. A female eastern milksnake will deposit her eggs where they will be secure and moist: under a large flat rock or pile of rotting tree bark, or inside a rotting log, stump, or old pile of sawdust (Ernst and Ernst 2003). The eggs hatch during mid-August to early September. Newly hatched young average 221 mm (8.7 in.) in total length and have vivid red dorsal blotches edged in black. Juveniles become sexually mature in two or more years. Milksnakes are considered a long-lived species, with a wild-caught adult surviving to 21 years of age in captivity (Snider and Bowler 1992).

Remarks: Just a few years ago, Missouri's milksnakes were considered a subspecies (*L. t. syspila*) and referred to as the red milksnake, one of nine subspecies in southern Canada and the United States (Ernst and Ernst 2003). Now, several of the subspecies in the central part of North America have been consolidated into one taxon and called the eastern milksnake (*L. triangulum*; Ruane et al. 2014).

The common name for this group of medium-sized, harmless snakes—"milksnake"—was coined by early European settlers who saw them in or near barns and milk sheds and assumed the colorful snake was there to "steal milk" from a cow. This is a myth. This small constrictor was likely in search of mice to eat and had no interest in the cow or its milk. Too often this species has been mistakenly identified by Missourians as a coralsnake; the colorful and venomous species found in southeastern and southern states. Our eastern milksnake has a white or yellowish background color that touches the black of their dorsal blotches and white or gray on their nose. True coralsnakes have their yellow bands touching red bands and a black band on their nose. Coralsnakes do not occur in Missouri and the closest population is the Texas coralsnake (*Micrurus tener*) found in a few southwestern counties of Arkansas (Trauth et al. 2004).

Although populations of this species are considered secure in Missouri, this brightly colored snake is often illegally collected from rocky glades in the Ozarks due to demand for the pet trade. Unfortunately, illegal collectors not only take individuals from glades but often destroy the large, embedded rocks necessary for milksnakes and other reptiles to live. Eastern milksnakes are often found dead on Missouri's roads and highways.

Mississippi Green Watersnake

Nerodia cyclopion (Duméril, Bibron and Duméril)

TOM R. JOHNSON

Adult Mississippi green watersnake.

Description: The Mississippi green watersnake is a medium-sized, dark-colored, semi-aquatic snake once somewhat common in southeastern Missouri but now endangered and possibly extirpated in the state. This harmless, heavy-bodied snake has a green-brown ground color with numerous small, obscure olive-brown or dark-brown markings. Close examination of the head will reveal a short row of scales (sub-oculars) between the eye and the upper lip scales. No other watersnake native to Missouri has this characteristic. The belly is dark gray or brown with numerous yellow markings, most of them in the form of half-moons. Dorsal markings are more distinct on young snakes. The dorsal scales are keeled, and the anal plate is divided.

Adult Mississippi green watersnakes range in total length from 760 to 1,140 mm (29.9 to 44.9 in.) but have been known to reach 1,295 mm (51.0 in.; Powell et al. 2016).

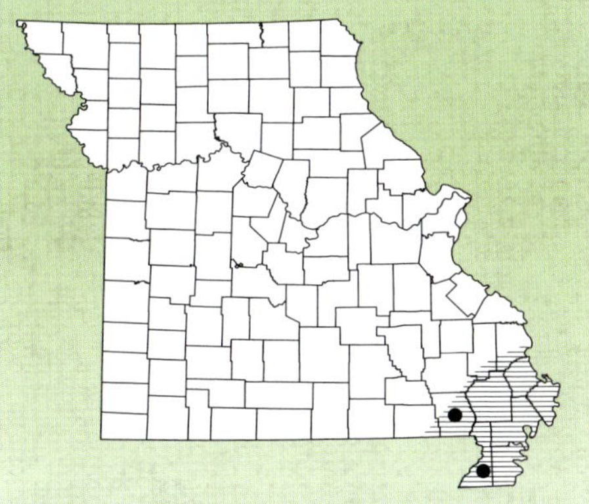

Distribution: Missouri: Presumed to occur in the natural cypress swamps of extreme southeastern Missouri. North America: Gulf Coast of southern eastern Texas to extreme western Florida, most of Louisiana, north through eastern and east-central Arkansas, southeastern Missouri and southern tip of Illinois (Powell et al. 2016).

TOM R. JOHNSON

JEFF BRIGGLER

Head of a Mississippi green watersnake (above). A Mississippi green watersnake ventral coloration (right).

Habits and Habitat: Very little is known about the natural history of this species in Missouri. It is probably active between late March and October. This watersnake is normally active on warm days, often basking on branches overhanging water (Hebrard and Mushinsky 1978). Overwintering sites are earth bank burrows, rock crevices, and levees with riprap rock (Ernst and Ernst 2003). Hurter (1911) reported individuals sunning on moss and aquatic plants in Missouri and Arkansas. Mississippi green watersnakes search for food during early evening or at night when the weather is hot. This species will bite and smear a foul-smelling musk onto its captor.

This is primarily a swamp-dwelling snake. Preferred habitats in Missouri are the cypress swamps, oxbows, river sloughs, and lakes of the Mississippi Alluvial Basin.

Food of the Mississippi green watersnake consists primarily of fish, but salamanders, frogs, and crayfish are also eaten. Kofron (1978) reported that Louisiana specimens ate mostly small fish (94 percent frequency). Garton et al. (1970) found western lesser sirens in three green watersnakes collected in southern Illinois.

SCOTT BALLARD

Juvenile Mississippi green watersnake.

Breeding: The reproductive biology of the Mississippi green watersnake has not been studied in Missouri. In general, courtship and mating occur in the spring, probably in April or May. Eggs are retained and the young develop within the females. The young are born during August and early September. In general, the number of young produced per female has a direct relation to the size of the female: the larger the female the greater the number of young that can be produced. Reported litter sizes are 7 to 37 young with an average of 18 (Dundee and Rossman 1989; Kofron 1978; Meade 1934). In Arkansas, litter size ranged from 8 to 34 (average 18) young (Trauth et al. 1994). Hurter (1911) captured a female near the Missouri-Arkansas border that gave birth to 19 young in mid-September in captivity. Young, which have vivid, dark crossbands and spots, are between 187 to 305 mm (7.4 to 12 in.) in total length (Ernst and Ernst 2003). Reproductive maturity is likely reached within three to four years (Ernst and Ernst 2003).

Remarks: The Mississippi green watersnake has been designated endangered and possibly extirpated in Missouri due to few confirmed reports of this species and the drastic reduction of native cypress swamps in southeastern Missouri. Commenting on the low population of this snake in both extreme southern Illinois and southeastern Missouri, Garton et al. (1970) stated: "Large fluctuations in population density are not unexpected in peripheral populations where ecological conditions presumably are marginal." This species was reported as quite common on the lower St. Francis River in the early 1900s (Hurter 1911). No specimens have been taken in Missouri since 1961. However, a young Mississippi green watersnake was observed by several employees of the Missouri Department of Conservation in extreme southern Dunklin County in 1994, but the snake was not collected or photographed (R. Thom pers. comm.). In addition, several dead specimens were taken from gravel roads on a national wildlife refuge in Mississippi County, Arkansas (Trauth 1990). This area is within a few miles of the Missouri-Arkansas border and close to Dunklin County. It is presumed from this report that this species is likely to be extant in southeastern Missouri and should continue to be listed as endangered in this state. Intensive surveys throughout the rivers and wetlands of southeastern Missouri are needed.

Plain-bellied Watersnake

Nerodia erythrogaster (Forster)

Adult plain-bellied watersnake from Stoddard County.

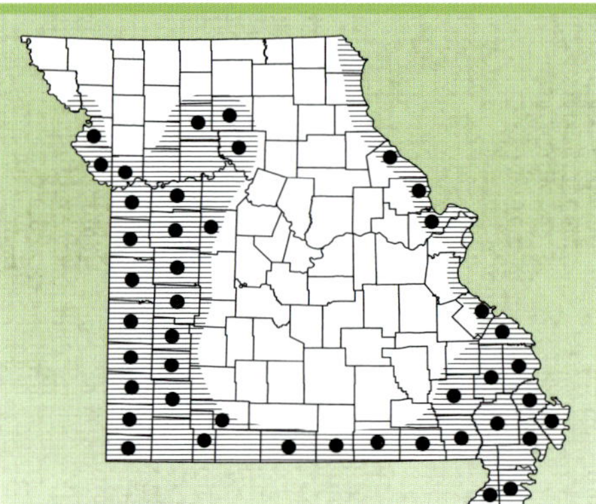

Description: The plain-bellied watersnake is a medium-sized, heavy-bodied, dark-colored, semi-aquatic snake with a plain yellow belly. Adults are gray, greenish gray, or brownish black, with little or no pattern along the dorsum. Populations in western Missouri more likely to have blotches along their dorsum and sides. The belly is plain yellow with some orange showing on occasional individuals. Young plain-bellied watersnakes are strongly patterned with brown dorsal and lateral blotches that may

Distribution: Missouri: Southeast corner of the state, north along the Mississippi River floodplain, and along the southern and western border of Missouri. North America: The plain-bellied watersnake ranges along the Atlantic coast from New Jersey into northern Florida, west through most of Texas and eastern New Mexico, northeast through Oklahoma and southeastern Kansas, western and southeastern Missouri, and southern tips of Illinois and Indiana. Several isolated populations occur throughout the Midwest (Powell et al. 2016).

TOM R. JOHNSON
TOM R. JOHNSON

A juvenile plain-bellied watersnake from Butler County (above). Ventral coloration of a plain-bellied watersnake from Stoddard County (left).

be joined to form transverse bars. These markings fade with age. The dorsal scales are keeled, and the anal plate is divided.

Adult plain-bellied watersnakes range in total length from 760 to 1,220 mm (29.9 to 48 in.) but have been known to reach 1,636 mm (64.4 in.; Powell et al. 2016).

Habits and Habitat: This species is apparently active from late March through October with peak emergence in April and early May. Individuals bask on logs in shallow water, on branches above the water, or along the shore. As with most of our watersnakes, this species is pugnacious. When cornered, an individual will strike or bite viciously. A foul-smelling musk also is excreted from glands in the base of the tail; it is often mixed with feces and smeared on the captor. They are known to overwinter in mammal burrows, rock piles and crevices, riprap rocks along levees, and decaying stumps and logs (Ernst and Ernst 2003). In the prairie regions of north-central Missouri, this species commonly uses crayfish burrows for overwintering (J. Briggler pers. obs.).

The plain-bellied watersnake prefers quiet bodies of water such as swamps, sloughs, oxbows, slow-moving rivers and streams, floodplains, seasonally flooded bottomland woods, drainage ditches, lakes, and ponds. Individuals may be encountered during the summer on land, far removed from any body of water.

This watersnake preys on fish, toads, frogs, tadpoles, salamanders, and crayfish. Most diets of watersnakes are dominated by fish; however, the plain-bellied watersnake appears

TOM R. JOHNSON

Plain-bellied watersnakes in the western part of Missouri have dorsal blotches.

to consume a larger volume of frogs and toads. An individual was captured in April in Linn County with a southern leopard frog in its mouth (J. Briggler pers. obs.).

Breeding: Courtship and mating mainly take place in the spring, especially during April and May. However, Blanchard (1924) discovered a pair mating in Pemiscott County in the grass and weeds at the edge of a swampy area in mid-June. As with most live-bearing species, the number of young depends on the size of females. Litter size varies from 2 to 37 young per female with most litters consisting between 10 and 20 young (Ernst and Ernst 2003). A Mississippi County female gave birth to 16 young in mid-September that ranged from 181 to 264 mm (7.1 to 10.4 in.) in total length (Anderson 1965). The young are born mainly during August and September. Young are more vividly patterned than adults with pronounced dorsal blotches. Reproductive maturity is obtained in about three years (Ernst and Ernst 2003). A plain-bellied watersnake lived in captivity for nearly 15 years (Snider and Bowler 1992).

Remarks: Once several subspecies were recognized in the plain-bellied watersnake group, with two subspecies occurring in Missouri: the yellow-bellied watersnake (*N. e. erythrogaster*), occurring in eastern Missouri counties along the Mississippi River and in southeastern Missouri, and the blotched watersnake (*N. e. transversa*), occurring in southwestern and western Missouri. A genetic study by Makowsky et al. (2010) did not support the recognition of these subspecies, so all are now classified as the plain-bellied watersnake.

Plain-bellied watersnakes are often misidentified as the venomous northern cottonmouth (*Agkistrodon piscivorus*) and killed because of unwarranted fear.

Broad-banded Watersnake

Nerodia fasciata confluens (Blanchard)

JEFF BRIGGLER

Adult broad-banded watersnake from Wayne County.

Description: The broad-banded watersnake is a medium-sized, semi-aquatic snake with irregular broad, dark blotches or bands. The bands may be brown, red brown, or black, separated by yellow or yellow gray. There are normally 11 to 17 bands; although irregular in shape, they are broadest on the back. Some of the bands may be connected. There is often a faint, dark line running diagonally from the eye past the corner of the mouth. The belly is yellow tan and boldly marked with black. The dorsal scales are keeled, and the anal plate is divided. Young broad-banded watersnakes are more brightly colored than adults.

Adult broad-banded watersnakes range in total length from 560 to 1,067 mm (22 to 42 in.) but have been known to reach 1,588 mm (62.5 in.; Powell et al. 2016).

Habits and Habitat: This species is normally active between late March and October. They are most active during daytime hours in the spring and often seen resting on

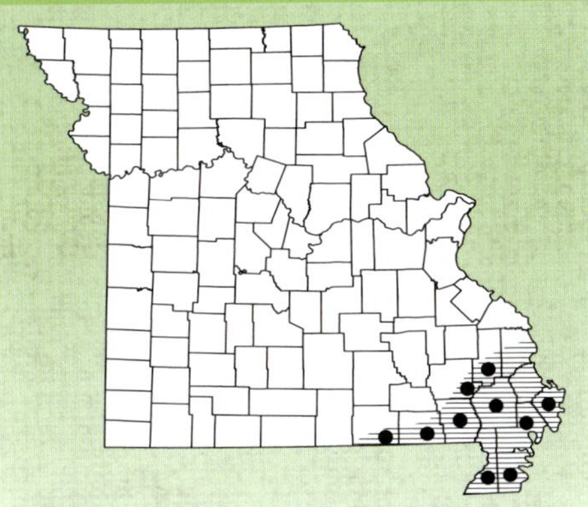

Distribution: Missouri: Mainly the Mississippi Alluvial Basin of southeastern Missouri. North America: Southeastern Missouri, extreme southern Illinois, along the western edges of Kentucky and Tennessee, western Mississippi, eastern and southern Arkansas, southeastern corner of Oklahoma, eastern third of Texas, all of Louisiana, and Gulf Coast area of Mississippi and Alabama (Powell et al. 2016).

JEFF BRIGGLER

TOM R. JOHNSON

Broad-banded watersnakes live in and along the edge of permanent swamps, such as this example in Butler County (above). Ventral coloration of a broad-banded watersnake (right).

thick algal mats in wetlands throughout southern Missouri. Although the broad-banded watersnake also basks on logs or branches overhanging water during the day, it is more active at night, especially during hot weather. If a specimen is encountered and not allowed to escape, it will flatten its head and neck and vigorously try to defend itself. Once captured it will bite viciously and smear its captor with a foul-smelling musk excreted from glands at the base of the tail. Like other watersnakes, broad-banded watersnakes are good swimmers and use lateral undulations to move quickly across the water's surface. They will dive underwater to escape predators or search for food. Overwintering is mainly in mammal burrows, rocky ledges, tree stumps, and logs. An individual was found overwintering in a rocky ledge over a quarter mile from water, and nine individuals were found overwintering in root channels under a rotten cypress trunk (Anderson 1965).

The broad-banded watersnake resides in and along the edge of cypress swamps, river sloughs, oxbow lakes, sluggish streams, and drainage ditches.

Food of this species includes mainly fish and amphibians (salamanders, frogs, toads, and their larvae), and occasionally crayfish.

Breeding: Courtship and mating occur during the spring with most activity occurring in April and May. Anderson (1965) observed a pair attempting to mate in mid-April in

TOM R. JOHNSON

A juvenile broad-banded watersnake from Butler County.

Tennessee. Litter size reported throughout their range varies from 6 to 83 with an average of 23 young (Kofron and Dixon 1980; Palmer and Braswell 1995). Estimated litter size for Arkansas populations was 20.6 (Trauth et al. 1994). The young are born alive during late July, August, or early September. Individuals collected from Missouri and Arkansas gave birth in captivity to 13, 7, and 19 young on July 13, July 15, and August 30, respectively (Anderson 1965). Hurter (1911) held in captivity an animal collected in the lowlands of southeastern Missouri that gave birth to 23 young in late August. The young are more brightly colored and patterned than adults and range in length from 180 to 266 mm (7.1 to 10.5 in.; Powell et al. 2016). Sexual maturity is generally reached within two to three years (Ernst and Ernst 2003).

Remarks: The type locality of this species was designated as Butler County by Blanchard (1923). This species has been and continues to be quite common within the remaining wetland habitats throughout the lowlands of southeastern Missouri (Hurter 1911; J. Briggler pers. obs.). As with other species of watersnakes, the broad-banded watersnake is often misidentified as the venomous northern cottonmouth (*Agkistrodon piscivorus*) and killed because of unwarranted fear. The true cottonmouth is more heavy-bodied with a larger, more chunky head, and a facial pit between the nostril and eye on either side of the head. The cottonmouth is darker and has a light line from each eye to the corner of the mouth.

Northern Diamond-backed Watersnake

Nerodia rhombifer rhombifer (Hallowell)

Adult northern diamond-backed watersnake from Butler County.

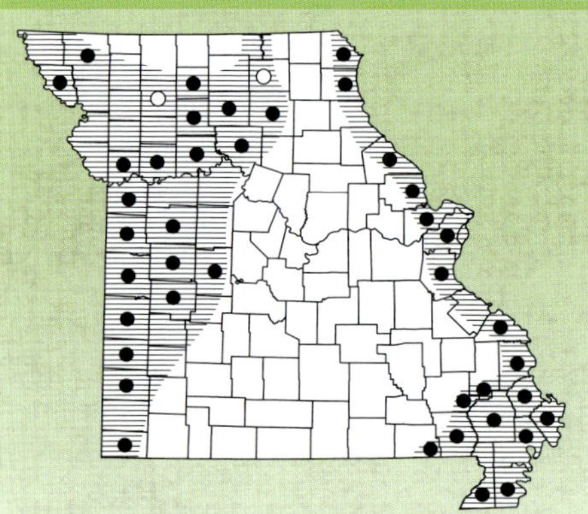

Distribution: Missouri: Southeastern corner, north along the Mississippi River floodplain, and in northern and western Missouri. North America: Southeastern corner of Iowa, western edge of Illinois, outer edges of Missouri, the western edge of Kentucky and Tennessee, eastern two-thirds of Arkansas, western and central Alabama, all of Mississippi and Louisiana, southern and eastern Kansas, most of Oklahoma, and the eastern two-thirds of Texas (Powell et al. 2016).

Description: The northern diamond-backed watersnake is a large, heavy-bodied snake with numerous diamond-shaped light markings along the back. This is Missouri's largest species of watersnake. The ground color may be gray, light brown, or dull yellow. There are 30 to 65 dark-brown blotches along the back, which are normally connected to form a chainlike pattern. The belly is yellow with several irregular rows of dark-brown spots or half-moons. The dorsal scales are strongly keeled, and the anal plate is divided.

Adult northern diamond-backed watersnakes range in total length from 760 to 1,220 mm (29.9 to 48 in.) but have been known to reach 1,753 mm (69 in.; Powell et al. 2016).

Habits and Habitat: This species is active from late March through October. During spring, early summer, and autumn, diamond-backed watersnakes often bask on branches, logs, tree roots, or along the

JEFF BRIGGLER

JEFF BRIGGLER

Northern diamond-backed watersnakes are found in and along wetlands, such as this example in Lincoln County (above). Ventral coloration of a northern diamond-backed watersnake from Butler County (left).

banks of waterways. During July, August, and September, diamond-backed watersnakes are distinctly nocturnal. They will readily take to water to escape predators and forage for food. They are capable of being underwater for over an hour (Baeyens et al. 1978). If molested, a diamond-backed watersnake will flatten its head and neck, bite viciously to defend itself, and secrete a strong-smelling musk from glands at the base of the tail. Overwintering takes place underground below the frost line in animal burrows and root systems (Ernst and Ernst 2003; LeClere 2013). Individuals have been observed emerging in April from crayfish burrows in Holt and Chariton counties (J. Briggler pers. obs.).

This large watersnake occurs in a variety of aquatic habitats but is most abundant in swamps, sloughs, oxbow lakes, marshes, and drainage ditches.

This species eats primarily fish and amphibians (frogs, toads, salamanders, and their larvae). They may also eat small snakes and turtles, earthworms, leeches, crayfish, and insects. Most fish captured are small sized, slow moving, or dead. Kofron and Dixon (1980) described a diamond-backed watersnake forming an enclosure with its body to catch small fish in shallow water. They also have another peculiar feeding behavior of lying underwater with mouth opened and tail anchored to a rock to await an approaching prey (Gillingham and Rush 1974).

TOM R. JOHNSON

Female northern diamond-backed watersnake giving birth from Pemiscot County.

Breeding: Courtship and mating take place in the spring, probably during April or early May, soon after emergence. Male diamond-backed watersnakes have a cluster of tiny raised bumps on their chin, which are likely used during courtship—the male rubs his chin along a receptive female's back to stimulate her. Several males may court a female at one time. Breeding can occur on the bank, in a shrub, or in the water. Females can produce 8 to 62 young (Gibbons and Dorcas 2004). Most females likely produce young every year (Aldridge et al. 1995; Betz 1963). Studies of diamond-backed watersnakes in Holt County found litter size averaged 27 young with a range of 16 to 54 young (Pilgrim 2001; Pilgrim et al. 2011). Young diamond-backed watersnakes are born during late August through early October. Anderson (1965) reported on a captive individual collected from Vernon County that gave birth to live young on two different dates in 1937. She gave birth to six young on September 9 and 24 young on October 5. A female collected in Pemiscot County gave birth to 35 young in late September; they averaged 294 mm (11.6 in.) in total length (Johnson 2000). The body pattern of the young is more distinct and often has some orange coloration on the belly compared to adults. Reproductive maturity occurs at two and one-half to three years of age (Betz 1963; Kofron 1979).

Remarks: The Mississippi River may have been a barrier to migration of the northern diamond-backed watersnake in the past. Populations east or west of the Mississippi River may be separate subspecies or full species, but additional research is needed (Brandley et al. 2010).

This watersnake is very common in the northern and western sections of Missouri, especially within bottomland prairie habitat that is associated with nearby wetland pools, drainage ditches, and rivers. Many individuals are killed by humans due to fear or because they are mistaken for a venomous snake, especially the cottonmouth. The cottonmouth does not have the chainlike dark pattern on the back. Cottonmouths have a distinct triangular head, a sensory pit between the eye and nostril on either side of the head, and a light line from each eye to the corner of the mouth.

Northern Watersnake

Nerodia sipedon sipedon (Linnaeus)

JEFF BRIGGLER

Adult northern watersnake from Callaway County.

Description: The northern watersnake, known locally as the banded watersnake or common watersnake, is a gray-to-brown snake with numerous dark-brown bands along the anterior one-third of the body; the bands become broken into alternating blotches posteriorly. Dorsal crossbands are red brown, dark brown, or nearly black. Color of the belly varies, but generally it is a combination of a cream-colored or yellow ground color and irregularly spaced half-moons or spots of orange or red, bordered

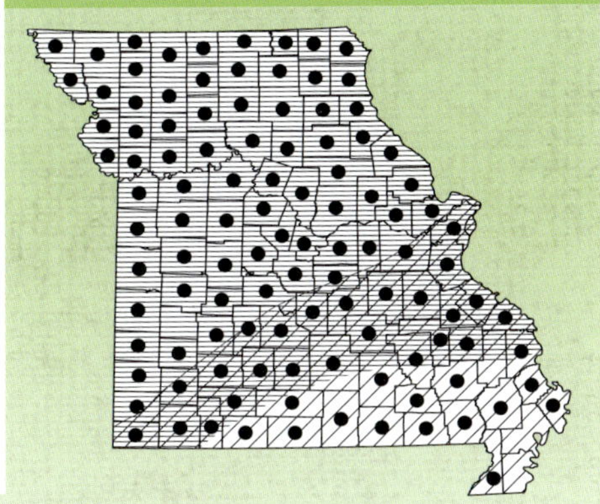

Distribution: Missouri: Northern two-thirds of the state (horizontal lines) and is replaced by the midland watersnake in the southern one-third (diagonal lines) with a wide area of intergradation. North America: The northern watersnake ranges from southern Canada and Maine south to northern South Carolina and Georgia, the Great Lakes states westward to the Central Plains states. The midland watersnake occurs from South Carolina to southern Alabama and Louisiana, to the southern half of Illinois and Missouri, and into Arkansas and eastern Oklahoma (Powell et al. 2016).

A juvenile midland watersnake eating a scuplin from Ozark County (above). Ventral coloration of a northern watersnake from Callaway County (right).

by gray or black. Dorsal scales are keeled, and the anal plate is divided. Male northern watersnakes are smaller and have longer tails than females.

Adult northern watersnakes range in total length from 560 to 1,067 mm (22 to 42 in.) but have been known to reach 1,500 mm (59.1 in.; Powell et al. 2016).

Habits and Habitat: Northern watersnakes are usually active from early April until October in Missouri. Individuals bask on branches overhanging water or on logs or rocks along the water's edge. This species becomes mostly nocturnal during hot weather; it is known to take shelter under rocks or other objects along the edge of rivers or ponds. Often riprap rock used to stabilize banks and levees of reservoirs is used for basking and shelter (Olds 2007). Overwintering sites are rock dams, levees, bank burrows, crayfish burrows, old logs and stumps, and crevices in rock ledges (Ernst and Ernst 2003; Olds 2007; Roe et al. 2003). In the Ozark Highlands, Anderson (1965) reported ledges with deep fissures generally near water are ideal overwintering sites. However, he observed snakes overwintering approximately 400 m from water in some cases. In Ozark County, an aggregation of at least 73 watersnakes was found in early March due to a collapsed rock and clay bank along an edge of a stream (Anderson 1965).

Like other members of the genus *Nerodia*, this species is pugnacious when captured and will flatten its head and neck and bite to defend itself. It discharges a strong-smelling musk from glands at the base of the tail as a means of defense.

Northern watersnakes live in a wide variety of aquatic habitats that are both natural and man-made: creeks, rivers, sloughs, farm ponds, lakes, marshes, and swamps. This is the most

JEFF BRIGGLER

Adult midland watersnakes are commonly observed basking on rocks and logs on Ozark rivers and streams, such as this example in Carter County.

common and widespread species of watersnake in Missouri, and it can be quite abundant along waterways with plenty of food and cover. In a single year, Bauman and Metter (1977) collected 1,370 northern watersnakes at a goldfish hatchery in central Missouri. Aldridge et al. (2005) estimated the average abundance of this species in May and June sampling was five adult snakes per kilometer (eight per mile) with a range of 1.6 to 8.2 per kilometer (2.6 to 13.2 per mile) in east-central Missouri. In a study of this species in southwestern Missouri (Lake Springfield), Roth and Greene (2006) estimated their home range to be 2.9 ha (7.2 ac) for males and 2.7 ha (6.7 ac) for females. Most activity occurs in the summer (July and August) and snakes generally remained within 5 m (16.4 ft) of aquatic habitat throughout the summer.

Small fish make up to 75 percent of the diet of northern watersnakes; they also eat amphibians. A northern watersnake captured in eastern Kansas had eaten a Woodhouse's toad (Smith and Powell 1993), and a young individual was observed eating a small-mouthed salamander larva in a temporary pool in west-central Illinois (McCallum 1995). Another individual was observed in Ozark County consuming a sculpin (*Cottus* spp.) on the North Fork of the White River (J. Briggler pers. obs.). This species will actively search for prey in the water. An adult watersnake was observed crawling along the rocky bottom of the spring-fed Current River in Missouri (J. Briggler pers. obs.). This individual stayed under water over 30 minutes and would poke its head within rocky crevices and slither beneath rocks presumably searching for prey. Northern watersnakes will also readily feed upon disposed carcasses of cleaned fish. They have learned through experience that an easy meal is provided by people cleaning their catch in heavy recreational use areas, such as the Lake of the Ozarks.

Breeding: Courtship and mating of this species mainly occur during the spring (April into June; Aldridge and Bufalino 2003; Aldridge et al. 2005; Bauman and Metter 1977). However, Anderson (1965) observed mating in late September in Jackson County. Males locate females by sensing an odor the females release and will actively search for them (Aldridge et al. 2005). The male swims with his head above the water and picks up the female odor by tongue flicking the air. He follows the smell and systematically searches the bushes, overhanging trees, and stream bank until the female is located. Courtship involves a male rubbing his chin along the back of the female and at the same time jerking spasmodically to stimulate her to copulate (Aldridge et al. 2005). More than one male may try to court and mate with a receptive female. Generally, only one male is involved, but occasionally two to four males will court a female (Aldridge et al. 2005). Courtship may last from several minutes to more than two hours (Aldridge et al. 2005). Females in various populations in Missouri are known to store sperm during the winter months and use it the following spring (Bauman and Metter 1977). Gestation is generally 9 to 12 weeks, but a study in central Missouri estimated the gestation period to be 58 days (Bauman and Metter 1977). Aldridge and Bufalino (2003) studied a population of this species in east-central Missouri and found that gravid females continue to feed throughout the breeding season, which is unusual for many snake species. The young are born during August and September in Missouri.

A litter contains 6 to 99 young with an average of 27 (Ernst and Ernst 2003; Minton 1972; Slevin 1951). Larger females typically produce more young (Bauman and Metter 1977). A Greene County specimen gave birth to 35 young in early September, and a Wright County specimen gave birth to 11 young in early September (Anderson 1965). These young measured from 189 to 272 mm (7.4 to 10.7 in.). Young northern watersnakes are light gray or tan with contrasting dark-brown or black crossbands and blotches. Males become reproductively mature in two years or less and most females take three years (Bauman and Metter 1977). Reported longevity of the northern watersnake is about 9.5 years (Snider and Bowler 1992).

Subspecies: Throughout much of the southern third of Missouri, the northern watersnake intergrades with—or is replaced by—a geographic race known as the midland watersnake, *Nerodia sipedon pleuralis* (Cope). The midland watersnake has a tan or red-brown ground color with brown or red-brown crossbands and blotches. Some individuals are almost orange with brown markings. The belly is usually yellow with irregularly spaced orange, red, or brown markings in the form of spots or half-moons. Northern watersnakes usually have 30 or more dark dorsal blotches (includes bands and blotches) along the back. Midland watersnakes usually have 30 or fewer dark dorsal blotches (includes bands and blotches).

This subspecies is the most commonly encountered snake on southern Ozark creeks and streams in Missouri. Anderson (1965) stated that the midland watersnake lives primarily in and near the clear, cool, gravel-bed creeks and rivers typical of the southern Missouri Ozarks. Other natural history and reproductive biology information of this subspecies is similar to that of the nominate subspecies described above.

Remarks: As with all watersnakes native to Missouri, northern and midland watersnakes are often misidentified as the venomous northern cottonmouth (*Agkistrodon piscivorus*) and killed because of unwarranted fear. People who fish in Missouri have persecuted watersnakes for years because of the mistaken belief that these reptiles eat game fish and are detrimental to their sport. In reality, just the opposite is the case. Game fish are too agile for watersnakes to catch—unless a fish is injured or diseased. Numerous studies (Pope 1944; Lagler and Salyer 1947; Raney and Roecker 1947; Trembley 1948) have shown that watersnakes, in association with natural bodies of water, improve fishing by preventing the spread of fish diseases, by reducing fish overpopulation, and by providing food for game species (large game fish eat young watersnakes). Watersnakes are a natural part of our outdoor heritage and a valuable component of the waters of our state.

Nickerson and Krager (1975) reported the collection of two aberrant specimens from Miller County that lacked the dorsal blotches and crossbands.

Northern Rough Greensnake

Opheodrys aestivus aestivus (Linnaeus)

JEFF BRIGGLER

Adult northern rough greensnake from Butler County.

Description: The northern rough greensnake is a slender, light-green arboreal snake. The dorsal color is plain light green; the belly is yellow or cream colored and devoid of any markings. Scales along the back and sides are weakly keeled, and the anal plate is divided. This snake has an elongated body and a long slender tail, allowing its body weight to be evenly spread across the branches and leaves of trees and bushes. The tail may account for 32 to 47 percent of their total length (Walley and Plummer 2000). Adult females are larger than adult males; males have a longer tail than females.

Adult northern rough greensnakes range in total length from 560 to 810 mm (22 to 31.9 in.) but have been known to reach 1,160 mm (45.7 in.; Powell et al. 2016).

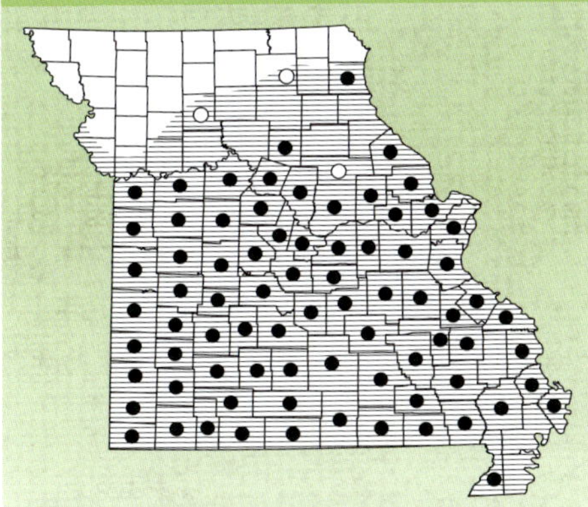

Distribution: Missouri: Throughout the southern two-thirds of the state and along the Mississippi River Hills in northeastern Missouri. North America: Southeastern U.S., from southern New Jersey to all of Florida, southern edge of West Virginia and Ohio, southern Indiana and Illinois, southern Missouri, southeastern Kansas, eastern Oklahoma, the eastern half of Texas, all Gulf Coast states, and the northeastern edge of Mexico (Powell et al. 2016).

Habits and Habitat: The active season of this species is late March through October, with peak activity in spring (April and May) and autumn (September and October). This species has not been well studied in

Northern rough greensnakes are often found in dense vegetation along the edge of rivers and streams, such as this example in Shannon County (above). A northern rough greensnake eating a grasshopper, Cole County (right).

Missouri, but extensive ecological studies have been conducted in northern Arkansas by Plummer and his colleagues. This species becomes active soon after sunrise in the morning, and they will select a site to spend the night from one-half to one hour before sunset (Plummer 1981). The rough greensnake is diurnal and spends most of its time among the branches and leaves of trees and bushes. It remains motionless among leaves and branches and relies on its coloration to remain undetected. If a slight wind moves the branches, the greensnake will bend and wave in response. Individuals may be observed crossing roads, trails, and creeks during the active season. Individuals are often observed crossing Katy Trail State Park in Cole County, especially in October (J. Briggler pers. obs.). Most of these individuals are migrating from the Missouri River floodplain into the adjacent upland, rocky, wooded hillsides. They are capable swimmers and have been observed swimming across an Ozark stream on several occasions (T. Johnson pers. obs.). At night, these snakes coil among the leaves near the tip of branches (Plummer 1981).

This arboreal species resides in the dense trees and bushes along creeks, rivers, ponds, reservoirs, and forest edges. Plummer (1981) found that rough greensnakes select resting places among branches from 1 to 4 m (3.3 to 13.1 ft) above the ground. Studies in Arkansas found that 86 percent of rough greensnakes' daytime micro habitat were within 3 m (9.8 ft) of a pond or creek, and 88 percent of their nighttime resting sites were within 3 m (9.8 ft) of a body of water (Plummer 1981). Observations of *Opheodrys aestivus* in North Carolina indicated that this species may not be as arboreal as has been reported by researchers in other sections of the species' range: out of 484 North Carolina greensnakes captured, 93 percent were found on the ground (Palmer and Braswell 1995). Anderson (1965) stated that in

TOM R. JOHNSON

A hatchling northern rough greensnake from Cole County.

Missouri this species can occur in buckbrush, bittersweet vines, wild blackberry vines, and willow trees with branches overhanging water. Plummer (1981), however, noted that willow and ironwood were used only slightly more than other species of trees and concluded that the rough greensnake did not seem to prefer any particular species of tree, bush, or plant. This lack of tree, bush, or plant preference was verified by a study in Oklahoma (Goldsmith 1984). Narrow strips of trees, bushes, and vines were more likely to hold rough greensnakes than a more open woodlot (Goldsmith 1984; Plummer 1981, 1997). Individuals may live on rocky, forested hillsides and river bluffs, and specimens have been found under flat rocks on occasion. The northern Arkansas study area had a population estimate that ranged from 100 to 800 rough greensnakes per hectare (40.5 to 323.7 per acre; Plummer 1997). A substantial population decrease of this species was caused by a long dry period that greatly reduced their insect and spider prey and, at the same time, greensnake predation had increased. Overwinter dormancy takes place in deep leaf litter, inside rotten logs and stumps, and in the ground at a depth of over 200 mm (7.9 in.; Ernst and Ernst 2003).

Rough greensnakes are considered a gentle and harmless species. When first captured a specimen may gape open its mouth and reveal a purplish-black lining, but this is a bluff and most specimens will not try to bite. Some newly captured specimens, as a form of defense, will produce a musk from glands at the base of their tail that may be mixed with feces.

The diet of this species is predominantly insects and spiders. Plummer (1981) listed spiders, hairless caterpillars, grasshoppers, crickets, beetles, dragonflies, and damselflies as comprising 85 percent of the diet of rough greensnakes studied in Arkansas. Rough greensnakes may eat young treefrogs, land snails, praying mantids, wood roaches, katydids, and walking sticks. Slender, arboreal, and insectivorous snakes like this species must eat more prey and eat more often to sustain themselves and reproduce, compared to snakes that eat vertebrate prey (Plummer 1985b). Predators of this species include hawks, bluejays, kingsnakes, and domestic cats. Plummer (1990b) found that gravid females were preyed upon by speckled kingsnakes and southern black racers. A rough greensnake had been eaten by an eastern coachwhip in Arkansas (Trauth and McAllister 1995).

Northern rough greensnakes are commonly found among the green vegetation of shrubs and trees.

Breeding: Male rough greensnakes are reproductively mature at one year old, and females are mature at two years old (Aldridge et al. 1990; Mitchell 1994). In an Arkansas study, rough greensnake females were found to first breed at two or three years old (Plummer 1985b). The Arkansas population study showed that, in most years, the sex ratio was 1:1 (Plummer 1985a). Sperm production and maturation is at its peak as a male reaches the age of 20 to 21 months (Aldridge, et al. 1990). Although males may be physiologically ready to breed in the fall, females are not receptive to autumn breeding, or, if they breed, the sperm will remain in the females' oviduct and can be used to fertilize their eggs the following spring (Plummer 1985b). Mating takes place in the spring but some autumn breeding was reported in Illinois, North Carolina, and Virginia (Ernst and Ernst 2003). A pair of adult rough greensnakes were captured within 30 m (98.4 ft) of each other in early October in western Illinois and were held together for observation (Morris and Vail 1991). The male immediately began jerky movements toward the female and rubbed his chin along the female's back. He also moved his cloacal area against the female's body to search for her cloaca and begin copulation. The pair were together for 37 minutes, and the female ended the episode when she constricted her abdomen and forced the male's hemipenis out of her cloaca. It should be noted that, although the female was kept in captivity an additional 11 months, no eggs were produced (Morris and Vail 1991). Males locate a female for mating by sight. Once found, the female will be followed and quickly approached with jerking movements by the male. Copulation will quickly follow and the pair are together for less than an hour (Ernst and Ernst 2003). Mating may occur both among branches of trees and on the ground. Most egg laying will take place in June and July (with an average of 35 days after mating; Ernst and Ernst 2003). Gravid females will move out of their arboreal haunts to select nesting sites: deep leaf litter, brush piles, behind loose bark of a standing dead tree, or in rotten logs or stumps. In Arkansas, some gravid females moved into an oak-hickory forest and laid their eggs in hollow, living trees at an average height of 1,220 mm (48.0 in.) above the ground (Plummer 1990a). In addition, it was found that females may use the same nesting site the following year, and several females used the same hollow tree trunk for nesting (Plummer 1981, 1990a). Each female produces from 1 to 14 eggs with an average of 5.9 eggs (Fitch 1985). Larger females will produce more eggs than smaller females. Their eggs are white, have a leathery feel, and stick to each other. A study in eastern Kansas found that only one clutch of eggs is produced per female during the summer (Fitch 1985). Studies in a few states have reported females producing two clutches of eggs per season (Ernst and Ernst 2003). There have been several documented cases of egg laying and hatching in Missouri with most eggs being laid in June and July and hatching in August and September. Anderson (1965) reported the following egg clutches: a female from Jackson County laid six eggs in early July, which hatched in early September; and a female captured in Johnson County laid four eggs in late June, which hatched in late September. Other reports include a female captured in Franklin County that laid six eggs in mid-July, with hatching taking place in late August. The hatchlings had a mean total length of 195 mm (7.7 in.). A female captured in Lincoln County in mid-July laid six eggs, which hatched in early August. The young had a mean total length of 171 mm (6.7 in.; Smith and Powell 1993). A female in Miller County laid seven eggs in mid-July (Johnson 2000). The eggs averaged 24 by 11 mm (0.9 by 0.4 in.) and hatched in late August after 42 days of incubation. Incubation period for clutches in Missouri ranged from 22 to 91 days. The newly hatched rough greensnakes averaged 184 mm (7.2 in.) in total length and were olive drab dorsally and white ventrally. Hatchlings' color will change to a bright green after they shed the first time (within a week after hatching). In the wild, the eggs can take up to two months to hatch. The many field studies of rough greensnakes in Arkansas found that females seemed to select nesting sites with high humidity, but by the time their eggs began to hatch some of the eggs had been developing under a wide range of humidity levels; with some nests being quite dry by the time of hatching. It was concluded that hatchling survivorship and the size of the hatchlings were not affected by such a wide range of humidity during egg incubation (Plummer and Snell 1988). Plummer (1985b) reported that this species may live as long as eight years in the wild.

Remarks: Rough greensnakes have an interesting internal adaptation that protects them from exposure to ultraviolet radiation. Because they are diurnal and arboreal, a great deal of time is spent in direct sunlight. Their internal organs—especially their reproductive organs—must be protected from the sun's dangerous ultraviolet radiation. This species, and several arboreal snakes in other parts of the world, have evolved a black peritoneum. This black envelope covers their internal organs and provides a shield to prevent damage from ultraviolet radiation (Plummer 1993). Likely, few other snakes in Missouri have this internal adaptation. A study of worldwide genetic relationships of snakes in the family Colubridae found that the North American rough greensnake originated from arboreal snakes found in tropical East Asia (Rabb and Marx 1973).

Any wildlife that depends on insects and spiders for their survival is vulnerable to insecticide contamination. This is especially true for rough greensnakes (Ernst and Ernst 2003). Rough greensnakes have been found dead on Missouri's roads and highways during their active season. Dead rough greensnakes will lose the yellow pigmentation within their skin and turn a blue color within several hours of death.

Smooth Greensnake

Opheodrys vernalis (Harlan)

Adult smooth greensnake.

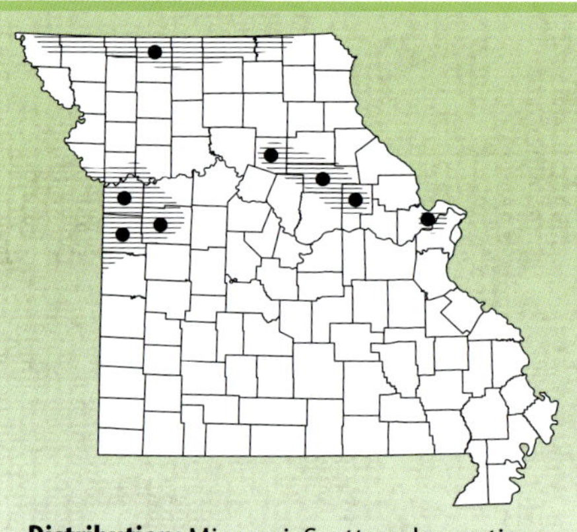

Distribution: Missouri: Scattered counties in the northern half of the state. North American: New England south to northern Virginia, west to Kansas, and north to Canada. Isolated populations occur in Texas. Populations on the western part of their range become more disjunct (Powell et al. 2016).

Description: The smooth greensnake is a small, gentle, secretive species. The dorsal color is bright green without any markings. The belly is plain white or with some yellow color along the sides. The lip scales on the head are generally yellow. Individuals in the upper Midwest may be a uniform light tan instead of green, but this color morph has not been reported in Missouri. Juveniles are similar to adults but have more of a blue gray coloration. The dorsal scales are smooth, and the anal plate is divided.

Adult smooth greensnakes range in total length from 303 to 510 mm (11.9 to 20.1 in.) but have been known to reach 797 mm (31.4 in.; Powell et al. 2016).

Habits and Habitats: Little is known about the smooth greensnake in Missouri, especially since it has not been seen in 50 years or more. The active season of this species presumably lasts from April into October. Most of the records for Missouri are from mid-April into June (J. Briggler unpubl. data). Smooth greensnakes search for food

TOM R. JOHNSON

Head of a smooth greensnake.

as they move about among grasses and shrubbery during the day and hide beneath logs and boards or within mats of grass and animal burrows at night. Their green coloration makes it almost impossible to see one among the grasses, and often they gently sway with the windblown grasses. When captured, a specimen may struggle to escape but seldom attempts to bite.

This species prefers grasslands. In Iowa, it occurs in grassy, moist meadows, native prairies, and upland grassland habitat surrounding lakes and reservoirs (LeClere 2013). It is known to occur in grassy meadows and pastures in Missouri (Anderson 1965). Hurter (1911) noted an individual was collected from a garden in St. Charles, but surrounding habitat was not mentioned. Easterla and Meadows (1993) found an individual under a board in a wet, grassy area below a pond spillway in Harrison County. This species is known to overwinter underground in abandoned mammal burrows, road embankments, and ant mounds (Degenhardt et al. 1996; Lang 1969; LeClere 2013). Ant mounds seem to be a preferred overwintering site, and crayfish burrows may also be used.

The smooth greensnake feeds on a variety of insects, slugs, snails, earthworms, and spiders. Individuals collected in Wisconsin and held in captivity ate hairless caterpillars, small crickets, and grasshoppers (Johnson 1987).

Breeding: Nothing is known about the reproductive biology of this species in Missouri. In other states this species normally mates in the spring but also breeds in early autumn. The eggs are laid in rotten logs and stumps, mounds of decaying vegetation, leaf litter, sawdust piles, and mammal burrows (Ernst and Ernst 2003). More than one female may deposit eggs in the same nest site. There may be from 3 to 15 eggs per clutch (LeClere 2003). The average clutch size for Illinois and Wisconsin populations is six (Smith 1961; Vogt 1981). The eggs are white, elongated, and have average measurements of 23 by 13 mm (0.9 by 0.5 in.; Moriarty and Hall 2014). Egg-laying and incubation time varies greatly with this species. Gravid females retain the developing eggs for a period during early summer, and the eggs consequently hatch soon after being laid; this is not the case for most other North American snake species. Incubation period is generally less than a month but can be only a few days after the eggs are laid (Ernst and Ernst 2003; Vogt 1981). Hatchling smooth greensnakes are olive tan to blue gray in color and range from 120 to 150 mm (4.7 to 5.9 in.) in length. One individual lived to slightly over six years of age in captivity (Snider and Bowler 1992).

Remarks: The smooth greensnake can be distinguished from the rough greensnake (*Opheodrys aestivus*) by its smaller size, shorter tail, and the smooth (unkeeled) dorsal scales. It is also known as a grass snake due to its grassland habits and green coloration.

This species of conservation concern is likely extirpated from the state with the last confirmed observation from 1970 in Harrison County (Easterla and Meadows 1993). The destruction of its grassland habitat and the use of agricultural insecticides are primarily responsible for the decline of this species in Missouri. Although the smooth greensnake is presumably gone from the state (Johnson 2000), this secretive species is known to occur in several Iowa border counties. Therefore, small, relict populations may still persist in Missouri. Besides keeping a watchful eye for a bright green colored snake in extreme northern Missouri, additional efforts are needed to determine this species' status in Missouri.

Great Plains Ratsnake

Pantherophis emoryi (Baird and Girard)

Adult Great Plains ratsnake from Taney County.

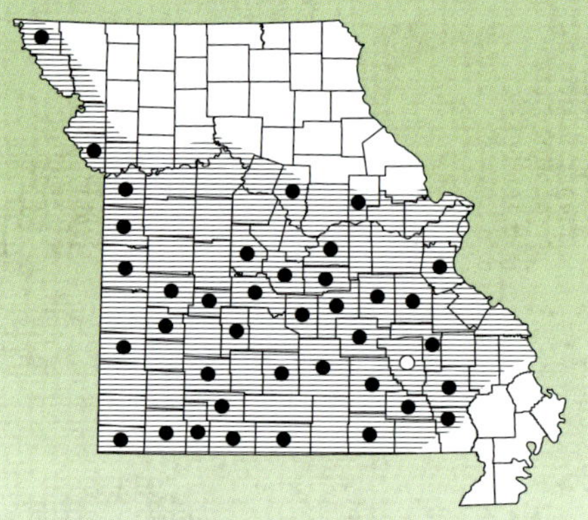

Distribution: Missouri: Northwestern, central, western, and southern parts of the state. North America: extreme southern edge of Nebraska, east to southwestern Illinois, southwest through most of Arkansas, and into central Texas and Mexico, and north through southern Great Plains states. There are numerous isolated populations in west-central Colorado and east-central New Mexico (Powell et al. 2016).

Description: The Great Plains ratsnake is a medium-sized gray snake with numerous dark-brown blotches. Its ground color is light gray or brown gray. There are 25 to 45 dorsal blotches of dark brown bordered with black. A dark-brown stripe between the eyes extends through each eye, along the sides of the head and onto the neck. There is also a spearpoint-shaped marking on the top of the head. The belly is white with bold, squarish black markings and black or dark-gray stripes along the underside of the tail. The body scales are weakly keeled, and the anal scale is divided.

Adult Great Plains ratsnakes range in total length from 610 to 910 mm (24 to 35.8 in.) but have been known to reach 1,530 mm (60.2 in.; Powell et al. 2016).

Habits and Habitat: This species is secretive and normally nocturnal during the active season (from late March to October). Daylight hours are spent hiding under rocks, logs, and boards or underground

JEFF BRIGGLER
JEFF BRIGGLER

Great Plains ratsnakes can often be found in open, rocky wooded hillsides, such as this example in Christian County (above). Ventral coloration of a Great Plains ratsnake (left).

in small mammal burrows and within rock crevices. In Missouri, the Great Plains ratsnake lives in open woodlands, rocky, wooded hillsides, glades, and caves. They are often found in the twilight zone of caves coiled up in rock crevices or resting on a cave gate to prey upon the evening exodus of bats. In a study in Texas, Great Plains ratsnakes were found to use man-made structures extensively, such as rock gulley plugs, and brush and mulch piles (Sperry and Taylor 2008). They are known to overwinter in rodent burrows, rock crevices, caves, old and hollow logs or stumps, and rock-lined wells (Ernst and Ernst 2003; J. Briggler unpubl. data). Drda (1968) found a Great Plains ratsnake overwintering in a cave in Jefferson County. The ratsnake will often vibrate its tail when alarmed. In Texas, its home range is about 4 to 27 ha (9.9 to 66.7 ac), with an average of 10 ha (24.7 ac), and in Kansas, it is about 2 to 30 ha (4.9 to 74.1 ac) with an average of 15 ha (37.1 ac; Klug et al. 2011; Sperry and Taylor 2008).

Its diet consists mainly of rodents, bats, and small birds but also occasionally frogs, lizards, and small snakes (Ernst and Ernst 2003). Larger prey are killed by constriction, but smaller items are likely grabbed by the mouth and swallowed directly.

Breeding: Breeding is presumed to take place soon after these snakes emerge from their overwintering retreat. In Kansas, breeding extends from late March through May (Fitch 1999). Courtship and mating involve some chasing by the male, followed by the male

TOM R. JOHNSON

Hatchling Great Plains ratsnake.

crawling over the female in a jerking and writhing fashion. Once their bodies are aligned mating will occur, which may last up to half an hour (Collins et al. 2010).

One clutch is laid per season, containing 3 to 37 eggs (Stebbins 2003). The average number of eggs per female was 7.2 and 9.5 for two studies in Kansas (Fitch 1999; Meshaka and Schmidt 2017). Anderson (1965) reported a female from Cedar County laid seven eggs in late June that averaged 46 by 24 mm (1.8 by 0.9 in.) in length and width. Seven clutches from 1994 to 2003 were obtained in captivity from individuals collected in Missouri (R. Krager pers. comm.). The average number of eggs per female was 8.9, with a range from 5 to 14 eggs and an average of 41 by 23 mm (1.6 by 0.9 in.) in length and width. These eggs were laid from mid-June to late July.

Eggs are smooth, white, and adherent; they are laid in mammal burrows and rotting logs and stumps between late May and early July (Ernst and Ernst 2003; Fitch 1999; Meshaka and Schmidt 2017). Hatching takes place in August and September with an incubation period of 51 to 77 days (Ernst and Ernst 2003). For six clutches from Missouri females, their eggs hatched from late August to late September with an incubation period of 55 to 74 days (B. Krager pers. comm.). Hatchling Great Plains ratsnakes are similar in appearance to the hatchlings of western ratsnakes (*Pantherophis obsoletus*). Most Great Plains ratsnakes mature in two to three years. A wild-caught individual lived to 21 years of age in captivity (Snider and Bowler 1992).

Remarks: House snake is a local name for the Great Plains ratsnake because it is occasionally found around abandoned farm buildings. Some authors call it the Emory's ratsnake; named in honor of Brigadier General William Hemsley Emory. This species can be confused with the prairie kingsnake (*Lampropeltis calligaster*). A careful reading of both species' accounts will facilitate proper identification of an individual. The prairie kingsnake is more commonly encountered than the Great Plains ratsnake in Missouri.

Little is known about the natural history of the Great Plains ratsnake in Missouri and range wide. It is a relatively uncommon snake in Missouri and due to its secretive and nocturnal habits, it might be more abundant than presumed.

Western Ratsnake

Pantherophis obsoletus (Say)

JEFF BRIGGLER

Adult western ratsnake from Callaway County.

Description: The western ratsnake is one of Missouri's largest and most familiar species. Its dorsal color is shiny black, but young adults and individuals in southeastern Missouri may have dark-brown or black blotches. Small patches of red skin may be visible between the scales along the sides of some individuals. The upper lip, chin, and ventral part of its neck are usually white. The belly is white with a black checkerboard pattern anteriorly, changing to a mottling of gray, brown, and white or yellow posteriorly. Scales along the back are weakly keeled, and the anal scale is divided. Hatchling western ratsnakes differ markedly in color from adults. The ground color is tan to gray; there are distinct black or dark-brown blotches along the back and sides. The head of young snakes has a black band that crosses from eye to eye and extends down at an angle toward the mouth. There is a gradual change in the ground color from gray to black as the young snakes mature. The adult color is present after two years of growth.

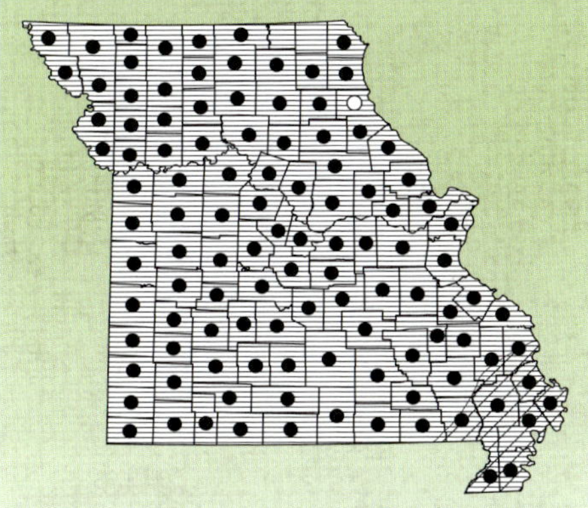

Distribution: Missouri: Statewide; may hybridize with the gray ratsnake (*P. spiloides*, diagonal lines) in southeastern corner of the state (see subspecies section). North America: West of the Mississippi River from southern and eastern Iowa south to Louisiana, west to eastern Texas, and north to southeastern Nebraska (Powell et al. 2016).

A hatchling western ratsnake from Cole County (above). Western ratsnakes hatching from Cole County (right).

Adult western ratsnakes range in total length from 1,067 to 1,830 mm (42 to 72 in.) but have been known to reach 2,184 mm (86 in.; Powell et al. 2016).

Habits and Habitat: The active season of the western ratsnake is between early April and October. Fitch (1963b) has conducted intensive field studies of this species in eastern Kansas; much of the following information is based on his work. This species is active when air temperatures are between 15.6°C and 31.1°C (60°F to 88°F). During spring, early summer, and autumn, western ratsnakes are active during the day; in hot weather, they become nocturnal. The western ratsnake is a forest-dwelling species; it prefers rocky, wooded hillsides or sections of woods along streams and rivers. These snakes take shelter in brush piles, hollow trees, farm buildings, and old houses where there is often an abundance of mice. Western ratsnakes are excellent climbers and will bask on horizontal branches in large trees. They are found along beams in barns, attics, and walls of man-made structures if not sealed tight. This species will climb into bluebird houses to eat eggs and young. They have been observed searching for bird nests along fencerows and raiding red-winged blackbird nests along the edge of ponds (R. Daniel pers. comm.). Predation of bird nests is more common in old field and edge habitat compared to forests (Thompson and Burhans 2003). A study in central Missouri used video cameras to monitor bird nests and found western ratsnakes were thorough nest predators (Stake et al. 2005). They probably located nests by searching randomly, smelling birds, and watching birds to learn where nests are located.

This species overwinters in mammal burrows, rock outcrops, caves, abandoned rock quarries, old stone wells or cisterns, foundations of old buildings, rotted tree stumps, hollow

TOM R. JOHNSON

JEFF BRIGGLER

Western ratsnake eating a house mouse (above). Western ratsnakes are excellent climbers, such as this adult from Crawford County (left).

trees, and in the root systems of dead and rotting standing trees. Winter quarters may be shared with other species of snakes. A small sandstone cave in Johnson County was used by both the western ratsnake and eastern yellow-bellied racer (*Coluber constrictor flaviventris*; Sexton and Hunt 1980). A few western ratsnakes were found overwintering in a cave in Jefferson County (Drda 1968). The population studied by Fitch (1963b) in eastern Kansas had an average home range of 10.1 to 12.1 ha (25 to 30 ac) with a population density of about 0.8 to 1.2 snakes per hectare (0.3 or 0.5 per acre).

The food of this species consists of rodents, small rabbits, bats, bird eggs, small birds, and on occasion lizards. Prey is killed by constriction. Easterla (1967) documented a western ratsnake tightly wrapped around a gray myotis (*Myotis grisescens*) bat in a Camden County cave. Cary et al. (1981) reported observing a western ratsnake constrict and devour an Indiana myotis (*Myotis sodalis*) bat in Texas County. A Jackson County specimen was found that had eaten a prairie vole (*Microtus o. ochrogaster*; Smith and Powell 1993). Predation of nestlings and adult European starlings (*Sturnus vulgaris*), as well as swallowing several chicken eggs have been observed in Callaway County (J. Briggler pers. obs.). Young ratsnakes eat frogs, lizards, and insects.

Breeding: Courtship and mating usually occur in the spring but may also happen in summer or autumn. Anderson (1965) observed a breeding pair in April. Eggs are laid in rotten stumps or logs, tree holes, sawdust, and mulch piles or under rocks. Four to 40 eggs (average 15) may be laid, usually in June or early July (Fitch 1970). The eggs adhere to each other when laid and average 46 by 23.5 mm (1.8 by 0.9 in.; Anderson 1965). Eggs hatch in the autumn, usually taking from 42 to 84 days (Ernst and Ernst 2003). Numerous egg clutches have been documented in Missouri. A clutch of 27 eggs was found in St. Louis County in late October (Sexton et al. 1976). Two completely formed young hatched from one egg in a clutch of eight eggs laid by a Lincoln County female (Schuette 1978). Another clutch of eight eggs was found in mid-July in a mulch pile in Lincoln County and hatched in late October (Schuette 1997). A clutch of 12 eggs was found in a hollow standing white oak tree being removed in early July in Cole County. These eggs were covered in moist soil that had accumulated within the hollow tree and later hatched in captivity in early October (J. Briggler unpubl. data). Newly hatched young are between 300 and 370 mm (11.8 and 14.6 in.) in total length. Females generally mature in four years and males in three years. Western ratsnakes are known to live more than 20 years (Snider and Bowler 1992).

Subspecies: Genetic and morphological data no longer support multiple subspecies of the former ratsnake (*P. obsoletus*) complex (Burbrink 2001; Burbrink et al. 2000). Burbrink divided this ratsnake group into three distinct species; eastern ratsnake (*P. alleghaniensis*), western ratsnake (*P. obsoletus*), and gray ratsnake (*P. spiloides*). The western ratsnake occurs statewide in Missouri, but the gray ratsnake that occurs east of the Mississippi River has been reported in southeastern Missouri (Anderson 1965). It is likely the population in southeastern Missouri is a hybrid of the western ratsnake and gray ratsnake (T. Johnson pers. obs.).

Remarks: Western ratsnakes, formerly known as black ratsnakes, are common in Missouri and often observed around homes and buildings, especially if rodents are abundant. They are also commonly known as the black snake, because of their coloration, or chicken snake, because of their frequent raids of chicken houses. Western ratsnakes will vibrate their tail when alarmed. If captured or cornered, they will bite in self-defense and release a pungent, unpleasant musk from glands at the base of the tail when frightened. Western ratsnakes help reduce damage to crops and stored grain by eating mice and rats; they are a valuable, natural rodent control. This service far outweighs the occasional theft of a few hen's eggs or baby chickens. Hawks are a major predator of adult western ratsnakes.

Western Foxsnake

Pantherophis ramspotti Crother, White, Savage, Eckstut, Graham, and Gardner

TOM R. JOHNSON

Adult western foxsnake from Holt County.

Description: The western foxsnake is a moderately large snake with distinct brown blotches. Ground color is gray, tan, or yellow tan. The western foxsnake averages 37 dorsal blotches compared to an average of 43 dorsal blotches in the eastern foxsnake. There is a series of smaller dark-brown blotches along the sides. Head is normally yellow, orange, or orange brown and usually unmarked. The belly is cream to yellow and boldly checkered with black. Dorsal scales are weakly keeled; the anal scale is divided. Young western foxsnakes lack the adults' yellow ground color; they appear gray with bold dark-brown or black blotches. Head is boldly marked with a black mask running across the head, through the eyes and slanting back to the angle of the jaw. There are also black markings on the top of the head and black bars along the upper lips. These dark head markings disappear when the snakes reach adulthood. Hatchling foxsnakes are very similar in appearance to hatchling western

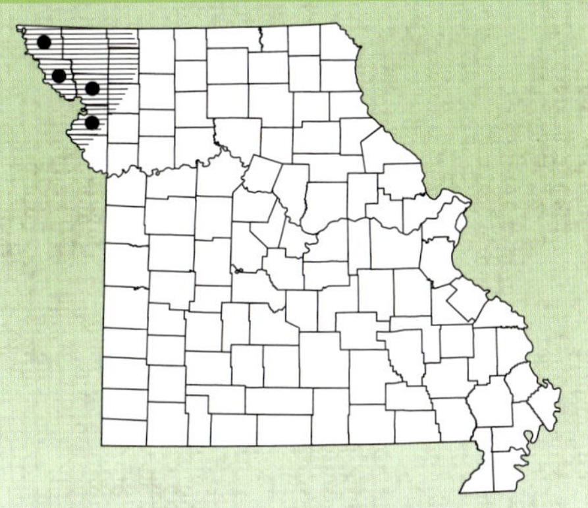

Distribution: Missouri: Northwestern corner. North America: West of the Mississippi River from southern Minnesota, throughout most of Iowa, southeastern South Dakota, eastern Nebraska, and northwestern Missouri. An area of overlap between the western and eastern foxsnake occurs along the Mississippi River (Powell et al. 2016).

JEFF BRIGGLER

JEFF BRIGGLER

Western foxsnakes are often found in wet prairies, such as this example in Holt County (above). A hatchling western foxsnake from Holt County (right).

ratsnakes; however, ventral scale counts are about 216 on western foxsnakes and about 221 on young western ratsnakes.

Adult western foxsnakes range in total length from 610 to 1,370 mm (24 to 53.9 in.) but have been known to reach 1,550 mm (61.0 in.; Powell et al. 2016).

Habits and Habitat: This is a marsh-dwelling member of the ratsnake group. The western foxsnake inhabits open grasslands and borders of woods that adjoin wet prairies, marshes, and river bottomlands in northwest Missouri. It is most abundant in the few remaining patches of natural, bottomland prairie. Approximately 50 foxsnakes were observed in cattail patches in deeper water habitat in Holt County (Griffin et al. 2007). Western foxsnakes are active between late April and early November with most reports occurring in May and June in Missouri. They are mostly diurnal and can be seen sunning or crossing roads in the morning or late afternoon. Unfortunately, many dead individuals have been reported from roads, especially those adjacent to wetlands.

Western foxsnakes take shelter beneath dense herbaceous vegetation under boards, logs, brush piles, mammal and crayfish burrows, and occasionally in a tree hollow a few feet off the ground (Shew et al. 2012; J. Briggler pers. obs.). We have captured adult foxsnakes resting under old pieces of corrugated tin along the edge of a natural wetland in Atchison and Holt counties in June. Foxsnakes defend themselves by vibrating their tail, coiling with head and neck raised, and repeatedly striking at an intruder. They are known to spend the winter underground within rock crevices, building foundations, and animal burrows (Ernst and Ernst 2003; Moriarty and Hall 2014). We have observed western foxsnakes using crayfish and mammal burrows in Missouri. The average home range varies from 12 to 54 ha (29.7 to 133.4 ac), and a study in northwest Missouri found males tend to have larger home ranges than females (Shew et al. 2012).

Its diet mainly includes rodents, small birds, bird eggs, and frogs. Its prey is killed by constriction. Young eat small frogs, lizards, earthworms, and insects. An American badger (*Taxidea taxus*) was observed digging up and eating snake eggs, including western foxsnake, and turtle eggs in Holt County (Davis and Welchert 2013).

Breeding: Little is known about the reproductive biology of the western foxsnake in Missouri and range wide. The species is presumed to court and breed in April, soon after emerging from overwintering dens. However, most courtship activity in northwestern Missouri was observed from late May into mid-June (Shew 2004; Shew et al. 2012). Courtship and mating are likely similar to the eastern foxsnake as described by Gillingham (1974) in Wisconsin. In late June or July, a female lays 10 to 20 white, leathery eggs, which adhere to each other and range from 37 to 46 mm (1.5 to 1.8 in.) in length (LeClere 2013). Their eggs are laid in rotten stumps or logs, sawdust piles, or leaf litter. Shew (2004) suspected egg laying in northwestern Missouri to occur in burrows or rocky crevices along roads or levees in Holt County. A female western foxsnake captured in Holt County laid nine eggs, which averaged 45 by 22 mm (1.8 by 0.9 in.; R. Seigel pers. comm.). In Iowa, nests with 8, 12, and 25 eggs have been found (LeClere 2013). Hatching usually occurs in August or September, with an incubation period between 35 and 75 days (Ernst and Ernst 2003). An incubation period of 42 days was documented in Iowa (LeClere 2013). Newly hatched young foxsnakes closely resemble those of western ratsnakes and are about 230 to 310 mm (9.1 to 12.2 in.) in total length. Most individuals probably reach sexual maturity by the age of three.

Remarks: Two subspecies were formerly recognized in the foxsnake group; western foxsnake (*Elaphe vulpina vulpina*) and eastern foxsnake (*E. v. gloydi*). Based on genetic data, these two subspecies were elevated to full species status, and this change is reflected in their range along the Mississippi River (Crother et al. 2011). New species names are the western foxsnake (*Pantherophis ramspotti*), mainly occurring west of the Mississippi River, and the eastern foxsnake (*P. vulpinus*) mainly occurring east of the Mississippi River. Only the western foxsnake was formerly known to occur in Missouri (Johnson 2000). It is now presumed that foxsnakes within the St. Louis area are eastern foxsnakes; individuals in northeastern Missouri are hybrids, and populations in northwestern Missouri are western foxsnakes. We choose to follow the current distribution outlined in Powell et al. (2016) and Crother et al. (2011). However, further genetic studies are needed for foxsnakes along the Mississippi River in Missouri.

The name foxsnake might have been given to this species because of the musky odor given off by newly captured specimens. Scent glands at the base of the tail produce an odor that is said to resemble the scent of a red fox. However, others believe it was named after Rev. Charles Fox, who collected snakes many years ago (Powell 1990). Due to habitat loss, persecution, and collection pressures, the western foxsnake is a species of conservation concern in Missouri.

Eastern Foxsnake

Pantherophis vulpinus (Baird and Girard)

TOM R. JOHNSON

Adult eastern foxsnake from Clark County.

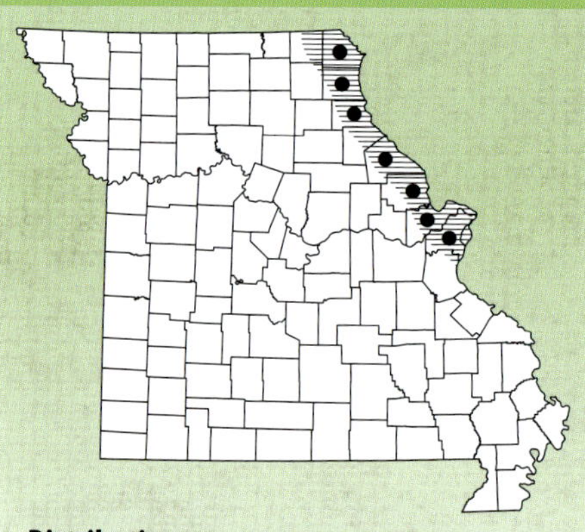

Distribution: Missouri: Along the Mississippi River floodplain north from St. Louis. It is likely that a hybridization zone occurs with the western foxsnake in the northeastern counties. North America: Primarily east of the Mississippi River in the Upper Peninsula of Michigan through most of Wisconsin, western Indiana, northern Illinois, and in the St. Louis area of Missouri. An area of overlap between the eastern and western foxsnake occurs along the Mississippi River (Powell et al. 2016).

Description: The eastern foxsnake is a moderately large snake with a gray, tan, or yellowish-tan ground color and distinct dark blotches. A row of dark-brown, rectangular-shaped blotches run along the back, with a series of smaller dark-brown blotches on each side. The blotches become bars or rings on the tail. The eastern foxsnake averages 43 dorsal blotches compared to an average of 37 dorsal blotches on the western foxsnake. The head is normally yellow, orange, or orange brown with few distinct markings. The belly is yellow and boldly checkered with black. Dorsal scales are weakly keeled, and the anal scale is divided. Young eastern foxsnakes lack the adults' yellow ground color; they appear gray with bold dark-brown blotches that are narrowly bordered in black. The head is boldly marked with a transverse line (black mask) running across the head, through the eyes, and slanting back to the angle of the jaw. There are also black markings on the top of the

JEFF BRIGGLER

TOM R. JOHNSON

Eastern foxsnakes are found in grassland habitats near marshes, such as this example in Clark County (above). Ventral coloration of an eastern foxsnake from Clark County (left).

head and black bars along the upper lips. These dark head markings fade as the snakes reach adulthood. Hatchling foxsnakes are very similar in appearance to western ratsnakes; however, ventral scale counts are about 216 on eastern foxsnakes and about 221 on young western ratsnakes.

Adult eastern foxsnakes range in total length from 610 to 1,370 mm (24.0 to 53.9 in.) but have been known to reach 1,705 mm (67.1 in.; Powell et al. 2016).

Habits and Habitat: Similar to the western foxsnake, the eastern foxsnake lives in open grassland, borders of woods, and along edges of agriculture fields adjacent to wet prairies, marshes, and river bottomlands. This species is generally active from May into October but has been reported as early as mid-April and as late as early November in Missouri. Peak observations occur in May and June (J. Briggler unpubl. data). Eastern foxsnakes can be seen crossing roads or sunning during the active season. A study in St. Charles County found several eastern foxsnakes sunning in areas of mowed grass adjacent to levee roads along wet ditches (Sexton 1991). Road-killed foxsnakes are often observed.

Eastern foxsnakes are secretive and spend a good amount of time hiding within dense vegetation and animal burrows or under objects (logs, boards, and brush piles). We have captured adult foxsnakes resting under old pieces of corrugated tin along the edge of a natural wetland in Clark County in June. Foxsnakes are generally a slow snake and seldom make a quick retreat when approached. However, they will defend themselves by vibrating their tail, coiling with head and neck raised, and repeatedly striking at an intruder. Foxsnakes spend the winter underground within rock crevices, building foundations, old wells, and animal burrows (Ernst and Ernst 2003; Gillingham 1974). We have observed foxsnakes using mammal burrows and riprap rock structures built specifically for overwintering sites in Missouri.

Prey consists mainly of small mammals (mice, voles, chipmunks) and occasionally birds and their eggs. Voles (*Microtus* spp.) are reportedly the primary food item in Indiana (Minton 2001). A foxsnake will grab prey with its mouth and kill it by constriction. Young are known to eat frogs, lizards, and insects.

Breeding: Little is known about the reproductive biology of the eastern foxsnake in Missouri. Similar to the western foxsnake, this species is presumed to court and breed in April, soon after emerging from overwintering dens, and continue into summer. Gillingham (1974) studied the courtship and mating behavior of the foxsnake in Wisconsin. He observed a male courting a female with a chase that may last 6 to 40 minutes. Once the female is still, the male moves along her side, rapidly flicking his tongue, then nudges her at midbody with his head. Jerky movements follow, and the male then moves his head forward along the female's body. Their tails begin twitching and the male moves his body over her. Once he has mounted her, their tails intertwine and mating occurs. During copulation, the male usually grasps the female behind the head with his mouth and continues to do so until copulation is completed (from 10 to 40 minutes). In late June or July, a female lays 8 to 27 eggs, which adhere to each other and average 45 by 28 mm (1.8 by 1.1 in.; Smith 1961; Vogt 1981). The eggs are laid in moist locations, such as rotten stumps or logs, sawdust piles, or leaf litter. Sexton (1991) captured two gravid females in late June in St. Charles County. One individual was held in captivity and laid 12 eggs in early July. The hatchlings emerged in late August after an incubation period of 50 days. Hatching takes place mainly in August but can occur into September. Newly hatched young resemble those of western ratsnakes and are 230 to 310 mm (9.1 to 12.2 in.) in total length. Sexual maturity is reached within three years with males maturing earlier than females.

Remarks: Only the western foxsnake was formerly known to be in Missouri (Johnson 2000), but Crother et al. (2011) determined that Missouri has both the western and eastern foxsnake (see remarks section of the western foxsnake for more detailed information). Additional genetic studies are needed to better understand the distribution of eastern and western foxsnakes along the Mississippi River floodplain of eastern and northeastern Missouri.

The eastern foxsnake may be mistaken for a copperhead, but the round, dark-brown blotches of the foxsnake differ from the distinctly hourglass-shaped markings of the copperhead. Others have mistaken a foxsnake for a rattlesnake because it vibrates its tail when alarmed. This species is valuable as a controller of destructive rodents. The eastern foxsnake is listed as species of conservation concern due to loss of wetlands and adjacent grassland habitat, collection pressures, persecution, and road fatalities.

Bullsnake

Pituophis catenifer sayi (Schlegel)

TOM R. JOHNSON

Adult bullsnake from Dade County.

Description: The bullsnake is Missouri's largest native snake species. It is a medium-to-large, tan, yellow, or white snake with around 41 dark-brown or black blotches along the back. Markings along the back and sides are generally black on the neck and tail and brown at midbody. There usually is a black line from the eye to the angle of the jaw and black bars along the upper lip. The head is large and the snout terminates with an enlarged, projecting rostral scale. The belly is yellow and strongly checkered with square or rectangular black markings. The scales along the back are keeled, and the anal plate is single.

Adult bullsnakes range in length from 948 to 1,830 mm (37.3 to 72 in.) but have been known to reach 2,667 mm (105 in.; Powell et al. 2016).

Habits and Habitat: In Missouri, bullsnakes are active from late March into November. Observations have occurred from late March to mid-November, with peaks in

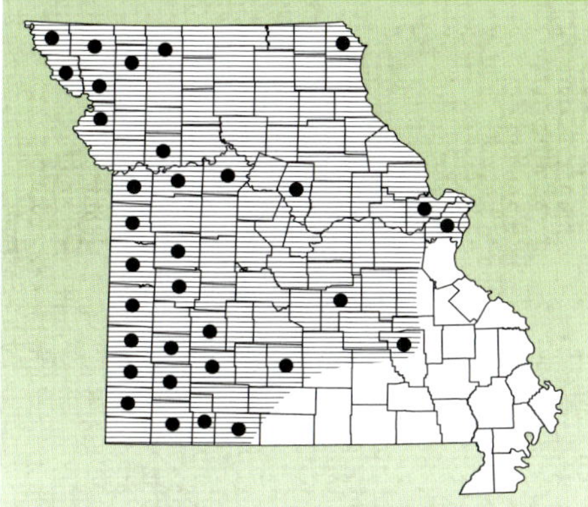

Distribution: Missouri: Primarily a prairie species and nearly statewide except for the southeastern quarter of the state. North America: Southern Alberta and Saskatchewan, throughout the Great Plains states, as well as New Mexico and Texas, southern Minnesota, southwestern Wisconsin, west-central Illinois and western Indiana (Powell et al. 2016).

Bullsnakes require high quality prairie habitat, such as this prairie in Barton County (above). Ventral coloration of a bullsnake (right).

May and June (J. Briggler unpubl. data). This species basks in the sun or searches for food by day; at night, it takes shelter in mammal burrows, clumps of vegetation, or rock piles, or under objects. It will readily climb into low bushes or trees. An individual was observed climbing a tree in Lincoln County (B. Schuette pers. comm.) Its pointy head and tough, enlarged snout allows this snake to burrow in loose or sandy soils.

If approached or cornered, a bullsnake will coil, vibrate its tail, and hiss loudly with the mouth partly open. The loud hiss is made by air being forced from the lung and out the windpipe, which has a special flap of skin that vibrates as air passes over it. Once captured, a bullsnake will bite to defend itself, but some specimens will calm down quickly and can be handled with ease.

In Missouri, bullsnakes occur on native prairies, grasslands, pastures, old fields, savannas and along some river bluffs. They are most commonly observed in native prairies and grasslands, especially in southwestern Missouri. This species overwinters in small mammal burrows, beneath rock piles, or in rock crevices on rocky hillsides. In studies from Kansas and Wisconsin, individuals returned to the same overwintering site year after year (Fitch 1958b; Kapfer et al. 2008). A telemetry study conducted in Wisconsin found the home range for males was 41 ha (101.3 ac) and females was 23 ha (56.8 ac; Kapfer et al. 2008).

The bullsnake diet includes mice, rats, ground squirrels, pocket gophers, small rabbits, birds, and bird eggs. Smaller prey is simply swallowed once grabbed with its mouth, but larger prey is killed by constriction.

TOM R. JOHNSON

Hatchling bullsnake from Barton County.

Breeding: Courtship and mating mainly take place during April and May. Adult males probably locate females by smell. Once a female is located, the male will court her by following behind, moving alongside and making jerky body movements. Once the female becomes passive, the male moves alongside and drapes his body over hers. Their cloacae meet when the male moves his tail under that of the female and mating occurs. During this time, the male grasps the female by the neck with his mouth.

Eggs are primarily laid during June or July. A study conducted in Nebraska found egg laying to occur from late June into mid-July for 38 clutches (Iverson et al. 2012). Two individuals collected in Missouri and held in captivity deposited eggs in mid- and late July (Anderson 1965). Throughout their range, bullsnakes laid 4 to 21 eggs with an average of about 11 (Iverson et al. 2012). Egg clutches reported for two Missouri specimens were 12 and 19 eggs (Anderson 1965). Females do not reproduce every year (Iverson et al. 2012). The elongated, leathery, white eggs average 52 by 32 mm (2 by 1.3 in.) and adhere to each other when laid (Iverson et al. 2012). Female bullsnakes normally lay their eggs in sandy or loose soil dug by the female, in abandoned small mammal burrows, or beneath large rocks, stumps, or logs (Ernst and Ernst 2003). The eggs hatch mainly in late August through September. Two clutches from Missouri hatched in late July and late September (Anderson 1965). Incubation period for 28 clutches in Nebraska was 51 to 70 days, with an average of 57 days (Iverson et al. 2012). Newly hatched bullsnakes average 365 mm (14.4 in.) in total length. Their color is similar to that of adult bullsnakes except their blotches are somewhat lighter. Throughout their range, males are reported to mature within two years, while females generally take about three or more years. Bullsnakes generally live into their 20s, with one captive individual surviving over 33 years (Snider and Bowler 1992).

Remarks: The bullsnake is a species of conservation concern due mainly to loss of native prairie and grassland habitat. Also, many individuals are unfortunately found dead on roads or killed by agricultural machines during haying. This large, nonvenomous species is a valuable neighbor to farmers because of the large number of crop-destroying rodents it eats. Consequently, the bullsnake is one of the most economically beneficial snake species in Missouri, and rural Missourians should make every effort to protect it.

Graham's Crawfish Snake

Regina grahamii Baird and Girard

JEFF BRIGGLER

Adult Graham's crawfish snake from Clark County.

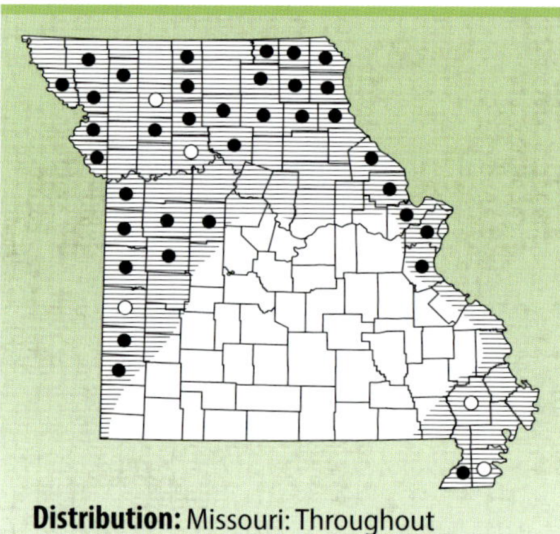

Distribution: Missouri: Throughout Missouri, except for the Ozarks. North America: Disjunct populations from Iowa and Illinois, south to Louisiana and Texas (Powell et al. 2016).

Description: Graham's crawfish snake is a medium-sized, dull-colored, semi-aquatic snake found in prairie streams, marshes, and ponds. The ground color is brown or yellow brown above, with a broad yellow stripe along the lower sides of the first three scale rows that may be bordered by a thin, dark-brown line. Some individuals may have a faint tan dorsal stripe. The chin and throat are often cream to yellow. The belly is cream colored or light yellow with a faint central row of gray or light-brown dots and a zigzag pattern of dark brown along both edges. Scales along the back and sides are keeled, and the anal plate is divided.

Adult Graham's crawfish snakes range in total length from 457 to 710 mm (18 to 28 in.) but have been known to reach 1,194 mm (47 in.; Powell et al. 2016).

Habits and Habitat: This harmless, reclusive species is active from late March or April through early November in Missouri (Johnson 1987; Seigel 1992). Peak

JEFF BRIGGLER

JEFF BRIGGLER

Graham's crawfish snakes are commonly found along the grassy edges of marshes, such as this example in Lincoln County (above). Ventral coloration of a Graham's crawfish snake (left).

activity occurs in April, May, and October with fewer observations during the summer in northwestern Missouri (Seigel 1992). Graham's crawfish snakes usually bask among branches overhanging water during spring and early summer. In the summer, individuals may estivate to escape the high temperatures and dry conditions of summer or become nocturnal. The habitat of this species is along slow-moving prairie streams, sloughs, marshes, swamps, ponds, and ditches, often with abundant aquatic vegetation. Graham's crawfish snakes were the most commonly captured snake in a natural, wet prairie marsh from mid-May into early August in Gentry County (Frese 2001). This species is reported to take shelter under rocks or logs along the water's edge, in nearby crayfish burrows, or burrows it constructs itself in moist, soft soil (Ernst and Ernst 2003; Kofron and Dixon 1980). In a telemetry study in Gentry County, Graham's crawfish snakes were never found under cover objects; they relied heavily on shallow water areas (average depth 133 mm; 5.2 in.) utilizing vegetation above the waterline for cover (Frese 2005). Sexton (1991), however, did capture crawfish snakes beneath cover boards adjacent to ditches in St. Charles County. At a Gentry County marsh, its home range averaged 1.5 ha (3.7 ac) and ranged from 0.14 to 3.8 ha (0.35 to 9.4 ac; Frese 2005). During winter, this species takes shelter in crayfish burrows. Dirt laden individuals have been observed emerging in April from crayfish burrows located in wet, bottomland prairies in Chariton, Holt, and Linn counties (J. Briggler pers. obs.).

Graham's crawfish snakes feed primarily on soft-bodied crayfish (recently molted individuals that are easier to digest). Other foods include frogs, salamanders, and small fish (Anderson 1965; Ernst and Ernst 2003; Seigel 1992). An adult Graham's crawfish snake was observed eating a northern crayfish (*Faxonius virilis*) on a rock above the waterline in Cass County (K. and A. Pogatshnik pers. comm.).

Breeding: This live-bearing species mates during April and May in the water. On two occasions, Anderson (1965) found groups of male Graham's crawfish snakes competing to mate with a single female in shallow water in northwestern Missouri. Mating of two pairs were observed in mid-May in Cass County (K. and A. Pogatshnik pers. comm.). These snakes were mating in the late afternoon with their heads above the water surface in a small pool with a water depth of 55 cm (22 in.) just below the exit of a large culvert. Mating was observed for about an hour with a bobbing behavior (extension of head and upper body) while bodies and tail regions were intertwined. At one point, the two mating pairs and an additional male joined into a mating ball (K. and A. Pogatshnik pers. comm.). Females give birth to live young from late July through September. A study in Kansas found a litter may contain 4 to 39 young with an average of 16 (Fitch 1985). A population studied in Holt County had females that likely reproduced each year of their life and had an average brood size of 11.6 young (range 5–16 young; Seigel 1992). Two individuals from Jackson County had four and six yolked follicles when dissected (Smith and Powell 1993). The young range in length from 190 to 254 mm (7.5 to 10 in.; Conant and Collins 1991) and have a color similar to that of adults. Males mature at one year of age, but females take up to three years (Hall 1969).

Remarks: Graham's crawfish snakes are difficult to keep in captivity due to their specialized diet (soft-bodied crayfish). Because of its close association with aquatic habitats, Graham's crawfish snake may be misidentified as a cottonmouth and killed because of unwarranted fear, although its general color and appearance look nothing like a cottonmouth. If captured, these snakes generally do not bite but usually release a foul-smelling musk mixed with feces.

When encountered, Graham's crawfish snakes are generally locally abundant in Missouri, especially on remnant, natural marshes with abundant aquatic vegetation and crayfish. Seigel (1992) captured 134 individuals over a five-year period within a wetland complex in Holt County, and Frese (2001) reported total captures of 58 individuals in a natural marsh in Gentry County. Keeping this species common in Missouri will require protecting and restoring wet prairie habitat adjacent to marshes, swamps, and streams. Such habitat supports high densities of crayfish necessary for food and crayfish burrows for shelter.

Variable Groundsnake

Sonora semiannulata semiannulata Baird and Girard

TOM R. JOHNSON

Adult variable groundsnake from Taney County.

Description: The variable groundsnake is a small species with smooth, shiny scales and highly variable coloration. The ground color of this gentle, secretive snake can be gray, light brown, or red brown. Individuals may have dark-brown or black bands along their entire length, a few bands or blotches along the forward part of the body, or be totally without pattern. Missouri individuals can have an orange or red tinge between crossbands. The belly is white or cream colored. The underside of the tail has numerous transverse, dark-gray bars. A melanistic (black pigmented) adult female groundsnake was captured on a glade in Taney County in 1981 (Miller 1983). Scales along the back are smooth, and the anal plate is divided.

Adult variable groundsnakes range in total length from 215 to 306 mm (8.5 to 12 in.) but has been known to reach 483 mm (19 in.; Powell et al. 2016).

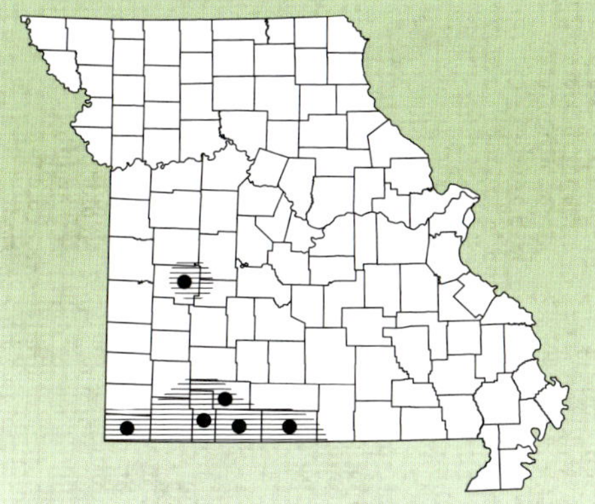

Distribution: Missouri: Restricted to open, rocky hillsides of the southwestern corner of the state, with an isolated population in St. Clair County. North America: Southern California to central Texas, and north to southern Kansas and southwestern Missouri. There are isolated populations in Utah, Idaho, and Colorado (Powell et al. 2016).

Variable groundsnakes live on open, rocky glades, such as this example in Stone County (above). Ventral coloration of a variable groundsnake (right).

Habits and Habitat: Variable groundsnakes are generally active from April until early November. This species usually remains under rocks during the day but may be active on the surface at night. They are frequently encountered in the spring, especially after rainfall. During hot weather, groundsnakes burrow underground to find cooler temperatures and higher humidity. In Missouri, this species resides on rocky, wooded hillsides that face south or southwest or on open, rocky glades. Groundsnakes are usually found in association with flat-headed snakes (*Tantilla gracilis*; Anderson 1965). Overwintering sites are likely beneath large rocks and within rocky crevices that are below the frost line. Individuals have been reported underground at depths of 60 cm (23.6 in.) in Oklahoma and up to 2.4 m (8 ft) in Kansas (Collins et al. 2010; Kassing 1961).

They consume scorpions, spiders (including black widows), centipedes, and soft-bodied insects. They are also known to eat mealworms and crickets in captivity (Anderson 1965; Ernst and Ernst 2003).

Breeding: Courtship and mating normally take place during April or May but autumn mating has been reported (Anderson 1965). A pair in St. Clair County was observed mating under a rock in late September (Anderson 1965). Courtship involves the male rubbing his

JEFF BRIGGLER

Variable groundsnakes have variation in color and pattern, such as this example in Taney County.

chin along the back of a female and occasionally biting her on the neck. Mating occurs when the tail of the male curls beneath that of the female. Their eggs are laid in loose, damp soil under rocks during June and July. Range wide, each female lays one to six eggs, with an average of about four (Bartlett and Tennant 2000; Goldberg 2001; Kassing 1961). Anderson (1965) documented an average of 4.8 eggs (range four to six eggs) from four females in Missouri. These egg clutches were laid in late June. The white eggs have a smooth, leathery shell and are about 20 by 6 mm (0.8 by 0.2 in.; Anderson 1965). Hatching takes approximately two months after the eggs are laid, with a reported incubation period of 49 to 70 days (Ernst and Ernst 2003). A Missouri clutch with four eggs was laid in late June and hatched in mid-August with an incubation period of 51 to 52 days (Anderson 1965). The young, which look similar to the adults, are 98 to 109 mm (3.9 to 4.3 in.) in length (Anderson 1965). Individuals of both sexes become mature from 12 to 18 months of age.

Remarks: Missouri's population of variable groundsnakes depends on the availability of open rocky glades and forested rocky hills. As long as this type of habitat is protected, the species should continue to survive. When threatened, a variable groundsnake usually attempts to flee to a nearby hiding place, tucks its head beneath its body, or even plays dead (opens mouth, sticks out tongue, and rolls onto back) to confuse predators (Tennant 1984; Werler and Dixon 2000). Variable groundsnakes are not known to bite people.

Dekay's Brownsnake

Storeria dekayi (Holbrook)

Adult Dekay's brownsnake from Callaway County.

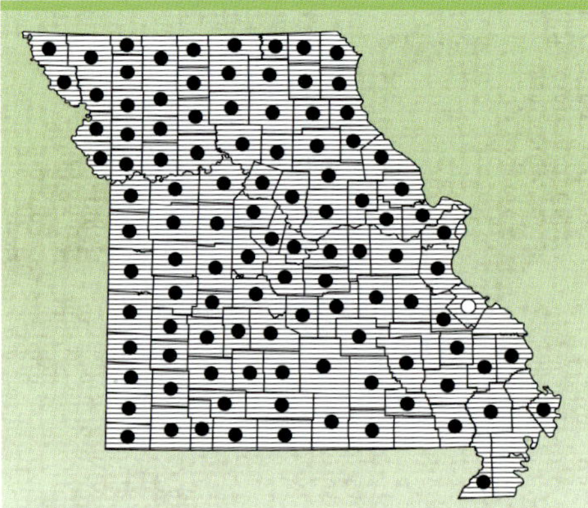

Distribution: Missouri: Presumed to occur statewide but appears to be less common in southeastern part of the state. North America: A widespread species that occurs from Nova Scotia through New England states south to panhandle of Florida, west to eastern Texas and Mexico, and north to the Great Lakes states (Powell et al. 2016).

Description: The Dekay's brownsnake is a small, secretive species that can be difficult to identify. The color may vary from gray to brown to red brown. There is usually a tan dorsal stripe bordered by two rows of small brown spots. These small spots are normally connected by a narrow brown line. The top of the head is usually dark; a dark spot is present under each eye and on each side of the neck. The belly is cream colored, yellow, or pink. The scales along the back are keeled, and the anal plate is divided.

Adult Dekay's brownsnakes range in total length from 230 to 330 mm (9.1 to 13 in.) but have been known to reach 527 mm (20.7 in.; Powell et al. 2016).

Habits and Habitat: In Missouri, brownsnakes are active from late March to early November. This species takes shelter under logs, rocks, boards, or other objects. It is most commonly found in moist forests or woodlands, especially along the edges of ravines, creeks, and river floodplains in

JEFF BRIGGLER

TOM R. JOHNSON

Defensive posture of a Dekay's brownsnake from Boone County (above). A newborn Dekay's brownsnake from Cole County (left).

Missouri. It can also be found in prairies and grasslands with nearby wetlands and woodlands and in urban areas (city parks, cemeteries, flower beds) where shelter, moisture, and food are abundant. They can be observed crossing trails and roads during the spring and fall. During hot weather, brownsnakes may be active at night. A specimen was captured crossing a Gasconade County blacktop road during a rainy night in September (T. Johnson pers. obs.). During the winter, individuals can be found beneath large rocks, and within old wells, stone walls, rock crevices, rock piles, logs, stumps, and building foundations (Ernst and Ernst 2003). A large aggregation (64 individuals) was uncovered by a pipeline company excavating in rocky, porous soil along the edge of a pasture in Jackson County in April (Anderson 1965).

Brownsnakes feed primarily on earthworms, slugs, land snails, and soft-bodied insects. While studying snakes in northwestern Missouri, R. Seigel (pers. comm.) found this species' diet to be 75 percent slugs and 25 percent earthworms. Rossman and Myer (1990) were the first to report on the specialized ability of brownsnakes to eat land snails. They suspected that the blunt head and elongated teeth of both brownsnakes and red-bellied snakes (*Storeria occipitomaculata*) helped extract the bodies of live land snails from their shells. If a brownsnake was able to get a firm hold of the snail's body with its teeth, it would exert a continuous pull

JEFF BRIGGLER

Ventral coloration of a Dekay's brownsnake from Boone County.

on the snail and twist its head and forward part of the body at the same time. It often took 10 minutes or more for the snail to become fatigued, which allowed the snake to extract the body and swallow it. Brownsnakes were found to be an important food of young prairie massasauga rattlesnakes in a northwestern Missouri wet prairie (Seigel 1986).

Breeding: DeKay's brownsnakes generally mate in the spring soon after emerging from their overwintering retreats, but autumn mating can also occur. This species gives birth to live young; 3 to 41 young are born between mid-July and September (Fitch 1985; Morris 1974). The largest litter size reported is 41 young from a female captured in east central Illinois (Morris 1974). Average number of young reported range wide is 13 (Ernst and Ernst 2003). Several gravid females have been caught in Missouri and held in captivity until birth occurred. Anderson (1965) documented a female from Jackson County giving birth to 23 young in early August, and a Webster County female giving birth to 13 young in mid-August. In addition, two gravid females caught in Lincoln County gave birth to 14 young in mid-July and 18 young in late July (Schuette 1998; Schuette 1999). In a study in Gentry County, young of the year were first captured in mid-August in a natural marsh (Frese 2001). Newly born brownsnakes from these four Missouri specimens ranged from 82 to 106 mm (3.2 to 4.2 in.) in total length with an average of about 94 mm (3.7 in.). The young are darker in color, which often obscures the dorsal pattern, and have a narrow yellow collar. Reproductive maturity normally occurs at two to three years of age (Ernst and Ernst 2003). An individual captured as an adult from the wild survived an additional seven years (Snider and Bowler 1992).

Remarks: Formerly two subspecies of brownsnakes were recognized in Missouri: the midland brownsnake (*Storeria dekayi wrightorum*), and the Texas brownsnake (*Storeria dekayi texana*; Johnson 2000). Genetic and morphologic data analyzed by Pyron et al. (2016) rejected the recognition of subspecies.

This species is not known to bite but when captured will flatten its head and neck and release a musky secretion from glands at the base of the tail.

Red-bellied Snake

Storeria occipitomaculata (Storer)

Red-bellied snake from Callaway County.

Description: The red-bellied snake is a small, woodland species that is either gray or red brown, normally with four narrow, dark stripes, a faint light mid-dorsal stripe, or a combination of these stripes present. Occasionally, individuals may have a distinct red or orange dorsal stripe. The head is usually darker than the body, and the nape of the neck has three light spots, which occasionally fuse to form a tan "collar" behind the head. The belly is orange, red, pink, or occasionally yellow. Scales along the back and sides are keeled, and the anal plate is divided.

Adult red-bellied snakes range in total length from 203 to 254 mm (8 to 10 in.) but have been known to reach 422 mm (16.6 in.; Powell et al. 2016).

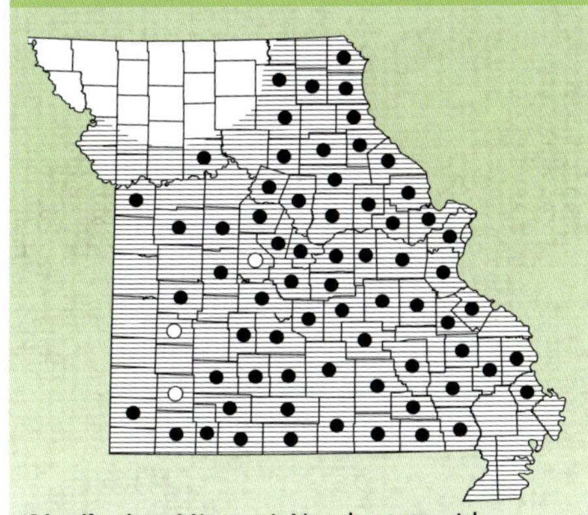

Distribution: Missouri: Nearly statewide in occurrence, but likely absent from the northwestern and southeastern corners. North America: Southern Canada, south to east Texas and Louisiana, east along the Gulf Coast states to northern Florida, and north along the Atlantic Coast to Canada (Powell et al. 2016).

Habits and Habitat: This secretive species is active from late March through early November. It spends a great deal of time hiding beneath rocks, boards, scattered tree bark, logs, and other objects; it will,

Red-bellied snakes are found in moist forest with an abundance of rocks, logs, and leaf litter, such as this example in Ripley County (above). Ventral coloration of a red-bellied snake from Barry County (right).

JEFF BRIGGLER

JEFF BRIGGLER

however, bask in the sun. A red-bellied snake was observed basking on a log in Pulaski County during mid-May (T. Johnson pers. obs.). This is a species of moist forests where there is ample shelter, such as rocks, logs, pieces of tree bark, and leaf litter. They are commonly encountered on rocky, north-facing, wooded hillsides in Missouri. Overwintering is spent beneath large rocks, in ant mounds and mammal burrows, within rock crevices, and possibly in rotting logs (Ernst and Ernst 2003). Although this species is secretive in Missouri, individuals can be found migrating across Katy Trail State Park along the Missouri River in central Missouri. On warm, sunny October days individuals migrate from the floodplain habitat to the rocky, wooded hillsides to overwinter (J. Briggler unpubl. data). An 18-year drift-fence study in a rocky, hilly forested habitat in the central Ozarks of Missouri captured a total of 3,497 red-bellied snakes (Rota et al. 2017).

This small snake is not known to bite and is docile and gentle when handled. A freshly captured specimen will secrete a musky odor from glands at the base of the tail. Jordan (1970) and Watermolen (1991) reported on individuals that, when provoked, went into convulsions, rolled over several times, opened their mouth, and remained motionless on their

JEFF BRIGGLER

Occasionally red-bellied snakes have a distinct red or orange dorsal stripe, such as this example in Shannon County.

back—true death-feigning behavior. This species was often captured on the family farm in central Wisconsin, and several specimens were observed flaring their lips soon after being captured (T. Johnson pers. obs.). This behavior has been reported by Gosner (1942) and de Queiroz (1997) and may be attributed to the fact that this species feeds on soft-bodied, slippery prey such as slugs, snails, and earthworms. The lip flaring may help them control the slime while swallowing a slug or worm.

Red-bellied snakes eat slugs, land snails, earthworms, and occasionally soft-bodied insects. See the Dekay's brownsnake species account for a description of snail-eating behavior.

Breeding: Courtship and mating take place in the spring, summer, or autumn. The young are born between late July and early September. Litter size has been reported to be up to 21 young (Blanchard 1937; Tennant 1984) with an average of eight. In Kansas, 75 percent of females were gravid and contained an average of seven well-developed embryos with a range from 4 to 13 (Busby et al. 2014). A St. Clair County female gave birth to six young in mid-August (Anderson 1965). These young measured 85 to 94 mm (3.3 to 3.7 in.) in total length. Reproductive maturity is reached in about two years (Blanchard 1937). Red-bellied snakes are known to live four to five years (Snider and Bowler 1992).

Remarks: Several subspecies were once recognized for the red-bellied snake group, and Missouri had the northern red-bellied snake (*S. o. occipitomaculata*). Pyron et al. (2016) concluded subspecies recognition was not warranted for this monotypic species (*S. occipitomaculata*).

This small, inoffensive snake is sometimes mistaken for a young copperhead and killed because of unwarranted fear.

Flat-headed Snake

Tantilla gracilis Baird and Girard

Adult flat-headed snake from Maries County.

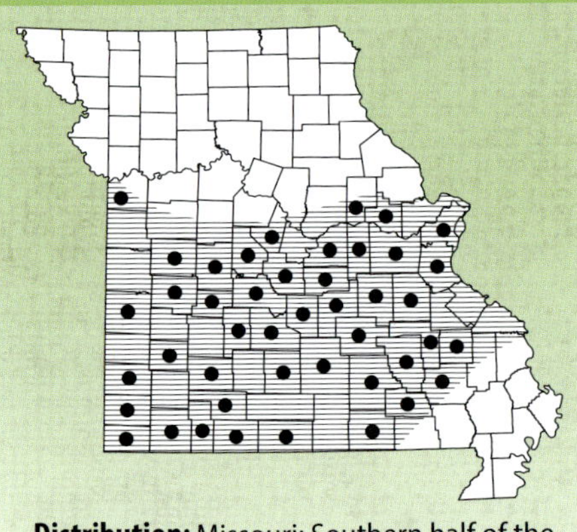

Distribution: Missouri: Southern half of the state, especially the Missouri Ozarks, but absent from the lowlands of southeastern Missouri. North America: Southwestern edge of Illinois, south to northwestern Louisiana and eastern two-thirds of Texas, and north to eastern Kansas (Powell et al. 2016).

Description: The flat-headed snake is Missouri's smallest species of snake. Its dorsal color is gray, tan, light brown, or slightly red brown. The head is normally darker than the body. The belly is salmon pink. Scales along the back and sides are smooth, and the anal plate is divided.

Adult flat-headed snakes range in total length from 180 to 203 mm (7.1 to 8 in.) but are known to reach 249 mm (9.8 in.; Powell et al. 2016).

Habits and Habitat: This burrowing species is normally active from late March through October with most surface activity from April into June (Roberson 1980). It spends a great deal of time in slightly moist soil under rocks or in underground burrows. Flat-headed snakes prefer open rocky hillsides, especially glades, throughout the Ozark Highlands. They are also found occasionally on exposed rocky road cuts, under rocks along logging roads and walking trails that have adequate sun

Flat-headed snakes live on open, rocky glades, such as this example in Shannon County (above). Head of a flat-headed snake from Shannon County (left).

exposure. They are often encountered where loose soil is in association with limestone. During hot weather, flat-headed snakes burrow into the ground, use other small animal burrows, or enter rock crevices to escape the extreme temperatures. The same shelters are also used during the winter.

This species is not known to bite when captured, and it is likely unable to inflict a bite to humans due to its small size.

The food of this snake includes scorpions, spiders, centipedes, and a variety of insects and their soft-bodied larvae (Carpenter 1958; Cobb 2004; Force 1935; Roberson 1980). Flat-headed snakes have a pair of enlarged, shallow-grooved teeth in the rear of the mouth that likely assist in capturing and subduing prey.

Breeding: Courtship and mating mainly take place during late April and May. Females lay from one to four eggs and deposit them in moist soil under rocks, within rock crevices, and within small animal burrows during June and mid-July. Clutch size averaged 3.1 eggs (range two to four eggs) in Arkansas (Trauth et al. 1994). Occasionally several females may deposit their eggs in the same location (Cobb 1990). The cream or white eggs are elongated and

TOM R. JOHNSON

Ventral coloration of a flat-headed snake.

measure about 18 mm (0.7 in.) in length and 6 mm (0.2 in.) in width. Hatching takes place during August and September. Two clutches were found in mid-July in St. Clair County in barely moist sand between two layers of limestone (Anderson 1965). One clutch contained two eggs and the other three eggs. These eggs hatched in captivity in mid-to-late September. The hatchlings averaged 82.5 mm (3.2 in.) in total length (Anderson 1965). Reproductive maturity is reached within 12 to 18 months in males and 18 to 30 months for females in Oklahoma (Force 1935). Roberson (1980) estimated that it took up to three years to attain maturity for females in Missouri.

Remarks: The preferred habitat of this species is the open, rocky areas, known as glades, throughout the Ozark Highlands. Glades have become overgrown with cedars due to lack of periodic fires, and the populations of many animals that live on glades—including the flat-headed snake—have been reduced. Considerable effort to restore these glades by land managers in the last 30 years has resulted in improved habitats for the flat-headed snake and other glade-dependent species. Maintaining the open nature of glade habitat and minimizing human disturbance of rocks will help keep this species common in Missouri.

Orange-striped Ribbonsnake

Thamnophis proximus proximus (Say)

JEFF BRIGGLER

Adult orange-striped ribbonsnake from Callaway County.

Description: The orange-striped ribbonsnake is a long, slender, colorful member of the gartersnake group. There are normally two wide, black stripes along the back and a narrow black stripe on each side. The mid-dorsal stripe is orange or yellow. A narrow yellow stripe is present along each side on the third and fourth scale rows. The head is black and usually has an orange, yellow, or white spot on the top. The belly is cream colored or light green and unmarked. Scales along the back and sides are keeled, and the anal plate is single.

Adult orange-striped ribbonsnakes range in total length from 510 to 760 mm (20.1 to 29.9 in.) but have been known to reach 1,268 mm (49.9 in.; Powell et al. 2016).

Habits and Habitat: In Missouri, this slender snake is active from mid-March through October. When temperatures are mild, it is active during the day but may become nocturnal during hot weather. When encountered, orange-striped ribbonsnakes

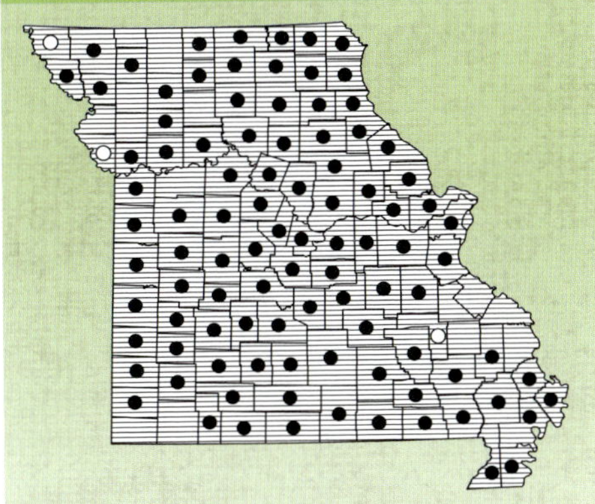

Distribution: Missouri: Statewide. North America: Scattered populations in southern Wisconsin, Indiana, and Illinois, along the eastern and southern edge of Iowa, the eastern edge of Nebraska, most of Kansas and Oklahoma, all of Missouri and Arkansas, western Tennessee and Mississippi, the northern two-thirds of Louisiana, and eastern Texas (Powell et al. 2016).

Orange-striped ribbonsnakes are found on the edges of heavily vegetated wetlands, such as this example in Macon County (above). Head, stripe, and belly patterns of an orange-striped ribbonsnake from Callaway County (right).

are quick and agile and once startled can be difficult to capture. This species will not hesitate to enter water.

This species is seldom far from water—including streams, lakes, ponds, and marshes—and often found in the border vegetation. It lives near swamps, marshes, ponds, roadside ditches, sloughs, streams, and rivers. Winter dormancy is spent below the surface in rocky crevices, springs, mammal and crayfish burrows, rotten logs and stumps, and anthills (Ernst and Ernst 2003). In the prairie regions of Missouri, this species relies heavily on crayfish burrows. It is not uncommon for up to five individuals to shelter in the same crayfish burrow during the winter months in Missouri (J. Briggler pers. obs.).

Small frogs, toads, salamanders, and small fish are the major prey of this species. However, earthworms were found in the stomachs of orange-striped ribbonsnakes in Holt County (R. Seigel pers. comm.). Orange-striped ribbonsnakes have been observed in a Linn County marsh during June feeding on tadpoles and young frogs (Johnson 2000).

Breeding: Mating takes place during April and early May. Mating has been observed in early April in bottomland prairie habit in north-central Missouri (J. Briggler pers. obs.). Not all females reproduce each year. Seigel and Ford (1987) found that 81 percent of adult

JEFF BRIGGLER

An orange-striped ribbonsnake attempting to eat a gray treefrog from Callaway County.

female orange-striped ribbonsnakes in a Missouri population reproduced within a given breeding season. A litter may contain 4 to 36 young with an average of 10 to 15 (Bowers 1967; Carpenter 1958; Powell 1982). Anderson (1965) reported that a St. Clair County female gave birth to nine young in late July. Pilgrim (2001) and Pilgrim et al. (2011) gathered data from many gravid females from Holt County, and the mean number per litter was 15 (range 6 to 26). The young are mainly born from July into September. In a study in Gentry County, young of the year were first captured in late August in a natural marsh (Frese 2001). The young from the St. Clair County female averaged 240 mm (9.4 in.) in total length with a range from 230 to 251 mm (9.1 to 9.9 in.; Anderson 1965). Males generally become reproductively mature in one to two years, and females may take up to three years (Clark 1974; Tinkle 1957).

Remarks: This species is quite common and widespread in Missouri. In the upland forest and woodlands of the Ozarks, individuals are less commonly encountered. However, they are abundant in bottomland prairies throughout the western and northern parts of the state. Bottomland prairies have saturated soils and are in close proximately to open water. Such habitat supports the ribbonsnake's prey and has an abundance of crayfish burrows necessary for winter dormancy.

Plains Gartersnake

Thamnophis radix (Baird and Girard)

Adult plains gartersnake from Chariton County.

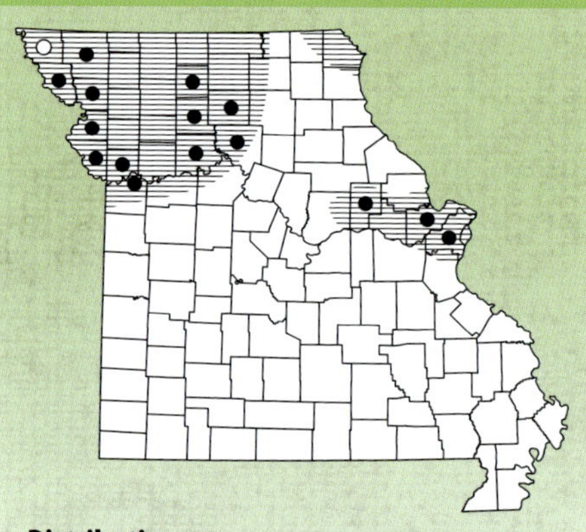

Distribution: Missouri: Occurs mainly in the north-central and western sections, with isolated populations in the eastern part of the state, especially in the St. Louis area. North America: Plains gartersnakes range from south-central Canada south and east to eastern Indiana, Illinois, Iowa, and Missouri, and south through eastern Montana, eastern Wyoming, and Colorado to northern Texas, with an isolated population in Ohio (Powell et al. 2016).

Description: The plains gartersnake is a medium-sized snake of wet prairies and marshes. Its ground color is green gray, olive, or brown, but occasionally it can be very dark. The dorsal stripe is bright yellow, orange, or orange yellow. The light stripe on each side is on the third and fourth scale rows. It may be yellow, green, or blue. The area between light stripes usually has an alternating double row of black spots. The light-green upper lip has black bars along the edges of the scales. The belly is gray or green gray with a row of black spots along each side. The scales along the back and sides are keeled, and the anal plate is single.

Adult plains gartersnakes range in total length from 380 to 710 mm (15 to 28 in.) but have been known to reach 1,093 mm (43 in.; Powell et al. 2016).

Habits and Habitat: This striking gartersnake is normally active from late March to late October. Plains gartersnakes spend warm,

JEFF BRIGGLER

JEFF BRIGGLER

Plains gartersnakes live mainly in wet prairie habitat, such as this example in Holt County (above). Juvenile plains gartersnake from Chariton County (left).

sunny days basking or searching for prey. They take shelter under logs, boards, rocks, or other objects. Winter is spent underground in abandoned rodent tunnels, anthills, and crayfish burrows in Missouri (J. Briggler pers. obs.).

As with all gartersnakes, the plains gartersnake will smear a musky secretion from glands at the base of the tail when captured or molested. Most of the individuals we have encountered have not tried to bite when captured.

In Missouri, this species lives in or near wet prairies, meadows, marshes, lakes, ponds, and streams and is most commonly encountered in grassland habitat with abundant crayfish burrows close to water that harbor many aquatic prey species.

This species' food includes small frogs and their tadpoles, salamanders, earthworms, small fish (minnows, mosquito fish), and occasionally, small rodents.

Breeding: Courtship and mating occur in the spring (April and May) soon after they emerge from overwintering sites, although autumn mating has been reported (Collins et al. 2010). Males find females by smell. Young in Missouri are born from late July through early

JEFF BRIGGLER

Head of a plains gartersnake from Chariton County

September, with 5 to over 60 young in a litter (Fitch 1985). Most clutches have 10 to 20 young (Ernst and Ernst 2003). The average number of offspring in an Illinois study was 18.4 (range 4–31; Stanford and King 2004). A mean of nine young per female was found in a total of 45 females from a population of plains gartersnakes taken in Holt County (R. Seigel pers. comm.). Additional studies in Holt County captured seven gravid females that produced an average of 16.5 offsprings (Pilgrim et al. 2011).

A female collected by Anderson (1965) in Holt County died in early July and contained 17 large follicles. Another female collected from northwest Missouri in Buchanan County gave birth to 15 young in early August (Anderson 1965). A female collected in St. Charles County gave birth to 29 young in late July (Anderson 1965). Newly born plains gartersnakes are 151 to 190 mm (5.9 to 7.5 in.) in length (Powell et al. 2016). Reproductive maturity is obtained in two to three years for females; males likely mature slightly earlier (Seibert and Hagen 1947; Stanford and King 2004). Plains gartersnakes can live eight or more years (Snider and Bowler 1992).

Remarks: This species may have declined in Missouri with the significant loss of wet prairie habitat throughout northern Missouri. In the remaining wet prairie habitat, this species appears to be relatively secure. Protection and restoration of remnant, wet prairie habitat will ensure that the plains gartersnake remains on the Missouri landscape.

Eastern Gartersnake

Thamnophis sirtalis sirtalis (Linnaeus)

JEFF BRIGGLER

Adult eastern gartersnake from Osage County.

Description: The eastern gartersnake is one of the most common and widely distributed species of snakes in North America. The general color is quite variable. The ground color may be black, brown, or olive. There are usually three distinct light stripes that are yellow, brown, green, or blue. One light stripe runs along the back, and there is a stripe on either side that involves the second and third scale rows. The area between the stripes on each side usually has a double row of alternating dark spots. The belly is green with two rows of faint black spots that are somewhat hidden by the overlapping scales. Scales along the back are keeled, and the anal plate is single.

Adult eastern gartersnakes range in total length from 457 to 660 mm (18 to 26 in.) but have been known to reach 1,372 mm (54 in.; Powell et al. 2016).

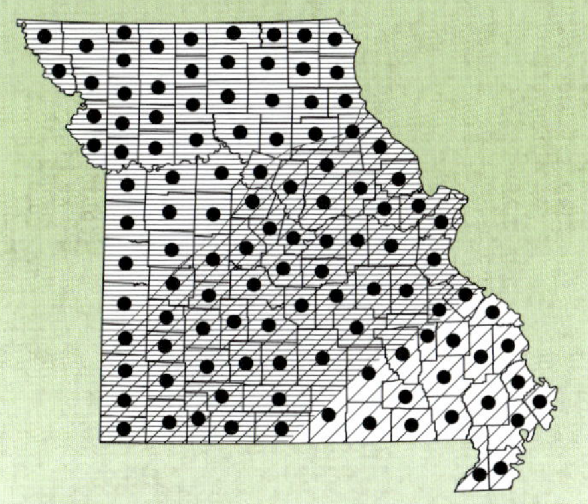

Distribution: Missouri: Southeastern half of the state (diagonal lines), intergrading with and replaced by the red-sided gartersnake in the northwestern half (horizontal lines). North America: The eastern gartersnake ranges from southern Ontario through the eastern half of the United States to southern Florida and west to southeastern Texas. The red-sided gartersnake occurs from south-central Canada south through Montana and western Minnesota south to Oklahoma and northern Texas (Powell et al. 2016).

Habits and Habitat: This common species is normally active from March through early November but may be observed in

Head, stripe, and belly patterns of a red-sided gartersnake from St. Louis County (above). An eastern gartersnake eating a southern leopard frog (right).

any month during a mild winter. This species is more tolerant of cold temperatures than most other snake species, and, therefore, it is one of the first snakes to emerge from winter dormancy and one of the last to enter.

Eastern gartersnakes move about mainly during the day searching for food. However, they are seen in the evening searching wetlands for amphibians. When approached by a predator, an individual quickly moves to a hiding place such as an animal burrow, a pile of rocks or boards, or thick brush. If a body of water is near, this species will not hesitate to enter the water to escape. When cornered, an eastern gartersnake will flatten its head and body or coil its body and strike repeatedly at an intruder. Once captured it will try to bite its captor and smear a foul-smelling musk from glands at the base of the tail.

During the winter, eastern gartersnakes take shelter in animal burrows or rotting stumps or congregate in deep cracks in south-facing limestone bluffs or rocky hillsides. In the prairie regions of western and northern Missouri, crayfish burrows and ant mounds are the main overwintering sites. Numerous eastern gartersnakes have been encountered in northern Missouri emerging from crayfish burrows in bottomland prairies in late March and into April. Although crayfish burrows are typically occupied by only one eastern gartersnake, it is not uncommon to see two or even three individuals emerging from one crayfish burrow (J. Briggler pers. obs.). Also, they can be found near people. Eastern gartersnakes will use holes in open fields, pastures, lawns, old rock-lined wells, cisterns, and foundations of abandoned

JEFF BRIGGLER

JEFF BRIGGLER

Adult red-sided gartersnake from Callaway County (above). When threatened, eastern gartersnakes will raise and flatten the forward part of the body and head, such as this example from Barry County (left).

homes for overwintering. It is common to see many eastern gartersnakes emerging in April from eastern mole (*Scalopus aguaticus*) tunnels in Jefferson City lawns (J. Briggler pers. obs.).

Eastern gartersnakes live in a variety of habitats but favor areas near water, such as ponds or marshes, swamps, and in damp woods or forested areas along creeks and rivers. Empty lots, brushy fencerows, roadside ditches, and old abandoned farms also are places where this species occurs.

The eastern gartersnake diet includes earthworms, frogs, tadpoles, toads, and salamanders. They occasionally will eat snails, slugs, insects, minnows, small snakes of other species, and small mice. Typically, amphibians make up most of their diet. Two red-sided gartersnakes captured in Jackson and Johnson counties had each eaten a plains leopard frog (Smith and Powell 1993). Another individual was captured in Linn County with a southern leopard frog in its mouth (J. Briggler pers. obs.).

Breeding: Courtship and mating normally occur in the spring, soon after emergence from overwintering; this species may occasionally mate in the autumn. Males locate females by scent, and several males may try to court and mate with one female. There is much writhing as the males try to rub their bodies along the sides and back of the female. Mating occurs when a male curls his tail beneath that of the female so that their cloacae meet. On several occasions in early April, multiple males (from 6 to 18) have been seen chasing a female or in a cluster with a female in bottomland prairie habitat in northern Missouri (J. Briggler unpubl. data). The young are born in mid-summer and early fall. A litter may contain 4 to 85 young (Fitch 1985). A female eastern gartersnake captured in southern Illinois gave birth to 103 young (Dyrkacz 1975). Anderson (1965) reported litter sizes of 18, 27, 29, 22, and 42 young for captive females between early July and early August. A study in Holt County in northwest Missouri observed mean litter sizes of 24 young with a range from 6 to 47 (Pilgrim 2001; Pilgrim et al. 2011). Young eastern gartersnakes average 125 to 230 mm (4.9 to 9.1 in.) in total length at birth (Powell et al. 2016). Young are brightly colored compared to most adults. Males become reproductively mature in one to two years, and females usually mature in two to three years (Rossman et al. 1996). A wild-caught individual lived 14 years in captivity (Snider and Bowler 1992).

Subspecies: At least two subspecies of *Thamnophis sirtalis* occur in Missouri: the nominate subspecies, the eastern gartersnake, *T. s. sirtalis* (Linnaeus), described above, and the red-sided gartersnake, *T. s. parietalis* (Say). The major difference between the two subspecies is the presence of red or orange-red skin between the scales along the sides of the red-sided gartersnake. Their natural history is similar. In southwestern Missouri, there are gartersnakes with a broad, orange stripe along their backs that are more characteristic of Texas gartersnakes (*T. sirtalis annectens*) found in Texas and western Oklahoma. Additional genetic and morphological studies are needed to better understand the distribution and taxonomy of the eastern gartersnake group in southwestern Missouri.

Remarks: The eastern gartersnake is one of the most frequently encountered snakes in the state; it can occur in large numbers, especially at overwinter sites. Eastern gartersnakes were the most abundant snake species and represented 38 percent of the snakes captured along the edge of wetlands on a protected area in St. Charles County (Sexton 1991). By far, it is the most abundant snake encountered in bottomland prairie habitat with high densities of crayfish burrows. In a drift-fence survey of a bottomland prairie in north-central Missouri, 119 eastern gartersnakes were captured during one spring (J. Briggler unpubl. data). This represented 55 percent of the captures of nine snake species. Another drift-fence survey conducted during one spring at three small bottomland prairies in Holt County (Durbian et al. 2006) captured 395 red-sided gartersnakes, representing 40 percent of the 13 species of snakes. This species is considered abundant and secure in Missouri due to utilization of a wide variety of habitats and annual production of large litter sizes. Protection and restoration of their preferred bottomland prairie habitat, however, will ensure this species remains abundant on the Missouri landscape.

Lined Snake

Tropidoclonion lineatum (Hallowell)

Adult lined snake from Miller County.

Description: The lined snake is a small, striped, secretive, burrowing snake similar in appearance to gartersnakes. Its head is small and slightly flattened. Its ground color is gray, brown, or tan gray, and the body has three white, light-gray, or yellow stripes: one along the dorsum and one on each side involving the second and third scale rows. In some populations, the dorsal stripe is orange or orange yellow (Ernst and Ernst 2003). There is also a dark-gray or brown stripe, or series of spots along each side. A white or light-yellow spot is usually present at the base of the head and in line with each eye. The chin and forward part of the neck is white. The belly is white, cream, or pale green, with two distinct rows of dark-gray or black spots, or half-moons along the midline. A common trait of burrowing snakes is the presence of a relatively short tail: the tail length of lined snakes is less than 20 percent of their total length (Smith 1961). Scales along the back are keeled, and the anal plate is single. The tail of adult males is longer than the tail of adult females.

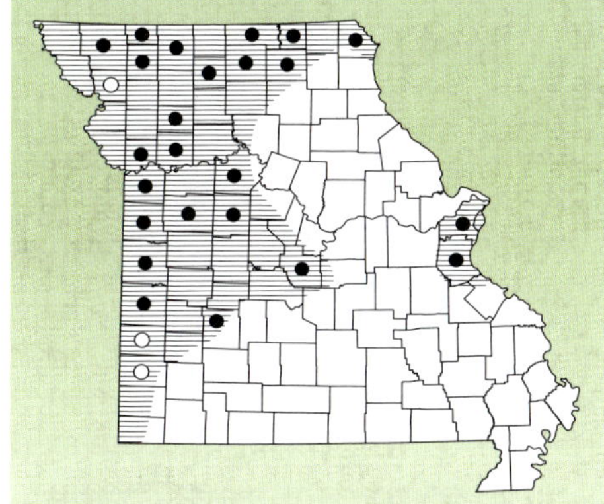

Distribution: Missouri: St. Louis and Kansas City areas, the northeastern and northwestern corners, and the central and western parts of the state. North America: From southeastern South Dakota, southwestern Minnesota and Iowa, and southwest to eastern Texas. Isolated or disjunct populations in southeastern Iowa, central Illinois, western Kansas, Nebraska, southwestern Wisconsin, Colorado, Texas, and New Mexico (Powell et al. 2016).

TOM R. JOHNSON

TOM R. JOHNSON

Small grassland patches and roadsides are some of the best remaining habitat for lined snakes, such as this example in Scotland County (above). Ventral coloration of a lined snake (right).

Adult lined snakes range in total length from 224 to 380 mm (8.8 to 15 in.) but have been known to reach 544 mm (21.4 in.; Powell et al. 2016).

Habits and Habitat: This species is normally active from April through October. During the day it hides under rocks, logs, and other debris, becoming active at night. Lined snakes retain a uniform body temperature by selecting shelters of varying shades and thicknesses. A thin, dark-colored shelter like tar paper can quickly increase the temperature of a snake hiding underneath. A snake hiding under a light-colored, flat rock will be heated more gradually (R. Krohmer pers. comm.).

When captured, lined snakes release an unpleasant-smelling musk from glands at the base of the tail. This species is rarely known to bite when handled.

Lined snakes prefer a slightly moist soil within a variety of habitats in Missouri: native prairies, rocky glades, empty lots in towns and suburbs, in or near old trash dumps, along highway right-of-ways where there is abundant debris for shelter, and in open, rocky woodlands. Overwinter dormancy is spent underground in small mammal burrows or under objects, such as rocks or logs embedded in the ground. There are a few reports of lined snakes on the surface during a warm spell in winter (Ernst and Ernst 2003).

Earthworms are the principle food of this species. A pair of enlarged teeth are present at the posterior of each maxillary bone, which may help lined snakes with swallowing earthworms. A few soft-bodied insects and slugs may also be eaten, but the small size of the head limits lined snakes to small prey (Ernst and Ernst 2003). As with most small species of snakes, lined snakes can be preyed upon by many animals, such as kingsnakes, racers, birds of prey, and shrikes. Sexton (1990) observed a bluejay attack and try to eat a lined snake in the St. Louis area. The snake measured 273 mm (10.7 in.) in snout-to-vent length.

Breeding: The reproductive biology of a population of lined snakes was studied in St. Louis County (Krohmer and Aldridge 1985a, 1985b), and the following information is derived from their papers. Lined snakes mate in late summer right after the young are born or they breed in autumn. Males have sperm stored in their vasa deferentia at a snout-to-vent length of 200 mm (7.9 in.) during their first year. One-year-old males are thus capable of breeding by late summer or early autumn. Females become sexually mature during midsummer of their second year at a mean minimum snout-to-vent length of 221 mm (8.7 in.). Females that had bred in autumn store the sperm, which then fertilize their ovum the following spring. Populations in other areas may breed in the spring (Collins et al. 2010). In Kansas, each female produces one litter of 2 to 12 young per season, with an average of six or seven (Fitch 1985). Observations of lined snakes from Oklahoma reported a captive female giving birth to 12 young and then eating nine of her young within 24 hours (Force 1931). Males in captivity readily ate earthworms, but gravid females were not interested in eating. Young lined snakes are born in late July into early September with most births occurring in August. Of the 23 births witnessed by Force (1931), all took place in August, and the number of young produced was from 2 to 12 per brood. The young began to shed their skin for the first time within one hour of birth and shed several more times during the next few days. Measurements of the 145 newly born lined snakes were from 70 to 130 mm (2.8 to 5.1 in.). Observations were made of the time it took for each young to be born: on average, it took from 10 seconds to 3 minutes; and intervals between the birth of each young snake was from under two minutes to 45 minutes (Force 1931). Larger females produce more offspring than small females (Krohmer and Aldridge 1985b). Newly born lined snakes are dark gray with faint white or yellow stripes (Ernst and Ernst 2003) and have an average total length of 100 to 120 mm (3.9 to 4.7 in.).

Remarks: Population status is largely unknown for this species throughout its range, including Missouri. This is mostly due to its secretive habit of spending considerable time beneath the ground or under cover. Rarely is this species observed by people, and reports are usually of dead individuals killed along highways. This species is most commonly observed in open, rocky habitat in the St. Louis and Kansas City areas. The best chance of seeing this secretive snake is at night, especially after rainfall, when they are searching for their favorite food (earthworms) on the surface.

Rough Earthsnake

Haldea striatula (Linnaeus)

Adult rough earthsnake from Douglas County.

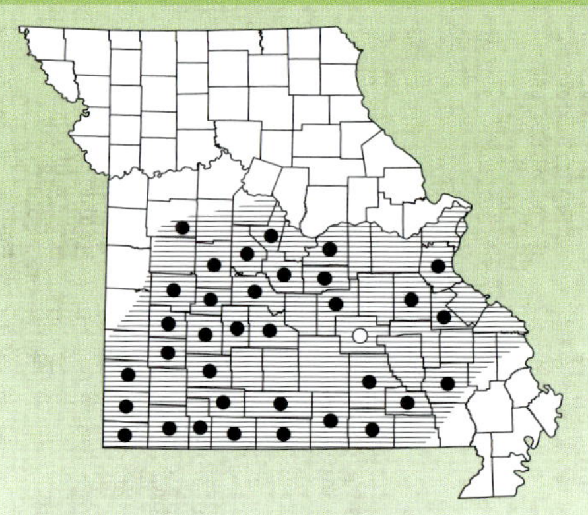

Distribution: Missouri: Throughout the Missouri Ozarks. North America: Southeastern states along the Atlantic Coast from southeastern Virginia to the northern edge of Florida, west to central Texas, and northeast to southern half of Missouri (Powell et al. 2016).

Description: The rough earthsnake is a small, plain-looking, secretive snake of open, rocky woodlands. Color is uniform gray, brown, or red brown. Individuals must be examined closely to verify identification. Characteristics to look for include five labial scales along the upper lip, a horizontal scale (loreal scale) anterior to each eye and a single scale (internasal scale) between the rostral scale on the snout. The belly is cream colored or light gray and unmarked. Scales along the back are keeled, and the anal plate is normally divided but occasionally single.

Adult rough earthsnakes range in total length from 180 to 254 mm (7.1 to 10 in.) but have been known to reach 324 mm (12.8 in.; Powell et al. 2016).

Habits and Habitat: Rough earthsnakes are active from late March into early November. This highly secretive species prefers rocky hillsides. It seldom ventures aboveground, remaining hidden under logs, flat rocks, or leaf litter. Individuals may

JEFF BRIGGLER

JEFF BRIGGLER

Rough earthsnakes are commonly found on open, rocky glades in southwestern Missouri, such as this example in Barry County (above). Head of a rough earthsnake from St. Francois County (left).

become active at night during warm, damp weather. This species is not known to bite, but when captured, it will release a foul-smelling musk from glands at the base of the tail. Winter dormancy occurs within rotting stumps and logs and probably small mammal burrows and rock crevices (Ernst and Ernst 2003; Trauth et al. 2004).

In Missouri, this species is most abundant on rocky, open, wooded hillsides (glades), along the edge of woodlands, and in open areas where there is abundant cover in the form of rocks or other ground debris. They are also found in areas disturbed by humans, such as trash dumps, rocky areas cut along roadsides, rock quarries, and abandoned buildings.

Bradford (1973) studied the movement pattern of the rough earthsnake in Missouri. Overall, individuals did not travel great distances, with a range from 1.8 to 22.3 m (5.9 to 73.2 ft). Although this study further confirmed individuals did seek shelter under rocks, surprisingly, most individuals were found in clumps of grass.

Earthworms are the principle food of this species, although they occasionally eat slugs, snails, and insects.

Breeding: Mating generally takes place during April through early June. However, Bradford (1973) described mating behavior for a pair of rough earthsnakes held in captivity in late

Rough earthsnakes can be quite abundant under rocks on glades, such as this example in Barry County.

March. Females give birth to their young from late July through September; a litter will contain 2 to 13 young (Wright and Wright 1957), with an average of nine. There are several documented reports of litter size for Missouri specimens. Bradford (1973) recorded the birth of a litter of four young in late September. Anderson (1965) documented a female from Ozark County that gave birth to four young in mid-August and a female from St. Clair County that gave birth to three young in late August. Young from these litters were 82 to 86 mm (3.2 to 3.4 in.; Bradford 1973) and 76.2 to 85 mm (3 to 3.3 in.) in total length (Anderson 1965). Young are darker in color than adults and usually have a white band on the back of the head. Sexual maturity is reached by the second year for males at a snout-to-vent length of 142 mm (5.6 in.) and the third year for females at a snout-to-vent length of 177 mm (7 in.) in Missouri (Bradford 1973). A wild-caught individual lived slightly over seven years in captivity (Snider and Bowler 1992). Bradford (1973) reported individuals up to five years of age for females and four years of age for males in Missouri.

Remarks: The rough earthsnake was formerly placed in the genus *Virginia*. More recent genetic data did not support this species being in this genus and, therefore, it was placed back in the formerly recognized *Haldea* genus (McVay and Carstens 2013).

Western Smooth Earthsnake

Virginia valeriae elegans Kennicott

JEFF BRIGGLER

Adult western smooth earthsnake from Butler County.

Description: The western smooth earthsnake is a small, slightly stout, plain-colored snake. It is often confused with the rough earthsnake. This species is gray, brown, or red brown, with tiny black dots. Sometimes a faint middorsal stripe is present. The belly is cream colored or light yellow and unmarked. For positive identification, the following characteristics must be determined: the presence of six upper, labial scales, a horizontal scale (loreal scale) in front of each eye, and two scales (internasal scales) between the nostrils on the snout. Scales along the back are smooth or weakly keeled, and the anal plate is divided.

Adult western smooth earthsnakes range in total length from 180 to 254 mm (7.1 to 10 in.) but have been known to reach 393 mm (15.5 in.; Laposha and Powell 1982; Powell et al. 2016).

Habits and Habitat: This species is normally active from late March through early November, with peaks in April and May,

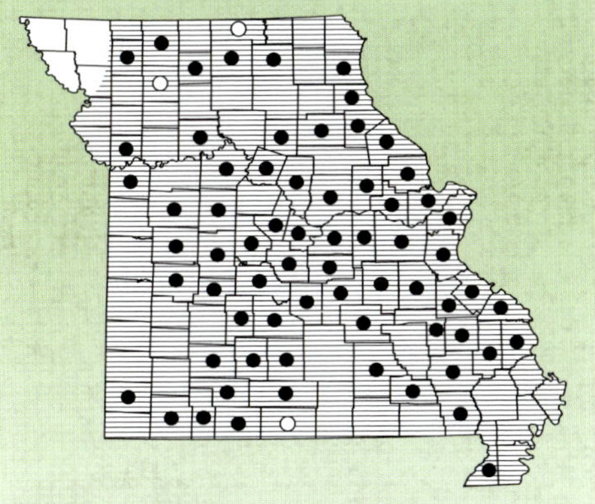

Distribution: Missouri: Nearly statewide, except northwestern corner. North America: South-central and southeastern Iowa through Missouri, northwestern corner of Kansas, eastern Oklahoma, eastern, western, and southern edges of Illinois, central and southern Indiana, central and western Kentucky, western Tennessee, most of Mississippi, and northwestern and southeastern Louisiana, and eastern half of Texas (Powell et al. 2016).

Western smooth earthsnakes are found in open woodlands with rocks, logs, and other ground cover, such as this example in Jackson County (above). Ventral coloration of a western smooth earthsnake from Lincoln County (right).

JEFF BRIGGLER

JEFF BRIGGLER

and again in September and October. This small, secretive snake lives in leaf litter and under rocks, logs, and other objects on rocky, wooded hillsides and in moist woods. It is more frequently encountered in the wooded areas with moist, loose soils.

The western smooth earthsnake is likely to be active at night during warm, humid weather. During the winter, individuals retreat underground into rocky crevices, small animal burrows, rotting stumps and logs, or trash dumps or burrow deep into loose soils (Ernst and Ernst 2003). An 18-year drift-fence study in a rocky, hilly forested habitat in the central Ozarks of Missouri captured a total of 5,630 western smooth earthsnakes (Rota et al. 2017).

Prey includes earthworms, slugs, and some soft-bodied insects. Anderson (1965) reported a captive specimen from Missouri ate earthworms and occasionally cutworms (moth larvae).

Breeding: Courtship and mating take place mainly in April and May (Bradford 1973) and possibly in the autumn. The young are generally born in August through September. Litter size ranges from 2 to 14 young (Fitch 1985) with an average of six to seven. A St. Clair County female gave birth to two young in mid-August (Anderson 1965). Anderson (1965) reported two additional litter sizes from St. Clair County. A female contained two nearly full-term embryos upon her death in late August. Another contained seven developing embryos, but there was no date provided. Finally, a Jackson County female that died of injury in early August contained eight well-developed embryos. Newborn young are generally 80 to 115 mm (3.1 to 4.5 in.) in total length (Powell et al. 2016). Young are darker in color than adults. Sexual maturity is likely reached within two or three years. A wild-caught individual lived to be a little over six years of age in captivity (Snider and Bowler 1992).

Remarks: The western smooth earthsnake is generally less frequently encountered than the rough earthsnake in Missouri, but the smooth earthsnake can be quite abundant in areas with suitable habitat. It is most commonly found under objects (rocks, leaves, logs) in moist, wooded areas, especially areas with loose soil. These areas provide ideal cover and loose soil for burrowing and locating their favorite prey (earthworms).

Family Viperidae

Vipers and Pit Vipers

Subfamily Crotalinae: Pit Vipers (Copperheads, Cottonmouths, & Rattlesnakes) VENOMOUS

The viper family occurs nearly worldwide and contains 328 species (Pough et al. 2018). The Crotalinae subfamily range includes eastern Asia from India to Japan, and from Canada through the United States, Mexico, and Central America to southern Argentina (Pough et al. 2018). Characteristics of pit vipers include a stout body, broad head, a pair of hollow fangs located at the front of the upper jaw, a heat-sensitive pit between the nostril and eye on each side of the head, eyes with a vertical pupil, and a single row of scales along the underside of the tail. The heat-sensitive pits of these snakes are used to detect warm-blooded prey. Even in total darkness, pit vipers can sense the location of a mouse or rat within a 0.5 to 0.6 m (1.6 to 2 ft) radius and can strike with great accuracy. Rattlesnakes (genera *Crotalus* and *Sistrurus*) are considered the most evolutionarily advanced group of snakes. These reptiles have evolved a distinct rattle on the tip of the tail that consists of hollow segments (specialized scales), which are loosely connected and hit against each other when the tail is rapidly vibrated, causing a buzzing sound. The six species of venomous snakes native to Missouri are pit vipers, represented by three genera: *Agkistrodon* (the copperhead and cottonmouth), *Crotalus* (rattlesnakes) and *Sistrurus* (pygmy and massasauga rattlesnakes).

Western Pygmy Rattlesnake

These venomous species are dangerous to people and should be avoided. Even freshly killed specimens can inflict a dangerous bite due to reflex action. A local or regional American Red Cross office can furnish up-to-date information on venomous snakebite first aid. In the event of a snakebite caused by a venomous species, the victim should be transferred to a hospital emergency room as soon as possible. Most bites occur when people are trying to kill or handle venomous snakes.

Accidental bites can be avoided by staying away from areas where there may be a concentration of venomous snakes. Wear protective footwear in habitats where dangerous snakes may occur. Never place your hands under rocks or logs, and do not step over rocks or logs. Step on them first, then over. When walking in a forest, step lively. Look the ground over when you stop to stand or sit, particularly around large rocks or logs. Pit vipers are most active in late evening and at night; be extra careful during these times. Wear rubber boots when fishing in streams or swampy areas that may harbor cottonmouths. Finally, avoid any snake you cannot identify. Most species of venomous snakes are shy and normally avoid people. When encountered in the wild, they usually try to escape detection by remaining motionless. Often, an individual that is provoked will try to escape rather than defend itself. Once cornered, however, these snakes will do their best to defend themselves. Learning to distinguish venomous from nonvenomous species by their color and pattern is recommended.

VENOMOUS

Eastern Copperhead

Agkistrodon contortrix (Linnaeus)

JEFF BRIGGLER

Adult eastern copperhead from Shannon County.

Description: The eastern copperhead is a medium-sized, stout-bodied snake with a sensory pit (heat sensing) between each nostril and eye. It is gray, copper, tan, or pinkish tan with hour-glass-shaped bands of dark brown. The markings are often edged in white. The top of the head can be gray or tan and without any markings. The eyes have vertical pupils. Their belly is cream colored with large, dark-gray or brown blotches along the edges that extend partly onto the sides of the body. Young eastern copperheads and some adults have a yellow tail tip. Scales along the back are weakly keeled, and the anal plate is single.

Adult eastern copperheads range in total length from 610 to 900 mm (24 to 35.4 in.) but have been known to reach 1,346 mm (53 in.; Powell et al. 2016).

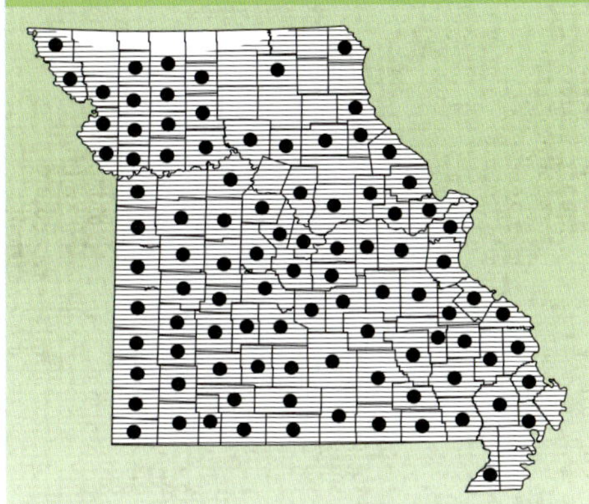

Distribution: Missouri: Throughout most of Missouri except for a few counties along the northern border. North America: The eastern copperhead ranges from Massachusetts south to northern Florida, west to eastern Texas, north to eastern Kansas, and east to southeast New York (Powell et al. 2016).

Habits and Habitat: This species is active from mid-March through early November, with most activity in Missouri occurring from mid-April to mid-October. A long-term

Head of an eastern copperhead with the vertical pupil that is indicative of venomous snake species in Missouri, Miller County (above). Ventral coloration of an eastern copperhead (right).

telemetry study conducted in Jefferson County documented copperheads on the surface as early as March 11 and as late as November 20 (W. Drda pers. comm.).

Fitch (1960) studied the eastern copperhead in eastern Kansas; some of his data is used below. The optimal temperature for this species is 27°C (80.6°F). On warm, sunny days, copperheads bask, especially in the morning. During late June, July, and August, this species becomes nocturnal. Its home range in Missouri varies from 2.4 to 14.6 ha (5.9 to 36.1 ac) with considerably smaller home ranges for gravid females (W. Drda pers. comm.). Copperheads rely on their camouflage pattern when resting in dead leaves and will usually remain motionless when encountered. This species is not particularly aggressive and will seldom strike unless provoked. They will occasionally climb into small bushes and trees searching for prey or to bask (Beaupre and Roberts 2001; Ernst and Ernst 2011).

Eastern copperheads live in a variety of upland and lowland habitats. Natural habitat includes open, rocky, wooded hillsides, brushy areas along creeks, and around swamps that offer abundant ground cover (scattered rocks, crevices along rock ledges, logs, and brush

A newborn eastern copperhead from Cole County (above). Eastern copperhead eating a white-footed mouse from Pettis County (left).

piles). They also reside near abandoned farm buildings, old sawmills, trash dumps, and old fields.

Most eastern copperheads enter overwintering sites from mid-to-late October in Missouri (W. Drda pers. comm.). Individuals overwinter singly or congregate at favorite overwintering sites, which are typically south-facing, rocky ledges in Missouri. They also use abandoned rock quarries, rocky lake spillways, animal burrows, hollow logs and stumps, and caves. Drda (1968) observed several copperheads overwintering on the floor in a small cave in Jefferson County.

Copperheads eat a wide variety of prey, but mice and voles make up the bulk of their diet. They also will eat frogs, lizards, small birds, insects (especially cicadas), and occasionally other small snakes (Ernst and Ernst 2011; Gloyd and Conant 1990). Young copperheads use their yellow tail as a lure to attract small frogs or lizards.

Breeding: Courtship and mating mainly take place in the spring but also in late summer into early autumn (Aldridge and Duvall 2002). Anderson (1965) observed a pair in Missouri mating in late April. A "combat dance" between two males can occur when a receptive female is nearby. These two males will raise 30 percent to 40 percent of their bodies off the ground, sway back and forth, and lean into each other in an attempt to push the opponent's head and neck to the ground (Ernst and Ernst 2003). Eventually the loser will crawl away from the area. Gravid females may congregate at a den site or at a nearby site that contains ample shelter (rocky areas) and basking potential while awaiting the birth of their young (Anderson 1965). The young are born in August, September, and early October. Litters normally contain from 3 to 11 young in Missouri. Anderson (1965) reported an average litter size for 20 captive Missouri females of 5.6 (range 3 to 11) young that were born from early September to early October. An additional seven females that were observed in the wild gave birth to three to nine young with an average of six. The mean birth date was mid-September. A female collected in Stoddard County gave birth to four young in mid-August (Anderson 1965). Most studies have indicated that female copperheads produce young every other year (Fitch 1960), but Anderson (1965) reported annual reproduction of two individuals collected in Missouri.

Young copperheads are born with functional fangs and venom glands. When first born they are generally dark and glossy in appearance until their first shed in 3 to 10 days (Anderson 1965). After this first shed, they are much paler in color, showing the general adult pattern, and have a yellow tail tip. Newborn Missouri copperheads measure between 177 and 215 mm (7 and 8.5 in.; Anderson 1965). Reproductive maturity takes three or more years for females and about two years for males (Fitch 1960). There are several reports of wild-caught adult copperheads surviving over 20 years in captivity, with one individual surviving 29 years, 10 months, and 6 days (Snider and Bowler 1992).

Remarks: Numerous subspecies of the copperhead were formerly recognized. Missouri had two and possibly three subspecies of copperheads—the Osage copperhead, *Agkistrodon c. phaeogaster* Gloyd, and the southern copperhead, *Agkistrodon c. contortrix* (Linnaeus), were found in the state, with influence from the northern copperheads (*A. c. mokasen*) along the Mississippi River bluffs in eastern Missouri (J. Ettling pers. comm.; Gloyd and Conant 1990; Johnson 2000). Recent genetic studies concluded that many of these subspecies of the copperhead were not valid (Burbrink and Guiher 2015).

Copperheads are the most commonly encountered venomous species in Missouri and thus account for most reported snake bites. Copperheads in Missouri bite an estimated 100 or more people each year, but death from copperhead bite is almost nonexistent, with a fatality rate of about 0.01 percent (Campbell and Lamar 1989). With common sense and a knowledge of copperhead habits and habitats, most bites by this species can be avoided, but if bitten, medical attention should be sought immediately.

VENOMOUS

Northern Cottonmouth

Agkistrodon piscivorus (Lacépède)

JEFF BRIGGLER

Adult northern cottonmouth from Ozark County.

Description: The northern cottonmouth is a heavy-bodied, dark-colored, semi-aquatic snake that is dark olive brown to nearly black. Dark crossbands are normally indistinct. The head is noticeably wider than the neck. The top of the head is dark brown or black. A black stripe may be present from the snout through the eye and onto the neck. Most individuals have a white upper lip. Northern cottonmouths in the Missouri Ozarks can be nearly black; coloration within a population can be quite variable (Gloyd and Conant 1990). There is a large sensory pit (heat sensing) located between the nostril and eye on either side of the head. The eyes have a vertical pupil, but the dark coloration of the eyes may make the pupils difficult to see. The belly is cream colored and heavily mottled with dark-brown or black blotches and smudges. Scales along the back and sides are heavily keeled, and the anal plate is single. Young northern cottonmouths are lighter in color and the 10 to 15 broad crossbands are distinct.

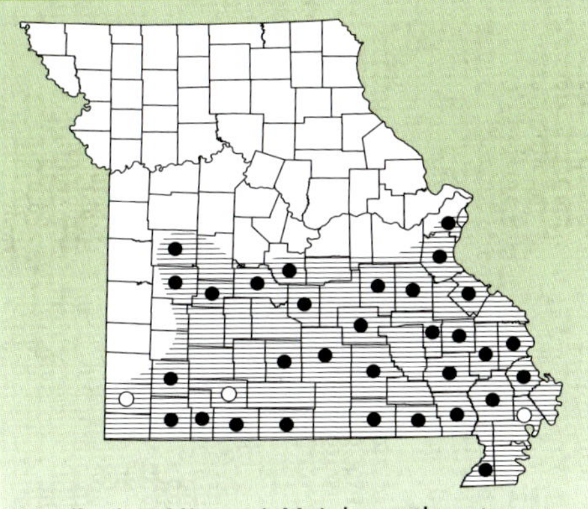

Distribution: Missouri: Mainly southeastern corner and scattered populations in the Ozarks. No northern cottonmouths are known to occur north of the Missouri River in this state (see remarks section). North America: Southeastern Virginia, south to Georgia, west to eastern Texas, and north to southern Missouri (Powell et al. 2016).

Defensive posture of a northern cottonmouth (above). Ventral coloration of a northern cottonmouth (right).

Newborn cottonmouths have a greenish yellow tail tip.

Adult northern cottonmouths range in total length from 760 to 1,220 mm (29.9 to 48 in.) but have been known to reach 1,892 mm (74.5 in.; Powell et al. 2016).

Habits and Habitat: In Missouri, the northern cottonmouth is normally active from late April through October. Individuals bask in the sun during the spring and autumn, but for the most part, this species is nocturnal. If encountered near brush or other cover, the northern cottonmouth will usually try to escape. If found in an open area, this snake remains in a loose coil and stands its ground. Cottonmouths often gape their mouth in a defensive posture, showing the white lining that is the origin of the common name. A cottonmouth is known to vibrate its tail and will not hesitate to strike if touched. Also, a strong-smelling musk is released from glands at the base of the tail any time an individual becomes defensive. When swimming through the water, a cottonmouth will hold its head above the surface with its back out of the water. This species often overwinters in rock crevices along limestone bluffs. In some areas, cottonmouths migrate from lowland wetland habitats to upland, forested bluffs in autumn, and return to wetlands in spring. These denning sites are often shared with racers, western ratsnakes, copperheads, and timber rattlesnakes (Burkett 1966; Ditmars 1931; Drda 1968; Smith 1961). Cottonmouths in the lowlands of southeastern

JEFF BRIGGLER

Newborn northern cottonmouth from Wayne County.

Missouri, which lack rocky outcrops, presumably use hollow logs and stumps, root channels of old trees, and animal burrows to overwinter. Radio telemetry studies in southwestern Missouri found home range sizes of 4.5 ha (11.1 ac) for males, 1.2 ha (3 ac) for non-gravid females, 0.3 ha (0.7 ac) for gravid females, and 0.4 ha (1 ac) for juveniles (B. Greene unpubl. data). Additional telemetry studies in Texas and Arkansas found that the majority of movement by cottonmouths is within about 10 m (32.8 ft) of the water's edge, with females making the farthest movement (up to 94 m; 308.4 ft) away from water (Metcalf et al. 2007; Roth 2005). Home range averaged 0.45 ha (1.1 ac) with a range from 0.13 to 1.14 ha (0.3 to 2.8 ac) in an Arkansas population (Metcalf et al. 2007).

In Missouri, the northern cottonmouth occurs in two distinct habitats. In southeastern Missouri, this species occurs in the cypress swamps, sloughs, oxbow lakes, and drainage ditches of the Mississippi Lowlands. In the Ozarks, this species occurs in scattered populations in cool, spring-fed, rocky creeks and small rivers. In the latter habitat, cottonmouths are highly nocturnal and secretive. Missouri is the northwestern limit of the range of the cottonmouth. Low winter temperatures are reported to be an important limiting factor in the distribution of this species along its northern range (Burkett 1966).

Cottonmouths eat a variety of animals—fish, frogs, other snakes, lizards, small turtles, rodents, and small birds, as well as small dead animals killed on roads (Gloyd and Conant 1990). Fish and frogs are most often eaten. Although most prey are captured within and around wetlands, cottonmouths will forage in upland, dry sites away from water on occasion.

Breeding: Courtship and mating may occur anytime within the active season, but in Missouri it typically occurs from April through June and August through October. Any time during their active season, male cottonmouths may instigate a "combat dance" when they encounter each other. Two cottonmouths were observed in early August in a combat dance on a wildlife refuge in northern Bollinger County (J. Stanard pers. comm.). The chance of this happening is enhanced when there is a female nearby (see timber rattlesnake account for a description of a combat dance). Perry (1978) also observed combat fighting in early

September in Missouri. In a study of a small, upland Ozark stream in northern Arkansas, Hill and Beaupre (2008) observed mating in mid-August and numerous male and female pairs during spring, late summer, and early autumn. A litter may contain 2 to 15 young with an average of six or seven (Burkett 1966). Litter size for a population studied in southwestern Missouri was 5.6 offspring per female (B. Greene unpubl. data). Anderson (1965) reported a female from Dent County that gave birth to eight young, and two Miller County females gave birth to four and five young. A study of an Ozark population in northern Arkansas found litter size averaged 4.1 young, with a range of three to five for 10 females (Hill and Beaupre 2008). In some areas in Missouri, gravid female cottonmouths move onto wooded bluffs or rocky hillsides during the summer to give birth. Crane and Greene (2008) found that gravid females in southwest Missouri will move into the open, rocky outcrops within the forested landscape prior to giving birth. These locations provide ideal temperatures and shelter for gravid females. Cottonmouths in Missouri mainly give birth during August and September, but births may occur into early October. In southwest Missouri, most births occur from late August to mid-September (B. Greene unpubl. data). In Missouri, it appears that cottonmouths give birth every other year, and if food is scarce, it may be every third year (Crane and Greene 2008). Newborn of this species are from 181 to 299 mm (7.1 to 11.8 in.) in length (Ernst and Ernst 2011). Young cottonmouths are light colored, with distinct markings somewhat similar to the hourglass-shaped dorsal markings of a copperhead. The tip of the tail is greenish yellow. The yellow tail has been reported to be used as a moving lure to attract frogs, lizards, and other prey (Wharton 1960). Sexual maturity is likely reached within two or three years, with females taking longer than males. A wild-caught adult lived slightly over 21 years in captivity (Snider and Bowler 1992).

Remarks: Formerly, three subspecies of the cottonmouths were known from North America, with the western cottonmouth (*Agkistrodon piscivorus leucostoma*) occurring in Missouri (Johnson 2000). Recent genetic data resulted in the elevation to species level for two species (northern cottonmouth and Florida cottonmouth), with no recognized subspecies (Burbrink and Guiher 2015).

A small population of this species was reported in Livingston County by Anderson (1965). It is presumed that this was an introduced population. Currently this species no longer occurs in that county or anywhere north of the Missouri River in this state (Gloyd and Conant 1990). There have been many common names associated with cottonmouths, causing some public confusion. The names cottonmouth, water moccasin, lowland moccasin, trapjaw, and gapper have been used for this species. Some Missourians use the common name cottonmouth for this species and water moccasin for the various nonvenomous watersnakes (*Nerodia* spp.). Often, any dark-colored snake found in or near an aquatic habitat is considered a venomous cottonmouth and either killed or unreasonably feared. The northern cottonmouth has a limited range in Missouri; it is likely that most semi-aquatic snakes seen or killed are in fact nonvenomous watersnakes. Also, a misconception is that cottonmouths are commonly found in trees along the edge of the water and drop on people or into boats. Most likely individuals are seeing nonvenomous watersnakes that do use shrubs and small trees along the water's edge. The cottonmouth prefers to be on the ground along the water's edge and rarely if ever climbs trees (Graham 2013). We hope that this book will help Missourians properly identify both venomous and nonvenomous species and refrain from killing snakes.

The bite of the northern cottonmouth is dangerous and can be fatal if medical treatment is not sought immediately. Within the known range of the cottonmouth in the United States, about 7.3 percent of the venomous snakebites can be attributed to this species (Parrish 1963). Cottonmouths bites are generally 4–5 percent of the venomous snakebites that occur in Missouri (Missouri Poison Center).

VENOMOUS

Timber Rattlesnake

Crotalus horridus Linnaeus

Adult timber rattlesnake from Perry County.

Description: The timber rattlesnake is Missouri's largest species of venomous snake. The timber rattlesnake is heavy bodied and has a prominent rattle at the end of its tail. Its ground color may be yellow, tan, brown, or gray, with dark-brown markings. The head normally has a dark-brown line from each eye to the angle of the jaw. Dark markings along the body are rounded anteriorly, changing to bands or V-shaped lines along the midbody to the tail. There usually is a rust-colored dorsal stripe. The

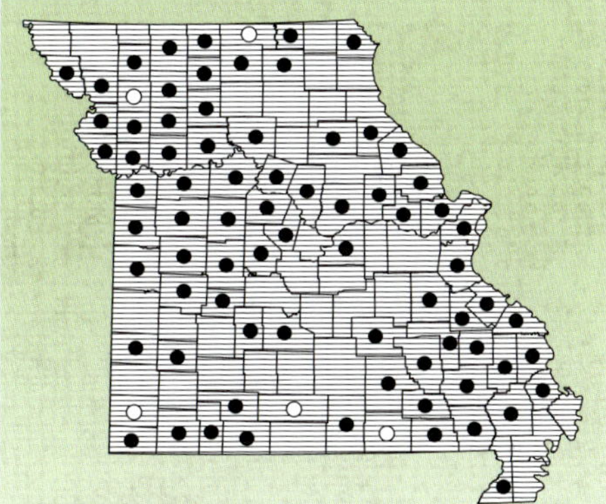

Distribution: Missouri: Generally considered to be statewide, with few records in the prairie regions in southwestern and northeastern corners, as well as the lowlands in southeastern. There are likely several locations where the species has been eliminated. North America: New England south to northern Florida, west through southern Ohio, Indiana, Illinois, north to western and southwestern Wisconsin, southeastern Minnesota, eastern and southern Iowa, most of Missouri, eastern edge of Kansas, eastern half of Oklahoma and Texas, and east to northern Florida. There are several regional color variations in this species (Brown 1993; Powell et al. 2016).

Timber rattlesnakes live on rough, rocky, wooded habitats on south-facing slopes, such as this example in Cass County (above). Rattles from two mature timber rattlesnakes (right).

tail is black and often referred to as "velvet-tailed." The top of the head is gray, light tan, or yellow, and unmarked. There is a large sensory pit (heat sensing) located between the nostril and eye on either side of the head. The belly is tan or light gray and sprinkled with small gray or brown specks. Scales along the back are keeled, and the anal plate is single. Most of the scales along the underside of the tail are in one row. The large rattle is straw colored. Young timber rattlesnakes are lighter in color than adults.

Adult timber rattlesnakes range in total length from 910 to 1,520 mm (35.8 to 59.8 in.) not including the rattle. They are known to reach up to 1,890 mm (74.4 in.) in length (Powell et al. 2016).

Habits and Habitat: In Missouri, timber rattlesnakes are mainly active from April through mid-October. During spring and autumn individuals bask in the sun on south-facing rock ledges and bluffs. They emerge from overwintering dens during periods of sunny weather and warm air temperatures, generally 23.8°C to 26.6°C (75°F to 80°F). During the hot summer, this species becomes mostly crepuscular or nocturnal.

Timber rattlesnakes are shy. If encountered, they may remain motionless to avoid detection, relying on their color to camouflage them among low bushes, branches, and dead leaves. Once disturbed, an individual may quickly move away and try to escape or coil in a defense posture and rattle.

During the active season, timber rattlesnakes spend a great deal of time coiled in a secure site such as near a fallen log or under a large flat rock. Inactive timber rattlesnakes are either

TOM R. JOHNSON

A newborn timber rattlesnake from Boone County.

lying in wait for a rodent to venture close or resting while a meal is digested. Observations of timber rattlesnakes in Pennsylvania and New York showed that an individual would rest its head on the edge of a downed log or along the base of a tree, look up and wait for a rodent to come close enough to strike and kill (Brown and Greenberg 1992; Reinert et al. 1984). This species is known to climb trees and into bushes that are usually less than 5 m (16.4 ft) high, presumably to seek prey or a safe place for shedding (Coupe 2001; Saenz et al. 1996; Wittenberg 2012). Some have been reported as high as 14.5 m (47.6 ft) in trees in Texas (Rudolph et al. 2004). Many rattlesnake species have an interesting way to collect and drink rainwater. An individual from St. Louis County was observed collecting falling water in its tightly coiled body and drinking it (Anderson and Drda 2005). Home range sizes are influenced by gender and reproductive status. Males travel quite often and generally have home ranges greater than 75 ha (185.3 ac). Females travel less often and have a smaller home range, generally 15 to 30 ha (37.1 to 74.1 ac). Gravid females have a home range of less than 10 ha (24.7 ac; Anderson 2010; Wittenberg 2012; J. Ettling pers. comm.). Gravid female timber rattlesnakes typically remain in rocky, south-facing, wooded habitats during July and early August, usually in the same area where they overwinter. During this time, while awaiting the birth of their young, the females feed very little, if at all (Keenlyne 1972). Male timber rattlesnakes are known to engage in a "combat dance" when they encounter each other, especially when a female is near. This behavior includes facing each other with their heads and forward part of their body above the ground, intertwining their necks and trying to push each other onto the ground to establish dominance (Anderson 1965; Collins et al. 2010).

This species prefers a habitat of mature forests and heavily wooded, rocky, south- and southwest-facing hillsides near streams and rivers. In general, timber rattlesnakes may be associated with large areas of rugged, hilly, heavily forested terrain with ledges and bluffs. In the northern half of Missouri, there are rock ledges and river bluffs where this species and others (such as copperheads and western ratsnakes) congregate in overwintering dens. In the southern half of the state, where there are many suitable overwintering sites, denning in large numbers is less common. Inactive rock quarries are often used as overwintering sites, especially if natural sites are lacking.

Timber rattlesnakes eat a variety of rodents (mice, rats, voles, shrews, and squirrels), small rabbits, and occasionally birds (Anderson 1965; Wittenberg 2012). An adult specimen from St. Louis County was found to have eaten an eastern chipmunk (*Tamias striatus*; Smith and Powell 1993).

Breeding: In Missouri, timber rattlesnakes begin courtship and mating mainly during the summer or early autumn, which is similar to their behavior in other states (Ernst and Ernst 2003). In mid-August, two rattlesnakes were observed in a combat dance in Shannon County, with a female nearby (C. Jensen pers. comm.). However, some mating has been reported in the spring (Anderson 1965; Keenlyne 1978). A mating pair was observed in late April in Clay County (Anderson 1965). A male courting a female will move alongside and on top of a receptive female, rub his head and body against hers and stimulate her with jerky movements (Anderson 1965). Copulation occurs when the male curls his tail under the female's tail and their cloacal openings meet. Female timber rattlesnakes in Missouri produce a litter of young mainly every third year, but some individuals reproduce every two or four years (W. Drda pers. comm.). A litter generally contains 5 to 14 young (average litter size is 8 or 9; Collin et al. 2010; Fitch 1985). Gravid females studied in a north-central Missouri population produced an average of seven offspring with a significant male bias (Muellman et al. 2017). A female caught in Jackson County gave birth to eight young in early September (Anderson 1965). Another female, collected from Bollinger County and held in captivity, gave birth to 19 young in early September (Anderson 1965). The young are born during August, September, and early October. They are born with a single rattle segment on their tail and each time the skin is shed a new segment is added to the base of the rattle. A timber rattlesnake may shed three to five times during an active season. As the rattle becomes longer, the terminal segments tend to weaken and break off. Besides general wear, a variety of insects or insect larvae (such as beetle larvae or camel crickets) may take advantage of a resting or sleeping rattlesnake and chew on a few segments of the rattle—which have no feeling—further weakening the rattle (T. Johnson pers. obs). Thus, counting the number of segments is not an accurate method for determining the age of a rattlesnake. Newborn timber rattlesnakes range in length from 250 to 330 mm (9.8 to 13 in.). The female stays with the young for 7 to 10 days prior to dispersing, and newborns are known to follow adult odor trials, including their mother, to find an overwintering site (Ernst and Ernst 2003; Muellman et al. 2018). Males mature at four to six years of age while females take 4 to 13 years (Ernst and Ernst 2012). A study by Brown and Simon (2018) in New York confirmed a life span of at least 41 years for males in the wild and estimated a life span of at least 51 years in a wild female.

Remarks: A subspecies of the timber rattlesnake, the canebrake rattlesnake (*Crotalus horridus atricaudatus* Latreille), was once recognized in the southeastern and southern portions of the range of the species, including southeastern Missouri. Studies have since shown that the characteristics separating the two subspecies tend to be clinal, and the subspecies *C. h. atricaudatus* is considered invalid (Clark et al. 2003; Pisani et al. 1972). The species, *Crotalus horridus*, is monotypic at this time.

Many people in Missouri mistakenly call the timber rattlesnake the diamond-backed rattlesnake. There has been some localized decline of the timber rattlesnake population in Missouri mainly due to habitat loss, persecution, and poaching.

The bite of a timber rattlesnake is dangerous and can cause human death. If bitten, seek medical attention immediately. Because this species normally occurs in rough country and is mostly nocturnal during the summer, few Missourians ever encounter it. There are few cases of people being bitten by this species in Missouri.

VENOMOUS

Eastern Massasauga

Sistrurus catenatus (Rafinesque)

JEFF BRIGGLER

Adult eastern massasauga from the Saint Louis Zoo.

Description: The eastern massasauga is a medium-sized rattlesnake associated with marsh habitats in the northeastern part of the state. Its general color is gray or brownish gray with 29 to 39 (average 34) dark-brown or black dorsal blotches (Anderson 1965). There are normally three alternating rows of round, dark blotches along the sides. The head is noticeably wider than the neck and has nine large scales on the top. There are two dark stripes edged in white along each side of the head. A sensory pit (heat sensing) is located between the nostril and eye on each side of the head. The tail is light with dark bands. The tail terminates in a distinct rattle. The belly is chiefly black or dark gray with irregular light markings. Scales along the back are keeled, and the anal plate is single.

Adult eastern massasaugas range in total length from 472 to 760 mm (18.6 to 29.9 in.) but have been known to reach 1,003 mm (39.5 in.; Powell et al. 2016).

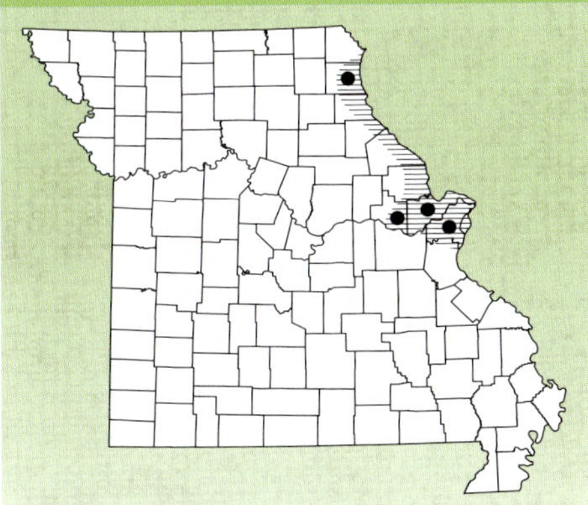

Distribution: Missouri: Along the Mississippi and Missouri River floodplain from St. Louis to northeastern Missouri. Likely extirpated from the state. North America: Isolated populations in New York and Pennsylvania, west to eastern Missouri and Iowa, and north to Michigan and northern Ontario (Powell et al. 2016).

A juvenile eastern massasauga from the Saint Louis Zoo (above). Head and rattle of an eastern massasauga from the Saint Louis Zoo (right).

Habits and Habitat: In Missouri, this species is mainly active from April through October. Specimens collected in Missouri were captured from mid-April to the end of September. This species spends considerable time basking during sunny, warm spring days. Basking locations are often in open spaces among vegetation, on top of matted vegetation, or coiled on or near the entrance of crayfish burrows. They are mainly active during the daytime but become nocturnal during the heat of the summer (July and August). Winter is spent mainly in crayfish burrows.

Massasaugas are normally not aggressive unless disturbed repeatedly. They generally remain motionless to avoid detection. They will readily attempt to escape down a nearby crayfish burrow or into thick vegetation if necessary. However, this rattlesnake will sound its rattles and defend itself when threatened.

In Missouri, eastern massasaugas were restricted to marshes or moist prairie habitats located within or near large river floodplains. Historically, numerous individuals occurred in bottomland prairies that were annually inundated by high water of the Mississippi River in St. Charles County (Hurter 1911).

Eastern massasaugas eat a wide variety of small animals: small mammals, snakes, frogs, birds, and, occasionally, invertebrates (Ernst and Ernst 2011; Holycross and Mackessy 2002). In Wisconsin, most of the adult diet consists of voles and mice (Keenlyne and Beer 1973). Newborns in Illinois primarily ate short-tailed shrews (*Blarina carolinensis*; Shepard et al. 2004). However, in laboratory feeding trials, there was a preference for small snakes. Young often lure prey, such as frogs, by waving their yellow tip tails (Schuett et al. 1984).

Breeding: Little is known about the reproductive biology of the eastern massasauga in Missouri. Courtship and mating usually occur in late summer and autumn, with some mating in spring (Aldridge et al. 2008). In southern Illinois, litter sizes range from 3 to 11 young with an average of 8.3 (Aldridge et al. 2008). Most births take place during August and September. Two females collected from St. Charles County gave birth to 8 and 11 young in early and late August (Anderson 1965). Newborn massasaugas average 220 mm (8.7 in.) in total length (Ernst and Ernst 2011). A study in Pennsylvania found newborns returned to their birthing area to overwinter (Jellen and Kowalski 2007). The young are paler than adults. The tail of newborn young has a yellow tip and a single "button" (the first element of the young snake's rattle). Sexual maturity is reached at three to four years (Ernst and Ernst 2011). Massasaugas are known to live up to 20 years in captivity (Snider and Bowler 1992).

Remarks: The taxonomy of massasauga populations has undergone considerable changes range wide, including Missouri. Until recently, there were three recognized subspecies of the massasauga in North American: eastern massasauga (*Sistrurus catenatus catenatus*), western massasauga (*Sistrurus catenatus tergeminus*) and desert massasauga (*Sistrurus catenatus edwardsii*). It was first believed that Missouri included a zone of intergradation between the eastern massasauga (*S. c. catenatus*) and western massasauga (*S. c. tergeminus*) based upon morphology (Anderson 1965). Given the similarities in ecology and habitat use, it was latter assumed populations were all eastern massasaugas in Missouri (Johnson 2000). More recent genetic data did not fully support the previous subspecies distinctions (Kubatko et al. 2011). Therefore, the eastern massasauga was elevated to full species status (*S. catenatus*) and the western subspecies was elevated to full species status (*S. tergeminus*) with two recognized subspecies: prairie massasauga (*S. t. tergeminus*) and desert massasauga (*S. t. edwardsii*). Missouri's extant populations in north-central and northwestern Missouri are recognized as the prairie massasauga and the historic populations along the Mississippi River floodplain in eastern Missouri are recognized as the eastern massasauga (Gibbs et al. 2011).

This species is listed as endangered in Missouri due to very few records and drastic reduction of bottomland prairie habitat. It is also designated as a federal threatened species throughout its range. The last confirmed report of the eastern massasauga in Missouri was in 1941 and, therefore, the species is likely extirpated from the state. Although this secretive snake has not been seen for many decades in the state, it is important to keep a watchful eye for this species. If an individual is observed, photograph it and report to the Missouri Department of Conservation.

Although this small rattlesnake has highly toxic venom, there are few records of human deaths over its entire range. In most cases, massasaugas are mild-tempered and rarely strike unless repeatedly harassed. If bitten, seek medical attention immediately.

Prairie Massasauga

VENOMOUS

Sistrurus tergeminus tergeminus (Say)

Adult prairie massasauga from Chariton County.

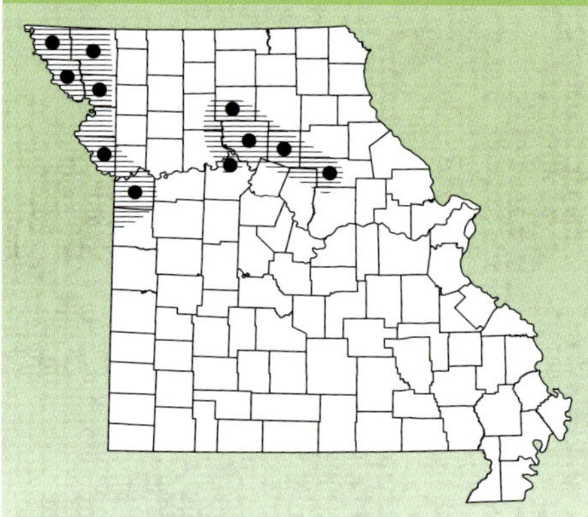

Distribution: Missouri: Isolated populations in north-central and northwestern corner of the state. North America: Southwestern Iowa and southeastern Nebraska, south into northern Missouri, eastern half of Kansas, western Oklahoma, and northern Texas (Powell et al. 2016).

Description: The prairie massasauga is a medium-sized rattlesnake associated with bottomland prairie habitats in north-central and northwestern Missouri. Its general color may be light to dark gray or gray brown with 34 to 50 (average 40) dark-brown or black dorsal blotches (Anderson 1965) and three alternating rows of smaller dark spots along the sides. The head is noticeably wider than the neck, with nine large scales on the top and two dark stripes edged in white along each side of the head. A sensory pit (heat sensing) is located between the nostril and eye on each side of the head. The tail is light with dark bands and terminates in a distinct rattle. The belly is normally light gray with darker-gray mottling, although some individuals will have a mostly black belly in the north-central Missouri populations. Scales along the back are keeled, and the anal plate is single.

Adult prairie massasaugas range in total length from 472 to 760 mm (18.6 to 29.9 in.) but have been known to reach 883 mm (34.8 in.; Powell et al. 2016).

Prairie massasaugas live in wet prairies, such as this example in Linn County (above). Ventral coloration of a prairie rattlesnake from Linn County (left).

Habits and Habitat: The prairie massasauga is the most studied snake in Missouri. This species is mostly active from April through October but surface activity can occur in March and into November if temperatures are warm (J. Briggler unpubl. data). They are primarily diurnal, especially in April and May, followed by October, but become more nocturnal during the heat of the summer (Seigel 1986). They spend considerable time basking during sunny, warm days in an open space among vegetation, coiled on top of a crayfish burrow or ant mound, or sometimes on top of matted vegetation several feet above ground. In Missouri, this species overwinters primarily in crayfish burrows and ant mounds, although other animal burrows can be used. Most individuals rely on crayfish burrows that reach groundwater below the frost line. Often massasaugas will move from moist prairie habitats after emergence to drier upland sites (Seigel 1986). In Missouri, massasaugas are restricted to marshes or moist prairie habitats located within or near large river floodplains or scattered moist prairie swales within upland grassland habitats. Due to extensive habitat loss, the range of this species has been reduced to a few isolated areas in north-central and northwestern Missouri.

Combat fighting between two male prairie massasaugas from Chariton County (above). A newborn prairie massasauga coiled on top of a crayfish burrow from Linn County (right).

The massasauga population in Holt County has an average home range from 7.4 to 17.1 ha (18.3 to 42.3 ac), with males utilizing larger areas (Durbian et al. 2008). Also, Durbian et al. (2008) recommended a 100 ha (247.1 ac) area to be the minimal amount of habitat necessary for sustaining a population.

This rattlesnake is generally mild mannered and hesitant to strike unless harassed. When approached, massasaugas usually remain motionless to avoid detection or quickly seek shelter in thick vegetation or crayfish burrows. However, this small rattlesnake will become defensive, sound its rattle, and strike when cornered or harassed.

This snake's diet consists mainly of small rodents. In Missouri, adult massasaugas prey primarily on voles and mice (Seigel 1986). Young massasaugas also eat other snake species, including Dekay's brownsnakes and eastern gartersnakes (Seigel 1986).

Breeding: Courtship and mating may occur in spring or autumn in Missouri. Although many studies have documented reproduction primarily from late summer into autumn for massasaugas throughout their range, Missouri populations appear to mainly mate after spring emergence. Mating pairs are commonly seen in Missouri populations in the spring, especially April. Welchert and Mills (2014) documented three mating pairs in late April in northwestern Missouri, and more than 20 mating pairs in the north-central Missouri populations have been observed from April 4 to April 28 (J. Briggler unpubl. data; T. Crabill pers. comm.). In addition, male combat fighting has been witnessed multiple times in north-central Missouri populations in April (J. Briggler pers. obs.). In each case, the males raise their head and upper part of the body with tails intertwined. They begin to sway their raised bodies back and forth and then press against each other's body until one male is pinned to the ground. The loser will retreat, and the winner will approach the nearby female for mating. Numerous studies have been conducted on litter size of Missouri's massasauga populations. The average litter size is six to seven young, with a range from 1 to 13 (Seigel 1986; J. Briggler unpubl. data; T. Crabill unpubl. data), but a litter size as high as 17 has been reported (Pilgrim 2001; Pilgrim et al. 2011). The size of a litter increases with the size of the female. Initial data suggested female massasaugas in Missouri produce a litter of young every other year (Seigel 1986), but more recent long-term data indicate that massasaugas can reproduce annually with an abundance of prey and favorable weather. Most births take place in August and September. Newborns were observed in mid-August to late September in Holt County (Seigel 1986). Anderson (1965) observed seven newborns with an adult female in Holt County in late August, and a female from Chariton County gave birth to 12 young in late August as well. Newborn massasaugas range in length from 206 to 275 mm (8.1 to 10.8 in.). The newborn's tail has a yellow tip and a single "button." The young have a lighter ground color and darker blotches than adults. Sexual maturity is reached between two to four years. Massasaugas can live at least 20 years (Snider and Bowler 1992).

Remarks: Massasauga is a Native American word meaning "great river mouth"; thus, referring to the wet, lowland habitats associated with this snake along rivers (Minton 1972). The taxonomy of massasauga rattlesnake populations has undergone considerable changes, and extant populations occurring in Missouri are now listed as the prairie massasauga (see remarks section of the eastern massasauga for more detailed information). There continues to be studies as to the recognition of the two subspecies of the western massasauga complex (*S. tergeminus*). A genetic study by Ryberg et al. 2015 did not support a subspecies distinction between the prairie massasauga and desert massasauga. Their study supported at least five genetically distinct populations, with Missouri prairie massasauga populations being one. This is further supported by McCluskey and Bender (2015). They found distinct genetic clusters within the isolated Missouri prairie massasauga populations compared to the more continuous distribution in Kansas.

The prairie massasauga once occurred in large numbers across much of north-central and northwestern Missouri. Five isolated populations are currently known in the state. Due to few populations, low numbers, and greatly reduced natural, bottomland prairie habitat, the species is classified as state endangered. Missouri populations of prairie massasaugas may be genetically distinct and deserve continued protection. If an individual massasauga is located, photograph and report it to the Missouri Department of Conservation. Protection and management of bottomland prairies is vital to sustaining this species in Missouri.

Western Pygmy Rattlesnake

VENOMOUS

Sistrurus miliarius streckeri Gloyd

Adult western pygmy rattlesnake from Shannon County.

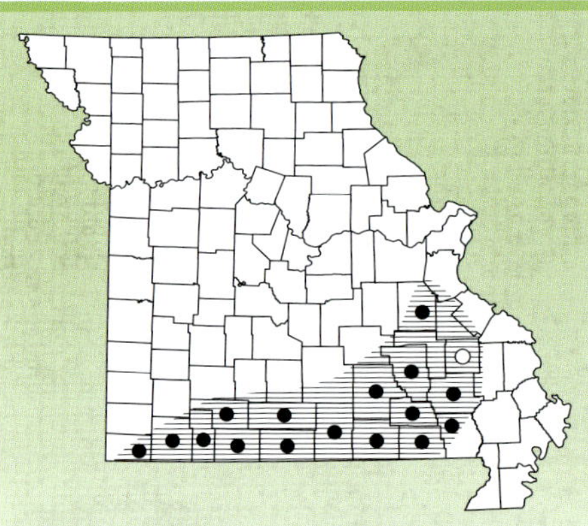

Distribution: Missouri: Extreme southern Missouri along the border with Arkansas, and in the eastern Missouri Ozarks and St. Francois Mountains. North America: Southern Missouri and southwestern Tennessee, south to the Gulf Coast states of Mississippi, Louisiana, and Texas, and north to eastern Oklahoma (Powell et al. 2016).

Description: The western pygmy rattlesnake is a small, colorful rattlesnake with a slender tail and tiny rattle. This is one of the smallest species of rattlesnake in North America. Its general color is brown gray with small dark-brown or black blotches. There are 20 to 30 dorsal blotches, which are round or in short bars. There are one or two alternating rows of small, rounded, dark blotches along the sides. A narrow orange-brown dorsal stripe is usually present. The head has a distinct black stripe that angles from the eye to the corner of the mouth and a sensory pit (heat sensing) located between each nostril and eye. The slender tail has six to eight dark bands and terminates in a small rattle. Its belly is dusky cream colored with numerous dark, irregularly spaced transverse bars. The scales along the back and sides are keeled, and the anal plate is single.

Adult western pygmy rattlesnakes range in total length from 380 to 540 mm (15 to 21.3 in.) but have been known to reach 832 mm (32.8 in.; Powell et al. 2016).

JEFF BRIGGLER
JEFF BRIGGLER

Western pygmy rattlesnake are found in rocky, scrubby areas with plenty of sunlight, such as this example in Shannon County (above). Head of a western pygmy rattlesnake from Shannon County (left).

Habits and Habitat: The western pygmy rattlesnake is normally active from early April to mid-October. During late spring and early summer this small snake will bask in rocky, open areas, near brush piles, or along the side of roads near forests and glades. During July and August this species tends to be nocturnal and can be observed crossing roads and highways at night.

This species usually takes shelter under rocks, logs, brush piles, or shrubs during spring, early summer, and autumn. Two adults were found under a large rock in mid-April on a Barry County glade (J. Briggler unpubl. data). A study in southwestern Missouri found that pygmy rattlesnakes selected closed canopy cover near their retreat sites, such as small logs and shrubs (Maag 2017). They also tended to avoid bare ground and leaf litter in favor of sites with vegetative cover (Maag 2017). Little information is published on winter dormancy locations for pygmy rattlesnakes, but in the Ozarks, they likely use animal burrows, logs and stumps, large rocks, and crevices within rock ledges.

The disposition of this species when encountered varies. Some pygmy rattlesnakes will try to defend themselves vigorously by coiling, sounding their rattles, jerking their head, and striking at any movement. Others will remain motionless and try to escape only when touched with a stick or snake hook. The sound made by the rattle of a western pygmy

Western pygmy rattlesnakes give birth to live young, such as this female from southwestern Missouri (above). A newborn western pygmy rattlesnake from Ozark County (right).

rattlesnake is a faint buzz reminiscent of the sound made by an insect. It can be heard from only a little over 1 m (3.3 ft) away.

In Missouri, the western pygmy rattlesnake is most often observed in or near glades, in second-growth forests near rock ledges, and along the area between forests and open lands. Its preferred habitat is south-facing, rocky, and partially wooded hillsides. A population of this species was studied in southwestern Missouri; it was found that males moved more than females and both sexes moved during September when they seemed to relocate to overwintering sites (Holder 1988; Holder and Moll 1988). Another study in southwest Missouri found that pygmy rattlesnakes are relatively sedentary and occupied very small home ranges of less than 2.6 ha (6.4 ac; Maag 2017). Gravid females moved considerably less often than both males and non-gravid females. Once these females give birth, they generally increase their home range by five-fold (0.2 to 1.04 ha; 0.5 to 2.6 ac; Maag 2017).

Food of this species includes a variety of lizards, small snakes, mice, and occasionally small frogs and insects. Individuals collected in Barry County ate prairie lizards, five-lined skinks, little brown skinks, and small mice in captivity (Anderson 1965; Johnson 2000). Juveniles are known to wave their yellow-tip tail to lure frogs and lizards into range to kill and eat (Ernst and Ernst 2011; Rabatsky and Waterman 2005).

Breeding: Little information is available on the reproductive biology of this species. Courtship and mating appear to mainly take place in late summer and early autumn but could happen in early spring after emergence from wintering sites. Maag (2017) observed four females mating with one or more males from late August to early October in southwest Missouri. A male courts a female by waving his tail and tongue flicking as he is crawling along her side to mate (Montgomery and Schuett 1989). Throughout their range, litter size varies from 2 to 32 young, with an average of six (Ernst and Ernst 2011). Northern populations, such as in Missouri, produce smaller litter sizes (Fitch 1985). Three to seven young are usually produced per litter in Missouri (Anderson 1965). A Taney County female was found with seven developing embryos in early July (Anderson 1965), and a Stone County female contained seven well-developed embryos in mid-August (Hurter 1911). The young are born from late August through September. A Wayne County female gave birth in late August to three young (Anderson 1965). A Carter County female gave birth in late September to five young (Anderson 1965). The young from these two litters averaged 127 mm (5 in.) in total length, with a range from 120 to 130 mm (4.7 to 5.1 in.). Newly born western pygmy rattlesnakes are paler than adults and have a yellow tail tip. Females have been reported to defend their newborns by positioning themselves between the young and the threat. Maag (2017) reported births occurring in mid-August, with females attending litters for up to three days. In Missouri, most female western pygmy rattlesnakes appear to breed every other year (Holder 1988; Holder and Moll 1988), but Maag (2017) observed four females in Missouri mating within a few weeks after giving birth, suggesting the potential for annual reproduction. Sexual maturity is estimated at two to three years of age (Messenger 2010). A wild-caught individual lived slightly over 16 years in captivity (Snider and Bowler 1992).

Remarks: Some people in extreme southern Missouri use the local name "ground rattler" for this species. This snake is rarely encountered and very few individuals have been bitten by them in Missouri. Even though this snake is small and rarely bites people, its venom is quite toxic. They should be respected and left alone.

NONNATIVE SPECIES

Nonnative species are those species introduced into areas outside of their native range. Typically, they arrive at a new location by human activities, either accidental (e.g., escaped pet, or hitchhiking in plants and other debris) or deliberate. In Missouri, nonnative amphibian and reptile species are often reported from the wild. Species more commonly reported in Missouri include African clawed frogs (*Xenopus laevis*), Cuban treefrogs (*Osteopilus septentrionalis*), greenhouse frogs (*Eleutherodactylus planirostris*), American alligators (*Alligator mississippiensis*), spectacled caimans (*Caiman crocodilus*), African spurred tortoises (*Centrochelys sulcate*), desert tortoises (*Gopherus agassizii*), yellow-bellied sliders (*Trachemys scripta scripta*), midland painted turtles (*Chrysemys picta marginata*), Mediterranean geckos (*Hemidactylus turcicus*), Italian wall lizards (*Podarcis siculus*), brown anoles (*Anolis sagrei*), green iguanas (*Iguana iguana*), Burmese pythons (*Pythron bivittatus*), and boa constrictors (*Boa constrictor*). In most cases, nonnative amphibian and reptile species do not survive Missouri's cold winter months. However, some have become established (e.g., Italian wall lizard and Mediterranean gecko) in Missouri. It can be difficult to predict with certainty which nonnative species will become established and negatively affect native species. Generally, nonnative species that become established are usually invasive and harm native species through predation, competition for resources (e.g, food, space), and as a vector of disease. To protect Missouri's native amphibians and reptiles, laws require nonnative species be confined to prevent accidental releases and deter intentional releases.

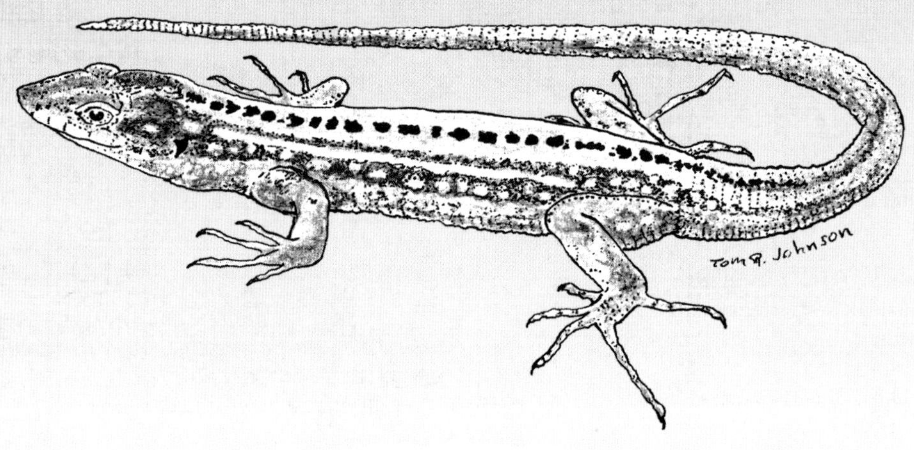

Italian Wall Lizard

INTRODUCED

Mediterranean Gecko

Hemidactylus turcicus (Linnaeus)

JEFF BRIGGLER

Adult Mediterranean gecko from Saint Louis County.

Description: The Mediterranean gecko is a small lizard with a large head, bulging eyes with vertical pupils, broad toepads, and flattened appearance. Ground color is white to pink with darker-brown or gray spots. The tail has broad cross bands. Warty looking tubercles are prominent on the body. The belly is white to translucent. At night, they appear ghostly white. Juveniles have more boldly marked tail bands (Powell et al. 2016).

Adult Mediterranean geckos range in total length from 100 to 127 mm (3.9 to 5 in.; Powell et al. 2016).

Habits and Habitats: Active at night on the outside of buildings mainly from April through October if warm temperatures permit (Collins et al. 2010). Mediterranean geckos are associated with human structures in urban areas in Missouri. Populations are known to occur on university buildings, greenhouses, and homes (Manning and Briggler 2003; Paulissen and Buchanan

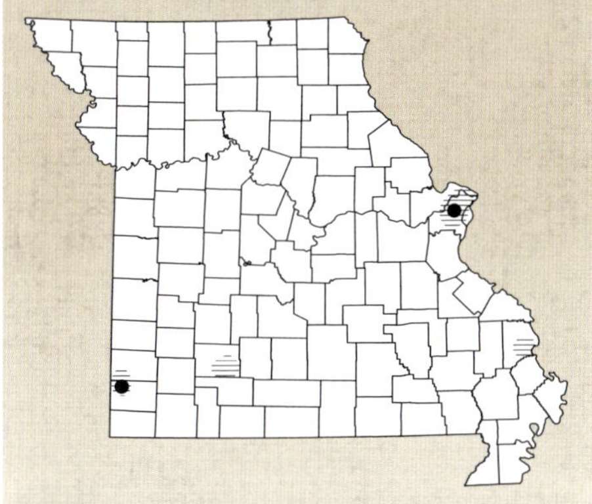

Distribution: Missouri: Introduced in urban areas in the southern half of Missouri with populations known in St. Louis, Springfield, Joplin, and Cape Girardeau. North America: Introduced throughout Florida and west into Texas. Numerous scattered populations in at least 13 additional states and southern California and Arizona (Powell et al. 2016).

Mediterranean geckos occur in and around buildings in urban areas, such as this individual climbing an interior wall next to a window at the Saint Louis Zoo.

1991). Their "sticky" toepads allow them to climb effectively. They remain hidden by day within cracks and gaps of buildings and under siding. At night, these geckos can be viewed with flashlights along building walls, around windows and doors, and on porch ceilings, especially near lights that attract insects. Mediterranean geckos are vocal at night, emitting squeaking and chirping sounds. If captured, they will bite, and their tails will drop off readily to distract a predator. Overwintering locations are likely cracks and crevices of building foundations and inside heated buildings. It is somewhat common to see them active during winter within heated homes and buildings.

Mediterranean geckos feed on a wide variety of insects (e.g., caterpillars, beetles, ants), spiders, and other invertebrates.

Breeding: Reproductive biology is unknown in Missouri. Reproductive season is generally from April to August throughout much of its introduced range in the United States. Females produce two eggs per clutch and often two to three clutches during a summer (Rose and Barbour 1968). In the northern part of their invaded range, individuals likely only produce one or two clutches per season (Paulissen and Buchanan 1991). Often several females will lay eggs together. The eggs are highly adhesive to each other and to the surface where deposited. They are known to deposit eggs in wall crevices in Tennessee (Wessels et al. 2018) and in peat moss in a greenhouse in Arkansas (Paulissen and Buchanan 1991). Eggs are also found on open surfaces within homes. Rose and Barbour (1968) found eggs on the open floor in a closet. The eggs were later covered with debris by the adult. Eggs have also been found in filing cabinets in Florida (Rose and Barbour 1968) and on top of a mantle in Missouri (K. Goellner pers. comm.). The white eggs, which average 11 by 9 mm (0.4 by 0.3 in.) are ovoid and hard-shelled (Rose and Barbour 1968). Eggs hatch about 40 days after laying during summer or early autumn. Maturity is likely reached within one year for males and two years for females (Rose and Barbour 1968).

Remarks: Because this species is commonly found in and around homes, it is also known as the Mediterranean house gecko. They are native to the Middle East and Mediterranean area. They have successfully invaded many areas worldwide and are the most widely introduced lizard in the United States. This lizard is easily dispersed via their astonishing ability to hitchhike via human transport (e.g., ship cargo, vehicles, plant trade). They are only found in and around buildings in urban areas in Missouri and do not appear to be a threat to native amphibian and reptile species. Their range in Missouri appears to be expanding to other urban areas.

INTRODUCED

Italian Wall Lizard

Podarcis siculus (Rafinesque)

JEFF BRIGGLER

Adult Italian wall lizard from Jasper County.

Description: The Italian wall lizard is generally a brown or gray lizard that usually has a green back. Patterns along the back and sides are highly variable, often with spots, stripes, or net-like patterns. The tail is mainly brown or gray. The belly consists of six rows of smooth scales that are usually white or cream (Powell et al. 2016).

Adult Italian wall lizards range in total length from 140 to 240 mm (5.5 to 9.4 in.; Powell et al. 2016).

Distribution: Missouri: Introduced population in Joplin's downtown area, Jasper County. North America: Scattered introduced populations in New York, Connecticut, New Jersey, Pennsylvania, Missouri, Kansas, and California (Briggler et al. 2015; Powell et al. 2016; Kolbe et al. 2013).

Habits and Habitat: This species is mainly active from early April to early November. Wall lizards can be seen basking on rock walls, sidewalks, and building foundations from mid-morning to early afternoon. When approached, they will retreat quickly within cracks and crevices of rock structures (e.g., walls, sidewalks, rock piles), mulch beds, trash piles, and vegetation. They will frequently enter human dwellings, such as garages and basements. Overwintering sites in Missouri mainly occur within and

Italian wall lizards occur on and within rock structures in urban areas, such as this rock retaining wall in Jasper County.

JEFF BRIGGLER

beneath rock structures (e.g., walls, sidewalks, building foundations), as well as small burrows in yards and hollow trees (Briggler et al. 2015). This species is native to southern Europe and along the Mediterranean Sea coast. In its native range, this species lives in shrubby vegetation, rocky areas, rural gardens, pasturelands, and urban areas. In Missouri, this species resides in an older neighborhood with abandoned structures and rock and concrete walls that provide ideal shelter (Briggler et al. 2015).

The diet of the Italian wall lizard has not been studied in Missouri. Small invertebrates such as grasshoppers, crickets, caterpillars, beetles, ants, bees, moths, spiders, and earthworms probably make up the bulk of its diet (Burke and Mercurio 2002). These lizards are often observed in grassy lawns by Joplin residents when mowing; it is presumed the lizards are foraging for food.

Breeding: No information is known on the reproductive biology of this species in Missouri. Courtship and mating probably occur in April and early June. Females likely dig burrows or use areas beneath or within rock structures to deposit eggs. Clutch sizes range from 2 to 12 eggs with most clutches containing five or six eggs. Eggs are presumably laid in June or early July. Females may produce more than one clutch a year (Collins et al. 2010). In captivity, females have been known to deposit from three to five clutches a year. Eggs normally hatch in five to seven weeks and offspring typically measure 30 to 35 mm (1.2 to 1.4 in.) in snout-to-vent length (Arnold and Ovenden 2002). Males likely mature within one year; females likely take up to two years.

Remarks: The Italian wall lizard, also called "ruin" lizard, is named for its association with and ability to climb rocky structures. This species has been accidentally and intentionally introduced throughout the world. It was first established in the 1950s in Kansas and since that time it has been introduced accidentally or deliberately via the pet trade (Kolbe et al. 2013). This lizard was accidentally introduced in 2001 in an urban area of Joplin from a collection of animals from Topeka, Kansas (Briggler et al. 2015). Once established, populations can become quite abundant in a short time. The population introduced in Missouri originated from nine individuals that now number several hundred over a six-square-block area (J. Briggler unpubl. data). So far, this population is restricted to an urban area that supports few native lizards, but nearby natural habitat (rocky glades) needs to be monitored to ensure this invasive lizard does not become established and compete with native lizards of Missouri.

Species of Possible Occurrence

The following species were, at one time, listed as part of the Missouri herpetofauna, or have populations within 25 miles of the state line. Although there are several species of salamanders (southern two-lined salamander, *Eurycea cirrigera*; three-lined salamander, *E. guttolineata*; northern zigzag salamander, *Plethodon doralis*; northern slimy salamander, *P. glutinosus*; and Mississippi slimy salamander, *P. mississippi*) that occur across the Mississippi River in adjacent Illinois, Kentucky, and Tennessee, the Mississippi River is likely a geographic barrier for these limited dispersal species. In addition, little to no habitat for several of these species occurs in Missouri. Therefore, these salamander species are unlikely to occur in Missouri and not addressed below. However, due to their proximity to Missouri's borders, some of the following species may eventually be found or rediscovered in this state.

Amphibians

TOM R. JOHNSON

Spotted Dusky Salamander – *Desmognathus conanti* Rossman

The spotted dusky salamander occurs in small colonies in extreme southern Illinois, western Kentucky and Tennessee, and northeastern Arkansas (Crowley's Ridge; Niemiller and Reynolds 2011; Powell et al. 2016; Trauth et al. 2004). The sandy hills of Crowley's Ridge in southeastern Missouri could harbor a small population of this species; thus far it has not been found in that area.

TOM R. JOHNSON

Hurter's Spadefoot – *Scaphiopus hurterii* Strecker

The Hurter's spadefoot's range extends into north-central and northwestern Arkansas (Powell et al. 2016; Trauth et al. 2004). This secretive species is an explosive breeder that is most likely to be found on nights with heavy rainfall along the southern and southwestern border with Arkansas and Oklahoma, especially in sandy areas within river floodplains. To date, the Hurter's spadefoot has not been found in Missouri.

TOM R. JOHNSON

Bird-voiced Treefrog – *Hyla avivoca* Viosca

The bird-voiced treefrog is a beautiful little treefrog that has been reported across the Mississippi River in southern Illinois, western Kentucky, and western Tennessee (Smith 1961, 1966; Niemiller and Reynolds 2011; Trauth et al. 2004). Many attempts to locate this species during the breeding season in the remaining tupelo-gum swamps of southeastern Missouri have failed.

Reptiles

TOM R. JOHNSON

American Alligator – *Alligator mississippiensis* (Daudin)

The American alligator was introduced beginning in 1972 into much of its former range in Arkansas by the Arkansas Game and Fish Commission. At that time, its population was rapidly declining throughout Arkansas, and the introduction was initiated to increase its range and abundance, as well as control beaver populations (Trauth et al. 2004). Individuals were introduced into three Arkansas counties that border the Missouri Bootheel, and occasional sightings are still reported (Watt et al. 2002). Although American alligators are occasionally seen in Missouri, these reports are most likely intentional or accidental releases based upon single specimens. However, due to the proximity to Arkansas and suitable habitat along the St. Francis River and other swamplands in extreme southeastern Missouri, it is possible this species could become established, especially with milder winters.

MIKE JEFFORDS

Eastern Fence Lizard – *Sceloporus undulatus* (Bosc and Daudin *in* Sonnini and Latreille)

The eastern fence lizard is found across the Mississippi River in all border counties in the southern half of Illinois, Kentucky, and Tennessee (Niemiller and Reynolds 2011; Powell et al. 2016). This species can be impossible to distinguish from the prairie lizard in Missouri based upon pattern, so molecular methods are needed to positively confirm identification. Although the Mississippi River likely poses a significant barrier, suitable habitat (upland forest) exists in Missouri.

TOM R. JOHNSON

Prairie Earless Lizard – *Holbrookia maculata perspicua* Axtell

Two prairie earless lizards were reported from a state park in Johnson County (Nickerson and Krager 1972). This is a range extension of about 115 miles northeast of the nearest natural occurring population in Kansas, and the species was subsequently listed as part of the Missouri herpetofauna (Johnson and Bader 1974). No additional specimens have been collected in Missouri; this lizard is not considered a part of the Missouri fauna.

DAREN RIEDLE

Western Milksnake – *Lampropeltis gentilis* (Baird and Girard)

Recent species evaluations of numerous subspecies of the milksnake group (*Lampropeltis triangulum*) has resulted in the potential of the western milksnake occurring in western Missouri from Kansas City south to the Arkansas border (Ruane et al. 2014). Powell et al. (2016) shows the range of this species occurring in the southern Kansas City area of Missouri and questions the identification of milksnakes in southwestern Missouri. This species apparently can only be distinguished from the eastern milksnake (*L. triangulum*) by using molecular methods and geographical distribution (Powell et al. 2016). Although this species likely exists along the border with Kansas, no published studies using molecular data have confirmed western milksnakes in Missouri.

TOM R. JOHNSON

Gray Ratsnake – *Pantherophis spiloides* (Duméril, Bibron and Duméril)

Recent molecular data has recognized several species of the former ratsnake complex with the western ratsnake (*P. obsoletus*) occurring west of the Mississippi River and gray ratsnake occurring east of the Mississippi River with areas of hybridizations occurring within southeastern Missouri (Burbrink 2001; Burbrink et al. 2000; Powell et al. 2016). Additional molecular studies are needed to determine if populations in southeastern Missouri truly represent hybrids with western ratsnakes, or if they are entirely gray ratsnakes.

JEFF BRIGGLER

Queensnake – *Regina septemvittata* (Say)

Three specimens of the semi-aquatic queensnake were collected in 1927 in southern Stone County (Anderson 1965). These specimens were collected by G.K. Noble with assistance by B.C. Marshall. There is some question as to the validity of the collection location since B.C. Marshall had no recollection of encountering this species in Missouri (Anderson 1965; Conant 1960). No additional specimens have been taken in the state. However, populations of queensnakes are still extant in northern Arkansas (Stanley and Trauth 2007). This species is not recognized as occurring in Missouri, but habitat similar to those of populations in Arkansas exists along numerous rivers and streams in southern Missouri.

ROBERT HAMILTON / ALAMY STOCK PHOTO

Common Ribbonsnake – *Thamnophis sauritus* (Linnaeus)

The common ribbonsnake is known from southern Illinois, western Kentucky, and western Tennessee (Niemiller and Reynolds 2011; Powell et al. 2016). Ribbonsnakes are good swimmers, and the Mississippi River is unlikely to prevent this species from occurring in southeastern Missouri.

Guide to Missouri Salamander Larvae

This section is furnished to help readers identify Missouri's salamander larvae that can be found in ponds, streams, and other aquatic habitats. Not all of our native salamanders breed in streams or wetlands (i.e., species in the genus *Plethodon*). Species in the genus *Plethodon* do not have an aquatic larvae stage.

Salamander larvae may not be as difficult to identify as Missouri's various toad or frog tadpoles, but many look similar. Also, several subtle characteristic changes take place during their aquatic development. The illustrations depict a variety of stages of development (i.e., some were made from older specimens than others). Very young salamander larvae may have to be raised in captivity for a few weeks or months so that their characteristics can be compared with the illustrations and descriptions. Costal grooves may be difficult to see on living specimens of many species, so questionable specimens may have to be preserved and costal grooves counted under magnification. Lengths provided are approximations for a complete (uninjured) specimen and indicate total length (combined head, body, and tail lengths). Gill size should not be used for identification because the size and bulk of gills are directly related to the amount of dissolved oxygen in their aquatic habitat. For example, small-mouthed salamander larvae (*Ambystoma texanum*) from a shallow pool with an abundance of rotting vegetation will have very large gills, while dark-sided salamander larvae (*Eurycea longicauda melanopleura*) taken from a spring-fed creek will have much smaller gill structures. Live or freshly preserved specimens can be identified, but very old, preserved specimens will lack much of the subtle pigmentation needed for identification. It is important to know the location or origin of the specimen(s) to be identified. Several Missouri salamanders have a restricted distribution; for example, the Oklahoma salamander (*Eurycea tynerensis*) is found only in the southwestern corner of the state. Thus, knowing the specimen's collection data will help narrow the possibilities by eliminating those not found in that part of the state. Information for this section was gleaned from Altig and Ireland (1984), Altig and McDiarmid (2015), Bishop (1943), Dundee and Rossman (1989), Ireland and Altig (1983), Smith (1961), and Trauth et. al. (2004).

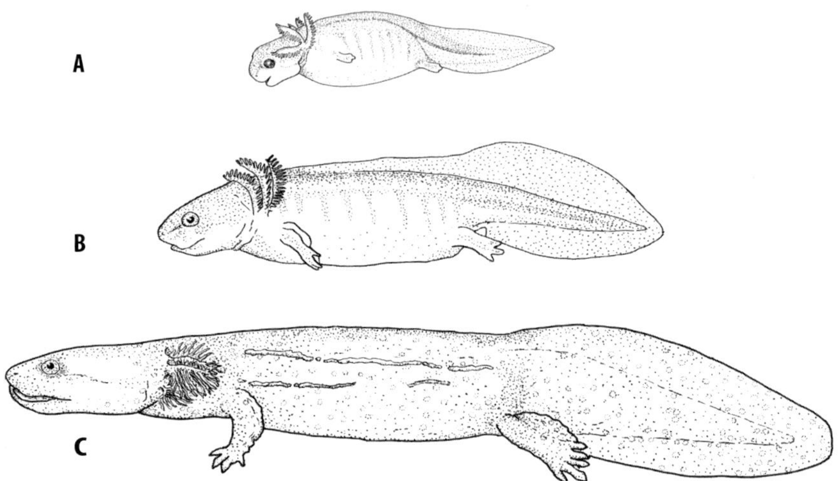

Eastern Hellbender and **Ozark Hellbender** (*Cryptobranchus alleganiensis alleganiensis* and *C. a. bishopi*). Stream-type larvae. No costal grooves. Total length at hatching averages 27.5 mm (1.1 in.). Larval stage may last up to 24 months. Stubby limbs and extended abdomen, with a large, visible yolk sack during the first two to three months of development. Four toes on front limbs; five toes on hind limbs. Pen and ink drawings depict (A) a larva soon after hatching; (B) a larva about two weeks after hatching; and (C) a hellbender larva at about four months post hatching. Limbs develop a prominent flap of skin, which extends backward and appears paddle-like. High and well-developed caudal fin. Hellbender larvae at three months usually dark brown to nearly black with numerous tiny, light spots on body and tail. Translucent ventral in newly hatched larvae; gray with older specimens. Larvae may retain external gills until 1½ to 2 years post hatching and at a total length of 100 to 130 mm (3.9 to 5.1 in.).

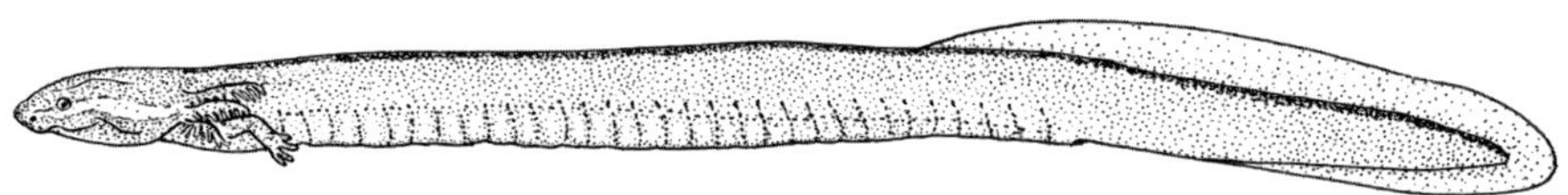

Western Lesser Siren (*Siren intermedia nettingi*). Costal grooves 31 to 38. Eel-like with feathery, external gills. Four toes on forelimbs; no hind limbs. Well-developed caudal fin, lacking in markings. Dark-brown or brownish-gray head and body. Some red coloration on head of live specimens. Light stripe extends from snout to gills. Lives in swamps and flooded ditches. Length: 64 to 100 mm (2.5 to 3.9 in.).

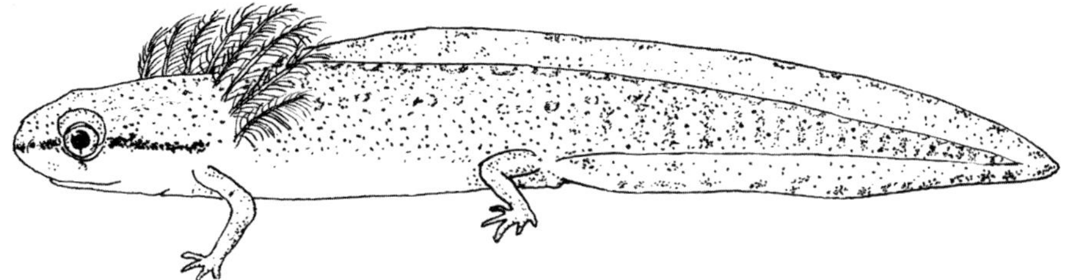

Central Newt (*Notophthalmus viridescens louisianensis*). Pond-type larvae. No costal grooves. Four toes on forelimbs; five toes on hind limbs. Eye diameter larger than distance between nostril and eye. Four external gill rackers on each side; gill structures large and full. Head width about the same as body width. Dark line on sides of head from nostril through eye to just below gills. Medium height dorsal caudal fin extending forward to behind head. Black markings along edge of caudal fin. Living larvae pale tan. Lives in swamps and woodland ponds. Length: 22 to 28 mm (0.9 to 1.1 in.).

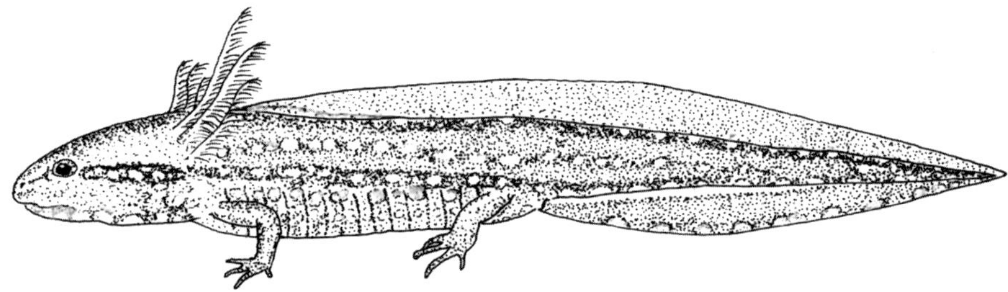

Ringed Salamander (*Ambystoma annulatum*). Pond-type larvae. Fifteen costal grooves. Four toes on forelimbs; five toes on hind limbs. Numerous irregular and poorly defined bilaterally paired cream-yellow (in life) dorsal blotches; ventrolateral dark bands of pigment from forelimbs to vent, separated from dorsal pigment by an irregular light band. Immaculate chin and throat. Dorsal fin extends from just behind the head. Uniform dark pigmentation over dorsal surface and lower sides. Often has a line of very small, white or silvery spots (iridophores) along lower edge of sides, between front and hind limbs. Transformation at a snout-to-vent length of 40 to 60 mm (1.6 to 2.4 in.).

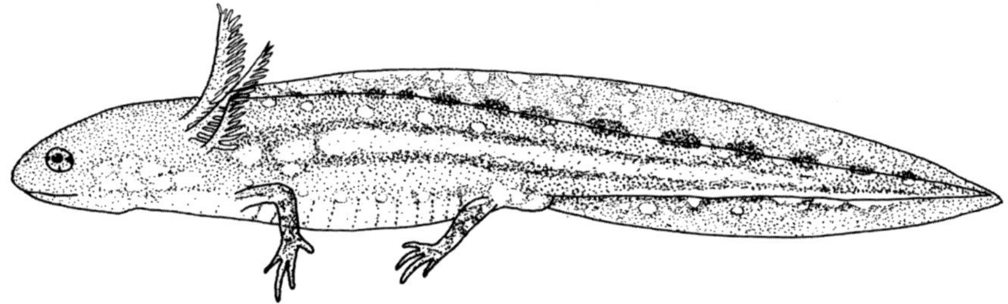

Spotted Salamander (*Ambystoma maculatum*). Pond-type larvae with 11 to 13 costal grooves. Four toes on forelimbs; five toes on hind limbs. Chin and throat lack pigment. Full caudal fin. Dorsal caudal fin begins immediately behind head, and ventral fin begins at vent. Olive drab to dark, grayish brown, with dark-gray to black small markings along dorsum to the tip of the tail. Faint light stripe along sides onto tail. Light areas yellow in living specimens. Resides in fishless, woodland ponds. Length: Average 51 mm (2 in.).

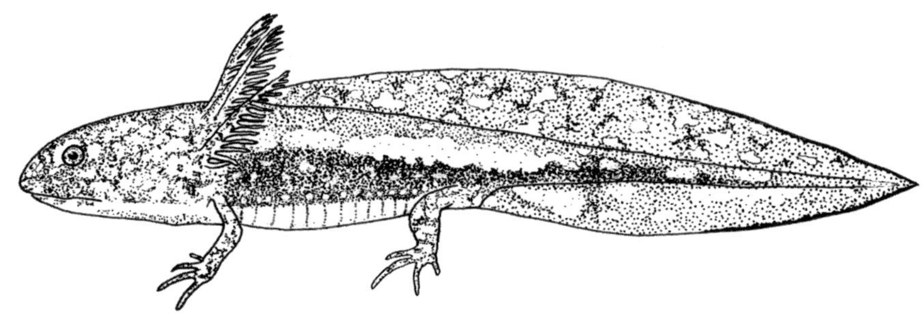

Marbled Salamander (*Ambystoma opacum*). Pond-type larvae with 11 to 12 costal grooves. Four toes on forelimbs; five toes on hind limbs. Slender or elongated forelimbs. Some chin and throat pigmentation. Dark gray with numerous light mottling on head and caudal fin. Wide, distinct, light lateral stripe and wide, dark stripe below, both extending onto tail. Light areas yellow or cream colored on living specimens. Distinct line of small iridophores along lower sides on living specimens. Abundance of flecking on the head, body, and tail of larvae about to transform. Develops in fishless, woodland ponds, swamps, and ephemeral pools. Length: 39 to 54 mm (1.5 to 2.1 in.).

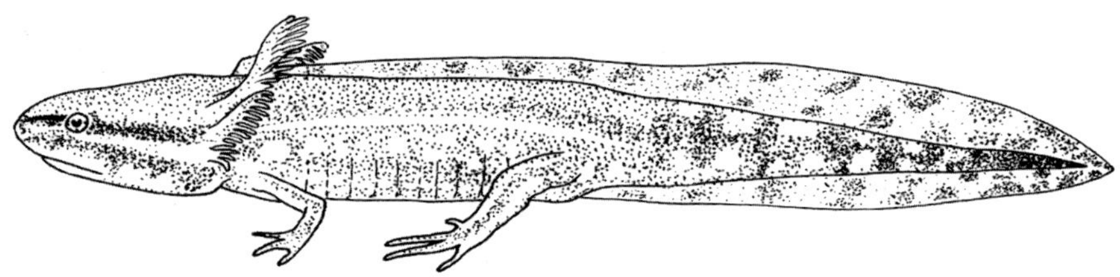

Mole Salamander (*Ambystoma talpoideum*). Pond-type larvae with 10 to 11 costal grooves. Four toes on forelimbs; five toes on hind limbs. Medium height dorsal caudal fin; narrow ventral caudal fin. Dark gray, with a dark and light stripe along sides of head. Thin, light stripe along sides does not extend onto tail; line may be cream or yellow in life. Large, dark-gray or black blotches or bands on tail; scattered, dark-gray markings on caudal fin. Distinct, medial dark-gray stripe on ventral. Specimen for drawing had an extended larval period. Develops in bottomland forest swamps and pools. Length: 50 to 70 mm (2 to 2.8 in.).

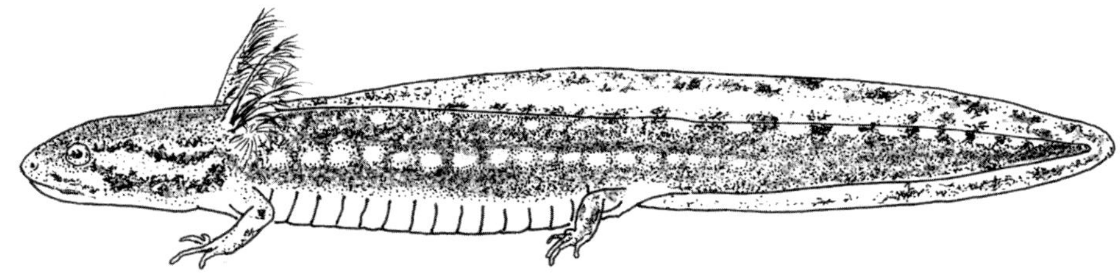

Small-mouthed Salamander (*Ambystoma texanum*). Pond-type larvae with 13 to 15 costal grooves. Four toes on forelimbs; five toes on hind limbs. Clear area with little or no pigmentation along the lower part of dorsal caudal fin. Squarish or rectangular dark markings along dorsum separated by light bars. Immaculate chin and throat. Blunt head. Wide, light area along each side of head from snout to under gills; line of faint, light, round areas along the sides. Newly transformed specimens may range from 35 to 75 mm (1.4 to 3 in.) in total length.

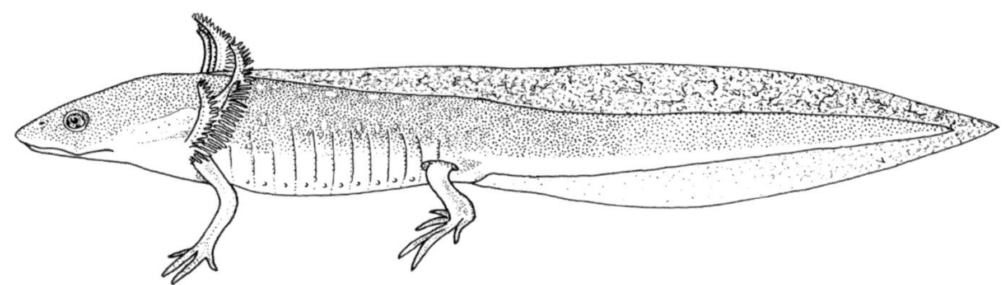

Eastern Tiger Salamander (*Ambystoma tigrinum*). Pond-type larvae with 11 to 14 costal grooves. Four toes on forelimbs; five toes on hind limbs. Flattened toes. Full caudal fin with dorsal fin extending forward to back of head. Head wider than body. Lateral view of head flattened. Uniform, dark gray to nearly black; small, irregular, dark flecks cover dorsal caudal fin. Small, light spots along dorsum on either side of caudal fin. May be a thin, light stripe along sides or a line of small round, light spots on either side of back. Tiny, white or silvery iridophores along lower section of sides, especially on young larvae. Develops in fishless prairie or woodland ponds and marshes. Length: 100 to 177 mm (3.9 to 7 in.).

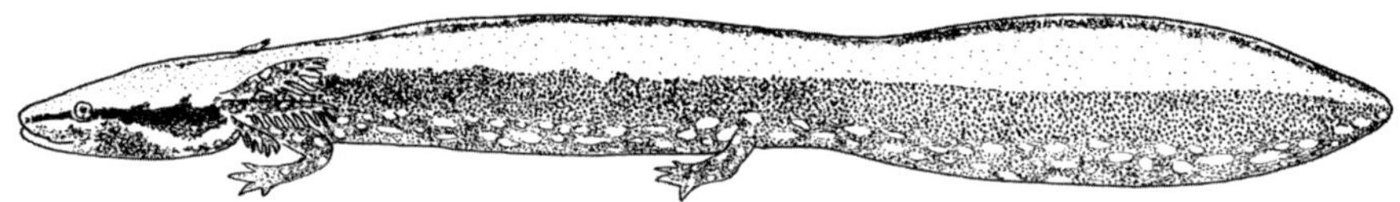

Common Mudpuppy and **Red River Mudpuppy** (*Necturus maculosus maculosus* and *N. m. louisianensis*). Stream-type larvae with 15 to 16 costal grooves that are difficult to see. Four toes on forelimbs; four toes on hind limbs. Wide and flattened head. Full and prominent gills. Well-developed caudal fins; dorsal fin begins at junction of tail and body. Dark line running from the nostril through each eye to gills. Dark dorsal stripe from the head to near the tip of tail. Dark-brown dorsal stripe bordered on either side by a wide, light stripe. Sides of body and tail dark brown with small, irregular light areas, which are largest near the tip of tail. Gray venter; central area without pigment. Found in medium-size streams, rivers, and large reservoirs. Length: 39 to 75 mm (1.5 to 3 in.).

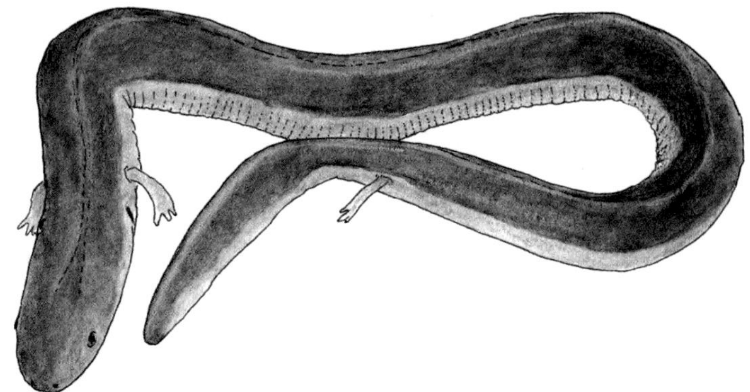

Three-toed Amphiuma (*Amphiuma tridactylum*). No costal grooves. Eel-like body; somewhat pointed head. Tiny, light gray spots on body and tail. Tiny front and hind limbs; three toes on each, appearing almost useless. Gills likely absent, as much of the larval stage is spent in the egg. Egg stage may last up to five months. Some authors believe external gills are resorbed soon after hatching. Hatchlings average 64 mm (2.5 in.) in total length. Dark brown to nearly black above; medium gray below. Line drawing depicts a late larva of 115 mm (4.5 in.) in total length. Found in natural, cypress swamps.

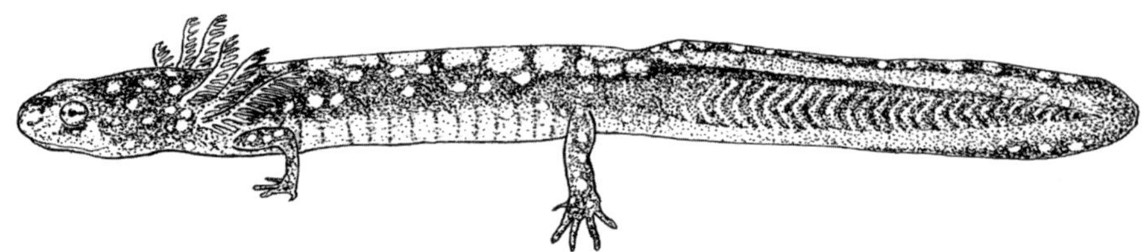

Eastern Long-tailed Salamander and **Dark-sided Salamander** (*Eurycea longicauda longicauda* and *E. l. melanopleura*). Stream-type larvae with 13 or 14 costal grooves. No pigmentation in center of throat (gular region). Four toes on forelimbs; five toes on hind limbs. Newly hatched larvae nearly black. No dark pigmentation on bottom of hind feet. Dark gray or brownish black. Large, light dorsal spots or well-developed light dorsal stripe (pale tan in life) on body. Venter without pigmentation. Reduced ventral caudal fin with some light spots near the tail tip. Develops in cave springs and small creeks. Length: 22 to 75 mm (0.9 to 3 in.) in total length.

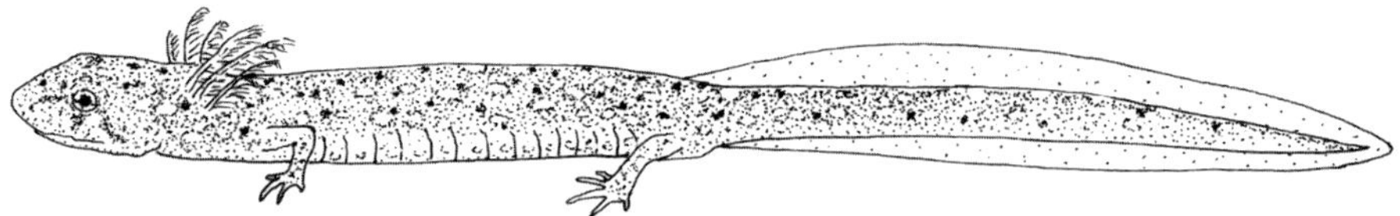

Cave Salamander (*Eurycea lucifuga*). Stream-type larvae with gills and 14 to 15 costal grooves. Four toes on forelimbs; five toes on hind limbs. Dorsal caudal fin begins just above hind limbs, and ventral fin begins just past vent. Early larvae usually darkly pigmented with small, faint light spots. Dark-gray stippling across the central throat (gular area) of later larvae. Irregular dark line extends from each eye to the corners of the mouth and a smaller dark line below each eye. Small black spots along the back and sides. Dark pigment on the bottom of hind feet. Larvae may take up to two years to transform. Total length at transformation averages 58 mm (2.3 in.). Can be easily confused with eastern long-tailed or dark-sided salamander larvae (see their description above). Larvae develop in cave and headwater streams, and springs.

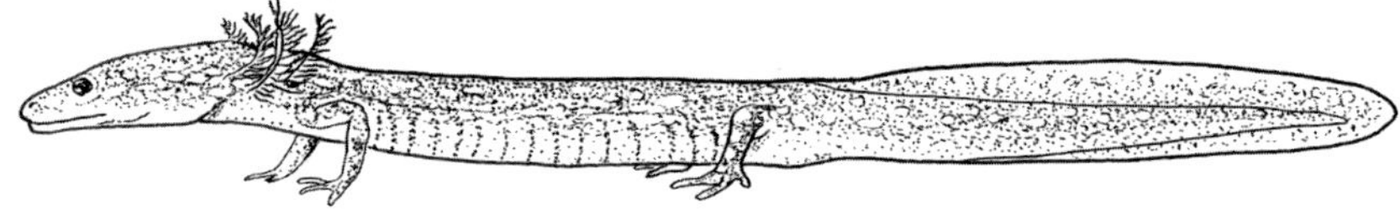

Grotto Salamander (*Eurycea spelaea*). Stream-type larvae with 16 to 19 costal grooves. Four toes on forelimbs; five toes on hind limbs. Head distinctly wider than body and widest just in front of gills. Blunt snout. Fully formed and functional eyes. Moderately developed caudal fins; dorsal fin does not extend forward onto body. Brown to dark gray or sometimes pinkish tan with numerous streaks along sides and onto tail. Two or three longitudinal lines of tiny, bright-white spots (iridophores) usually present along the sides, often extending onto tail. Light-gray and unmarked venter. Found in cave springs and springs flowing out from the base of bluffs (grottoes). Length: 32 to 55 mm (1.3 to 2.2 in.).

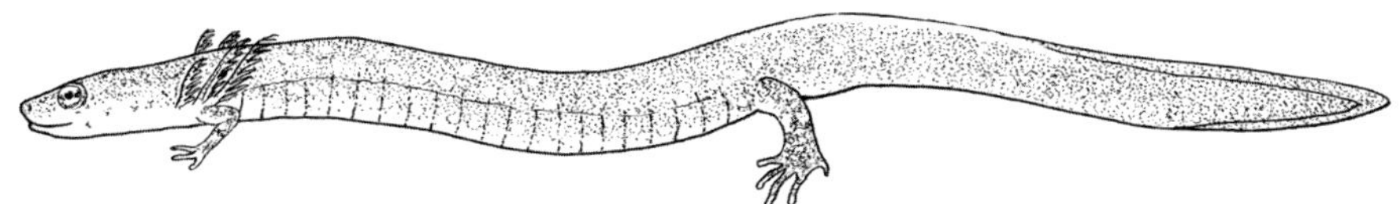

Oklahoma Salamander (*Eurycea tynerensis*). Stream-type larvae with 19 to 21 costal grooves. Four toes on forelimbs; five toes on hind limbs. When front and hind limbs pressed inward (adpressed), the costal groove count is seven or more. Slender body; head not wider than neck. Light tan; thin yellow or golden stripe (on live specimens) on dorsal section of tail. Poorly developed caudal fin. Several rows of tiny, white spots along sides of body. Gray to light-yellow venter on living specimens. Tiny, faint, dark markings or stippling cover head, body, and tail. Missouri population mostly neotenic. Lives in small, clear, gravelly, spring-fed streams. Length: 39 to 67 mm (1.5 to 2.6 in.).

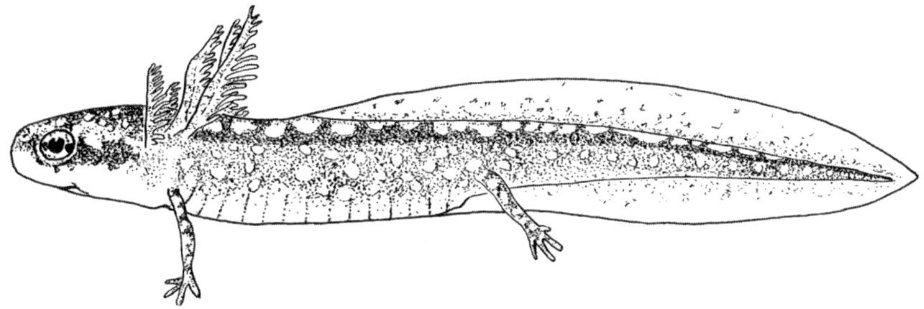

Four-toed Salamander (*Hemidactylium scutatum*). Stream-type larvae with 13 to 14 costal grooves. Four toes on forelimbs; four toes on hind limbs. Dark line from nostril through eye onto sides of head. Well-developed caudal fin with little pigmentation; dorsal fin extends forward to near midbody. Dark-brown head and body with tan or cream mottling and large, light spots along dorsum. Develops in small, headwater streams, ephemeral pools with thick mosses along edge, pools in fens, and fishless sinkhole ponds. Length: 16 to 35 mm (0.6 to 1.4 in.).

Guide to Missouri Toad and Frog Tadpoles

Identifying anuran larvae is difficult. Even professional herpetologists have difficulty separating many species, and some biologists simply avoid working with tadpoles. This guide should help the reader identify the tadpoles of Missouri's toads and frogs. Based on the works of Altig (1970), Altig and McDiarmid (2015), Gosner (1960), and others, a standardized method of tadpole identification has been established. The reader should refer to these works for more details.

The descriptions and illustrations that follow depict a development stage of each species at a point just prior to—or at the point of—their hind limbs appearing (around stage 30; Gosner 1960). This guide is based on preserved specimens, but some descriptions of live specimens are included when deemed valuable. Familiarity with tadpole anatomy (tadpole mouthparts, shape, tail or tail fin shape, markings, etc.) is helpful in identification. In this guide, emphasis is placed on examining mouthparts for the accurate identification of tadpoles, which requires preserved tadpoles. In each species description, the number and placement of labial teeth rows on the upper and lower lips are represented by two numbers separated by a slash. For example, 2/3 indicates there are two rows (anterior labial teeth rows) on the upper lip and three rows (posterior labial teeth rows) on the lower lip. Total tadpole length is given for each species only as a general reference. Several environmental factors may cause tadpoles to develop at different rates, so exact measurements are of limited value. It is helpful to know the collecting locations of specimens to be identified. For example, tadpoles of American bullfrogs (*Lithobates catesbeianus*) can be expected in all parts of Missouri, whereas the northern leopard frog (*Lithobates pipiens*) is limited to a few extreme northern counties. For more details on tadpole identification refer to the following works: Altig 1970, Altig and McDiarmid 2015, and Gosner 1960.

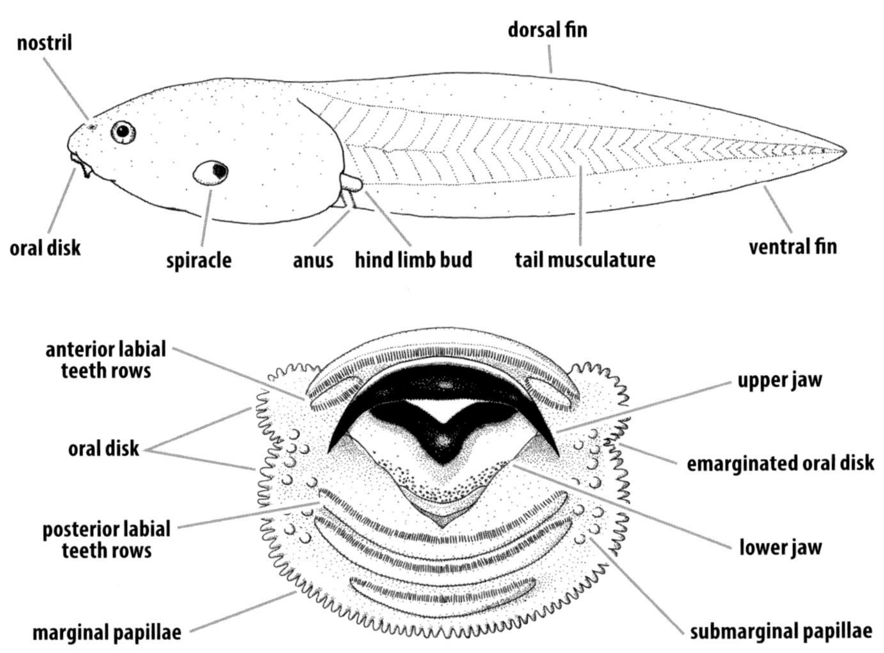

Anatomy of a tadpole body (top) and mouthparts (bottom).

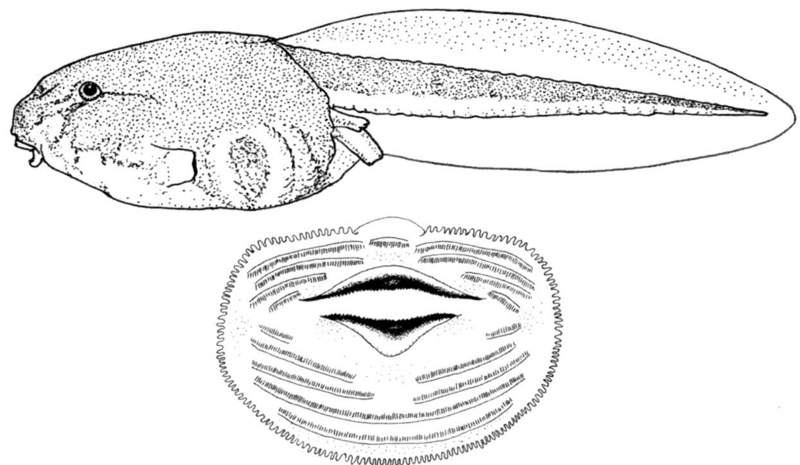

Eastern Spadefoot (*Scaphiopus holbrookii*). Medium size, heavy bodied, with dorsally placed eyes and closely placed nostrils. Light lip line below eye from nostril to just forward of spiracle. Live specimens dark-brown to shiny, brassy-brown body and tail musculature, with numerous small, brassy to gold flecks (iridophores). Clear tail fins, rounded at tip of tail. Round oral disk, small marginal papillae; few submarginal papillae. Labial teeth row formula may range from 4/4, 5/6 to 6/6; most anterior labial teeth rows with medial gap. Posterior labial teeth rows 1 to 3 with medial gap. Total length prior to transformation around 28 mm (1.1 in.). Develop mainly in temporary pools.

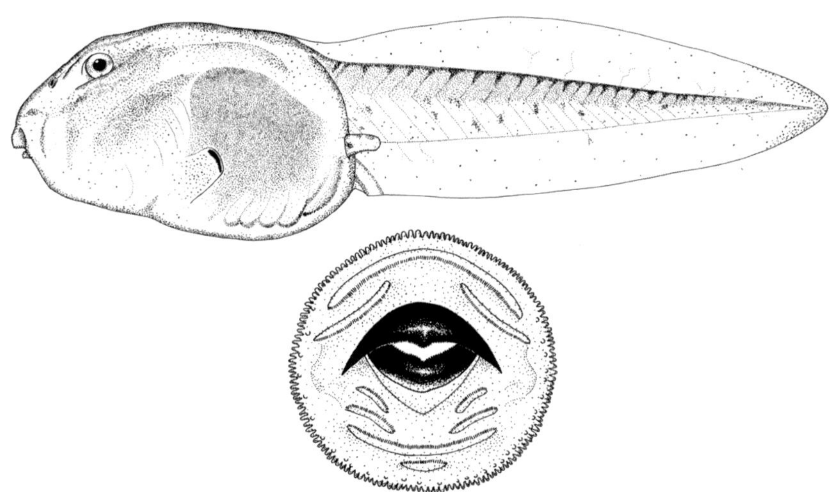

Plains Spadefoot (*Spea bombifrons*). Medium size, deep bodied. Body broadest just behind the eyes. Live tadpoles uniformly tan or brown. Dorsally positioned eyes; small and closely spaced nostrils. Clear and medium height tail fins. Tail musculature delineated with pigment. Visible intestinal coil at posterior area. Round oral disk; small, nonpigmented marginal papillae completely bordering disk. Under certain conditions, some tadpoles become cannibalistic and mouthparts change. Labial teeth row formula may range from 2/4 to 4/4. Upper jaw with cuspate; notched lower jaw. Total length 40 to 48 mm (1.6 to 1.9 in.). Develop mainly in temporary pools.

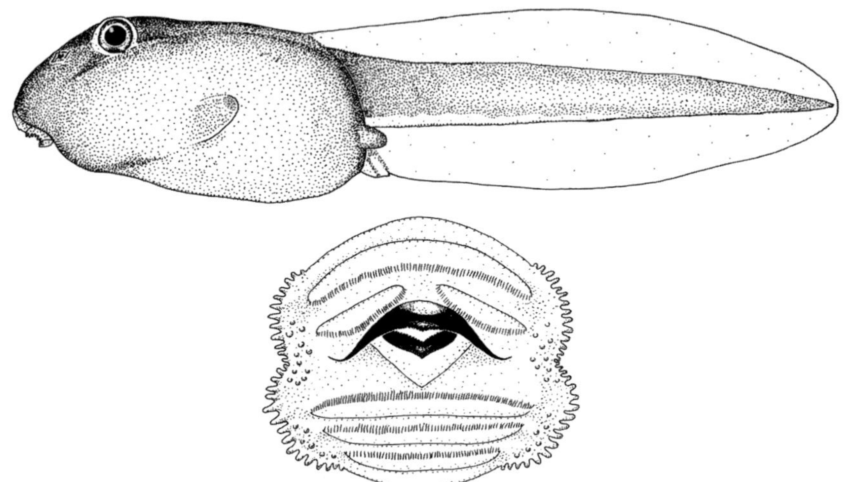

Eastern American Toad and **Dwarf American Toad** (*Anaxyrus americanus americanus* and *A. a. charlesmithi*). Very small, dark. Somewhat flattened body and sloping snout in lateral view. Small, dorsally positioned eyes; small nostrils. Low tail fin, without pigment and rounded at tip. Tail musculature without pigment along ventral edge. Oral disk with lateral emargination; without marginal papillae anteriorly and posteriorly. Serrated jaws. Upper jaw is slightly cuspate; notched lower jaw. Labial teeth row formula 2/3. Total length 18 to 24 mm (0.7 to 0.9 in.). Develop in temporary pools, shallow ponds, and lake edges.

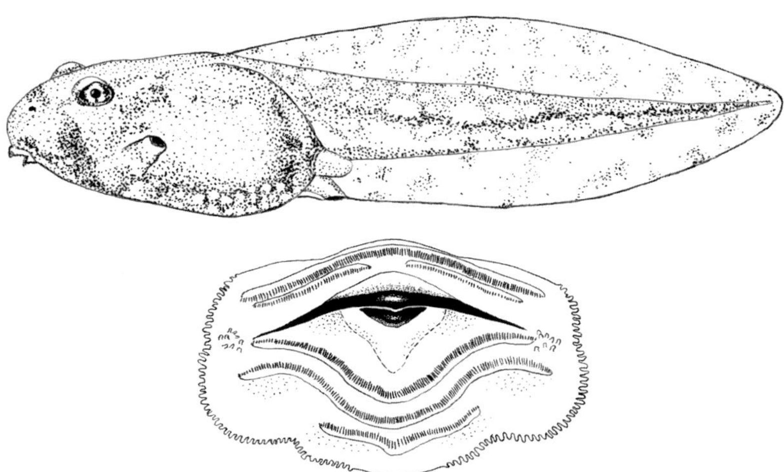

Great Plains Toad (*Anaxyrus cognatus*). Dark brown to brassy brown with a wide, dark bar between nostril and eye down to belly. Long and sloping snout as seen from the side. Brown body and dark-brown tail with numerous brassy iridophores. Rounded tip of tail fin. Labial teeth row formula 2/3. Second anterior labial teeth row with medial gap; third posterior labial teeth row less than half the width of second posterior labial teeth row. Small marginal papillae. Less than 10 submarginal papillae at either side of jaws; slight lateral emargination; wide anterior gap; medium posterior gap. Average total length 22 mm (0.9 in.). Develop in flooded ditches and shallow pools.

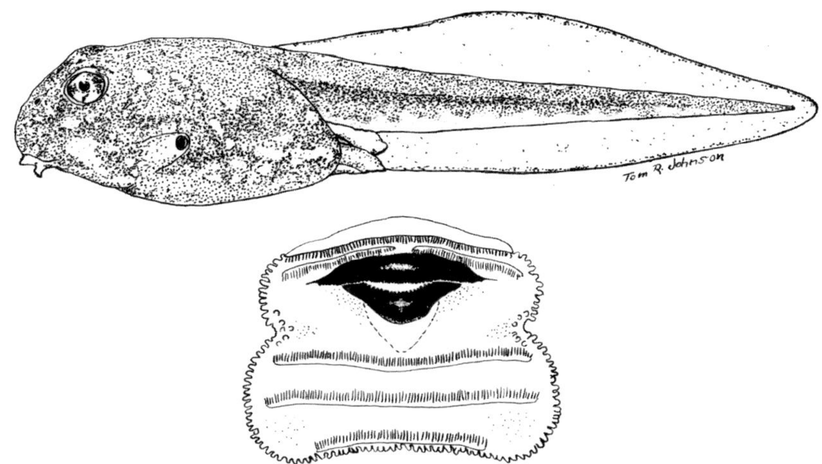

Fowler's Toad (*Anaxyrus fowleri*). Small size. Dark brown to black; older tadpoles may show a bronze brown on body and tail. Tail fin clear or with some pigment along dorsal edge; dorsal tail fin rounded at the middle; slightly rounded tip of tail fin. Tail with dark midline and light area ventrally. Small light area below eye; small light areas on lower sides. Labial teeth row formula 2/3. Oral disk with wide anterior gap; emarginated below lower jaw line; small marginal papillae. Few submarginal papillae. Jaws with very fine serrations. Small gap in second anterior labial teeth row. Average total length around 30 mm (1.2 in.). Develop in pools and shallow areas of streams and backwater areas.

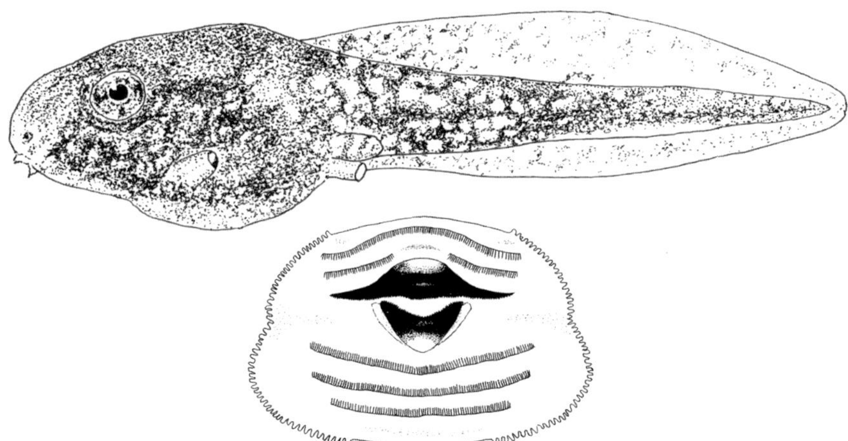

Rocky Mountain Toad (*Anaxyrus woodhousii woodhousii*). Small size. Dark-brown to black body and tail; abundant and scattered iridophores—some in clusters producing a frosted appearance on body and onto the tail musculature. Light-gray to white venter. Rounded head profile. Large eyes relative to size of tadpole. Medium height dorsal tail fins; low ventral tail fin. Tail fins darkened with stippling; rounded at tail tip. Oral disk with medium to wide anterior and posterior gap; slight lateral emargination; small marginal papillae. Few to no submarginal papillae. Labial teeth row formula 2/3. Second anterior labial teeth row with medial gap. Posterior teeth rows without gap; first posterior teeth row widest of the three. Finely serrated upper and lower jaws. Total length prior to transformation from 25 to 35 mm (1 to 1.4 in.).

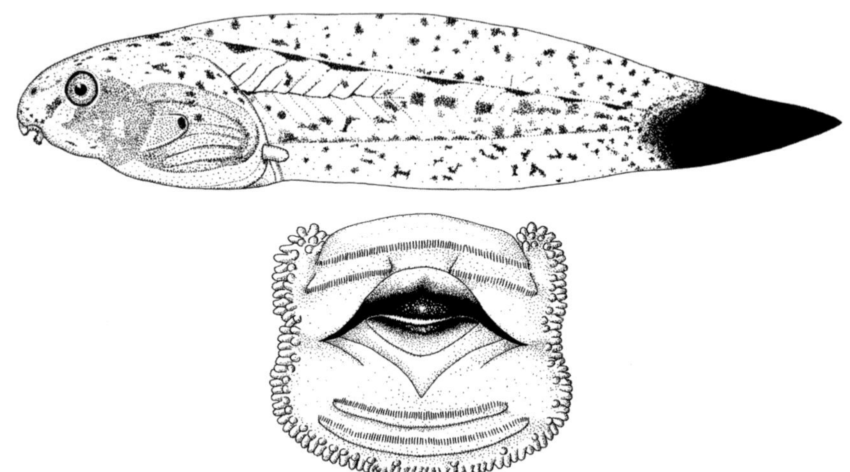

Blanchard's Cricket Frog (*Acris blanchardi*). Dorsally depressed body. Dorsally positioned eyes; large nostrils. Up to one half of spiracle tube free from body wall. Visible intestinal coil. Brown head and body. Similar height tail fins with faint mottling; banded dorsum of tail musculature. Usually some dark blotches on tail and tail fins; large black section on tail tip. Labial teeth row formula 2/2. Oral disk with slight lateral emargination; wide dorsal gap; second anterior lateral teeth row with gap, and no gap on second posterior labial teeth row. Medium-size marginal papillae. Average total length 36 mm (1.4 in.). Develop in a wide variety of aquatic habitats: marshes and swamps, stream backwaters and shallow areas, and farm ponds.

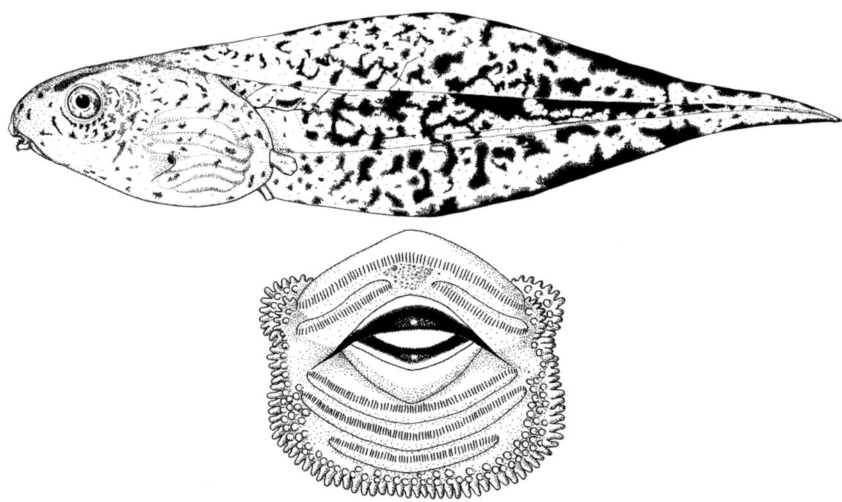

Cope's Gray Treefrog and **Gray Treefrog** (*Hyla chrysoscelis* and *H. versicolor*). Body-to-tail ratio of 1:2. Visible intestinal coil. High tail fins heavily mottled with black. Often orange or red tail fin in live specimens. Tail usually has a well-developed flagellum. Oral disk has slight lateral emargination. Labial teeth row formula 2/3; third posterior teeth row medium in length. Serrated jaws with narrow lower jaw. Total length 32 to 38 mm (1.3 to 1.5 in.). Usually develop in fishless ponds and pools in or near forested areas.

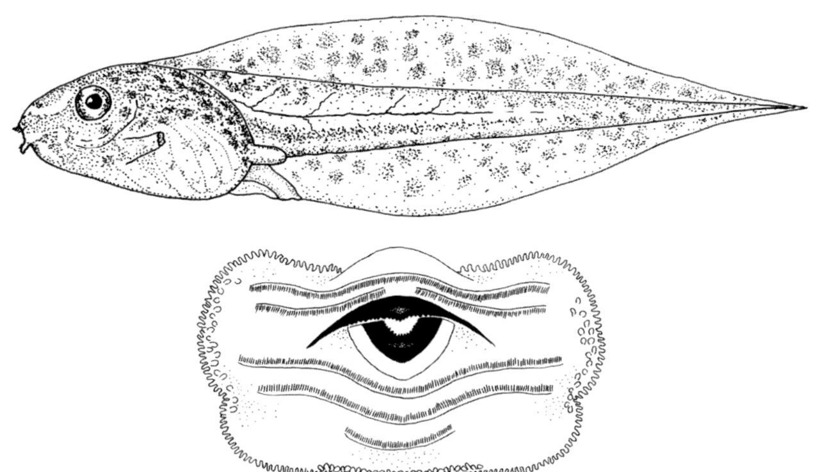

Green Treefrog (*Hyla cinerea*). Body-to-tail ratio of 1:2.1. Medium height tail fins. In living specimens, body, tail, and tail fins olive to pale green with small, brown blotches or stippling, and one or two light marks behind head. Pale-green to yellow line in front of eyes. Labial teeth row formula 2/3; second anterior labial teeth row with small medial gap. Oral disk has wide anterior gap; emarginated along posterior area below second posterior teeth row. Biserial (divided) posterior marginal papillae below third posterior teeth row. Throat with some stippling. Average total length 48 mm (1.9 in.). Develop in swamps and river backwaters.

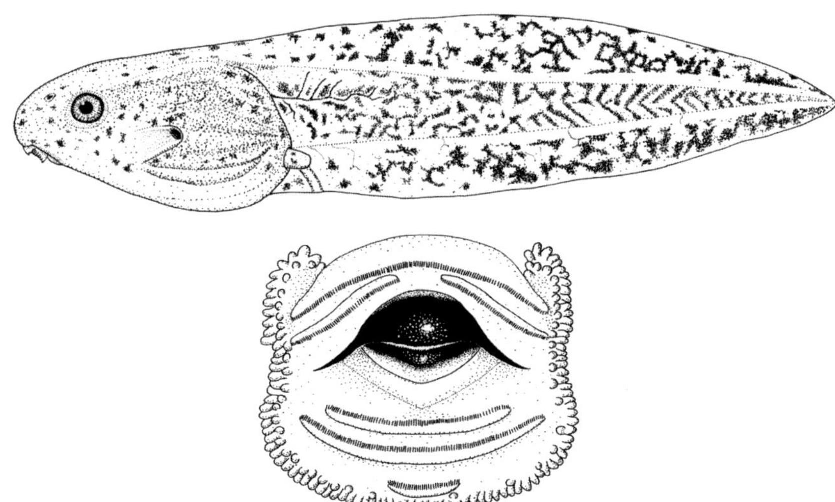

Spring Peeper (*Pseudacris crucifer*). Body-to-tail ratio of 1:2. Moderately visible intestinal coil. Oral disk has wide anterior gap; slight lateral emarginations; submarginal papillae close to marginal papillae. Large upper jaw. Labial teeth row formula 2/3; second anterior teeth row with medium gap; third posterior teeth row short. Body with scattered small, diffused dark markings. Mottled tail musculature; medium height tail fins. Dorsal fin slightly higher than ventral fin. Tail fins with large, irregular blotches and clear zone near musculature. Total length 27 to 35 mm (1.1 to 1.4 in.). Develop in fishless forest ponds and pools.

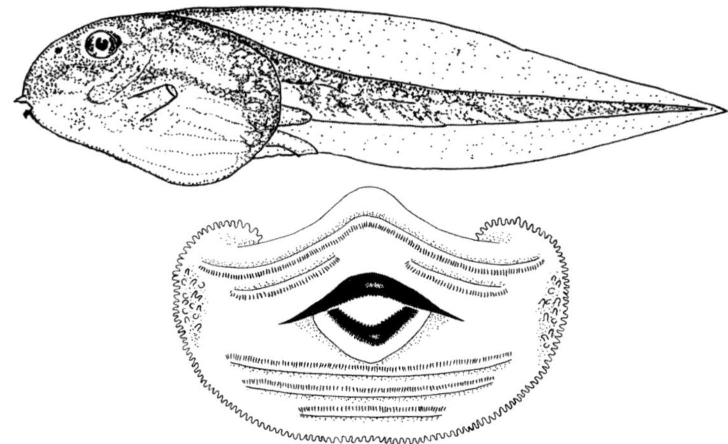

Boreal Chorus Frog, **Cajun Chorus Frog**, and **Upland Chorus Frog** (*Pseudacris maculata*, *P. fouquettei*, and *P. feriarum*). A lack of information on ways to distinguish these three species required them to be described together in this section. See their species accounts to separate by distribution. Body-to-tail ratio of 1:2. Usually visible intestinal coil. Tail musculature with some pigmentation or with only dorsal pigmentation. Dorsal tail fin higher than ventral fin; both with faint pigmentation. Scattered or extensive iridophores on head, mid, and lower section of body. Labial teeth rows formula 2/2 or 2/3. Wide oral disk; wide anterior gap; narrow upper jaw. Elongated first anterior labial teeth row; second anterior teeth row with medium medial gap; third posterior labial tooth row (if present) shorter than first and second rows. Average total length 35 mm (1.4 in.). Develop in temporary pools and fishless ponds.

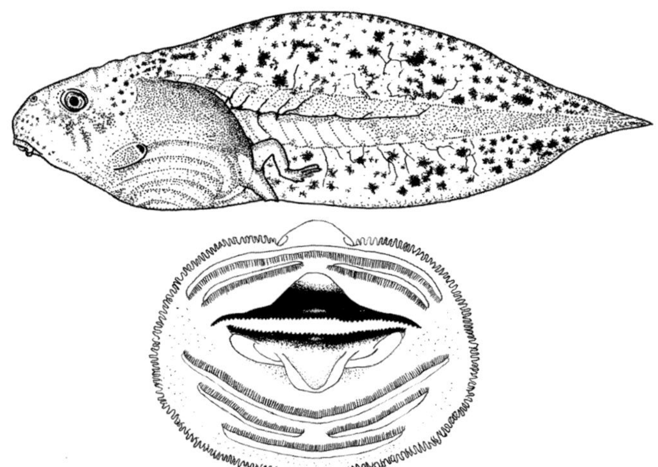

Illinois Chorus Frog (*Pseudacris illinoensis*). Medium size; head slopes in lateral view. Intestinal coil not visible in life; visible in preserved specimens. Live specimens may be tan or brown with a scattering of small, black dots on back of head and eyes with orange cast. High tail fins and a rounded body; high dorsal tail fin begins just above spiracle; tail fins covered with small, scattered, black irregular markings; tail fins terminate in a fine point. Slightly elongated oral disk with no emargination; labial teeth rows formula 2/3. Second anterior and second posterior labial teeth rows each with small gap. Wide upper jaws. Maximum total length 65 mm (2.6 in.). Develop in temporary pools and flooded ditches.

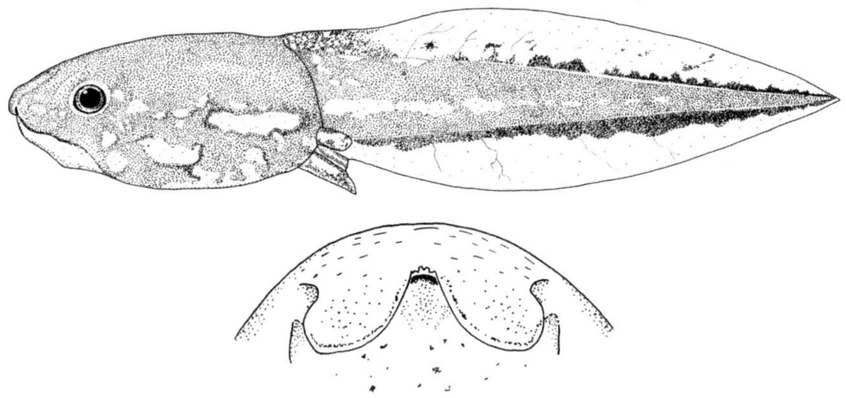

Eastern Narrow-mouthed Toad (*Gastrophryne carolinensis*). Small with noticeably flattened body. Laterally positioned eyes. Oral disk replaced by labial flaps with a median notch. Nostrils not present until late in development. A single spiracle present and located immediately ventral to anus. Intestinal coil not visible. Low tail fins pigmented along dorsal and ventral edges of tail musculature. Tail ends with fine point. Large, light blotches on sides and belly, extending onto tail. Total length from 25 to 30 mm (1 to 1.2 in.). Develop in temporary puddles and pools.

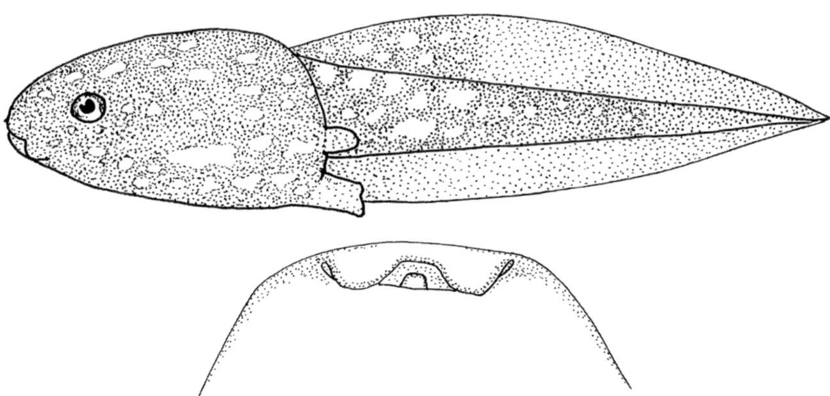

Western Narrow-mouthed Toad (*Gastrophryne olivacea*). Small with flattened appearing body. Eyes positioned on sides of head. Oral disk replaced by labial flaps with a median notch making lower jaw visible. One spiracle located ventrally alongside anus. Nostrils not present until near end of development. Tan to gray body and tail musculature with scattered, light blotches on head, body, tail, and dorsal tail fin. Little or no tail strip. Body and tail can be lightened by an abundance of iridophores. Total length from 30 to 35 mm (1.2 to 1.4 in.). Develop in puddles and temporary pools.

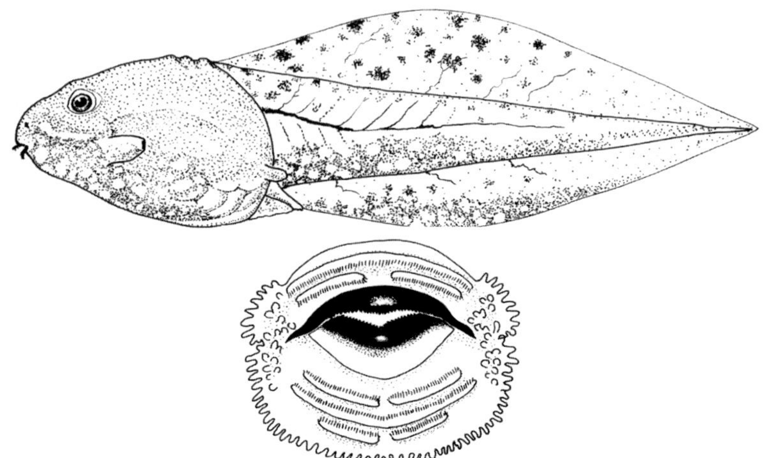

Northern Crawfish Frog (*Lithobates areolatus circulosus*). Medium size. Body-to-tail ratio of 1:1.9. Usually or partially visible intestinal coil. Irises can be orange in some populations. Usually with a narrow light line below each eye to indicate a future white upper lip line. Wide oral disk dorsal gap; emarginated at corners of mouth; enlarged lower jaw. Large submarginal papillae at the corners of the jaw may number up to 20 per side. Labial teeth row formula 2/3; second anterior labial teeth row with narrow medial gap; first and third posterior labial teeth rows with or without narrow medial gap. Tadpoles may be light gray to greenish gray or dark gray with numerous small, irregular markings on body, tail, and tail fins—all depending on the water conditions of their aquatic habitat. Lighter tadpoles would be from clear water, and darker, more pigmented ones from a cloudy pond or pool. If markings are present on body and tail, they are usually not well defined and appear diffused or "fuzzy." Average total length 70 mm (2.8 in.). Develop in fishless ponds and prairie pools.

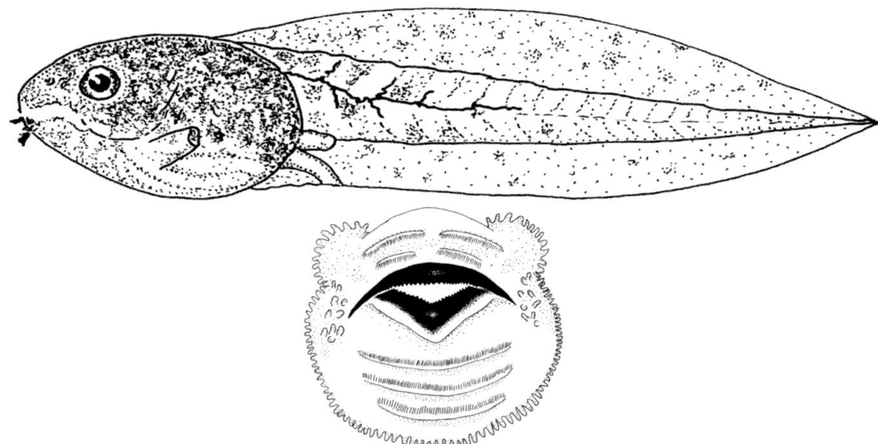

Plains Leopard Frog (*Lithobates blairi*). Moderate size. Live specimens tan to brownish gray of somewhat uniform appearance; eyes may have an orange iris. Usually a faint light line from mouth to below each eye. Body-to-tail ratio of 1:1.5. Oral disk with large anterior gap; lateral emargination; large posterior marginal papillae; few but large submarginal papillae at jaw edges. Medium-size lower jaw. Labial teeth row formula 2/3 but variable; both anterior labial teeth rows may have a narrow to medium gap. Posterior labial teeth rows without gap (but can be variable). Head and tail musculature with faint-to-moderate dark markings. Tail fins may be clear or covered with small-to-medium dark markings. Average total length 60 mm (2.4 in.). Develop in marshes, flooded ditches and farm ponds.

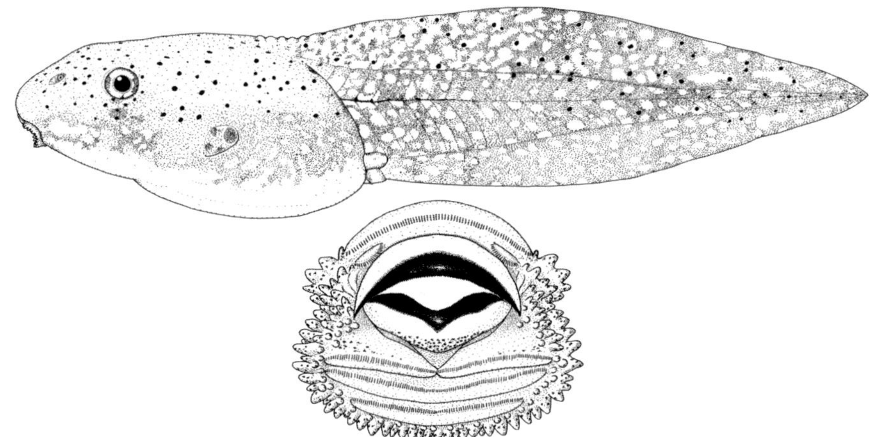

American Bullfrog (*Lithobates catesbeianus*). Large size. Live specimens with a wash of green or olive drab over body and tail, and some yellow or light orange on belly and base of tail. Body-to-tail ratio of 1:1.5. Some gray mottling on body, tail, and tail fins. Belly normally without pigmentation. Distinct black dots on upper body and dorsal tail fin, and scattered along the tail musculature. Intestinal coil not visible. Oral disk has large pigmented papillae and subpapillae; wide anterior notch; noticeably emarginated at jaw edges. Large and serrated upper and lower jaws. Labial teeth row formula 2/3 or 1/3; second anterior labial teeth row split into two short sections. Total length 78 to 121 mm (3.1 to 4.8 in.). Develop in any natural, permanent body of water. American bullfrog tadpoles in Missouri overwinter and transform the following summer.

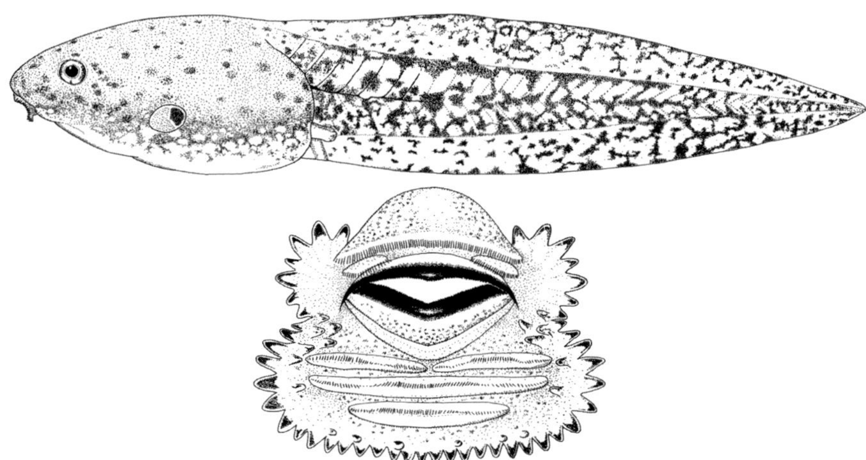

Green Frog (*Lithobates clamitans*). Medium to large size. Body-to-tail ratio of 1:1.8. Irregular dark-gray markings and diffused spots cover body, tail, and tail fins. Live specimens gray to gray brown with orange irises. No small, distinct, black dots on body and tail as found on the tadpoles of American bullfrogs (*Lithobates catesbeianus*). Oral disk with extensive lateral emarginations; wide anterior gap; large, somewhat flattened, and heavily pigmented marginal papillae. Labial teeth row formula 2/3, but somewhat variable. Second anterior labial teeth row with wide medial gap; first posterior labial teeth row with narrow medial gap. Upper jaw slightly cuspate. Total length 74 to 100 mm (2.9 to 3.9 in.). In Missouri, most green frog tadpoles overwinter in their pools, ponds, or swamp habitat, and transform the following summer.

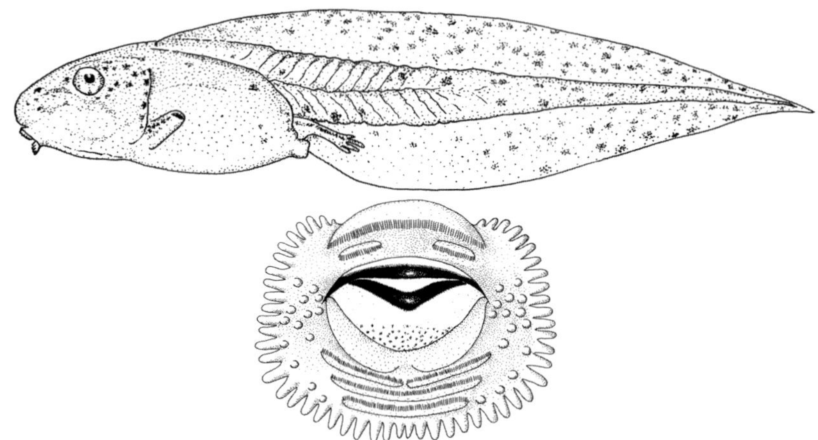

Pickerel Frog (*Lithobates palustris*). Medium size. Body-to-tail ratio of 1:1.7. Few scattered dark markings on body. Intestinal coil not visible or only slightly visible. Live tadpoles have dark-tan or brown body and tail musculature with a bronzy cast and a scattering of gold iridophores. Low tail fins; ventral fin somewhat deeper than height of dorsal fin. Tail moderately marked with dark mottling. Oral disk not emarginated; labial teeth row formula 2/3; wide dorsal gap above first anterior labial teeth row; second anterior teeth row small with wide medial gap. Large marginal papillae; 10 or fewer papillae below third posterior labial teeth row. Large submarginal papillae. Large lower jaw. Total length 48 to 55 mm (1.9 to 2.2 in.). Develop in woodland ponds, swamps, and old beaver ponds. Can easily be confused with wood frog tadpoles (*Lithobates sylvaticus*).

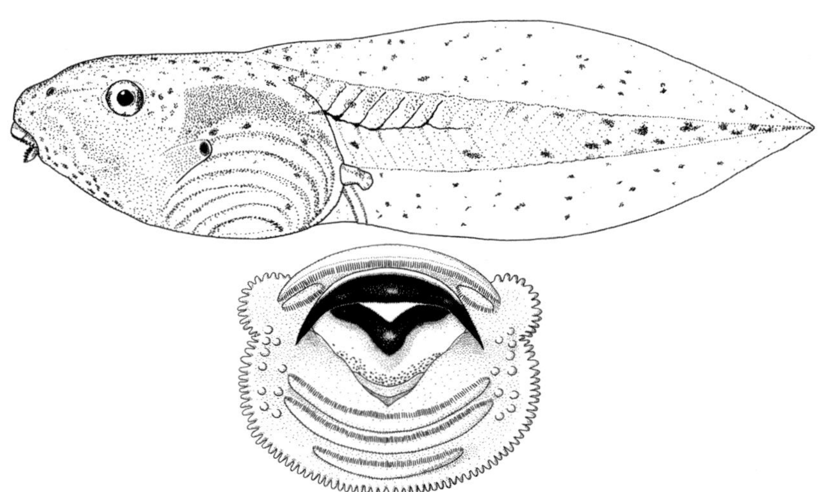

Northern Leopard Frog and **Southern Leopard Frog** (*Lithobates pipiens* and *L. sphenocephalus*). There are too few differences in the characteristics of these two tadpoles to determine which you may have. In addition, there are variations throughout their extensive North American range, which complicates identification. Knowing where a tadpole was collected will greatly help determine the species. Medium size. Body-to-tail ratio of 1:1.5. Moderately pigmented body and tail musculature; could be extensively covered with dark markings. Tail fins clear or with a scattering of small, dark markings. Throat with a scattering of small, dark markings. Visible intestinal coil. Labial teeth row formula 2/3 (but highly variable). Second anterior labial teeth row with wide gape, reducing the size of each section. Third posterior labial teeth row reduced in length. Oral disc has a wide anterior gap; laterally emarginated; large sublabial papillae, few in number. Large lower jaw. Light line below each eye. Total length 45 to 54 mm (1.8 to 2.1 in.). Develop in a variety of aquatic habitats: flooded ditches, farm ponds, marshes, swamps, river backwaters, and the shallow sections of reservoirs.

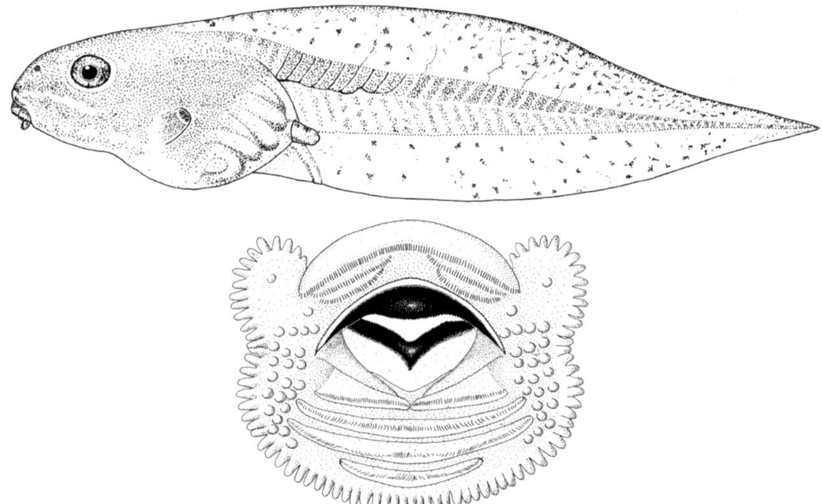

Wood Frog (*Lithobates sylvaticus*). Medium size. Body-to-tail ratio of 1:1.8. No distinct markings on body. Partially visible intestinal coil, hidden by subdermal pigmentation. Live specimens dark, brassy brown, with scattered, numerous tiny golden flecks (iridophores). Dorsally rounded tail fin, with dorsal and ventral fins of similar height and tapering to a fine point. Faint, small stippling on tail fins. A pale upper lip line usually present between nostril and eye. Oral disk has a wide dorsal gap and is laterally emarginated; large submarginal papillae. Labial teeth row formula 3/4. Anterior third labial teeth row with medium gap; no gap on any posterior labial teeth rows. Total length 42 to 48 mm (1.7 to 1.9 in.). Similar in size and appearance to pickerel frog tadpoles (*Lithobates palustris*). Develop in fishless, woodland ponds and temporary pools.

Glossary

Adpressed: To position forelimbs backward and hind limbs forward along the side of a salamander or its larvae and either count the number of costal grooves or measure the distance between digits pointing toward each other. Helpful technique for identification of long, slender salamanders.

Amplexus: Sexual embrace of amphibians in which the male grasps the female's body from above with his forelimbs. This allows the eggs to be fertilized externally by the male as they are being laid.

Anal plate: The last belly scale of a snake, which covers the anal opening. This scale may be single or divided in two.

Anura: An order of the class Amphibia comprising toads and frogs. All have large hind legs for jumping and swimming. A synonym of Salientia.

Arboreal: Related to trees, as in the behavior of some amphibians and reptiles of living in trees.

Barbels: Fleshy, tubular projections of skin on the chin and throat of some turtles.

Biserial: Divided structure; with two halves; as in tadpole marginal papillae in two, connected rows.

Boss: A swollen, rounded area on the middle of the snout. Also called a cranial boss.

Carapace: A turtle's upper shell.

Carnivorous: Flesh eating; consuming and digesting other animals.

Cirri: Downward projections from the nostrils on males of certain lungless salamanders. The naso-labial groove extends downward to the tip of each cirrus.

Clinal: A gradual change of one or more characteristics of an animal from one part of its geographic range to another. This may be observed, for example, from north to south or from a higher to a lower elevation.

Cloaca: The common chamber through which the urinary, digestive, and reproductive canals discharge their contents through the vent or anus.

Constriction: The method of killing a prey animal whereby a nonvenomous snake coils its body around the prey and squeezes.

Costal grooves: Vertical grooves on most salamanders located on the sides of the body between the fore- and hind limbs, which correspond to the segmental and abdominal muscles.

Cranial crests: Raised ridges on the head located between or behind the eyes on some toads.

Cuspate: Possessing an enlarged point at the middle of the upper jaw (in tadpoles or turtles).

Diploid: Having two sets of chromosomes per animal cell.

Diurnal: Active during the day.

Dorsal: The upper surface or back of an animal.

Dorsolateral fold: A line of raised glandular skin along an area between the back and sides on some frogs.

Ectothermic: Not producing body heat; gaining heat from surroundings or by basking, as do amphibians and reptiles.

Dorsum: The entire back of an animal.

Eft: The terrestrial immature stage of the life cycle of a newt, between the aquatic larva and the aquatic adult.

Emarginate: Having a notched or indented edge, as in the indentation on either side of the oral disk of some tadpoles.

Endothermic: Producing body heat through metabolizing food, as do birds and mammals.

Facial pit: The double-chambered, deep cavity located between each nostril and eye on the head of pit vipers. The chambers are separated by a thin membrane. Used to detect warm-blooded prey.

Flagellum: An extension of the tail of some tadpoles capable of moving independently of the rest of the tail.

Fossorial: Adapted for living underground.

Frontal scale: A single, large scale between the eyes and on top of the head of most lizards and snakes.

Glochidia: Larvae of certain freshwater mussels that attach mainly to gills of fish and some aquatic amphibian hosts as part of their life cycle.

Granule: Tiny granular scale on, for example, members of the Teiidae lizards.

Hemipenis: One of a pair of copulatory organs of male lizards and snakes housed in the base of the tail.

Herbivorous: Consuming and digesting plants.

Herpetology: The branch of science dealing with the study of amphibians and reptiles. From the Greek *herpeto-*, "creeping thing," and *-logos*, "to study."

Hind limb bud: A projection at the base of the tail of a tadpole indicating early development of hind legs.

Hinge: A joint between the anterior and posterior halves of the plastron of some turtles allowing the closure of the plastron against the carapace.

Hybridization: When two or more species interbreed in regions where their populations overlap.

Intergradation: When two or more subspecies interbreed in regions where their populations overlap.

Internasal scales: Scales between the nostrils on lizards and snakes.

Iridophore: Small, bright white, silvery, brassy, or golden flecks in the skin. Can be found on amphibian larvae and adults.

Keeled scales: Body scales along the sides and back of some snakes and lizards having small, raised ridges.

Labial scales: A row of scales that border the upper and lower lips on lizards and snakes.

Labial teeth: Tiny, horny projections on the lips of tadpoles that are arranged in rows like the teeth of a comb.

Larva: An early stage of many animals that undergo metamorphosis, differing markedly in form and appearance from the adult, as in the aquatic, gilled young of a salamander (plural: **larvae**).

Lateral: Referring to the side of an animal.

Lentic: Referring to aquatic habitats with no current (standing water); lakes, ponds, oxbows and river backwaters.

Liebesspiel: A German term meaning "love play" used to describe the aquatic courtship of salamanders, such as the Ambystomatidae (mole salamanders).

Loreal scale: One or two small scales on snakes or lizards located between the nasal and preocular scales.

Lotic: Referring to aquatic habitats with a detectable current (running water); rivers, streams, creeks, and springs.

Melanism: A color condition of an animal where black pigment covers most of the body.

Mental gland: A light-colored swelling below the chin on males of several species of plethodontid or lungless salamanders (usually more apparent during the breeding season).

Metamorphosis: A morphological change from a larva to an adult with the loss of larval characteristics and the acquisition of adult characteristics.

Mid-dorsal: The area in the center of the back.

Morph: The structure, shape, form, or color of an animal species that may be used by a biologist for identification.

Mullerian duct: Embryonic oviducts. Also, the vestigial oviducts present along the kidneys of some male true frogs (family Ranidae) such as the northern leopard frog.

Naso-labial groove: A groove extending downward from the nostril and across the lip in lungless salamanders.

Neotenic: Sexually mature and able to reproduce, but retaining many larval characteristics.

Nocturnal: Active at night.

Nuchal: A small, centrally located scute on the anterior edge of the carapace of some turtles, directly behind the neck.

Ocelli: Markings on an animal that resemble eyes; round markings of concentric rings.

Oral disk: Fleshy parts of a tadpole's mouth, including anterior and posterior labia and papillae.

Osteoderm: A bony plate or scale in the dermal layer of skin in some reptiles.

Oviparous: In reference to animals that lay eggs.

Ovoviviparous: In reference to animals that produce eggs that are retained in the oviduct of the female throughout the development of the young. The young are born "live," or completely developed.

Papillae: Small fleshy projections along the labial edges of a tadpole's mouth.

Parotoid glands: The large, wartlike glands located behind each eye on toads.

Plastron: A turtle's lower shell.

Pleural scute: Any of the scutes covering the ribs on a turtle's carapace.

Pond-type: Salamander larvae with characteristics of species that breed in ponds: large or wide head; large, external gills; high caudal fin.

Postmental scale: One or two scales behind the mental scale on the chin of some lizards (i.e., skinks).

Prefrontal scale: A scale or set of scales on the dorsum, just before the frontal scale and near each eye on the head of lizards and snakes.

Process: An extension of a common structure or organ, e.g., in the alligator snapping turtle, the extension of the tongue that resembles a red worm.

Resorb: The loss and assimilation of tissue (e.g., tail and gills) during metamorphosis from a larval stage in amphibians (e.g., tadpole to a toadlet or froglet).

Rostral scale: The scale at the tip of the snout on lizards or snakes.

Scutes: The large scales that cover the upper and lower shells of turtles. These scales are lacking in softshells (genus *Apalone*).

Serrate: Having projections in the shape of the teeth of a saw. Jaws of some tadpoles may be serrated.

Smooth scales: Body scales along the sides and back of some snakes that are devoid of ridges or keels and are smooth to the touch.

Snout-to-vent length: The measurement of an animal from the tip of the nose to the cloaca or vent. Usual method of measuring toads and frogs.

Spermatophore: A gelatinous cone with a cap consisting of sperm secreted by male salamanders. The sperm cap is picked up by a female with her cloacal lips after courtship.

Spiracle: A tubelike opening on the left side of most tadpoles that allows water to pass out of the gill chambers during aquatic respiration.

Stream-type: Salamander larvae with characteristics of species that breed in cool, rocky streams: small or narrow head; small, external gills; low caudal fin.

Subcaudal: A series of scales along the ventral surface of the tail of lizards and snakes.

Subspecies: A subgrouping within a species having a unique set of characteristics, usually based on geographical differences.

Supranasal scale: One or two scales just above the nasal opening on the head of lizards and snakes.

Supraocular scale: A large scale located immediately above each eye on the head of lizards and snakes.

Symbiotic: The living together of two unrelated organisms where one or both may gain some benefit (e.g., the green alga found in Ambystomatidae eggs).

Sympatric: Referring to two or more species occupying the same geographic area.

Tadpole: The aquatic larva of toads and frogs. Gills are present but hidden behind a flap of skin. The stage of life between the egg and the tailless young adult.

Tetraploid: Having four haploid sets of chromosomes per cell.

Troglobitic: An animal restricted to living in caves.

Tympanum: The round, flat external eardrum of most toads and frogs.

Type locality: Collection site of the original specimen used for the first published description of a species.

Vent: External opening to the cloaca; also called the anus.

Venter: The lower surface or belly of an animal.

Vermiculation: Arrangement of dark pigment in a wormlike pattern.

Literature Cited

ALDRIDGE, R.D. and A.P. BUFALINO.
2003. Reproductive female common watersnakes (*Nerodia sipedon sipedon*) are not anorexic in the wild. Journal of Herpetology 37(2):416–419.

ALDRIDGE, R.D., A.P. BUFALINO, and A. REEVES.
2005. Pheromone communication in the watersnake, *Nerodia sipedon*: A mechanistic difference between semi-aquatic and terrestrial species. The American Midland Naturalist 154:412–422.

ALDRIDGE, R.D. and D. DUVALL.
2002. Evolution of the mating season in the pitvipers of North America. Herpetological Monographs 16:1–25.

ALDRIDGE, R.D., W.P. FLANAGAN, and J.T. SWARTHOUT.
1995. Reproductive biology of the water snake *Nerodia rhombifer* from Veracruz, Mexico, with comparisons of tropical and temperate snakes. Herpetologica 51:182–192.

ALDRIDGE, R.D., J.J. GREENHAW, and M.V. PLUMMER.
1990. The male reproductive cycle of the rough green snake (*Opheodrys aestivus*). Amphibia-Reptilia 11:165–172.

ALDRIDGE, R.D., B.C. JELLEN, M.C. ALLENDER, M.J. DRESLIK, D.B. SHEPARD, J.M. COX, and C.A. PHILLIPS.
2008. Reproductive biology of the massasauga (*Sistrurus catenatus*) from south-central Illinois. Pp 403–412. *In:* W.K. Hayes, K.R. Beaman, M.D. Cardwell, and S.P. Bush (eds). The Biology of Rattlesnakes. Loma Linda University Press. California.

ALDRIDGE, R.D. and D.E. METTER.
1973. The reproductive cycle of the western worm snake, *Carphophis vermis*, in Missouri. Copeia 1973(3):472–477.

ALEXANDER, D.G.
1966. An ecological study of the swamp cricket frog, *Pseudacris nigrita feriarum* (Baird), with comparative notes on two other hylids of the Chapel Hill, North Carolina, region. PhD dissertation. University of North Carolina, Chapel Hill.

ALEXANDER, M.M.
1943. Food habits of the snapping turtle in Connecticut. Journal of Wildlife Management 7:278–282.

ALEXANDER, W.P.
1927. The Allegheny hellbender and its habitat. Buffalo Society of Natural Science 7:13–18.

ALLEN, E.R. and W.T. NEILL.
1950. The alligator snapping turtle, *Macrochelys temminckii*, in Florida. Ross Allen's Reptile Institution Special Publication 4. 15 pp.

ALTIG, R.
1967. Food of *Siren intermedia nettingi* in a spring-fed swamp in Illinois. The American Midland Naturalist 77:239–241.
1970. A key to the tadpoles of the continental United States and Canada. Herpetologica 26(2):180–207.

ALTIG, R. and P.H. IRELAND.
1984. A key to salamander larvae and larviform adults of the United States and Canada. Herpetologica 40(2):212–218.

ALTIG, R. and R.W. MCDIARMID.
2015. Handbook of Larval Amphibians. Cornell University Press. Ithaca, New York. xvi + 345.

ALTNETHER, S.
2003. Subterranean oviposition in the worm snake, *Carphophis vermis*. Missouri Herpetological Association Newsletter 16:16.

ANDERSON, C.D.
2010. Effects of movement and mating patterns on gene flow among overwintering hibernacula of the timber rattlesnake (*Crotalus horridus*). Copeia 2010(1):54–61.

ANDERSON, C.D. and W.J. DRDA.
2005. *Crotalus horridus* (timber rattlesnake). Behavior. Herpetological Review 36(4):456–457.

ANDERSON, P.
1942. Amphibians and Reptiles of Jackson County, Missouri. Bulletin of the Chicago Academy of Sciences 6:203–220.
1957. A second list of new herpetological records for Missouri. Natural History Miscellanea 161:1–5.
1965. The Reptiles of Missouri. University of Missouri Press. Columbia. xxiii + 330 pp.

ANDERSON, P.K.
1954. Studies in the ecology of the narrow-mouthed toad, *Microhyla carolinensis carolinensis*. Tulane Studies in Zoology 2:15–46.

ANGERT, A.L., D. HUTCHISON, D. GLOSSIP, and J.B. LOSOS.
2002. Microhabitat use and thermal biology of the collared lizard (*Crotaphytus collaris collaris*) and the fence lizard (*Sceloporus undulates hyacinthinus*) in Missouri glades. Journal of Herpetology 36(1):23–29.

ANTON, T.G., D. MAUGER, C.A. PHILLIPS, M.J. DREZLIK, J.E. PETZING, A.R. KUHNS, and J.M. MUI.
2003. *Clonophis kirtlandii* (Kirtland's snake). Aggregating behavior and site fidelity. Herpetological Review 34(3):248–249.

ARESCO, M.J.
2004. Reproductive ecology of *Pseudemys floridana* and *Trachemys scripta* (Testudines: Emydidae) in northwestern Florida. Journal of Herpetology 38:249–256.

ARNOLD, E.N. and D.W. OVENDEN.
2002. A Field Guide to the Reptiles and Amphibians of Britain and Europe. Harper Collins, London, United Kingdom. 272 pp.

ARNOLD, S.J.
1976. Sexual behavior, sexual interference and sexual defense in the salamanders *Ambystoma maculatum*, *Ambystoma tigrinum*, and *Plethodon jordani*. Zeitschrift fur Tierpsychologie 42:247–300.

AUSTIN, J.D. and K.R. ZAMUDIO.
2008. Incongruence in the pattern and timing of intra-specific diversification in bronze frogs and bullfrogs (Ranidae). Molecular Phylogenetics and Evolution 48(2008):1041–1053.

AVERILL-MURRAY, R.C.
2006. Natural history of the western hog-nosed snake (*Heterodon nasicus*) with notes on envenomation. Sonoran Herpetologist 19(9):98–101.

AXTELL, R.W. and N. HASKELL.
1977. An interhiatal population of *Pseudacris streckeri* from Illinois, with an assessment of its postglacial dispersion history. Natural History Miscellanea 202:1–8.

AYRES, D.E.
1973. Field behavior of the six-lined racerunner in Peoria County. Proceedings of the Peoria Academy of Science 6:23–30.

BAEYENS, D.A., C.T. MCALLISTER, and L.F. MORGANS.
1978. Some physiological and morphological adaptations for underwater survival in *Natrix rhombifera* and *Elaphe obsoleta*. Journal of the Arkansas Academy of Science 32:18–21.

BAIRD, T.A.
2004. Reproductive coloration in female collared lizards, *Crotaphytus collaris*, stimulates courtship by males. Herpetologica 60(3):337–348.

BAIRD, T.A., M.A. ACREE, and C.L. SLOAN.
1996. Age and gender-related differences in the social behavior and mating success of free-living collared lizards, *Crotaphytus collaris*. Copeia 1996(2):336–347.

BAKER, C.L., L.C. BAKER, and M.F. CALDWELL.
1947. Observation of copulation in *Amphiuma tridactylum*. Journal of the Tennessee Academy of Science 22:87–88.

BANNING, J.L., A.L. WEDDLE, G.W. WAHL III, M.A. SIMON, A. LAUER, R.L. WALTERS, and R.N. HARRIS.
2008. Antifungal skin bacteria, embryonic survival, and communal nesting in four-toed salamanders, *Hemidactylium scutatum*. Oecologia 156(2):423–429.

BARBEAU, T.R. and H.B. LILLYWHITE.
2005. Body wiping behaviors associated with cutaneous lipids in hylid tree frogs of Florida. Journal of Experimental Biology 208:2147–2156.

BARBOUR, R.W.
1971. Amphibians and Reptiles of Kentucky. University of Kentucky Press. Lexington. 334 pp.

BARDEN, R.B. and L.J. KEZER.
1944. The eggs of certain plethodontid salamanders obtained by pituitary gland implantation. Copeia 1944:115–118.

BARKO, V.A. and J.T. BRIGGLER.
2006. Midland smooth softshell (*Apalone mutica*) and spiny softshell (*Apalone spinifera*) turtles in the middle Mississippi River: Habitat associations, population structure, and implications for conservation. Chelonian Conservation and Biology 5:225–231.

BARKO, V.A., J.T. BRIGGLER, and D.E. OSTENDORF.
2004. Passive fishing techniques: A cause of turtle mortality in the Mississippi River. Journal of Wildlife Management 68:1145–1150.

BARTELT, P.E., M. MCDOWELL, and C. BRIDGES.
2001. Continued study of the movements and habitat use of leopard frogs (*Rana pipiens*) on the Union Slough National Wildlife Refuge. Final Report. U.S. Fish & Wildlife Service. 11 pp.

BARTLETT, R.D. and A. TENNANT.
2000. Snakes of North America. Western Region. Gulf Publishing Company, Houston, Texas.

BASTIEN, H. and R. LECLAIR, JR.
1992. Aging wood frogs (*Rana sylvatica*) by skeletochronology. Journal of Herpetology 26(2):222–225.

BATESON, Z.W., J.D KRENZ, and R.E. SORENSEN.
2011. Multiple paternity in the common five-lined skink (*Plestiodon fasciatus*). Journal of Herpetology 45(4):504–510.

BAUMAN, M.A. and D.E. METTER.
1977. Reproductive cycle of the northern watersnake, *Natrix s. sipedon* (Reptilia, Serpentes, Colubridae). Journal of Herpetology 11(1):51–59.

BAUR, B.E.
1986. Longevity of horned lizards of the genus *Phrynosoma*. Bulletin of the Maryland Herpetological Society 22:149–151.

BEANE, J.C. and P.R. TRAIL.
1991. *Scincella lateralis* (ground skink). Predation. Herpetological Review 22(3):99.

BEASLEY, B.J.
2010. Analysis of *Eurycea* hybrid zone in eastern Missouri. MS thesis. Missouri University of Science and Technology, Rolla. 63 pp.

BEASLEY, V.R., S.A. FAEH, B. WIKOFF, C. STAEHLE, J. EISOLD, D. NICHOLS, R. COLE, A.M. SCHOTTHOEFER, M. GREENWELL, and L.E. BROWN.
2005. Risk factors and declines in northern cricket frog (*Acris crepitans*). Pp. 75–86. *In:* M.J. Lannoo (ed.). Amphibian Declines: The Conservation Status of United States Species. University of California Press. Berkeley. 1094 pp.

BEATTY, J.J.
1973. A comparison of the life history of *Notophthalmus viridescens* Rafinesque in a temporary and permanent pond. MA thesis. University of Missouri, Columbia. 89 pp.

BEAUPRE, S.J. and K.G. ROBERTS.
2001. *Agkistrodon contortrix contortrix* (southern copperhead). Chemotaxis, arboreality, and diet. Herpetological Review 32:44–45.

BEE, M.A. and H.C. GERHARDT.
2002. Individual voice recognition in a territorial frog (*Rana catesbeiana*). Proceedings of the Royal Society, London 269:1443–1448.

BERGER-BISHOP, L.E. and R.N. HARRIS.
1996. A study of caudal allometry in the salamander *Hemidactylium scutatum*. Herpetologica 52(4):515–525.

BERINGER, J. and T.R. JOHNSON.
1995. *Rana catesbeiana* (bullfrog). Diet. Herpetological Review 26(2):98.

BERNSTEIN N.P. and R.J. RICHTSMEIER.
2007. Home range and philopatry in the ornate box turtle, *Terrapene ornata ornata*, in Iowa. The American Midland Naturalist 157:162–174.

BEST, T.L. and G.S. PFAFFENBERGER.
1987. Age and sexual variation in the diet of collared lizards (*Crotaphytus collaris*). The Southwestern Naturalist 32(4):415–426.

BETZ, T.W.
1963. The gross ovarian morphology of the diamond-

backed water snake, *Natrix rhombifera*, during the reproductive cycle. Copeia 1963:692–697.
Birchfield, G.L.
2002. Green frog (*Rana clamitans*) movement behavior and terrestrial habitat use in fragmented landscapes in central Missouri. PhD dissertation. University of Missouri, Columbia. 136 pp.
Bishop, S.C.
1941. The salamanders of New York. New York State Museum Bulletin 324:1–365.
1943. Handbook of Salamanders: The Salamanders of the United States, of Canada, and of lower California. Comstock Publishing Company, Ithaca, New York. 555 pp.
Black, J.H.
1970. Amphibians of Montana. Montana Fish, Wildlife, and Parks Department, Animals of Montana Series No. 1. 32 pp.
Blair, W.F.
1936. A note on the ecology of *Microhyla olivacea*. Copeia 1936(2):115.
1976. Some aspects of the biology of the ornate box turtle, *Terrapene ornata*. The Southwestern Naturalist 21(1):89–104.
Blanchard, F.N.
1923. A new North American snake of the genus *Natrix*. Michigan University Occasional Papers of the Museum of Zoology 140:1–7.
1924. A collection of amphibians and reptiles from southeastern Missouri and southern Illinois. Papers of the Michigan Academy of Science. 4:533–541.
1937. Data on the natural history of the red-bellied snake, *Storeria occipito-maculata* (Storer), in northern Michigan. Copeia 1937:151–162.
1942. The ring-necked snakes, genus *Diadophis*. Bulletin of the Chicago Academy of Sciences 7:1–144.
Blomquist, S.M. and M.L. Hunter, Jr.
2009. A multi-scale assessment of habitat selection and movement patterns by northern leopard frogs (*Lithobates* [*Rana*] *pipiens*) in a managed forest. Herpetological Conservation and Biology 4:142–160.
Bodie, J.R.
2001. Stream and riparian management for freshwater turtles. Journal of Environmental Management 62:443–455.
Bodie, J.R. and R.D. Semlitsch.
2000a. Spatial and temporal use of floodplain habitats by lentic and lotic species of aquatic turtles. Oecologia 122:138–146.
2000b. Size-specific mortality and natural selection in freshwater turtles. Copeia 2000:732–739.
Bodie, J.R, R.D. Semlitsch, and R.B. Renken.
2000. Diversity and structure of turtle assemblages: Associations with wetland characters across the floodplain landscape. Ecography 23:444–456.
Bonett, R.M. and P.T. Chippindale.
2004. Speciation, phylogeography and evolution of life history and morphology in plethodontid salamanders of the *Eurycea multiplicata* complex. Molecular Ecology 13:1189–1203.
Bouchard, S.S. and K.A. Bjorndal.
2006. Ontogenetic diet shifts and digestive constraints in the omnivorous freshwater turtle *Trachemys scripta*. Physiological and Biochemical Zoology 79(1):150–158.
Bounty, J. and J.L. Carr.
2017. Amphibians & Reptiles of Louisiana: An Identification and Reference Guide. Louisiana State University Press. Baton Rouge. 386 pp.
Bour, R.
1987. Type-specimen of the alligator snapper, *Macroclemys temminckii* (Harlan 1835). Journal of Herpetology 21(4):340–343.
Bowen, K.D., P.L. Colbert, and F.J. Janzen.
2004. Survival and recruitment in a human-impacted population of ornate box turtles, *Terrapene ornata*, with recommendations for conservation and management. Journal of Herpetology 38:562–568.
Bowers, J.H.
1967. A record litter of *Thamnophis sirtalis proximus* (Say). The Southwestern Naturalist 2:200.
Boyer, D.A. and A.A. Heinze.
1934. An annotated list of the amphibians and reptiles of Jefferson County, Missouri. Transactions of the St. Louis Academy of Science 28:185–200.
Bradford, J.
1973. Reproduction and ecology of two species of earth snakes: *Virginia striatula* and *Virginia valeriae*. PhD dissertation. University of Missouri, Columbia. 106 pp.
Bragg, A.N. and C.C. Smith.
1943. Observations on the ecology and natural history of Anura IV. The ecological distribution of toads in Oklahoma. Ecology 24:285–309.
Brandley, M.C., T.J. Guiher, R.A. Pyron, C.T. Winne, and F.T. Burbrink.
2010. Does dispersal across an aquatic geographic barrier obscure phylogeographic structure in the diamond-backed watersnake (*Nerodia rhombifer*)? Molecular Phylogenetics and Evolution 57(2):552–560.
Brandon, R.A.
1965. *Typhlotriton*, *T. nereus*, and *T. spelaeus*. Catalogue of American Amphibians and Reptiles, p. 20.
1966. A reevaluation of the status of the salamander, *Typhlotriton nereus* Bishop. Copeia 1966:555–561.
1970. *Typhlotriton* and *Typhlotriton spelaeus*. Catalogue of American Amphibians and Reptiles, p. 84.
1971. Correlation of seasonal abundance with feeding and reproductive activity in the grotto salamander (*Typhlotriton spelaeus*). The American Midland Naturalist 86(1):93–100.
Brandon, R.A. and S.R. Ballard.
1998. Status of Illinois chorus frogs in southern Illinois. Pp. 102–112. *In*: M.J. Lannoo (ed.). Status and Conservation of Midwestern Amphibians. University of Iowa Press. Iowa City.
Brandon, R.A. and D.J. Bremer.
1966. Neotenic newts, *Notophthalmus viridescens louisianensis*, in southern Illinois. Herpetologica 22(3):213–217.
Brandon, R.A. and D.J. Bremer.
1967. Overwintering of larval tiger salamanders in southern Illinois. Herpetologica 23(1):67–68.

Brauman, R.J. and R.A. Florillo.
1995. *Lampropeltus getula holbrooki* (speckled kingsnake). Oophagy. Herpetological Review 26(2):101–102.
Braun, A.P. and Q.E. Phelps.
2016. Habitat use by five turtle species in the middle Mississippi River. Chelonian Conservation and Biology 15(1):62–68.
Brecke, B. and J.J. Moriarty.
1989. *Emydoidea blandingii* (Blanding's turtle). Longevity. Herpetological Review 20(2):53.
Breden, F.
1988. Natural history and ecology of Fowler's toad, *Bufo woodhousei fowleri* (Amphibia: Bufonidae) in the Indiana Dunes National Lakeshore. Fieldiana Zoology 49:1–16.
Breitenbach, G.L.
1982. The frequency of communal nesting and solitary brooding in the salamander, *Hemidactylium scutatum*. Journal of Herpetology 16(4):341–346.
Briggler, J.
2004. Rediscovery of the dusty hog-nosed snake (*Heterodon nasicus gloydi*) in Missouri. Missouri Herpetological Association Newsletter 17:13.
Briggler, J.T.
1998. Amphibian use of constructed woodland ponds in the Ouachita National Forest. MS thesis. University of Arkansas, Fayetteville. 79 pp.
Briggler, J.T., J.E. Johnson, and D.D. Rambo.
2004. Demographics of a ringed salamander (*Ambystoma annulatum*) breeding migration. The Southwestern Naturalist 49(2):209–217.
Briggler, J.T. and J.W. Prather.
2006. Seasonal use and selection of caves by plethodontid salamanders in a karst area of Arkansas. The American Midland Naturalist 155(1):136–148.
Briggler, J.T. and W.L. Puckette.
2003. Observations on reproductive biology and brooding behavior of the Ozark zigzag salamander, *Plethodon angusticlavius*. The Southwestern Naturalist 48(1):96–100.
Briggler, J.T., R.L. Rimer, and G. Deichsel.
2015. First record of the northern Italian wall lizard (*Podarcis siculus campestris*) in Missouri. Reptiles and Amphibians 22(1):43–45.
Britton, J.M.
1981. Microhabitat distribution and its effect on prey utilization in sympatric populations of *Plethodon glutinosus* and *Plethodon dorsalis* in northwestern Arkansas. Journal of the Arkansas Academy of Science 35:26–28.
Brodie, E.D., Jr.
1968. Investigations on the skin toxin of the red-spotted newt, *Notophthalmus viridescens viridescens*. The American Midland Naturalist 80(1):276–280.
1977. Salamander antipredator postures. Copeia 1977:523–535.
Brodie, E.D. Jr., J.L. Hensel, and J.A. Johnson.
1974. Toxicity of the urodele amphibians *Taricha*, *Notophthalmus*, *Cynops*, and *Paramesotrition* (Salamandridae). Copeia 1974:506–511.
Brooks, G.R., Jr.
1967. Population ecology of the ground skink, *Lygosoma laterale* (Say). Ecological Monographs 37(2):71–87.
Brown, E.E.
1956. Nests and young of the six-lined racerunner *Cnemidophorus sexlineatus* Linnaeus. Journal of the Elisha Mitchell Scientific Society 72:30–40.
Brown, L.E.
1978. Subterranean feeding by the chorus frog *Pseudacris streckeri* (Anura: Hylidae). Herpetologica 34:212–216.
Brown, L.E., H.O. Jackson, and J.R. Brown.
1972. Burrowing behavior of the chorus frog, *Pseudacris streckeri*. Herpetologica 28:325–328.
Brown, W.S.
1993. Biology, status, and management of the timber rattlesnake (*Crotalus horridus*): A guide for conservation. Society for the Study of Amphibians and Reptiles, Herpetological Circular No. 22. 72 pp.
Brown, W.S. and D.B. Greenberg.
1992. Vertical-tree ambush posture in *Crotalus horridus*. Herpetological Review 23(3):67.
Brown, W.S. and M.G. Simon.
2018. Record life span in a population of timber rattlesnakes (*Crotalus horridus*). Herpetological Review 49(3):442–445.
Budhabhatti, J. and E.O. Moll.
1990. *Chelydra serpentina* (common snapping turtle). Feeding behavior. Herpetological Review 21(1):19.
Bufalino, A. and D.A. Easterla.
1996. A preliminary ecological investigation of three Great Plains anuran species in Holt County, Missouri. Transactions of Missouri Academy of Science 30:89.
Buhlmann, K.A.
1995. Habitat use, terrestrial movements, and conservation of the turtle, *Deirochelys reticularia*, in Virginia. Journal of Herpetology 29(2):173–181.
Buhlmann, K.A. and J.W. Gibbons.
2001. Terrestrial habitat use by aquatic turtles from a seasonally fluctuating wetland: Implications for wetland conservation boundaries. Chelonian Conservation and Biology 4:115–127.
Buhlmann, K.A., J. W. Gibbons, and D.R. Jackson.
2008. *Deirochelys reticularia* (Latreille 1801)-chicken turtle. Chelonian Research Monographs 5:14.1–14.6.
Buhlmann, K.A. and T.R. Johnson.
1995. *Deirochelys reticularia miaria*. Geographic distribution. Herpetological Review 26(4):209.
Buhlmann, K.A. and M.R. Vaughan.
1991. Ecology of the turtle *Pseudemys concinna* in the New River, West Virginia. Journal of Herpetology 25:72–78.
Bull, J.J., R.C. Vogt, and C.J. McCoy.
1982. Sex determining temperatures in turtles: A geographic comparison. Evolution 36:326–332.
Burbrink, F.T.
2001. Systematics of the eastern ratsnake complex (*Elaphe obsoleta*) Herpetological Monographs 15:1–53.
Burbrink, F.T. and T.J. Guiher.
2015. Considering gene flow when using coalescent methods to delimit lineages of North American pitvipers of the genus *Agkistrodon*. Zoological Journal of Linnean Society 173:505–526.

Burbrink, F.T., R. Lawson, and J.B. Slowinski.
2000. Mitochondrial DNA phylogeography of the polytypic North American rat snake (*Elaphe obsoleta*): A critique of the subspecies concept. Evolution 54:2107–2118.

Burke, R.L. and R.J. Mercurio.
2002. Food habits of a New York population of Italian wall lizards, *Podarcis sicula* (Reptilia, Lacertidae). The American Midland Naturalist 147(2):368–375.

Burkett, R.D.
1966. Natural history of cottonmouth moccasin, *Agkistrodon piscivorus* (Reptilia). University of Kansas Publications Museum of Natural History 17(9):435–491.
1969. An ecological study of the cricket frog, *Acris crepitans*, in northeastern Kansas. PhD dissertation. University of Kansas, Lawrence. 111 pp.

Burns, J.A., H. Zhang, E. Hill, E. Kim, and R. Kerney.
2017. Transcriptome analysis illuminates the nature of the intracellular interaction in a vertebrate-algal symbiosis. eLife 2017:6 doi:10.7554/eLife.22054.

Burt, C.E.
1935. Further records of the ecology and distribution of amphibians and reptiles in the middle west. The American Midland Naturalist 16(3):311–336.

Busby, W.H., G.R. Pisani, E. Townsend Peterson, and N. Barve.
2014. Ecological studies of the smooth earth snake and redbelly snake, and niche modeling of forest species in eastern Kansas. Kansas Biological Survey, Lawrence, Kansas. 68 pp.

Butterfield, B.P.
1988. Age structure and reproductive biology of the Illinois chorus frog (*Pseudacris streckeri illinoensis*) from northeastern Arkansas. MS thesis. Arkansas State University, Jonesboro, Arkansas. 26 pp.

Butterfield, B.P., W.E. Meshaka, and S.E. Trauth.
1989. Fecundity and egg mass size of the Illinois chorus frog, *Pseudacris streckeri illinoensis* (Hylidae), from northeastern Arkansas. Southwest Naturalist 34:556–557.

Cagle, F.R.
1937. Egg laying habits of the slider turtle (*Pseudemys troostii*), the painted turtle (*Chrysemys picta*), and the musk turtle (*Sternotherus odoratus*). Journal of Tennessee Academy of Science 12:87–95.
1942. Herpetological fauna of Jackson and Union counties, Illinois. The American Midland Naturalist 28(1):164–200.
1948. Observations on a population of the salamander, *Amphiuma tridactylum* Cuvier. Ecology 29:479–491.
1950. The life history of the slider turtle *Pseudemys scripta troostii* (Holbrook). Ecological Monographs 20:31–54.

Camp, C.D.
1988. Aspects of the life history of the southern red-backed salamander *Plethodon serratus* Grobman in the southeastern United States. The American Midland Naturalist 119:93–100.

Campbell, J.A. and W.W. Lamar.
1989. The Venomous Reptiles of Latin America. Cornell University Press. Ithaca, New York. 425 pp.

Capparella, A.P., T. Springer, and L.E. Brown.
2012. Discovery of new localities for the threatened Kirtland's snake (*Clonophis kirtlandii*) in central Illinois. Transactions of Illinois Academy of Science 105(3&4):101–105.

Caron, M.D. and D.E. Swann.
2009. Great Plains skink, *Plestiodon obsoletus* (Baird and Girard, 1852). Pp. 456–459. *In:* L.L.C. Jones and R.E. Lovich (eds.). Lizards of the American Southwest, A Photographic Field Guide. Rio Nuevo Publishers, Tucson, Arizona. 567 pp.

Carpenter, C.C.
1958. Reproduction, young, eggs and food of Oklahoma snakes. Herpetologica 14:113–115.

Carreno, C.A. and R.N. Harris.
1998. Lack of nest defense behavior and attendance patterns in a joint-nesting salamander, *Hemidactylium scutatum* (Caudata: Plethodontidae). Copeia 1998(1):183–189.

Cary, D.L., R.L. Clawson, and D. Grimes.
1981. An observation of snake predation on a bat. Transactions of Kansas Academy of Science 84(4):223–224.

Churchill, T.A. and K.B. Storey.
1996. Organ metabolism and cryoprotectant synthesis during freezing in spring peepers, *Pseudacris crucifer*. Copeia 1996(3):517–525.

Christiansen, J.L., N.P. Bernstein, C.A. Phillips, J.T. Briggler, and D. Kangas.
2012. Declining populations of Yellow Mud Turtles (*Kinosternon flavescens*) in Iowa, Illinois, and Missouri. The Southwestern Naturalist 57(3):304–313.

Christiansen, J.L. and J.W. Bickham.
1989. Possible historic effects of pond drying and winterkill on the behavior of *Kinosternon flavescens* and *Chrysemys picta*. Journal of Herpetology 23:91–94.

Christiansen, J.L. and R.R. Burken.
1979. Growth and maturity of the snapping turtle (*Chelydra serpentina*) in Iowa. Herpetologica 35(3):261–266.

Clark, A.M., P.E. Moler, E.E. Possardt, A.H. Savitzky, W.S. Brown, and B.W. Brown.
2003. Phylogeography of the timber rattlesnake (*Crotalus horridus*) based on mtDNA sequences. Journal of Herpetology 37:145–154.

Clark, D.R., Jr.
1970. Ecological study of the worm snake, *Carphophis vermis* (Kennicott). University of Kansas Publications. Museum of Natural History 19(2):87–194.
1974. The western ribbon snake (*Thamnophis proximus*): Ecology of a Texas population. Herpetologica 30:372–379.

Clark, D.R. Jr. and R.J. Hall.
1970. Function of the blue tail-coloration of the five-lined skink (*Eumeces fasciatus*). Herpetologica 26(2):271–274.

Clark, R.D.
1974. Activity and movement patterns in a population of Fowler's toad, *Bufo woodhousei fowleri*. The American Midland Naturalist 92:257–274.

Claussen, D.L., P.M. Daniel, S. Jiang, and N.A. Adams.
1991. Hibernation in the eastern box turtle,

Terrapene c. carolina. Journal of Herpetology 25(3):334–341.
Cline, G.R. and R. Tumlison.
2001. Distribution and relative abundance of the Oklahoma salamander (*Eurycea tynerensis*). Proceedings of the Oklahoma Academy of Science 81:1–10.
Cline, G.R., R. Tumlison, and P. Zwank.
1989. Biology and ecology of the Oklahoma salamander, *Eurycea tynerensis*: Literature review and comments on research needs. Bulletin of Chicago Herpetological Society 24:164–168.
Cobb, V.A.
1990. Reproductive notes on the eggs and offspring of *Tantilla gracilis* (Serpentes: Colubridae), with evidence of communal nesting. The Southwestern Naturalist 35:222–224.
2004. Diet and prey size of the flathead snake, *Tantilla gracilis*. Copeia 2004(2):397–402.
Cocroft, R.B.
1994. A cladistic analysis of chorus frog phylogeny (Hylidae: *Pseudacris*). Herpetologica 50(4):420–437.
Collins, J.T.
1991. Viewpoint: A new taxonomic arrangement for some North American amphibians and reptiles. Herpetological Review 22(2):42–43.
1993. Amphibians and reptiles in Kansas, 3rd edition. Pub. Edu. Series No. 13, University of Kansas. xx + 397 pp.
Collins, J.T., S.L. Collins, and T.W. Taggart.
2010. Amphibians, Reptiles, and Turtles in Kansas. Eagle Mountain Publishing, LC. Eagle Mountain, Utah. 312 pp.
Conant, R.
1943. Studies on North American water snakes—I *Natrix kirtlandii* (Kennicott). The American Midland Naturalist 29(2):313–341.
1951. The Reptiles of Ohio. University of Notre Dame Press. Notre Dame, Indiana. 284 pp.
1958. A Field Guide to Reptiles and Amphibians of Eastern and Central North America. Houghton Mifflin Co., Boston. xv + 366 pp.
1960. The queen snake, *Natrix septemvittata*, in the interior highlands of Arkansas and Missouri, with comments upon similar disjunct distributions. Proceedings of the Academy of Natural Sciences of Philadelphia 112:25–40.
1975. A Field Guide to Reptiles and Amphibians of Eastern and Central North America, 2nd edition. Houghton Mifflin, Boston. xviii + 429 pp.
Conant, R. and J.T. Collins.
1991. A Field Guide to Reptiles and Amphibians of Eastern and Central North America, 3rd edition. Houghton Mifflin Co., Boston, Massachusetts. xviii + 450 pp.
Converse, S.J., J.B. Iverson, and A. Savidge.
2002. Activity, reproduction and overwintering behavior of ornate box turtles (*Terrapene ornata ornata*) in the Nebraska sandhills. The American Midland Naturalist 148:416–422.
Conaway, C.H. and D.E. Metter.
1967. Skin glands associated with breeding in *Microhyla carolinensis*. Copeia 1967(4):672–673.
Cooper, W.E., Jr. and L.J. Vitt.
1988. Orange head coloration of the male broad-headed skink (*Eumeces laticeps*), a sexually selected social cue. Copeia 1988(1):1–6.
Corse, W.A. and D.E. Metter.
1980. Economics, adult feeding and larval growth of *Rana catesbeiana* on a fish hatchery. Journal of Herpetology 14(3):231–238.
Costanzo, J.P., R.E. Lee, Jr., and M.F. Wright.
1993. Physiological responses to freezing in the turtle *Terrapene carolina*. Journal of Herpetology 27(1):117–120.
Costanzo, J.P., M.F. Wright, and R.E. Lee, Jr.
1992. Freeze tolerance as an overwintering adaptation in Cope's grey treefrog (*Hyla chrysoscelis*). Copeia 1992(2):565–569.
Coupe, B.
2001. Arboreal behavior in timber rattlesnakes (*Crotalus horridus*). Herpetological Review 32(2):83–85.
Craig, J.M., D.A. Mifsud, A.S. Briggs, J. Boase, and G. Kennedy.
2015. Mudpuppy (*Necturus maculosus maculosus*) spatial distribution, breeding water depth, and use of artificial spawning habitat in the Detroit River. Herpetological Conservation and Biology 10(3):926–934.
Crane, A.L. and B.D. Greene.
2008. The effect of reproductive condition on thermoregulation in female *Agkistrodon piscivorus* near the northwestern range limit. Herpetologica 64(2):156–167.
Crawford, J.A., D.B. Shepard, and C.A. Conner.
2009. Diet composition and overlap between recently metamorphosed *Rana areolata* and *Rana sphenocephala*: Implications for a frog of conservation concern. Copeia 2009(4):642–646.
Crawford, J.A., J.A. Tunnage, and E.M. Wright.
2017. Breeding pond occupancy of the ringed salamander (*Ambystoma annulatum*) in east-central Missouri. The American Midland Naturalist 178(1):151–157.
Croshaw, D.A. and D.E. Scott.
2005. Experimental evidence that nest attendance benefits female marbled salamanders (*Ambystoma opacum*) by reducing egg mortality. The American Midland Naturalist 154(2):398–411.
Crother, B.I. (ed.).
2017. Scientific and Standard English Names of Amphibians and Reptiles of North America North of Mexico, with Comments Regarding Confidence in Our Understanding pp. 1–102. SSAR Herpetological Circular 43.
Crother, B.I., M.E. White, J.M. Savage, M.E. Eckstut, M.R. Graham, and D.W. Gardner.
2011. A reevaluation of the status of the foxsnakes *Pantherophis gloydi* Conant and *P. vulpinus* Baird and Girard (Lepidosauria). International Scholarly Research Network Zoology 2011, doi:10.5402/2011/436049.
Cunningham, C.A. and T. Trautwein.
1996. The occurrence, habitat use and breeding status of an aquatic salamander, *Amphiuma tridactylum*, in southeast Missouri. Report to the Missouri Department of Conservation, Small Grant Program, Jefferson City, Missouri. iv + 60 pp.

Daigle, C., P. Galois, and Y. Chagnon.
2002. Nesting activities of an eastern spiny softshell turtle, *Apalone spinifera*. Canadian Field-Naturalist 116:104–107.
Dalton, B. and A. Mathis.
2014. Identification of sex and parasitism via pheromones by the Ozark zigzag salamander. Chemoecology 24:189–199.
Daniel, R.E.
2011. Updated maximum size records for amphibians and reptiles from Missouri. Missouri Herpetological Association Newsletter 24:17–20.
Daniel, R.E. and B.S. Edmond.
2014. Geographic distribution: *Lampropeltis nigra*. Herpetological Review 45(3):465.
Daszak, P., A. Strieby, A.A. Cunningham, J.E. Longcore, C.C. Brown, and D. Porter.
2004. Experimental evidence that the bullfrog (*Rana catesbeiana*) is a potential carrier of chytridiomycosis, an emerging fungal disease of amphibians. Herpetological Journal 14:201–207.
Davis, J.L. and D.M. Welchert.
2013. *Pantherophis vulpinus* (western fox snake). Predation. Herpetological Review 44(3):525–526.
Degenhardt, W.G., C.W. Painter, and A.H. Price.
1996. Amphibians and Reptiles of New Mexico. University of New Mexico Press. Albuquerque. 431 pp.
Demuth, J.P. and K.A. Buhlmann.
1997. Diet of the turtle *Deirochelys reticularia* on the Savannah River site, South Carolina. Journal of Herpetology 31(3):450–453.
de Queiroz, A.
1997. Lip-flaring in Thamnophiine snakes and its possible association with feeding on soft-bodied, sticky prey. Herpetological Review 28(1):28–29.
Deyrup, M., L. Deyrup, and J. Carrel.
2013. Ant species in the diet of a Florida population of eastern narrow-mouthed toads, *Gastrophryne carolinensis*. Southeastern Naturalist 12(2):367–378.
Dinkelacker, S.A. and N.L. Hilzinger.
2014. Demographic and reproductive traits of western chicken turtles, *Deirochelys reticularia miaria*, in Central Arkansas. Journal of Herpetology 48(4):439–444.
Ditmars, R.L.
1931. Snakes of the World. MacMillan Company, New York, New York. 207 pp.
Dobie, J.L.
1971. Reproduction and growth in the alligator snapping turtle, *Macroclemys temmincki* (Troost). Copeia 1971(4):645–658.
Dodd, C.K., Jr.
1980. Notes on the feeding behavior of the Oklahoma salamander, *Eurycea tynerensis* (Plethodontidae). The Southwestern Naturalist 25(1):111–113.
2013a. Frogs of the United States and Canada, Volume 1. The John Hopkins University Press. Baltimore, Maryland. 1–460 pp.
2013b. Frogs of the United States and Canada, Volume 2. The John Hopkins University Press. Baltimore, Maryland. 461–982 pp.
Dole, J.W.
1971. Dispersal of recently metamorphosed leopard frogs, *Rana pipiens*. Copeia 1971:221–228.
Doody, J.S., R.J. Brauman, J.E. Young, and R.A. Fiorillo.
1996. *Farancia abacura* (mud snake). Death feigning. Herpetological Review 27(2):82–83.
Doroff, A.M. and L.B. Keith.
1990. Demography and ecology of an ornate box turtle (*Terrapene ornata*) population in south-central Wisconsin. Copeia 1990:387–399.
Dowling, H.G.
1956. Geographic relations of Ozarkian amphibians and reptiles. The Southwestern Naturalist 1:174–189.
Drake, D.L., T.L. Anderson, L.M. Smith, K.M. Lohraff, and R.D. Semlitsch.
2014. Predation of eggs and recently hatched larvae of endemic ringed salamanders (*Ambystoma annulatum*) by native and introduced aquatic predators. Herpetologica 70(4):378–387.
Drake, D.L., A. Drayer, and S.E. Trauth.
2007. *Bufo americanus* (American toad). Algal symbiosis. Herpetological Review 38(4):435–436.
Drake, D.L. and K.M. O'Donnell.
2014. Sampling of terrestrial salamanders reveals previously unreported atypical color morphs in the southern red-backed salamander *Plethodon serratus*. The American Midland Naturalist 171(1):172–177.
Drake, D.L., K. O'Donnell, and B. Ousterhout.
2012. *Plethodon serratus* (southern red-backed salamander). Cicada burrow use. Herpetological Review 43(2):318–319.
Drake, D.L. and B.H. Ousterhout.
2011. Summer breeding of the southern leopard frog, *Rana sphenocephala* (=*Lithobates sphenocephalus*), in southern Missouri. Missouri Herpetological Association Newsletter 24:21–22.
Drda, W.J.
1968. A study of snakes wintering in a small cave. Journal of Herpetology 1(1–4):64–70.
Dreslik, M.J.
1997. Ecology of the river cooter, *Pseudemys concinna*, in a southern Illinois floodplain lake. Herpetological Natural History 5:135–145.
Dreslik, M.J., A.R. Kuhns, C.A. Phillips, and B.C. Jellen.
2003. Summer movements and home range of the cooter turtle, *Pseudemys concinna*, in Illinois. Chelonian Conservation and Biology 4:706–710.
Dreslik, M.J., E.O. Moll, C.A. Phillips, and T.P. Wilson.
1998. The endangered and threatened turtles of Illinois. Illinois Audubon, Spring 1998:10–15.
Duellman, W.E. and L. Trueb.
1986. Biology of the Amphibians. McGraw-Hill, New York, New York. xvii + 670 pp.
Dundee, H.A.
1958. Habitat selection by aquatic plethodontid salamanders of the Ozarks, with studies on their life histories. PhD dissertation. University of Michigan, Ann Arbor. 185 pp.
1965. *Eurycea tynerensis*. Catalogue of American Amphibians and Reptiles. 22.1–22.2.

DUNDEE, H.A. and D.A. ROSSMAN.
1989. The Amphibians and Reptiles of Louisiana. Louisiana State University Press. Baton Rouge. 300 pp.
DURBIAN, F.E., A. CUNNINGHAM, D.E. EASTERLA, and B.N. LOMAS.
2006. Site specific surveys for new massasauga rattlesnake (*Sistrurus catenatus*) populations in Holt County, Missouri. United States Fish and Wildlife Service, Mound City, Missouri. 10 pp.
DURBIAN, F.E., R.S. KING, T. CRABILL, H. LAMBERT-DOHERTY, and R.A. SEIGEL.
2008. Massasauga home range patterns in the Midwest. Journal of Wildlife Management 72(3):754–759.
DURBIAN, F.E., J.L. LEHMER, and J.T. BRIGGLER.
2001. Distribution. *Emydoidea blandingii* (Blanding's turtle). Herpetological Review 32(2):117.
DYER, W.G.
1999. Newly reported snake hosts for two ochetosomatid digeneans in Illinois. Transactions of the Illinois State Academy of Science 92:243–245.
DYRKACZ, S.
1975. Life history: *Thamnophis s. sirtalis*. Litter size. Herpetological Review 6(1):20.
EASTERLA, D.A.
1967. Black rat snake preys upon gray myotis and winter observation of red bats. The American Midland Naturalist 77(2):527–528.
1970. Albinistic small-mouthed salamander from southeastern Missouri. Transactions of Missouri Academy of Science 4:93–94.
1971. A breeding concentration of four-toed salamanders, *Hemidactylium scutatum*, in southeastern Missouri. Journal of Herpetology 5(3–4):194–195.
1972. Herpetological records for northwest Missouri. Transactions of Missouri Academy of Science 6:158–160.
EASTERLA, D.A. and H. GREGORY.
1967. First record of the mole salamander for Missouri. Herpetologica 23(3):239–240.
EASTERLA, D.A. and M.D. MEADOWS.
1993. The northern prairie skink rediscovered in Missouri, the last recorded occurrence of the smooth green snake for Missouri, plus a Missouri range extension for the northern spring peeper. The Bluebird 60(3):105–109.
EASTERLA, D.A. and D.D. SLEEP.
1999. Records of reptiles from northwestern Missouri, including county range extensions for the northern prairie skink (*Eumeces septentrionalis*). Missouri Herpetological Association Newsletter 12:19–21.
EDGREN, R.A.
1942. A nesting rendezvous of the musk turtle. Chicago Naturalist 5:63.
1955. The natural history of the hog-nosed snakes, Heterodon: A review. Herpetologica 11:105–117.
EDMOND, B.S.
2011. Successful hatching from an unusually large racer (*Coluber constrictor*) clutch. Missouri Herpetological Association Newsletter 24:22–25.
2015. Two five-lined skink (*Plestiodon fasciatus*) clutches. Missouri Herpetological Association Newsletter 28:16–17.
EDMOND, B.S. and R.E. DANIEL.
2014. A new kingsnake for Missouri and some comments on the biogeography of southeast Missouri. Missouri Herpetological Association Newsletter 27:18–21.
ELLIOTT, L., C. GERHARDT, and C. DAVIDSON.
2009. The Frogs and Toads of North America: A Comprehensive Guide to their Identification, Behavior, and Calls. Houghton Mifflin Harcourt. Boston, Massachusetts. 343 pp.
ELLIOTT, W.R.
2007. Zoogeography and biodiversity of Missouri caves and karst. Journal of Cave and Karst Studies 69(1):135–162.
EMLEN, S.T.
1977. "Double clutching" and its possible significance in the bullfrog. Copeia 1977(4):749–751.
ERNST, C.H. and R.W. BARBOUR.
1972. Turtles of the United States. University of Kentucky Press. Lexington. x + 347 pp.
ERNST, C.H. and E.M. ERNST.
2003. Snakes of the United States and Canada. Smithsonian Books, Washington and London. 668 pp.
ERNST, C.H. and E.M. ERNST.
2011. Venomous Reptiles of the United States, Canada, and Northern Mexico. Volume 1 *Heloderma, Micruroides, Micrurus, Pelamis, Agkistrodon, Sistrurus.* The Johns Hopkins University Press. Baltimore, Maryland. 352 pp.
ERNST, C.H. and E.M. ERNST.
2012. Venomous Reptiles of the United States, Canada, and Northern Mexico, Volume 2 *Crotalus.* The John Hopkins University Press. Baltimore, Maryland. 391 pp.
ERNST, C.H. and J.E. LOVICH.
2009. Turtles of the United States and Canada, 2nd edition. The Johns Hopkins University Press. Baltimore, Maryland. 819 pp.
ETTLING, J.A., M.D. WANNER, A.S. PEDIGO, J.L. KENKEL, K.R. NOBLE, and J.T. Briggler.
2017. Augmentation programme for the endangered Ozark hellbender *Cryptobranchus alleganiensis bishopi* in Missouri. International Zoo Yearbook (51):79–86.
EWERT, M.A.
1969. Seasonal movement of the toads *Bufo americanus* and *B. cognatus* in northwestern Minnesota. PhD dissertation. University of Minnesota, Minneapolis. 193 pp.
1976. Nests, nesting and aerial basking of *Macroclemys* under natural conditions, and comparisons with *Chelydra* (Testudines: Chelydridae). Herpetologica 32(2):150–156.
1979. Geographic distribution: *Chrysemys picta belli* x *marginata* (painted turtle). Herpetological Review 10(3):101–102.
FAIR, W.S. and S.E. HENKE.
1999. Movements, home ranges, and survival of Texas horned lizards (*Phrynosoma cornutum*). Journal of Herpetology 33(4):517–525.

Farrar, J.
2001. Yellow mud turtles living on the edge. Nebraskaland 79(4):24–33.
Femmer, S.R.
1978. Distribution and life history studies of *Scaphiopus bombifrons* Cope and *Bufo cognatus* Say in Missouri. MA thesis. University of Missouri, Columbia. 58 pp.
Fenolio, D.B. and G.O. Graening.
2009. Report of a mass aggregation of isopods in an Ozark cave of Oklahoma with considerations of population sizes of stygobionts. Speleobiology Notes 1:9–11.
Fenolio, D.B., G.O. Graening, B.A. Collier, and J.F. Stout.
2006. Coprophagy in a cave-adapted salamander; the importance of bat guano examined through nutritional and stable isotope analyses. Proceedings of the Royal Society B: Biological Sciences 273:439–443.
Fenolio, D.B., G.O. Graening, and J.R. Stout.
2005. Seasonal movement patterns of pickerel frogs (*Rana palustris*) in an Ozark cave and trophic implications supported by stable isotope evidence. The Southwestern Naturalist 50(3):385–389.
Fenolio, D.B., M.L. Niemiller, R.M. Bonett, G.O. Graening, B.A. Collier, and J.F. Stout.
2014. Life history, demography, and the influence of cave-roosting bats on a population of the grotto salamander (*Eurycea spelaea*) from the Ozark Plateaus of Oklahoma (Caudata: Plethodontidae). Herpetological Conservation and Biology 9(2):394–405.
Ferrara, C.R., R.C. Vogt, and R.S. Sousa-Lima.
2013. Turtle vocalizations as the first evidence of posthatching parental care in chelonians. Journal of Comparative Psychology 172(1):24–32.
Figiel, C.R., Jr. and R.D. Semlitsch.
1995. Experimental determination of oviposition site selection in the marbled salamander, *Ambystoma opacum*. Journal of Herpetology 29(3):452–454.
Fitch, H.S.
1955. Habits and adaptations of the Great Plains skink (*Eumeces obsoletus*). Ecological Monographs 25(1):59–83.
1956a. A field study of the Kansas ant-eating frog, *Gastrophryne olivacea*. University of Kansas Publications. Museum of Natural History 8(4):275–306.
1956b. An ecological study of the collared lizard (*Crotaphytus collaris*). University of Kansas Publications. Museum of Natural History 8(3): 213–274.
1958a. Natural history of the six-lined racerunner (*Cnemidophorus sexlineatus*). University of Kansas Publications. Museum of Natural History 11(2):11–62.
1958b. Home ranges, territories, and seasonal movement of vertebrates on the natural history reservation. University of Kansas Publications. Museum of Natural History 11:63–326.
1960. Autecology of the copperhead. University of Kansas Publications. Museum of Natural History 13(4):85–288.
1963a. Natural history of the racer, *Coluber constrictor*. University of Kansas Publications. Museum of Natural History 15(8):351–468.
1963b. Natural history of the black rat snake (*Elaphe o. obsoleta*) in Kansas. Copeia 1963(4):649–658.
1970. Reproductive cycles in lizards and snakes. University of Kansas Museum of Natural History Miscellaneous Publications 52:1–247.
1975. A demographic study of the ringneck snake (*Diadophis punctatus*) in Kansas. University of Kansas Museum of Natural History Miscellaneous Publications 62:1–53.
1978. A field study of the prairie kingsnake (*Lampropeltis calligaster*). Transactions of the Kansas Academy of Science 81(4):353–363.
1982. Resources of a snake community in prairie-woodland habitat of northeastern Kansas. Pp. 83–97. In: N.J. Scott, Jr. (ed.). Herpetological communities. U.S. Fish and Wildlife Service, Wildlife Res. Report 13:1–239.
1985. Variation in clutch and litter size in New World reptiles. University of Kansas Museum of Natural History Miscellaneous Publications 76: 1–76.
1989. A field study of the slender glass lizard, *Ophisaurus attenuatus*, in northeastern Kansas. University of Kansas Museum of Natural History Occasional Papers. 125:1–50.
1999. A Kansas Snake Community: Composition and Changes over 50 years. Krieger Publishing Company. Malabar, Florida. 165 pp.
2003. Reproduction in snakes of the Fitch Natural History Reservation in northeastern Kansas. Journal of Kansas Herpetology 6:21–24.
Fitch, H.S. and R.R. Fleet.
1970. Natural history of the milk snake (*Lampropeltis triangulum*) in northeastern Kansas. Herpetologica 26(4):387–396.
Fitch, H.S. and H.W. Greene.
1965. Breeding cycle in the ground skink, *Lygosoma laterale*. University of Kansas Publications. Museum of Natural History 15(11):565–575.
Flower, S.S.
1925. Contributions to our knowledge of the duration of life in vertebrate animals. -II. Batrachians. Proceedings of the Zoological Society of London 95(1):269–289.
Fontenot, C.L., Jr.
1999. Reproductive biology of the aquatic salamander *Amphiuma tridactylum* in Louisiana. Journal of Herpetology 33(1):100–105.
2011. *Hyla cinerea* (green treefrog). Winter aggregation. Herpetological Review 42:84–85.
Fontenot, C.L., Jr. and L.W. Fontenot.
1989. *Amphiuma tridactylum* (three-toed amphiuma). Feeding. Herpetological Review 20(2):48.
Force, E.R.
1931. Habits and birth of young of the lined snake, *Tropidoclonion lineatum* (Hallowell). Copeia 1931(2):51–53.
1935. A local study of the opisthogylph snake *Tantilla gracilis* Baird and Girard. Papers of Michigan Academy of Science, Arts and Letters 20:645–659.

FORD, D.K.
1999. Foraging ecology and demography of *Sternotherus odoratus* in a southwestern Missouri population. MS thesis. Southwest Missouri State University, Springfield. 54 pp.
FORD, D.K. and D. MOLL.
2004. Sexual and seasonal variation in foraging patterns in the stinkpot, *Sternotherus odoratus*, in southwestern Missouri. Journal of Herpetology 38(2):296–301.
FORESTER, D.C., J.W. SNODGRASS, K. MARSALEK, and Z. LANHAM.
2006. Post-breeding dispersal and summer home range of female American toads (*Bufo americanus*). Northeastern Naturalist 13(1):59–72.
FRAZER, N.B., J.W. GIBBONS, and J.L. GREENE.
1991. Life history and demography of the common mud turtle *Kinosternon subrubrum* in South Carolina, USA. Ecology 72(6):2218–2231.
FREEDBERG, S., M.A. EWERT, B.J. RIDENHOUR, M. NEIMAN, and C.E. NELSON.
2005. Nesting fidelity and molecular evidence for natal homing in the freshwater turtle, *Graptemys kohnii*. Proceeding of Royal Society B: Biological Sciences 272:1345–1350.
FREIBURG, R.E.
1951. An ecological study of the narrow-mouthed toad (*Microhyla*) in northeastern Kansas. Transactions of the Kansas Academy of Science 54(3):374–386.
FRESE, P.W.
2000. Notes on the ecology of the northern crawfish frog (*Rana areolata circulosa*) in southwest Missouri. Missouri Herpetological Association Newsletter 13:20–21.
2001. Observations on the herpetofaunal community of a natural marsh in northern Missouri. Missouri Herpetological Association Newsletter 14:13–14.
2003a. *Eumeces septentrionalis septentrionalis* (northern prairie skink). Nesting behavior. Herpetological Review 34(2):143.
2003b. Tallgrass prairie amphibian and reptile assemblage. Fire mortality. Herpetological Review 34(2):159–160.
2005. Graham's crayfish snake spatial ecology and habitat use project summary. Final report to Missouri Department of Conservation, Jefferson City. Missouri. 3 pp.
FRESE, P.W. and E. BRITZKE.
2001. *Siren intermedia nettingi* (western lesser siren). Predation. Herpetological Review 32(2):99.
FRESE, P.W., A. MATHIS, and R. WILKINSON.
2003. Population characteristics, growth, and spatial activity of *Siren intermedia* in an intensively managed wetland. The Southwestern Naturalist 48(4):534–542.
FRESE, P.W. and A.M. SULLIVAN.
2000. *Rana areolata circulosa* (northern crayfish frog): vocalization. Herpetological Review 31(2):101.
FROST, D.R., A.G. KLUGE, and D.M. HILLIS.
1992. Species in contemporary herpetology: Comments on phylogenetic inference and taxonomy. Herpetological Review 23(2):46–54.
FUERST, G.S. and C.C. AUSTIN.
2004. Population genetic structure of the prairie skink (*Eumeces septentrionalis*): Nested clade analysis of post pleistocene populations. Journal of Herpetology 38(2):257–268.
GALBRAITH, D.A.
1993. Multiple paternity and sperm storage in turtles. Herpetological Journal 3(4):117–123.
GALBRAITH, D.A., B.N. WHITE, R.J. BROOKS, and P.T. BOAG.
1993. Multiple paternity in clutches of snapping turtles (*Chelydra serpentina*) detected using DNR fingerprint. Canadian Journal of Zoology 71:318–324.
GARTON, J.D. and H.R. MUSHINKSY.
1979. Integumentary toxicity and unpalatability as an antipredator mechanism in the narrow-mouthed toad, *Gastrophryne carolinensis*. Canadian Journal of Zoology 57:1965–1973.
GARTON, J.S.
1972. Courtship of the small-mouthed salamander, *Ambystoma texanum*, in southern Illinois. Herpetologica 28(1):41–45.
GARTON, J.S. and R.A. BRANDON.
1975. Reproductive ecology of the green treefrog, *Hyla cinerea*, in southern Illinois (Anura: Hylidae). Herpetologica 31(2):150–161.
GARTON, J.S., E.W. HARRIS, and R.A. BRANDON.
1970. Descriptive and ecological notes on *Natrix cyclopion* in Illinois. Herpetologica 26(4):454–461.
GATZ, A.J., JR.
1973. Algal entry into the eggs of *Ambystoma maculatum*. Journal of Herpetology 7(2):137–138.
GEHLBACH, F.R., R. GORDON, and J.B. JORDAN.
1973. Aestivation of the salamander, *Siren intermedia*. The American Midland Naturalist 89:455–463.
GEHLBACH, F.R. and S.E. KENNEDY.
1978. Population ecology of a highly productive aquatic salamander (*Siren intermedia*). The Southwestern Naturalist 23(3):423–430.
GEHLBACH, F.R. and B. WALKER.
1970. Acoustic behavior of the aquatic salamander, *Siren intermedia*. BioScience 20:1107–1108.
GELLER, G.A. and G.S. CASPER.
2019. Late-term embryos and hatchlings of Ouachita map turtles (*Graptemys ouachitensis*) make sounds within the nest. Herpetological Review 50(3):449–452.
GERHARDT, H.C., R.E. DANIEL, S.A. PERRILL, and S. SCHRAMM.
1987. Mating behaviour and male mating success in the green treefrog. Animal Behaviour 35(5):1490–1503.
GERHARDT, H.C., B. DIEKAMP, and M. PTACEK.
1989. Inter-male spacing in choruses of the spring peeper, *Pseudacris (Hyla) crucifer*. Animal Behaviour 38(6):1012–1024.
GERHARDT, H.C., M.B. PTACEK, L. BARNETT, and K.G. TORKE.
1994. Hybridization in the diploid-tetraploid treefrogs *Hyla chrysoscelis* and *Hyla versicolor*. Copeia 1994(1):51–59.
GIBBONS, J.W.
1987. Why do turtles live so long? BioScience 37(4):262–269.
2013. A long-term perspective of delayed emergence (aka overwintering) in hatchling turtles: Some they

do and some they don't, and some you just can't tell. Journal of Herpetology 47(2):203–214.
Gibbons, J.W. and M.E. Dorcas.
2004. North American Watersnakes, A Natural History. University of Oklahoma Press. Norman. 438 pp.
Gibbons, J.W. and J.L. Greene.
1978. Selected aspects of the ecology of the chicken turtle, *Deirochelys reticularia* (Latreille) (Reptilia, Testudines, Emydidae). Journal of Herpetology 12(2):237–241.
Gibbs, H.L., M. Murphy, and J.E. Chiucchi.
2011. Genetic identity of endangered massasauga rattlesnake (*Sistrurus* sp.) in Missouri. Conservation Genetics 12:433–439.
Gilbert, M., R. Leclair, Jr. and R. Fortin.
1994. Reproduction of the northern leopard frog (*Rana pipiens*) in floodplain habitat in the Richelieu River, P. Quebec, Canada. Journal of Herpetology 28(4):465–470.
Gilhen, J.
1970. An unusual Nova Scotian population of the northern ringneck snake, *Diadophis punctatus edwardsi* (Merrem). Occasional Paper of Nova Scotia Museum 9:1–12.
1984. Amphibians and Reptiles of Nova Scotia. Nova Scotia Museum, Halifax, Nova Scotia. 162 pp.
Gillingham, J.C.
1974. Reproductive behavior of the western fox snake, *Elaphe v. vulpina* (Baird and Girard). Herpetologica 30(3):309–313.
Gillingham, J.C. and T. Rush.
1974. Notes on the fishing behavior of water snakes. Journal of Herpetology 8:384–385.
Glorioso, B.M., A.J. Vaughn, and J.H. Waddle.
2010. The aquatic turtle assemblage inhabiting a highly altered landscape in southeast Missouri. Journal of Fish and Wildlife Management 1(2):161–168.
Gloyd, H.K. and R. Conant.
1990. Snakes of the *Agkistrodon* Complex: A Monographic Review. Society for the Study of Amphibians and Reptiles. Contributions to Herpetology, No. 6. Oxford, Ohio. vi + 614 pp.
Godley, J.S.
1983. Observations on the courtship, nests and young of *Siren intermedia* in southern Florida. The American Midland Naturalist 110(1):215–219.
Goldsmith, S.K.
1984. Aspects of the natural history of the rough green snake, *Opheodrys aestivus* (Colubridae). The Southwestern Naturalist 29(4):445–452.
Goldberg, S.R.
2001. Reproduction in the ground snake, *Sonora semiannulata* (Serpentes: Colubridae), from Arizona. The Southwestern Naturalist 46(3):387–391.
Gosner, K.L.
1942. Lip curling of the red-bellied snake. Copeia 1942(42):181–182.
1960. A simplified table for staging anuran embryos and larvae with notes on identification. Herpetologica 16(3):183–190.
Graeter, C.J.
2005. Habitat selection and movement patterns of amphibians in altered forest habitats. MS thesis. University of Georgia, Athens.
Graham, S.P.
2013. How frequently do cottonmouth (*Agkistrodon piscivorus*) bask in trees? Journal of Herpetology 47(3):428–431.
Graham, T.E. and A.A. Graham.
1992. Metabolism and behavior of wintering common map turtles, *Graptemys geographica*, in Vermont. Canadian Field-Naturalist 106:517–519.
Graves, B.M. and J.J. Krupa.
2005. *Bufo cognatus* Say, 1823. Great Plains toad. Pp. 401–404. *In:* M.J. Lannoo (ed.). Amphibian Declines: The Conservation Status of United States Species. University of California Press. Berkeley. 1094 pp.
Gray, R.H., L.E. Brown, and L. Blackburn.
2005. *Acris crepitans* Baird, 1954(b) northern cricket frog. Pages 441–443. *In*: M.J. Lannoo (ed). Amphibian Declines: The Conservation Status of United States Species. University of California Press. Berkeley, California. 1094 pp.
Green, D.M.
1981. Adhesion and the toe pads of treefrogs. Copeia 1981(4):790–796.
Griffin, A.D., F.E. Durbian, and R.L. Bell.
2007. *Elaphe vulpina* (western fox snake). Habitat use. Herpetological Review 38:206–207.
Grobman, A.B.
1990. The effect of soil temperatures on emergence from hibernation of *Terrapene carolina* and *T. ornata*. The American Midland Naturalist 124(2):366–371.
Groves, J.
1982. Egg-eating behavior of brooding five-lined skinks, *Eumeces fasciatus*. Copeia 1982(4):969–971.
Groves, J.D. and F. Groves.
1978. Spider predation on amphibians and reptiles. Bulletin of Maryland Herpetology Society 14: 44–46.
Gruberg, E.R. and R.V. Stirling.
1972. Observations on the burrowing habits of the tiger salamander (*Ambystoma tigrinum*). Herpetological Review 4(3):85–89.
Guttman, D., J.E. Bramble, and O.J. Sexton.
1991. Observations on the breeding immigration of wood frogs *Rana sylvatica* reintroduced in east-central Missouri. The American Midland Naturalist 125:269–274.
Hall, R.J.
1969. Ecological observations on Graham's watersnake (*Regina grahami* Baird and Girard). The American Midland Naturalist 81(1):156–163.
1972. Food habits of the Great Plains skink (*Eumeces obsoletus*). The American Midland Naturalist 87(2):258–263.
Hall, R.J. and H.S. Fitch.
1971. Further observation of the demography of the Great Plains skink (*Eumeces obsoletus*). Transactions of the Kansas Academy of Science 74(1):93–98.
Harding, J.H.
1997. Amphibians and Reptiles of the Great Lakes Region. University of Michigan Press. Ann Arbor.
Harrel, J.B. and G.L. Stringer.
1997. Feeding habits of the alligator snapping turtle

(*Macroclemys temminckii*) as indicated by Teleostean otoliths. Herpetological Review 28(4):185–187.
HARRIS, J.P., JR.
1959. The natural history of *Necturus*: III. Food and feeding. Field and Laboratory 27:105–111.
HARRIS, R.N.
2005. *Hemidactylium scutatum*. Pages 780–781. *In*: M.J. Lannoo (ed.). Amphibian Decline: The Conservation Status of United States Species. University of California Press. Berkeley, California. 1094 pp.
HARRIS, R.N. and D.E. Gill.
1980. Communal nesting, brooding behavior, and embryonic survival of the four-toed salamander *Hemidactylium scutatum*. Herpetologica 36(2):141–144.
HEALY, W.R.
1970. Reduction of neoteny in Massachusetts populations of *Notophthalmus viridescens*. Copeia 1970(3):578–581.
HEATWOLE, H., N. PORAN and P. KING.
1999. Ontogenetic changes in the resistance of bullfrogs (*Rana catesbeiana*) to the venom of copperheads (*Agkistrodon contortrix contortrix*) and cottonmouths (*Agkistrodon piscivorus piscivorus*). Copeia 1999(3):808–814.
HEBRARD, J.J. and H R. MUSHINSKY.
1978. Habitat use by five sympatric water snakes in a Louisiana swamp. Herpetologica 34(3):306–311.
HEDGES, S.B.
1986. An electrophoretic analysis of Holarctic hylid frog evolution. Systematic Zoology 35:1–21.
HEDRICK, A. and J.B. IVERSON.
2017. *Kinosternon flavescens* (yellow mud turtle). Female growth and longevity. Herpetological Review 48(1):178–179.
HEEMEYER, J.L. and M.J. LANNOO.
2012. Breeding migrations in crawfish frogs (*Lithobates areolatus*): Long-distance movements, burrow philopatry, and mortality in a near-threatened species. Copeia 2012(3):440–450.
HEEMEYER, J.L., P.J. WILLIAMS, and M.J. LANNOO.
2012. Obligate crayfish burrow use and core habitat requirements of crawfish frogs. Journal of Wildlife Management 76(5):1081–1091.
HENDERSON, R.W.
1970. Feeding behavior, digestion, and water requirements of *Diadophis punctatus arnyi* Kennicott. Herpetologica 26(4):520–526.
HENDRICKS, L.J. and J. KEZER.
1958. An unusual population of a blind cave salamander and its fluctuation during one year. Herpetologica 14:41–43.
HERBECK, L.A. and D.R. LARSEN.
1999. Plethodontid salamander response to silvicultural practices in Missouri Ozark forests. Conservation Biology 13(3):623–632.
HERBECK, L.A. and R.D. SEMLITSCH.
2000. Life history and ecology of the southern redback salamander, *Plethodon serratus*, in Missouri. Journal of Herpetology 34(3):341–347.
HERMAN T.A.
2013. Four-toed salamander *Hemidactylium scutatum* (Schlegel 1838). Pages 141–155. *In*: R.A. Pfingsten, J.G. Davis, T.O. Matson, G.J. Lipps, Jr., D. Wynn and D.J. Armitage (eds). Amphibians of Ohio. Ohio Biological Survey Bulletin New Series. Volume 17(1). 899 pp.
HERMAN, T.A. and J.L. BOUZAT.
2016. Range-wide phylogeography of the four-toed salamander: out of Appalachia and into the glacial aftermath. Journal of Biogeography 43(4):666–678.
HESS, Z.J. and R.N. HARRIS.
2000. Eggs of *Hemidactylium scutatum* (Caudata: Plethodontidae) are unpalatable to insect predators. Copeia 2000(2):597–600.
HIGHTON, R.
1962. Revision of North American salamanders of the genus *Plethodon*. Bulletin of Florida State Museum 6(3):235–367.
HIGHTON, R. and A. LARSON.
1979. The genetic relationships of salamanders of the genus *Plethodon*. Systematic Zoology 28(4):579–599.
HIGHTON, R., G.C. MAHA, and L.R. MAXSON.
1989. Biochemical evolution in the slimy salamanders of the *Plethodon glutinosus* complex in the eastern United States. Part I. Geographic protein variation. University of Illinois Biological Monographs 57:1–78.
HIGHTON, R. and T.P. WEBSTER.
1976. Geographic protein variation and divergence in populations of the salamander *Plethodon cinereus*. Evolution 30(1):33–45.
HILL, J.G., III and S.J. BEAUPRE.
2008. Body size, growth, and reproduction in a population of western cottonmouths (*Agkistrodon piscivorus leucostoma*) in the Ozark Mountains of northwest Arkansas. Copeia 2008(1):105–114.
HIME, P.M., A.N. DRAYER, and S.J. PRICE.
2014. *Necturus maculosus* (common mudpuppy). Larval guarding. Herpetological Review 45(3):474.
HINSHAW, S.H. and B.K. SULLIVAN.
1990. Predation of *Hyla versicolor* and *Pseudacris crucifer* during reproduction. Journal of Herpetology 24(2):196–197.
HOCKING, D.J., T.A.G. RITTENHOUSE, B.B. ROTHERMEL, J.R. JOHNSON, C.A. CONNOR, E.B. HARPER, and R.D. SEMLITSCH.
2008. Breeding and recruitment phenology of amphibians in Missouri oak-hickory forest. The American Midland Naturalist 160(1):41–60.
HOCKING, D.J. and R.D. SEMLITSCH.
2007. Effects of timber harvest on breeding-site selection by gray treefrogs (*Hyla versicolor*). Biological Conservation 138(3–4):506–513.
2008. Effects of experimental clearcut logging on gray treefrog (*Hyla versicolor*) tadpole performance. Journal of Herpetology 42(4):689–698.
HODGES, W.L.
2009. Texas horned lizard *Phrynosoma cornutum* (Harlan, 1825). Pp. 166–169. *In*: L.L.C. Jones and R.E. Lovich (eds.). Lizards of the American Southwest: A Photographic Field Guide. Rio Nuevo Publishers, Tucson, Arizona. 567 pp.
HOISINGTON, D. and C. DURBIN.
2019. Eastern yellow-bellied racer, *Coluber constrictor*

flaviventris, egg clutch discovered in drainage pipe. Missouri Herpetological Association Newsletter 32:11–13.
HOLDER, T.L.
1988. Movement and life history aspects of the pygmy rattlesnake in southwest Missouri. MS thesis. Missouri State University, Springfield.
HOLDER, T. and D. MOLL.
1988. Movement and natural history of *Sistrurus miliarius* in southwest Missouri. Missouri Herpetological Association Newsletter 1:2–3.
HOLOMUZKI, J.R.
1995. Oviposition sites and fish-deterrent mechanisms of two stream anurans. Copeia 1995(3):607–613.
HOLYCROSS, A.T. and S.P. MACKESSY.
2002. Variation in the diet of *Sistrurus catenatus* (massasauga), with emphasis on *Sistrurus catenatus edwardsii* (desert massasauga). Journal of Herpetology 36(3):454–464.
HOUSEAL, T.W., J.W. BICKHAM, and M.D. SPRINGER.
1982. Geographic variation in the yellow mud turtle, *Kinosternon flavescens*. Copeia 1982(3):567–580.
HOWZE, J.M. and L.L. SMITH.
2015. Spatial ecology and habitat use of the coachwhip in a longleaf pine forest. Southeastern Naturalist 14(2):342–350.
HULLINGER, A.R.
2018. Critical habitat assessment and recovery plan for the Kansas state threatened broad-headed skink. MS thesis. Fort Hays State University, Hays, Kansas. 79 pp.
HURTER, J.
1911. Herpetology of Missouri. Transactions of the St. Louis Academy of Science 20(5):59–274.
HUTCHERSON, J.E., C.L. PETERSON, and R.F. WILKINSON.
1989. Reproductive and larval biology of *Ambystoma annulatum*. Journal of Herpetology 23(2):181–183.
HUTCHISON, V.H.
1956. Notes on the plethodontid salamanders, *Eurycea lucifuga* (Rafinesque) and *Eurycea longicauda longicauda* (Green). Occasional Papers of the National Speleological Society 3:1–24.
1958. The distribution and ecology of the cave salamander, *Eurycea lucifuga*. Ecological Monographs 28(1):1–20.
IRELAND, P.H.
1974. Reproduction and larval development of the dark-sided salamander, *Eurycea longicauda melanopleura* (Green). Herpetologica 30(4):338–343.
1976. Reproduction and larval development of the gray-bellied salamander *Eurycea multiplicata griseogaster*. Herpetologica 32(3):233–238.
IRELAND, P.H. and R. ALTIG.
1983. Key to the gilled salamander larvae and larviform adults of Arkansas, Kansas, Missouri and Oklahoma. The Southwestern Naturalist 28(3):271–274.
IVANYI, C.S.
2009. Eastern collared lizard, *Crotaphytus collaris* (Say, 1823). Pp. 104–107. *In*: L.L.C. Jones and R.E. Lovich (eds.). Lizards of the American Southwest, A Photographic Field Guide. Rio Nuevo Publishers, Tucson, Arizona. 567 pp.
IVERSON, J.B.
1979. Reproduction and growth of the mud turtle, *Kinosternon subrubrum* (Reptilia, Testudines, Kinosternidae), in Arkansas. Journal of Herpetology 13(1):105–111.
1990. Nesting and parental care in the mud turtle, *Kinosternon flavescens*. Canadian Journal of Zoology 68(2):230–233.
1991. Life history and demography of the yellow mud turtle, *Kinosternon flavescens*. Herpetologica 47(4):373–395.
2001. Reproduction of the river cooter, *Pseudemys concinna*, in Arkansas and across its range. The Southwestern Naturalist 46(3):364–370.
IVERSON, J.B., C.A. YOUNG, T.S.B. AKRE, and C.M. GRIFFITHS.
2012. Reproduction by female bullsnakes (*Pituophis catenifer sayi*) in the Nebraska sandhills. The Southwestern Naturalist 57(1):58–73.
JACKSON D.C.
2000. Living without oxygen: lessons from the freshwater turtle. Comparative Biochemistry and Physiology Part A 125(3):299–315.
JACKSON, D.R.
2012. *Plestiodon laticeps* (broad-headed skink). Herbivory. Herpetological Review 43(1):138–139.
JELLEN, B.C. and M.J. KOWALSKI.
2007. Movement and growth of neonate eastern massasaugas (*Sistrurus catenatus*). Copeia 2007(4):994–1000.
JENSEN, E.L., P. GOVINDARAJULU, and M.A. RUSSELLO.
2015. Genetic assessment of taxonomic uncertainty in painted turtles. Journal of Herpetology 49(2):314–324.
JENSSEN, T.A. and W.D. KLIMSTRA.
1966. Food habits of the green frog, *Rana clamitans*, in southern Illinois. The American Midland Naturalist 76(1):169–182.
JOHNSON, B.K. and J.L. CHRISTIANSEN.
1976. The food and food habits of Blanchard's cricket frog, *Acris crepitans blanchardi* (Amphibia, Anura: Hylidae), in Iowa. Journal of Herpetology 10(1):63–74.
JOHNSON, C.
1966. Species recognition in the *Hyla versicolor* complex. Texas Journal of Science 18:361–364.
JOHNSON, J.R.
2004. Fall breeding of the southern leopard frog (*Rana sphenocephala*) in central Missouri. Missouri Herpetological Association Newsletter 17:14–16.
JOHNSON, J.R., J.H. KNOUFT, and R.D. SEMLITSCH.
2007. Sex and seasonal differences in a spatial terrestrial distribution of gray treefrog (*Hyla versicolor*) populations. Biological Conservation 140(2007):250–258.
JOHNSON, J.R., R.D. MAHAN, and R.D. SEMLITSCH.
2008. Seasonal terrestrial microhabitat use by gray treefrogs (*Hyla versicolor*) in Missouri oak-hickory forests. Herpetologica 64(3):259–269.
JOHNSON, J.R. and R.D. SEMLITSCH.
2003. Defining core habitat of local populations of

the gray treefrog (*Hyla versicolor*) based on choice of oviposition sites. Oecologia 137(2):205–210.

Johnson, R.W., R.R. Fleet, M.B. Keck, and D.C. Rudolph.
2007. Spatial ecology of the coachwhip, *Masticophis flagellum* (Squamata: Colubridae), in eastern Texas. Southeastern Naturalist 6(1):111–124.

Johnson, T.R.
1975. Fall choruses of *Hyla crucifer* in Missouri. St. Louis Herpetological Society Newsletter 2(12):5–6.
1977. The Amphibians of Missouri. University of Kansas. Museum of Natural History. Public Education Series 6:1–134.
1983. Courtship under water. Missouri Conservationist 44(2):22–24.
1987. The Amphibians and Reptiles of Missouri. Missouri Department of Conservation, Jefferson City, Missouri. 369 pp.
2000. The Amphibians and Reptiles of Missouri. Missouri Department of Conservation, Jefferson City, Missouri. 400 pp.

Johnson, T.R. and R.N. Bader.
1974. Annotated checklist of Missouri amphibians and reptiles. St. Louis Herpetological Society Special Issue 1:1–16.

Jordan, R., Jr.
1970. Death-feigning in a captive red-bellied snake, *Storeria occipitomaculata* (Storer). Herpetologica 26(4):466–468.

Kangas, D.A.
2007. Observations on the ecology and distribution of several aquatic turtles in northeast Missouri with attention to the yellow mud turtle (*Kinosternon flavescens*) and the Blanding's turtle (*Emydoidea blandingii*). Final report to the Missouri Department of Conservation. 188 pp.

Kangas, D.A., B. Miller, and D. Noll.
1980. A report on the 1980 studies of the Illinois mud turtle in Missouri. Final report to the Missouri Department of Conservation. 47 pp.

Kangas, D.A. and K.R. Palmer.
1982. Field studies of the distribution and ecology of the Illinois Mud Turtle in northeast Missouri. Final report to the Missouri Department of Conservation. 62 pp.

Kapfer, J.M., J.R. Coggins, and R. Hay.
2008. Spatial ecology and habitat selection of bullsnakes (*Pituophis catenifer sayi*) at the northern periphery of their range. Copeia 2008(4):815–826.

Kapus, E.J.
1964. Anatomical evidence for *Heterodon* being poisonous. Herpetologica 20:137–138.

Kassing, E.F.
1961. A life history study of the Great Plains ground snake, *Sonora episcopa episcopa* (Kennicott). Texas Journal of Science 13:185–203.

Keen, W.H.
1975. Breeding and larval development of three species of *Ambystoma* in central Kentucky (Amphibia: Urodela). Herpetologica 31(1):18–21.

Keenlyne, K.D.
1972. Sexual differences in the feeding habits of *Crotalus horridus horridus*. Journal of Herpetology 6:234–237.
1978. Reproductive cycles in two species of rattlesnakes. The American Midland Naturalist 100(2):368–375.

Keenlyne, K.D. and J.R. Beer.
1973. Food habits of *Sistrurus catenatus catenatus*. Journal of Herpetology 7(4):382–384.

Kerney, R., E. Kim, R.P. Hangarter, A.A. Heiss, C.D. Bishop, and B.K. Hall.
2011. Intracellular invasion of green algae in a salamander host. Proceeding of National Academy of Science 108(16):6497–6502.

Kezer, J.
1952. The eggs of *Typhlotriton spelaeus* Stejneger, obtained by pituitary gland implantation. National Speleological Society Bulletin 14:58–59.

Kiester, A.R., C.W. Schwartz, and E.R. Schwartz.
1982. Promotion of gene flow by transient individuals in an otherwise sedentary population of box turtles (*Terrapene carolina triunguis*). Evolution 36(3):617–619.

Kimmons, J.B. and D. Moll.
2010. Seed dispersal by red-eared sliders (*Trachemys scripta elegans*) and common snapping turtles (*Chelydra serpentina*). Chelonian Conservation and Biology 9(2):289–294.

Klemens, M.W.
1993. Amphibians and reptiles of Connecticut and adjacent regions. State Geological and Natural History Survey of Connecticut Bulletin 112. 112:1–318.

Klimstra, W.D.
1959. Food habits of the yellow-bellied king snake in southern Illinois. Herpetologica 15:1–5.

Klug, P.E., J. Fill, and K.A. With.
2011. Spatial ecology of eastern yellow-bellied racer (*Coluber constrictor flaviventris*) and Great Plains rat snake (*Pantherophis emoryi*) in a contiguous tallgrass-prairie landscape. Herpetologica 67(4):428–439.

Kofron, C.P.
1978. Foods and habitats of aquatic snakes (Reptilia, Serpentes) in a Louisiana swamp. Journal of Herpetology 12(4):543–554.
1979. Reproduction of aquatic snakes in south-central Louisiana. Herpetologica 35(1):44–50.

Kofron, C.P. and J.R. Dixon.
1980. Observations on aquatic colubrid snakes in Texas. The Southwestern Naturalist 25(1):107–109.

Kofron, C.P. and A.A. Schreiber.
1985. Ecology of two endangered aquatic turtles in Missouri: *Kinosternon flavescens* and *Emydoidea blandingii*. Journal of Herpetology 19(1):27–40.
1987. Observations on aquatic turtles in a northeastern Missouri marsh. The Southwestern Naturalist 32(4):517–521.

Kolbe, J.J., L.J. Harmon, and D.A. Warner.
1999. New state record lengths and associated natural history notes for some Illinois snakes. Transactions of the Illinois Academy of Science 92:133–135.

Kolbe, J.J., B.R. Lavin, R.L. Burke, L. Rugiero, M. Capula, and L. Luiselli.
2013. The desire for variety: Italian wall lizard (*Podarcis siculus*) populations introduced to the United States via the pet trade are derived from

multiple native-range sources. Biological Invasions 15(4):775–783.

Korschgen, L.J. and T.S. Baskett.
1963. Foods of impoundment- and stream-dwelling bullfrogs in Missouri. Herpetologica 19(2):89–99.

Korschgen, L.J. and D.L. Moyle.
1955. Food habits of the bullfrog in central Missouri farm ponds. The American Midland Naturalist 54(2):332–341.

Kreeger, F.B.
1942. The cloaca of the female *Amphiuma tridactylum*. Copeia 1942(4):240–245.

Krenz, J.D. and D.E. Scott.
1994. Terrestrial courtship affects mating locations in *Ambystoma opacum*. Herpetologica 50(1):46–50.

Krohmer, R.W. and R.D. Aldridge.
1985a. Male reproductive cycle of the lined snake, *Tropidoclonion lineatum*. Herpetologica 41(1):33–38.
1985b. Female reproductive cycle of the lined snake, *Tropidoclonion lineatum*. Herpetologica 41(1):39–44.

Kroll, J.C.
1976. Feeding adaptation of hognose snakes. The Southwestern Naturalist 20:537–557.

Krupa, J.J.
1986. Multiple egg clutch production in the Great Plains toad. Prairie Naturalist 18:151–152.
1988. Fertilization efficiency in the Great Plains toad (*Bufo cognatus*). Copeia 1988(3):800–802.
1994. Breeding biology of the Great Plains toad in Oklahoma. Journal of Herpetology 28(2):217–224.
1995. *Bufo woodhousii* (Woodhouse's toad). Fecundity. Herpetological Review 26:142–144.

Kubatko, L.S., H.L. Gibbs, and E.W. Bloomquist.
2011. Inferring species-level phylogenies and taxonomic distinctiveness using multilocus data in *Sistrurus* rattlesnakes. Systematic Biology 60:393–409.

Lagler, K.F.
1943. Food habits and economic relations of the turtles of Michigan with special reference to fish management. The American Midland Naturalist 29:257–312.

Lagler, K F. and J.C. Salyer, II.
1947 [1945]. Food and habits of the common water snake, *Natrix s. sipedon*, in Michigan. Michigan Academy of Science Arts Letters 31:169–180.

Lahti, M.E. and A.D. Leaché.
2009. Prairie Lizard *Sceloporus consbrinus* Baird and Girard, 1853. Pp. 210–213. *In:* L.L.C. Jones and R.E. Lovich (eds.). Lizards of the American Southwest: A Photographic Field Guide. Rio Nuevo Publishers, Tucson, Arizona. 567 pp.

Landreth, H.F. and M.T. Christensen.
1971. Orientation of the plains spadefoot toad, *Scaphiopus bombifrons*, to solar cues. Herpetologica 27(4):454–461.

Lang, J.W.
1969. Hibernation and movements of *Storeria occipitomaculata* in northern Minnesota. Journal of Herpetology 3:196–197.

Langford, G.J. and J.A. Borden.
2004. *Farancia abacura* (mud snake). Courtship behavior. Herpetological Review 35(4):400–401.

Laposha, N.A. and R. Powell.
1982. *Virginia valeriae* (smooth earth snake). Size. Herpetological Review. 13(3):97.

Lardie, R.L.
1975. Courtship and mating behavior in the Yellow Mud Turtle, *Kinosternon flavescens flavescens*. Journal of Herpetology 9(2):223–227.

Leclair, R., Jr. and J. Castanet.
1987. A skeletochronological assessment of age and growth in the frog *Rana pipiens* Schreber (Amphibia, Anura) from southwestern Quebec. Copeia 1987(2):361–369.

LeClere, J.B.
2013. A Field Guide to the Amphibians and Reptiles of Iowa. ECO Herpetological Publishing & Distribution, Rodeo, New Mexico. viii + 349 pp.

Legler, J.M.
1960. Natural history of the ornate box turtle, *Terrapene ornata ornata* Agassiz. University of Kansas Publications. Museum of Natural History 11(10):527–669.

Lehnhoff, L.
2004. Habitat use and spatial ecology of Blanding's turtle, *Emydoidea blandingii*, on Squaw Creek National Wildlife Refuge, Mound City, Missouri. MS thesis. Southwest Missouri State University, Springfield. 57 pp.

Lemmon, E.M., A.R. Lemmon, J.T. Collins, and D.C. Cannatella.
2008. A new North America chorus frog species (Amphibia: Hylidae: *Pseudacris*) from the south-central United States. Zootaxa 1675(3):1–30.

Lemmon, E.M., A.R. Lemmon, J.T. Collins, J.A. Lee-Yaw, and D.C. Cannatella.
2007. Phylogeny-based delimitation of species boundaries and contact zones in the trilling chorus frogs (*Pseudacris*). Molecular Phylogenetics and Evolution 44:1068–1082.

Lescher, T.C., J.T. Briggler, and Z. Tang-Martinez.
2013a. Relative abundance, population structure, and conservation of alligator snapping turtles (*Macrochelys temminckii*) in Missouri between 1993–1994 and 2009. Chelonian Conservation and Biology 12(1):163–168.

Lescher, T.C., Z. Tang-Martinez, and J.T. Briggler.
2013b. Habitat use by the alligator snapping turtle (*Macrochelys temminckii*) and eastern snapping turtle (*Chelydra serpentina*) in southeastern Missouri. The American Midland Naturalist 169(1):86–89.

Lindeman, P.V.
1998. Of deadwood and map turtles (*Graptemys*): An analysis of species status for five species in three river drainages using replicated spotting-scope counts of basking turtles. Chelonian Conservation and Biology 3:137–141.
2006. Zebra and Quagga Mussels (*Dreissena* spp.) and other prey of a Lake Erie population of common map turtles (Emydidae: *Graptemys geographica*). Copeia 2006(2):268–273.

Littlejohn, M.J.
1958. Mating behavior in the treefrog *Hyla versicolor*. Copeia 1958(3):222–223.

Livezey, R.L.
1952. Some observations on *Pseudacris nigrita triseriata* (Wied) in Texas. The American Midland Naturalist 47(2):372–381.

Lykens, D.V. and D.C. Forester.
1987. Age structure in the spring peeper: do males advertise longevity? Herpetologica 43(2):216–223.

Maag, D.W.
2017. The spatial ecology and microhabitat selection of the pygmy rattlesnake (*Sistrurus miliarius*) in southwestern Missouri. MS thesis. Missouri State University, Springfield. 49 pp.

Madison, D.M.
1997. The emigration of radio-implanted spotted salamanders, *Ambystoma maculatum*. Journal of Herpetology 31(4):542–551.

Madison, D.M. and L. Farrand, III.
1998. Habitat use during breeding and emigration in radio-implanted tiger salamanders, *Ambystoma tigrinum*. Copeia 1998(2):402–410.

Mahan, R.D. and J.R. Johnson.
2007. Diet of the gray treefrog (*Hyla versicolor*) in relation to foraging site location. Journal of Herpetology 41(1):16–23.

Mahmoud, I.Y.
1967. Courtship behavior and sexual maturity in four species of kinosternid turtles. Copeia 1967:314–319.
1968. Feeding behavior in kinosternid turtles. Herpetologica 24(4):300–305.
1969. Comparative ecology of the kinosternid turtles of Oklahoma. The Southwestern Naturalist 14(1):31–66.

Makowsky, R., J.C. Marshall, Jr., J. McVay, P.T. Chippindale, and L.J. Rissler.
2010. Phylogeographic analysis and environmental niche modeling of the plain-bellied watersnake (*Nerodia erythrogaster*) reveals low levels of genetic and ecological differentiation. Molecular Phylogenetics and Evolution 55(3):985–995.

Manning, G.J. and J.T. Briggler.
2003. *Hemidactylus turcicus* (Mediterranean house gecko). USA. Arkansas. Herpetological Review 34(4):384.

Marion, K.R.
1970. The reproductive cycle of the fence lizard, *Sceloporus undulatus*, in eastern Missouri. PhD dissertation. Washington University, St. Louis, Missouri. vi + 191 pp.

Martof, B.
1956. Growth and development of the green frog, *Rana clamitans*, under natural conditions. The American Midland Naturalist 55(1):101–117.

Masta, S.E., B.K. Sullivan, T. Lamb, and E.J. Routman.
2002. Molecular systematics, hybridization, and phylogeography of the *Bufo americanus* complex in eastern North America. Molecular Phylogenetics and Evolution 24(2):302–314.

Masta, S.E., N.M. Laurent, and E.J. Routman.
2003. Population genetic structure of the toad *Bufo woodhousii*: An empirical assessment of the effects of haplotype extinction on nested cladistic analysis. Molecular Ecology 12(6):1541–1554.

Mathis, A.
2003. Use of chemical cues in detection of conspecific predators and prey by newts, *Notophthalmus viridescens*. Chemoecology 13:193–197.

Mathis, A. and F. Vincent.
2000. Differential use of visual and chemical cues in predator recognition and threat-sensitive predator-avoidance response by larval newts (*Notophthalmus viridescens*). Canadian Journal of Zoology 78:1646–1652.

Matson, T.O.
2005. *Necturus maculosus* (Rafinesque, 1818) Mudpuppy. Pp. 870–871. *In*: M.J. Lannoo (ed.). Amphibian Declines: The Conservation Status of United States Species. University of California Press. Berkeley.

McAlister, W.H.
1963. Evidence of mild toxicity in the saliva of the hognose snake (*Heterodon*). Herpetologica 19(2):132–137.

McAllister, C.T.
1985. Food habits and feeding behavior of *Crotaphytus collaris collaris* (Iguanidae) from Arkansas and Missouri. The Southwestern Naturalist 30(4):597–600.

McAllister, C.T. and S.P. Tabor.
1985. *Gastrophryne olivacea* (Great Plains narrowmouth toad). Coexistence. Herpetological Review 16(4):109.

McCallum, M.L.
1995. *Nerodia sipedon sipedon* (northern water snake). Feeding. Herpetological Review 26(1):39–40.

McCallum, M.L. and S.E. Trauth.
2001. *Pseudacris streckeri illinoensis* (Illinois chorus frog). Terrestrial feeding. Herpetological Review 32(1):35.

McCluskey, E.M. and D. Bender.
2015. Genetic structure of western massasauga rattlesnakes (*Sistrurus catenatus tergeminus*). Journal of Herpetology 49(3):343–348.

McDowell, C.R., B.A. Wheeler, and S.E. Trauth.
2004. *Terrapene carolina triunguis* (three-toed box turtle). Aquatic behavior. Herpetological Review 35(3):265–266.

McGaugh, S.E., C.M. Eckerman, and F.J. Janzen.
2008. Molecular phylogeography of *Apalone spinifera* (Reptilia, Trionychidae). Zoologica Scripta 37(3):289–304.

McGregor, J.H. and W.R. Teska.
1989. Olfaction as an orientation mechanism in migrating *Ambystoma maculatum*. Copeia 1989(3):779–781.

McKelvy, A.D. and F.T. Burbrink.
2017. Ecological divergence in the yellow-bellied kingsnake (*Lampropeltis calligaster*) at two North American biodiversity hotspots. Molecular Phylogenetics and Evolution 106(2017):61–72.

McKinstry, D.M.
1978. Evidence of toxic saliva in some colubrid snakes of the United States. Toxicon 16(6):523–534.

McKnight, D.T., J.R. Harmon, J.L. McKnight, and D.B. Ligon.
2015. The spring-summer nesting and activity pattern of the western chicken turtle (*Deirochelys reticularia miaria*). Copeia 103(4):1043–1047.

McKnight, D.T., J.L. McKnight, T.L. Dean, and D.B. Ligon.
2013. *Acris blanchardi* (Blanchard's cricket frog). Feigning death. Herpetological Review 44(1):117.
McMurray, S.E., J.S. Faiman, A.D. Roberts, B. Simmons, and M.C. Barnhart.
2012. A Guide to Missouri's Freshwater Mussels. Missouri Department of Conservation, Jefferson City, Missouri. 94 pp.
McVay, J.D. and B. Carstens.
2013. Testing monophyly without well-supported gene trees: Evidence from multi-locus nuclear data conflicts with existing taxonomy in the snake tribe Thamnophiini. Molecular Phylogenetics and Evolution 68:425–431.
McWilliams, S.R. and M. Bachman.
1989. Foraging ecology and prey preference of pond-form larval small-mouthed salamander, *Ambystoma texanum*. Copeia 1989(4):948–961.
Meade, G.P.
1934. Some observations on captive snakes. Copeia 1934(1):4–5.
1937. Breeding habits of *Farancia abacura* in captivity. Copeia 1937(1):12–15.
1940. Observations on Louisiana captive snakes. Copeia 1940(3):165–168.
Mecham, J.S., M.J. Littlejohn, R.S. Oldham, L.E. Brown, and J.R. Brown.
1973. A new species of leopard frog (*Rana pipiens* complex) from the plains of the central United States. Occasional Papers of the Museum of Texas Tech University 18:1–11.
Meshaka, W.E., Jr., J.W. Gibbons, D.F. Hughes, M.W. Klemens, and J.B. Iverson.
2017. *Kinosternon subrubrum* (Bonnaterre 1789)—Eastern mud turtle. Chelonian Research Monographs 5(10):101.1–101.16.
Meshaka, W.E., Jr. and C.J. Schmidt.
2017. Reproductive traits in the Great Plains ratsnake, *Pantherophis emoryi* (Baird and Girard, 1853), in northern Kansas. Transactions of the Kansas Academy of Science 120(3–4):170–174.
Meshaka, W.E., Jr. and S.E. Trauth.
1995. Reproductive cycle of the Ozark zigzag salamander, *Plethodon dorsalis angusticlavius* (Caudata, Plethodontidae), in north central Arkansas. Alytes 12:175–182.
Messenger, K.R.
2010. Growth and age at reproductive maturity of the Carolina pigmy rattlesnake, *Sistrurus m. miliarius* (Reptilia: Serpentes). MS thesis. Marshall University, Huntington, West Virginia. 106 pp.
Metcalf, A.L. and E.L. Metcalf.
1985. Longevity in some ornate box turtles (*Terrapene ornata ornata*). Journal of Herpetology 19(1):157–158.
Metcalf, C.K., F. Pezold, and B.G. Crump.
2007. *Agkistrodon piscivorus leucostoma* (western cottonmouth): Activity. Herpetological Review 38(4):465–466.
Metcalf, E. and A.L. Metcalf.
1970. Observations on ornate box turtles (*Terrapene ornata ornata* agassiz). Transactions of the Kansas Academy of Science 73(1):96–117.
1979. Mortality in hibernating ornate box turtles, *Terrapene ornata*. Herpetologica 35(1):93–96.
Middendorf, G.A., III and W.C. Sherbrooke.
1992. Canid elicitation of blood-squirting in a horned lizard (*Phrynosoma cornutum*). Copeia 1992(2):519–527.
Milanovich, J.R., S.E. Trauth, and T. McKay.
2008. Diet of western slimy salamander, *Plethodon albagula* (Caudata: Plethodontidae), from two mountain ranges in Arkansas. Southeastern Naturalist 7(2):323–330.
Milanovich, J.R., S.E. Trauth, D.A. Saugey, and R.R. Jordan.
2006. Fecundity, reproductive ecology, and influence of precipitation on clutch size in the western slimy salamander (*Plethodon albagula*). Herpetologica 62(3):292–301.
Miller, B.T.
1983. *Sonora semiannulata* (groundsnake). Coloration. Herpetological Review 14(4):121–122.
Miller, J.K.
2001. Escaping senescence: demographic data from the three-toed box turtle (*Terrapene carolina triunguis*). Experimental Gerontology 36(2001):829–832.
Minton, S.A.
1972. Amphibians and Reptiles of Indiana. Indiana Academy of Science. Indianapolis. 346 pp.
Minton, S.A., Jr.
2001. Amphibians and Reptiles of Indiana, revised 2nd edition. Indiana Academy of Science, Indianapolis. 404 pp.
Minx, P.
1992. Variation in phalangeal formulae in the turtle genus *Terrapene*. Journal of Herpetology 26(2):234–238.
Missouri Cave Database.
2019. Missouri Speleological Survey Inc.
Mitchell, J.C.
1990. *Pseudacris feriarum* (upland chorus frog). Predation. Herpetological Review 21(4):89.
1994. The Reptiles of Virginia. Smithsonian Institution Press. Washington D.C. 352 pp.
Mitchell, J.C. and J.M. Anderson.
1994. Amphibians and Reptiles of Assateague and Chincoteague Islands. Virginia Museum of Natural History, Special Publication No. 2.
Mitchell, J.C., K.A. Buhlmann, and R.L. Hoffman.
1996. Predation of marbled salamander (*Ambystoma opacum* Gravenhorst) eggs by the milliped *Uroblaniulus jerseyi* (Causey). Banisteria (8):55–56.
Mohr, C.E.
1943. The eggs of the long-tailed salamander, *Eurycea longicauda* (Green), Proceedings of the Pennsylvania Academy of Science 17:86.
Moll, D.
1976. Food and feeding strategies of the Ouachita map turtle (*Graptemys pseudogeographica ouachitensis*). The American Midland Naturalist 96(2):478–482.
1979. Subterranean feeding by the Illinois mud turtle, *Kinosternon flavescens spooneri*. Journal of Herpetology 13(3):371–373.

Moll, E.O.
1973. Latitudinal and intersubspecific variation in reproduction of the painted turtle *Chrysemys picta*. Herpetologica 29(4):307–318.
2001. A fortieth for Clementine. Sonoran Herpetology 14(5):50–51.

Montgomery, W.B. and G.W. Schuett.
1989. Autumnal mating with subsequent production of offspring in the rattlesnake *Sistrurus miliarius streckeri*. Bulletin of the Chicago Herpetological Society 24:205–207.

Moriarty, E.C. and D.C. Cannatella.
2004. Phylogenetic relationships of the North American chorus frog (*Pseudacris*: Hylidae). Molecular Phylogenetics and Evolution 30(2):409–420.

Moriarty, J.J. and C.D. Hall.
2014. Amphibians and Reptiles in Minnesota. University of Minnesota Press. Minneapolis. 370 pp.

Morris, M.A.
1974. Observations on a large litter of the snake *Storeria dekayi*. Transactions of Illinois State Academy of Science 67(3):359–360.

Morris, M.A. and K. Vail.
1991. Courtship and mating of captive rough green snakes, *Opheodrys aestivus*, from Illinois. Bulletin of the Chicago Herpetological Society 26(2):34.

Morris, P.A.
1944. They Hop and Crawl. The Jacques Cattell Press. York, Pennsylvania. xiv + 253 pp.

Mosauer, W.
1932. On the locomotion of snakes. Science 76:583–585.

Mount, R.H.
1975. The Reptiles and Amphibians of Alabama. Auburn University, Auburn, Alabama. 347 pp.

Moyle, D.L.
1952. Life history of the bullfrog in central Missouri farm ponds. MA thesis. University of Missouri, Columbia. 89 pp.

Muellman, P.J., O.D. Cunha, and C.E. Montgomery.
2017. *Crotalus horridus* (timber rattlesnake). Reproduction. Herpetological Review 48(4):858.
2018. *Crotalus horridus* (timber rattlesnake) maternal scent trailing by neonates. Northeastern Naturalist 25(1):50–55.

Mulcare, D.J.
1965. The problem of toxicity in *Rana palustris*. Proceeding of Indiana Academy of Science 75:319–324.

Munyer, E.A.
1967. Behavior of an eastern hognose snake, *Heterodon platyrhinos*, in water. Copeia 1967(3):668–670.

Myers, C.W.
1958a. Notes on the eggs and larvae of *Eurycea lucifuga* Rafinesque. Quarterly Journal of the Florida Academy of Science 21(2):125–130.
1958b. Amphibia in Missouri caves. Herpetologica 14:35–36.

Neill, W.T.
1948. Hibernation of amphibians and reptiles in Richmond County, Georgia. Herpetologica 4(3):107–114.

Nelson, C.E.
1972. *Gastrophryne olivacea* (Hallowell). Western narrow-mouthed toad. Catalogue of American Amphibians and Reptiles 122.1–122.4.

Nelson, D.H. and J.W. Gibbons.
1972. Ecology, abundance and seasonal activity of the scarlet snake, *Cemophora coccinea*. Copeia 1972(3):582–584.

Nelson, P.W.
2005. The Terrestrial Natural Communities of Missouri. Missouri Natural Area Committee. 550 pp.

Newman, C.E. and C.C. Austin.
2016. Thriving in the cold: glacial expansion and post-glacial contraction of a temperate terrestrial salamander (*Plethodon serratus*). Molecular Ecology 25(24):6162–6174.

Nickerson, M.A.
2000. *Sternotherus odoratus* (common musk turtle). Aerial basking. Herpetological Review 31(4):238–239.

Nickerson, M.A. and R. Krager.
1972. Additional noteworthy records of Missouri amphibians and reptiles with a possible addition to the herpetofauna. Transactions of the Kansas Academy of Science 75(3):276–277.
1975. The Lake Erie water snake "phenotype" in central Missouri. Herpetological Review 6(3):75.

Nickerson, M.A. and C.E. Mays.
1973. The hellbenders: North American giant salamanders. Milwaukee Public Museum Publications in Biology and Geology 1. Milwaukee, Wisconsin. 106 pp.

Niemiller, M.L., D. Fenolio, G.O. Graening, and B.T. Miller.
2009. Observations on oviposition and reproduction of the cave salamander, *Eurycea lucifuga* (Caudata: Plethodontidae), from Arkansas and Tennessee, USA. Speleobiology Notes 2009(1):17–19.

Niemiller, M.L. and R.G. Reynolds (eds).
2011. The Amphibians of Tennessee. The University of Tennessee Press. Knoxville. 369 pp.

Niemiller, M.L., R.G. Reynolds, and B.T. Miller (eds).
2013. The Reptiles of Tennessee. The University of Tennessee Press. Knoxville. 366 pp.

Nigh, T.A. and W.A. Schroeder.
2002. Atlas of Missouri Ecoregions. Missouri Department of Conservation, Jefferson City, Missouri. 212 pp.

Nyman, S., R.F. Wilkinson, and J.E. Hutcherson.
1993. Cannibalism and size relations in a cohort of larval ringed salamanders (*Ambystoma annulatum*). Journal of Herpetology 27(1):78–84.

O'Conner, K.M., C.D. Rittenhouse, J.J. Millspaugh, and T.A.G. Rittenhouse.
2015. Demographics and density estimates of two three-toed box turtle (*Terrapene carolina triunguis*) populations within forest and restored prairie sites in central Missouri. PeerJ 3:e1256.

O'Donnell, K.M., F.R. Thompson III, and R.D. Semlitsch.
2014. Predicting variation in microhabitat utilization of terrestrial salamanders. Herpetologica 70(3):259–265.

2016. Prescribed fire alters surface activity and movement behavior of a terrestrial salamander. Journal of Zoology 298(4):303–309.

Olds, M.J.
2007. Habitat use and overwintering ecology of the northern watersnake (*Nerodia sipedon*) on an artificial levee in central Illinois. MS thesis. Eastern Illinois University, Charleston, Illinois. 51 pp.

Osbourn, M.S., S.E. Pittman, D.L. Drake, and R.D. Semlitsch.
2012. *Ambystoma annulatum* (ringed salamander) and *Ambystoma maculatum* (spotted salamander). Climbing behavior. Herpetological Review 43(3):458–459.

Organ, J.A.
1968. Courtship behavior and spermatophore of the cave salamander, *Eurycea lucifuga* (Rafinesque). Copeia 1968(3):576–580.

Owen, P.C. and J.E. Juterbock.
2013. Small-mouthed Salamander *Ambystoma texanum* (Matthes 1855). Pp. 141–155. *In:* R.A. Pfingsten, J.G. Davis, T.O. Matson, G.J. Lipps, Jr., D. Wynn and B.J. Armitage (eds.). Amphibians of Ohio. Ohio Biological Survey Bulletin New Series. Volume 17(1). 899 pp.

Packard, G.C., J.K. Tucker, and L.D. Lohmiller.
1998. Distribution of Strecker's chorus frog (*Pseudacris streckeri*) in relation to their tolerance for freezing. Journal of Herpetology 32(3):437–440.

Palmer, W.M. and A.L. Braswell.
1995. Reptiles of North Carolina. University of North Carolina Press. Chapel Hill and London. 412 pp.

Parmelee, J.R.
1993. Microhabitat segregation and spatial relationships among four species of mole salamanders (genus *Ambystoma*). Occasional Papers of the Museum of Natural History, University of Kansas 160:1–33.

Parris, M.J.
1998. Terrestrial burrowing ecology of newly metamorphosed frogs (*Rana pipiens* complex). Canadian Journal of Zoology 76:2124–2129.

Parrish, H.M.
1963. Analysis of 460 fatalities from venomous animals in the United States. American Journal of Medical Science 245(2):129–141.

Paulissen, M.A.
1987. Diet of adult and juvenile six-lined racerunners, *Cnemidophorus sexlineatus* (Sauria: Teiidae). The Southwestern Naturalist 32(3):395–397.

Paulissen, M.A. and T.M. Buchanan.
1991. Observations on the natural history of the Mediterranean gecko, *Hemidactylus turcicus* (Sauria; Gekkonidae) in northwestern Arkansas. Journal of the Arkansas Academy of Science 45:81–83.

Pearson, P.G.
1955. Population ecology of the spadefoot toad, *Scaphiopus h. holbrooki* (Harlan). Ecological Monographs 25(3):233–267.
1957. Further notes on the population ecology of the spadefoot toad. Ecology 38:580–586.

Perkins, R.M. and M.J.R. Lentz.
1934. Contribution to the herpetology of Arkansas. Copeia 1934(3):139–140.

Perrill, S.A. and R.E. Daniel.
1983. Multiple egg clutches in *Hyla regilla*, *H. cinerea* and *H. gratiosa*. Copeia 1983(2):513–516.

Perrill, S.A., H.C. Gerhardt, and R. Daniel.
1978. Sexual parasitism in the green treefrog (*Hyla cinerea*). Science 200:1179–1180.
1982. Mating strategy shifts in male green treefrogs (*Hyla cinerea*): An experimental study. Animal Behaviour 30(1):43–48.

Perrill, S.A. and M. Magier.
1988. Male mating behavior in *Acris crepitans*. Copeia 1988(1):245–248.

Perry, J.
1978. An observation of "dance" behavior in western cottonmouth, *Agkistrodon piscivorus leucostoma* (Reptilia, Serpentes, Viperidae). Journal of Herpetology 12(3):429–431.

Peterson, C.L.
1988. Breeding activities of the hellbender in Missouri. Herpetological Review 19(2):28–29.

Peterson, C.L., D.E. Metter, B.T. Miller, R.F. Wilkinson, and M.S. Topping.
1988. Demography of the hellbender (*Cryptobranchus alleganiensis*) in the Ozarks. The American Midland Naturalist 119(2):291–303.

Peterson, C.L., J.W. Reed, and R.F. Wilkinson.
1989. Seasonal food habits of *Cryptobranchus alleganiensis* (Caudata: Cryptobranchidae). The Southwestern Naturalist 34(3):438–441.

Peterson, C.L. and R.F. Wilkinson.
1996. Home range size of the hellbender (*Cryptobranchus alleganiensis*) in Missouri. Herpetological Review 27(3):126–127.

Peterson, C.L., R.F. Wilkinson, D. Moll, and T. Holder.
1991. Premetamorphic survival of *Ambystoma annulatum*. Herpetologica 47(1):96–100.

Peterson, C.L., R.F. Wilkinson, Jr., M.S. Topping, and D.E. Metter.
1983. Age and growth of the Ozark Hellbender (*Cryptobranchus alleganiensis bishopi*). Copeia 1983(1):225–231.

Peterman, W.E. and R.D. Semlitsch.
2013. Fine-scale habitat associations of a terrestrial salamander: The role of environmental gradients and implications for population dynamics. PLoS ONE 8(5): e62184.

Petranka, J.W.
1982a. Courtship behavior of the small-mouthed salamander (*Ambystoma texanum*) in Central Kentucky. Herpetologica 38(2):333–336.
1982b. Geographic variation in the mode of reproduction and larval characteristics of the small-mouthed salamander (*Ambystoma texanum*) in the east-central United States. Herpetologica 38(4):475–485.
1990. Observations on nest site selection, nest desertion, and embryonic survival in marbled salamanders. Journal of Herpetology 24(3):229–234.

1998. Salamanders of the United States and Canada. Smithsonian Institution Press. Washington D.C. 587 pp.

Petranka, J.W., M.E. Hopey, B.T. Jennings, S.D. Baird, and S.J. Boone.
1994. Breeding habitat segregation of wood frogs and American toads: The role of interspecific tadpole predation and adult choice. Copeia 1994(3):691–697.

Petranka, J.W., A.W. Rushlow, and M.E. Hopey.
1998. Predation by tadpoles of *Rana sylvatica* on embryos of *Ambystoma maculatum*: Implications of ecological role reversals by *Rana* (predator) and *Ambystoma* (prey). Herpetologica 54(1):1–13.

Pflanz, D.J. and R. Powell.
1990. Death feigning by a coachwhip from Missouri. Missouri Herpetological Association Newsletter 3:12.

Phillips, C.A.
1994. Geographic distribution of mitochondrial DNA variants and the historical biogeography of the spotted salamander, *Ambystoma maculatum*. Evolution 48(3):597–607.

Phillips, C.A. and O.J. Sexton.
1989. Orientation and sexual differences during breeding migrations of the spotted salamander, *Ambystoma maculatum*. Copeia 1989(1):17–22.

Phillips, J.B., K. Adler, and S.C. Borland.
1995. True navigation by an amphibian. Animal Behaviour 50:855–858.

Phillips, J.G., D.B. Fenolio, S.L. Emel, and R.M. Bonett.
2017. Hydrologic and geologic history of the Ozark Plateau drive phylogenomic patterns in a cave-obligate salamander. Journal of Biogeography 44(11):2463–2474.

Pierce, B.A. and J.R. Shayevitz.
1982. Within and among population variation in spot number of *Ambystoma maculatum*. Journal of Herpetology 16(4):402–405.

Pilgrim, M.A.
2001. Offspring size variation in five species of snakes (*Sistrurus miliarius*, *Sistrurus catenatus*, *Thamnophis sirtalis*, *Thamnophis proximus*, and *Nerodia rhombifer*): Implications for optimal offspring size theory. MS thesis. Southeastern Louisiana University, Hammond, Louisiana. 69 pp.

Pilgrim, M.A., T.M. Farrell, P.G. May, M.R. Vollman, and R.A. Seigel.
2011. Secondary sex ratios in six snake species. Copeia 2011(4):553–558.

Pisani, G.R., J.T. Collins, and S.R. Edwards.
1972. A re-evaluation of the subspecies of *Crotalus horridus*. Transactions of the Kansas Academy of Science 75(3):255–263.

Pitt, A.L. and M.A. Nickerson.
2006. *Cryptobranchus alleganiensis* (hellbender salamander). Larval diet. Herpetological Review 37(1):69.
2012. Reassessment of the turtle community in the North Fork of White River, Ozark County, Missouri. Copeia 2012(3):367–374.

Pittman, S.E., A.L. Jendrek, S.J. Price, and M.E. Dorcas.
2008. Habitat selection and site fidelity of Cope's gray treefrog (*Hyla chrysoscelis*) at the aquatic-terrestrial ecotone. Journal of Herpetology 42(2):378–385.

Platt, D.R.
1969. Natural history of the hognose snakes *Heterodon platyrhinos* and *Heterodon nasicus*. University of Kansas Publications. Museum of Natural History 18:253–420.

Platz, J.E. and A. Lathrop.
1993. Body size and age assessment among advertising male chorus frogs. Journal of Herpetology 27(1):109–111.

Plummer, M.V.
1977. Observations on breeding migrations of *Ambystoma texanum*. Herpetological Review 8(3):79–80.
1981. Habitat utilization, diet and movements of a temperate arboreal snake (*Opheodrys aestivus*). Journal of Herpetology 15(4):425–432.
1985a. Demography of green snakes (*Opheodrys aestivus*). Herpetologica 41(4):373–381.
1985b. Growth and maturity in green snakes (*Opheodrys aestivus*). Herpetologica 41(1):28–33.
1990a. Nesting movements, nesting behavior, and nest sites of green snakes (*Opheodrys aestivus*) revealed by radiotelemetry. Herpetologica 46(2):190–195.
1990b. High predation on green snakes, *Opheodrys aestivus*. Journal of Herpetology 24(3):327–328.
1993. Thermal ecology of arboreal green snakes (*Opheodrys aestivus*). Journal of Herpetology 27(3):254–260.
1997. Population ecology of green snakes (*Opheodrys aestivus*) revisited. Herpetological Monographs 11:102–123.
2002. Observations on hibernacula and overwintering ecology of eastern hog-nosed snakes (*Heterodon platirhinos*). Herpetological Review 33(2):89–90.

Plummer, M.V. and D.B. Farrar.
1981. Sexual dietary differences in a population of *Trionyx muticus*. Journal of Herpetology 15:175–179.

Plummer, M.V. and N.E. Mills.
1996. Observations on trailing and mating behaviors in hognose snake (*Heterodon platirhinos*). Journal of Herpetology 30(1):80–82.
2000. Spatial ecology and survivorship of resident and translocated hognose snakes (*Heterodon platirhinos*). Journal of Herpetology 34(4):565–575.

Plummer, M.V., N.E. Mills, and S.L. Allen.
1997. Activity, habitat, and movement patterns of softshell turtles (*Trionyx spiniferus*) in a small stream. Chelonian Conservation and Biology 2:514–520.

Plummer, M.V. and H.W. Shirer.
1975. Movement patterns in a river-population of the softshell turtle, *Trionyx muticus*. University of Kansas. Museum of Natural History 43:1–26.

Plummer, M.V. and H.L. Snell.
1988. Nest site selection and water relations of eggs in the snake, *Opheodrys aestivus*. Copeia 1988(1):58–64.

Pope, C.H.
1939. Turtles of the United States and Canada. Alfred A. Knopf, New York, New York. 343 pp.
1944. Amphibians and reptiles of the Chicago area.

Chicago Natural History Museum Press. Chicago, Illinois. 275 pp.
Pope, P.H.
1928. The longevity of *Ambystoma maculatum* in captivity. Copeia 1928:99–100.
1937. Notes on the longevity of an *Ambystoma* in captivity. Copeia 1937(2):140–141.
Porter, K.R.
1972. Herpetology. W.B. Saunders Co., Philadelphia, Pennsylvania. 524 pp.
Potter, M.L.
2008. AFLP fingerprint analysis of hybrid salamanders in the Missouri Caverns section of Onondaga Cave. MS thesis. Missouri University of Science and Technology, Rolla. 112 pp.
Pough, F.H., R.M. Andrews, M.L. Crump, A.H. Savitzky, K.D. Wells, and M.C. Brandley.
2018. Herpetology, fourth edition. Oxford University Press. New York, New York. 591 pp.
Powell, M.A. and R. Powell.
2011. Aquatic turtles feasting on periodical cicadas. Missouri Herpetological Association Newsletter 24:21.
Powell, R.
1982. *Thamnophis proximus* (western ribbon snake). Reproduction. Herpetological Review 13:48.
1990. *Elaphe vulpina* (Baird and Girard) fox snake. Catalogue of American Amphibians and Reptiles 470.1–470.3.
Powell, R., R. Conant, and J.T. Collins.
2016. Peterson Field Guide to Reptiles and Amphibians of Eastern and Central North America, 4th edition. Houghton Mifflin Harcourt Publishing Company, New York, New York. 494 pp.
Powell, R., T.R. Johnson and D.D. Smith.
1994. New records of amphibians and reptiles in Missouri for 1994. Missouri Herpetological Association Newsletter 7:5–9.
Powell, R. and J.S. Parmerlee, Jr.
1991. Notes on reproduction in *Clonophis kirtlandii* (Serpentes: Colubridae). Bulletin of the Chicago Herpetological Society 26(2):32.
Prather, J.W. and J.T. Briggler.
2001. Use of small caves by anurans during a drought period in the Arkansas Ozarks. Journal of Herpetology 35(4):675–678.
Ptacek, M.B.
1992. Calling sites used by male gray treefrogs, *Hyla versicolor*, and *Hyla chrysoscelis*, in sympatry and allopatry in Missouri. Herpetologica 48(4):373–382.
Pyron, R.A. and F.T. Burbrink.
2009. Systematics of the common kingsnake (*Lampropeltis getula*; Serpentes: Colubridae) and the burden of heritage in taxonomy. Zootaxa 2241:22–32.
Pyron, R.A., F.W. Hsieh, A.R. Lemmon, E.M. Lemmon, and C.R. Hendry.
2016. Integrating phylogenomic and morphological data to assess candidate species-delimitation models in brown and red-bellied snakes (*Storeria*). Zoological Journal of the Linnean Society 177(4):937–949.
Rabatsky, A.M. and J.M. Waterman.
2005. Ontogenetic shifts and sex differences in caudal luring in the dusky pygmy rattlesnake, *Sistrurus miliarius barbouri*. Herpetologica 61(2):87–91.
Rabb, G.B. and H. Marx.
1973. Major ecological and geographical patterns in the evolution of colubroid snakes. Evolution 27:69–83.
Raney, E.C. and R.M. Roecker.
1947. Food and growth of two species of watersnakes from western New York. Copeia 1947(3):171–174.
Raun, G.G.
1962. Observations on behavior of newborn hog-nosed snakes, *Heterodon p. platyrhinos*. Texas Journal of Science 14:3–6.
Raymond, L.R. and L.M. Hardy.
1990. Demography of a population of *Ambystoma talpoideum* (Caudata: Ambystomatidae) in northwestern Louisiana. Herpetologica 46(4):371–382.
Reagan, D.P.
1974. Habitat selection in the three-toed box turtle, *Terrapene carolina triunguis*. Copeia 1974 (2):512–527.
Redmer, M.
1992. *Rana sphenocephala* (southern leopard frog). Variation. Herpetological Review 23(2):58–59.
1995. *Ambystoma texanum* (smallmouth salamander). Maximum size. Herpetological Review 26(1):29.
2000. Demographic and reproductive characteristics of a southern Illinois population of the crayfish frog, *Rana areolata*. The Journal of the Iowa Academy of Science 107(3–4):128–133.
2002. Natural history of the wood frog (*Rana sylvatica*) in the Shawnee National Forest, southern Illinois. Illinois Natural History Survey Bulletin 36:163–194.
Regosin, J.V., B.S. Windmiller, and J.M. Reed.
2003. Influence of abundance of small-mammal burrows and conspecifics on the density and distribution of spotted salamanders (*Ambystoma maculatum*) in terrestrial habitats. Canadian Journal of Zoology 81(4):596–605.
Reigle, N.J., Jr.
1967. The occurrence of *Necturus* in the deeper waters of Green Bay. Herpetologica 23(3):232–233.
Reinert, H. K., D. Cundall, and L. M. Bushar.
1984. Foraging behavior of the timber rattlesnake, *Crotalus horridus*. Copeia 1984(4):976–981.
Reinhard, S., S. Voitel, and A. Kupfer.
2013. External fertilisation and paternal care in the paedomorphic salamander *Siren intermedia* Barnes, 1826 (Urodela: Sirenidae). Zoologischer Anzeiger 253(1):1–5.
Resetarits, W.J., Jr.
1986. Ecology of cave use by the frog, *Rana palustris*. The American Midland Naturalist 116(2):256–266.
Resetarits, W.J., Jr. and R.D. Aldridge.
1988. Reproductive biology of a cave-associated population of the frog *Rana palustris*. Canadian Journal of Zoology 66(2):329–333.
Reynolds, H.C.
1945. Some aspects of the life history and ecology of the opossum in Central Missouri. Journal of Mammalogy 26(4):361–379.

Richardson, M.L., P.J. Weatherhead, and J.D. Brawn.
2006. Habitat use and activity of prairie kingsnakes (*Lampropeltis calligaster calligaster*) in Illinois. Journal of Herpetology 40(4):423–428.
Richtsmeier, R.J., N.P. Bernstein, J.W. Demastes, and R.W. Black.
2008. Migration, gene flow, and genetic diversity within and among Iowa populations of ornate box turtles (*Terrapene ornata ornata*). Chelonian Conservation and Biology 7(1):3–11.
Riedle, J.D.
2014. Demography of an urban population of ring-necked snakes (*Diadophis punctatus*) in Missouri. Herpetological Conservation and Biology 9(2):278–284.
Riedle, J.D., T. Weiberg, F. King-Cooley, S. Johnson, T.D.H. Riedle, and D. Asahl.
2017. Observations of the population ecology of three-toed box turtles in small, urban forest fragments. Collinsorum 6(2–3):10–14.
Riemer, W.J.
1957. The snake *Farancia abacura*: An attended nest. Herpetologica 13(1):31–32.
Ringia, A.M. and K.R. Lips.
2007. Oviposition, early development and growth of the cave salamander, *Eurycea lucifuga*: Surface and subterranean influences on a troglophilic species. Herpetologica 63(3):258–268.
Risley, P.L.
1933. Contributions on the development of the reproductive system in the musk turtle, *Sternotherus odoratus* (Latreille). Zeitschrift fur Zellforschung und Mikroskopische Anatomie 18:493–543.
Ritke, M.E. and J.G. Babb.
1991. Behavior of the gray treefrog (*Hyla chrysoscelis*) during the non-breeding season. Herpetological Review 22(1)5–8.
Ritke, M.E., J.G. Babb, and M.K. Ritke.
1990. Life history of the gray treefrog (*Hyla chrysoscelis*) in western Tennessee. Journal of Herpetology 24(2):135–141.
Rittenhouse, T.A.G., M.C. Doyle, C.R. Mank, B.B. Rothermel, and R.D. Semlitsch.
2004. Substrate cues influence habitat selection by spotted salamanders. Journal of Wildlife Management 68(4):1151–1158.
Rittenhouse, T.A.G. and R.D. Semlitsch.
2007a. Postbreeding habitat use of wood frogs in a Missouri oak-hickory forest. Journal of Herpetology 41(4):645–653.
2007b. Distribution of amphibians in terrestrial habitat surrounding wetlands. Wetlands 27:153–161.
2009. Behavioral response of migrating wood frogs to experimental timber harvest surrounding wetlands. Canadian Journal of Zoology 87(7):618–625.
Roberson, K.W.
1980. The biology of *Tantilla gracilis hallowelli* in Missouri. MS thesis. University of Missouri, Columbia. 61 pp.
Robinette, J.W. and S.E. Trauth.
1992. Reproduction in the western mud snake, *Farancia abacura reinwardtii* (Serpentes: Colubridae), in Arkansas. Journal of the Arkansas Academy of Science 46:61–64.
Robinson, C. and J.R. Bider.
1988. Nesting synchrony—a strategy to decrease predation of snapping turtles (*Chelydra serpentina*) nests. Journal of Herpetology 22(4):470–473.
Roble, S.M.
1985. Observations on satellite males in *Hyla chrysoscelis*, *Hyla picta*, and *Pseudacris triseriata*. Journal of Herpetology 19(3):432–436.
Roe, J.H., B.A. Kingsbury, and N.R. Herbert.
2003. Wetland and upland use patterns in semi-aquatic snakes: Implication for wetland conservation. Wetlands 23:1003–1014.
Rogers, C.P.
1996. *Rana catesbeiana* (bullfrog). Predation. Herpetological Review 27(1):19.
Rose, F.L. and C.D. Barbour.
1968. Ecology and reproduction cycles of the introduced gecko, *Hemidactylus turcicus*, in the southern United States. The American Midland Naturalist 79(1):159–168.
Rose, R.
1978. Observations on natural history of the ornate box turtle (*Terrapene o. ornata*). Transactions of the Kansas Academy of Science 81(2):171–172.
Rossman, D.A., N.B. Ford, and R.A. Seigel.
1996. The Garter Snakes: Evolution and Ecology. University of Oklahoma Press. Norman. 336 pp.
Rossman, D.A. and P.A. Myer.
1990. Behavioral and morphological adaptation for snail extraction in the North American brown snakes (genus *Storeria*). Journal of Herpetology 24(4):434–438.
Rota, C.T., A.J. Wolf, R.B. Renken, R.A. Gitzen, D.K. Fantz, R.A. Montgomery, M.G. Olson, L.D. Vangilder, and J.J. Millspaugh.
2017. Long-term impacts of three forest management strategies on herpetofauna abundance in the Missouri Ozarks. Forest Ecology and Management 387:37–51.
Roth, E.D.
2005. Buffer zone applications in snake ecology: A case study using cottonmouths (*Agkistrodon piscivorus*). Copeia 2005(2):399–402.
Roth, T.C. II and B.D. Greene.
2006. Movement patterns and home range use of the northern watersnake (*Nerodia sipedon*). Copeia 2006(3):544–551.
Rowe, J.W.
1992. Dietary habits of the Blanding's turtle (*Emydoidea blandingi*) in northeastern Illinois. Journal of Herpetology 26(1):111–114.
Ruane S, R.W. Bryson, Jr., R.A. Pyron, and F.T. Burbrink.
2014. Coalescent species delimitation in milksnakes (genus *Lampropeltis*) and impacts on phylogenetic comparative analyses. Systematic Biology 63(2)231–250.
Rudolph, D.C.
1978. Aspects of the larval ecology of five plethodontid salamanders of the Western Ozarks. The American Midland Naturalist 100(1):141–159.
Rudolph, D.C., R.R. Schaefer, D. Saenz, and R.N. Conner.
2004. Arboreal behavior in the timber rattlesnake,

Crotalus horridus, in eastern Texas. Texas Journal of Science 56(4):395–404.
Russell, B.L.
2004. Geographic variation in larval life history traits of the mole salamander (*Ambystoma talpoideum*): A thesis. MS thesis. Southeast Missouri State University, Cape Girardeau. 35 pp.
Ryan, T.J.
2007. Hydroperiod and metamorphosis in small-mouthed salamanders (*Ambystoma texanum*). Northeastern Naturalist 14(4):619–628.
Ryberg, W.A., J.A. Harvey, A. Blick, T.J. Hibbitts, and G. Voelker.
2015. Genetics structure is inconsistent with subspecies designations in the western massasauga *Sistrurus tergeminus*. Journal of Fish and Wildlife Management 6(2):350–359.
Saenz, D., S.J. Burgdorf, D.C. Rudolph, and C.M. Duran.
1996. *Crotalus horridus* (timber rattlesnake). Climbing. Herpetological Review 27(3):145.
Sage, R.D.
1992. Distribution and genetic interactions among Missouri leopard frogs (*Rana pipiens* complex). Missouri Herpetological Association Newsletter 5:3–4.
Salthe, S.N.
1973. *Amphiuma tridactylum*. Catalogue of American Amphibians and Reptiles, pp. 149.1–149.3.
Salveter, A. and C. Overby.
1998. Early summer courtship and mating observations of eastern collared lizards, *Crotaphytus collaris collaris*, in Reynolds County, Missouri. Missouri Herpetological Association Newsletter 11:18–19.
Sammartano, D.V.
1994. Spatial, dietary, and temporal niche parameters of two species of box turtle (*Terrapene*) in microsympatry. MS thesis. Southwest Missouri State University, Springfield. 39 pp.
Schepis, D.J.
2013. Spatial patterns and multi-scale habitat selection of the mudsnake (*Farancia abacura*) at the northern limits of its range. MS thesis. Missouri State University, Springfield. 58 pp.
Schmidt, K.P.
1953. A Checklist of North American Amphibians and Reptiles, 6th edition. American Society of Ichthyologists and Herpetologists. viii + 280 pp.
Schroeder, E.E.
1975. The reproductive cycle in the male bullfrog, *Rana catesbeiana* in Missouri. Transactions of Kansas Academy of Science 77(1):31–35.
Schuett, G.W., D.L. Clark, and F. Kraus.
1984. Feeding mimicry in the rattlesnake *Sistrurus catenatus*, with comments on the evolution of the rattle. Animal Behaviour 32:625–626.
Schuette, B.
1978. Two black rat snakes from one egg. Herpetological Review 9(3):92.
1980. Two broad-headed skink nests. Journal of St. Louis Herpetological Society 7:13–14.
1992. A clutch of ground skink eggs from Cuivre River State Park. Missouri Herpetological Association Newsletter 5:14.
1997. Two clutches of black rat snakes from eastern Missouri. Missouri Herpetological Association Newsletter 10:70–71.
1998. Natural history notes on two reptiles from Lincoln County, Missouri. Missouri Herpetological Association Newsletter 11:18.
1999. A note on reproduction in the midland brown snake (*Storeria dekayi wrightorum*). Missouri Herpetological Association Newsletter 12:18.
2005. Reproductive notes for reptiles from Cuivre River State Park. Missouri Herpetological Association Newsletter 18:12.
Schuette, B., S. Farrow, and D. Sherb.
1992. An observation of mating behavior in the broadhead skink (*Eumeces laticeps*). Missouri Herpetological Association Newsletter 5:14–15.
Schwartz, C.W. and E.R. Schwartz.
1974. The three-toed box turtle in Central Missouri: Its population, home range and movements. Missouri Department of Conservation. Terrestrial Series 5:1–28.
Schwartz, E.R.
2000. Update on permanent residency, persistence, and longevity in a 35-year study of a population of three-toed box turtles in Missouri. Chelonian Conservation Biology 3:737–738.
Schwartz, E.R. and C.W. Schwartz.
1991. A quarter-century study of survivorship in a population of three-toed box turtles in Missouri. Copeia 1991(4): 1120–1123.
Schwartz, E.R., C.W. Schwartz, and A.R. Kiester.
1984. The three-toed box turtle in central Missouri. Part II. A nineteen-year study of home range, movements and population. Missouri Department of Conservation Terrestrial Series 12:1–29.
Seale, D.B.
1980. Influence of amphibian larvae on primary production, nutrient flux, and competition in a pond ecosystem. Ecology 61(6):1531–1550.
Seibert, H.C. and C.W. Hagen, Jr.
1947. Studies on a population of snakes in Illinois. Copeia 1947(1):6–22.
Seigel, R.A.
1986. Ecology and conservation of an endangered rattlesnake (*Sistrurus catenatus*) in Missouri, USA. Biological Conservation 35(1986):333–346.
1992. Ecology of a specialized predator: *Regina grahami* in Missouri. Journal of Herpetology 26(1):32–37.
Seigel, R.A., and N.B. Ford.
1987. Reproductive ecology. Pp. 210–252. *In:* R.A. Seigel, J.T. Collins, and S.S. Novak (eds.). Snakes: Ecology and Evolutionary Biology. Macmillan Publishing Company, New York, New York. 529 pp.
Semlitsch, R.D.
1983. Structure and dynamics of two breeding populations of the eastern tiger salamander, *Ambystoma tigrinum*. Copeia 1983(3):608–616.
1987. Relationship of pond drying to the reproductive success of the salamander *Ambystoma talpoideum*. Copeia 1987(1):61–69.

Semlitsch, R.D. and T.L. Anderson.
2016. Structure and dynamics of spotted salamander (*Ambystoma maculatum*) populations in Missouri. Herpetologica 72(2):81–89.
Semlitsch, R.D., T.L. Anderson, M.S. Osbourn, and B.H. Ousterhout.
2014a. Structure and dynamics of ringed salamander (*Ambystoma annulatum*) populations in Missouri. Herpetologica 70(1):14–22.
Semlitsch, R.D., K.M. O'Donnell, and F.R. Thompson III.
2014b. Abundance, biomass production, nutrient content, and the possible role of terrestrial salamanders in Missouri Ozark forest ecosystems. Canadian Journal of Zoology 92(12):997–1004.
Semlitsch, R.D., J.H.K. Pechmann, and J.W. Gibbons.
1988. Annual emergence of juvenile mud snakes (*Farancia abacura*) at aquatic habitats. Copeia 1988(1):243–245.
Semlitsch, R.D. and H.M. Wilbur.
1988. Effects of pond drying time on metamorphosis and survival in the salamander *Ambystoma talpoideum*. Copeia 1988(4):978–983.
Serb, J.M., C.A. Phillips, and J.B. Iverson.
2001. Molecular phylogeny and biogeography of *Kinosternon flavescens* based on complete mitochondrial control region sequences. Molecular Phylogenetics and Evolution 18(1):149–162.
Settle, R.A., J.T. Briggler, and A. Mathis.
2018. A quantitative field study of paternal care in Ozark hellbender, North America's giant salamanders. Journal of Ethology 36:235–242.
Sexton, O.J.
1979. Remarks on defensive behavior of hognose snakes, *Heterodon*. Herpetological Review 10(3):86–87.
1990. *Tropidoclonion lineatum* (lined snake). Bird attack. Herpetological Review 21(4):94.
1991. A survey of the amphibians and reptiles of the Marais Temps Clair Wildlife Area, St. Charles, County, Missouri. Final Report to Missouri Department of Conservation. Jefferson City, Missouri. 11 pp.
Sexton, O.J., R.M. Andrews, and J.B. Bramble.
1992. Size and growth rate characteristics of a peripheral population of *Crotaphytus collaris* (Sauria: Crotaphytidae). Copeia 1992(4):968–980.
Sexton, O.J., J. Bizer, D.C. Gayou, P. Freiling, and M. Moutseous.
1986. Field studies of breeding spotted salamanders, *Ambystoma maculatum*, in eastern Missouri, U.S.A. Milwaukee Public Museum, Contributions in Biology and Geology 67:1–19.
Sexton, O.J. and S.R. Hunt.
1980. Temperature relationships and movements of snakes (*Elaphe obsoleta, Coluber constrictor*) in a cave hibernaculum. Herpetologica 36(1):20–26.
Sexton, O.J. and K.R. Marion.
1974. Duration of incubation of *Sceloporus undulatus* eggs at constant temperature. Physiological Zoology. 47(2):91–98.
Sexton, O.J. and C. Phillips.
1986. A qualitative study of fish-amphibian interactions in three Missouri ponds. Transactions of the Missouri Academy of Science 20:25–35.
Sexton, O.J., C. Phillips, and J.E. Bramble.
1990. The effects of temperature and precipitation on the breeding migration of the spotted salamander (*Ambystoma maculatum*). Copeia 1990(3):781–787.
Sexton, O.J., N. Shannon, and S. Shannon.
1976. Late season hatching success of *Elaphe o. obsoleta*. Herpetological Review 7(4):171.
Shaffer, S.A., J.T. Briggler, R.A. Gitzen, and J.J. Millspaugh.
2017. Abundance and harvest proportion of river turtles in Missouri. Journal of Freshwater Ecology 32(1):541–555.
Shepard, D.B., C.A. Phillips, M.J. Dreslik, and B.C. Jellen.
2004. Prey preference and diet of neonate eastern massasaugas (*Sistrurus c. catenatus*). The American Midland Naturalist 152(2):360–368.
Sherbrooke, W.C.
1990. Rain-harvesting in the lizard, *Phrynosoma cornutum*: Behavior and integumental morphology. Journal of Herpetology 24(3)302–308.
Shew, J.J.
2004. Spatial ecology and habitat use of the western fox snake (*Elaphe vulpina vulpina*) on Squaw Creek National Wildlife Refuge. MS thesis. Southwest Missouri State University, Springfield. 51 pp.
Shew, J.J., B.D. Greene, and F.E. Durbian.
2012. Spatial ecology and habitat use of the western foxsnake (*Pantherophis vulpinus*) on Squaw Creek National Wildlife Refuge (Missouri). Journal of Herpetology 46(4):539–548.
Shipman, P.A. and J.D. Riedle.
2008. Status and distribution of the alligator snapping turtle (*Macrochelys temminckii*) in southeastern Missouri. Southeastern Naturalist 7(2):331–338.
Shirose, L.J. and R.J. Brooks.
1995. Growth rate and age at maturity in syntopic populations of *Rana clamitans* and *Rana septentrionalis* in central Ontario. Canadian Journal of Zoology 73(8):1468–1473.
Shoop, C.R.
1960. The breeding habits of the mole salamander, *Ambystoma talpoideum* (Holbrook) in southeastern Louisiana. Tulane Studies of Zoology 8:65–82.
1965. Aspects of reproduction in Louisiana *Necturus* populations. The American Midland Naturalist 74(2):357–367.
Shulse, C.
1994. A possible neotenic tiger salamander from Missouri. Missouri Herpetological Association Newsletter 7:9.
2006. Rediscovery of Kirtland's snake (*Clonophis kirtlandii* Kennicott) in Missouri. Missouri Herpetological Association Newsletter 19:13–14.
Shulse, C.D.
2011. Building better wetlands for amphibians: Investigating the roles of engineered wetland features and mosquitofish (*Gambusia affinis*) on amphibian abundance and reproductive success. PhD dissertation. University of Missouri, Columbia. 165 pp.

Shulse, C.D. and R.D. Semlitsch.
2014. Western mosquitofish (*Gambusia affinis*) bolster the prevalence and severity of tadpole tail injuries in experimental wetlands. Hydrobiologia 723:131–144.

Shulse, C.D., R.D. Semlitsch, and K.M. Trauth.
2013. Mosquitofish dominate amphibian and invertebrate community development in experimental wetlands. Journal of Applied Ecology 50(5):1244–1256.

Skorepa, A.C. and J.E. Ozment.
1968. Habitat, habits, and variation of *Kinosternon subrubrum* in southern Illinois. Transactions of Illinois State Academy of Science 61:247–251.

Slevin, J.R.
1951. A high birth rate for *Natrix sipedon sipedon* (Linne). Herpetologica 7(3):132.

Smith, B.G.
1907. The life history and habits of *Cryptobranchus allegheniensis*. Biological Bulletin 13:5–39.
1912. The embryology of *Cryptobranchus allegheniensis*, including comparisons with some other vertebrates. II. General embryonic and larval development, with special reference to external features. Journal of Morphology 23:455–579.

Smith, D.D. and P.A. Bredehoft.
2020. *Hyla Chrysoscelis/Hyla versicolor* (gray treefrog). Predation. Herpetological Review 51(2):299.

Smith, D.D. and R. Powell.
1993. Life history observations of amphibians and reptiles from Missouri. Missouri Herpetological Association Newsletter 6:27–30.

Smith, G.R., A. Todd, J.E. Rettig, and F. Nelson.
2003. Microhabitat selection by northern cricket frogs (*Acris crepitans*) along a west-central Missouri creek: Field and experimental observations. Journal of Herpetology 37(2):383–385.

Smith, H.M.
1934. The amphibians of Kansas. The American Midland Naturalist 15(4):377–528.
1946. Handbook of lizards: lizards of the United States and Canada. Comstock Publishing Company, Ithaca, New York. xxi + 557 pp.

Smith, H.M., C.W. Nixon, and P.E. Smith.
1948. A partial description of the tadpole of *Rana areolata circulosa* and notes on the natural history of the race. The American Midland Naturalist 39(3):608–614.

Smith, M.E. and S.M. Secor.
2017. Physiological responses to fasting and estivation for the three-toed amphiuma (*Amphiuma tridactylum*). Physiological and Biochemical Zoology 90(2):240–256.

Smith, P.W.
1948. Food habits of cave dwelling amphibians. Herpetologica 4(6):205–208.
1961. The Amphibians and Reptiles of Illinois. Illinois Natural History Survey Bulletin 28(1). 298 pp.
1966. *Hyla avivoca*. Catalogue of the American Amphibians and Reptiles 28.1–28.2.

Smith, P.W. and H.M. Smith.
1952. Geographic variation in the lizard *Eumeces anthracinus*. University of Kansas Science Bulletin. 34(11):679–694.

Snider, A.T. and J.K. Bowler.
1992. Longevity of reptiles and amphibians in North American collections. 2nd Edition. Herpetological Circular, Number 21. Society for the Study of Amphibians and Reptiles. Oxford, Ohio. 40 pp.

Somma, L.A.
1985. Notes on maternal behavior and post-brooding aggression in the prairie skink (*Eumeces septentrionalis*). Nebraska Herpetological Newsletter 6(4):9–12.
1991. *Eumeces septentrionalis* (prairie skink). Piscivory. Herpetological Review 22(2):58–59.

Spencer, W.A.
1964. The relationship of dispersal and migration to gene flow in the boreal chorus frog. PhD dissertation. Colorado State University, Fort Collins, Colorado. 116 pp.

Sperry, J.H. and C.A. Taylor.
2008. Habitat use and seasonal activity patterns of the Great Plains ratsnake (*Elaphe guttate emoryi*) in central Texas. The Southwestern Naturalist 53(4):444–449.

Spotila, J.R.
1976. Courtship behavior of the ringed salamander (*Ambystoma annulatum*): observations in the field. The Southwestern Naturalist 21(3):412–413.

Spotila, J.R. and R.J. Beumer.
1970. The breeding habits of the ringed salamander, *Ambystoma annulatum* (Cope), in northwestern Arkansas. The American Midland Naturalist 84(1):77–89.

Spotila, J.R. and P.H. Ireland.
1970. Notes on the eggs of the gray-bellied salamander, *Eurycea multiplicata griseogaster*. The Southwestern Naturalist 14(3):366–368.

Stake, M.M., F.R. Thompson III, J. Faaborg, and D.E. Burhans.
2005. Patterns of snake predation at songbird nests in Missouri and Texas. Journal of Herpetology 39(2):215–222.

Stanford, K.M. and R.B. King.
2004. Growth, survival, and reproduction in a northern Illinois population of the plains gartersnake, *Thamnophis radix*. Copeia 2004(3):465–478.

Stanley, J.W. and S.E. Trauth.
2007. Distribution of the queen snake (*Regina septemvittata*) in Arkansas. Journal of Arkansas Academy of Science 61:99–103.

Starkey, D.E., H.B. Shaffer, R.L. Burke, M.R.J. Forstner, J.B. Iverson, F.J. Janzen, A.G.J. Rhodin, and G.R. Ultsch.
2003. Molecular systematics, phylogeography, and the effects of Pleistocene glaciation in the painted turtle (*Chrysemys picta*) complex. Evolution 57(1):119–128.

Stebbins, R.C.
2003. Peterson Field Guide to Western Reptiles and Amphibians. 3rd Edition. Houghton Mifflin Company, Boston, Massachusetts. 560 pp.

Steen, D.A., L.L. Smith, G.J. Miller, and S.C. Sterrett.
2006. Post-breeding terrestrial movements of

Ambystoma tigrinum (eastern tiger salamanders). Southeastern Naturalist 5(2):285–288.

Storey, K.B. and J.M. Storey.
1987. Persistence of freeze tolerance in terrestrially hibernating frogs after spring emergence. Copeia 1987(3):720–726.

Sullivan, B.K., K.B. Malmos, and M.F. Given.
1996. Systematics of the *Bufo woodhousii* complex (Anura: Bufonidae): Advertisement call variation. Copeia 1996(2):274–280.

Swanson, D.L. and B.M. Graves.
1995. Supercooling and freeze intolerance in overwintering juvenile spadefoot toads (*Scaphiopus bombifrons*). Journal of Herpetology 29(2):280–285.

Taber, C.A., R.F. Wilkinson, Jr., and M.S. Topping.
1975. Age and growth of hellbenders in the Niangua River, Missouri. Copeia 1975(4):633–639.

Taylor, S.J., J.K. Krejca, M.L. Niemiller, M.J. Dreslik, and C.A. Phillips.
2015. Life history and demographic differences between cave and surface populations of the western slimy salamander, *Plethodon albagula* (Caudata: Plethodontidae), in central Texas. Herpetological Conservation and Biology 10(2):740–752.

Templeton, A.R., H. Brazeal, and J.L. Neuwald.
2011. The transition from isolated patches to a metapopulation in the eastern collared lizard in response to prescribed fires. Ecology 92(9):1736–1747.

Tennant, A.
1984. The Snakes of Texas. Texas Monthly Press. Austin. 561 pp.

Thesing, B.D., R.D. Noyes, D.E. Starkey, and D.B. Shepard.
2016. Pleistocene climatic fluctuations explain the disjunct distribution and complex phylogeographic structure of the southern red-backed salamander, *Plethodon serratus*. Evolutionary Ecology 30:89–104.

Thom, R.H. and J.H. Wilson.
1980. The natural divisions of Missouri. Transaction of Missouri Academy of Science 14:9–23.

Thomas, R.B.
1993. Growth, diet, and reproduction of the red-eared slider (*Trachemys scripta*) inhabiting a reservoir receiving a cold effluent. MS thesis. Southwest Missouri State University, Springfield.

Thomas, R.B., D. Moll, and J. Steiert.
1994. Evidence of a symbiotic relationship between cellulolytic bacteria and a freshwater herbivorous turtle. The Southwestern Naturalist 39(4):386–388.

Thompson, F.R. III and D.E. Burhans.
2003. Predation of songbird nests differs by predator and between field and forested habitats. Journal of Wildlife Management 67(2):408–416.

Tinkle, D.W.
1957. Ecology, maturation, and reproduction of *Thamnophis sauritus proximus*. Ecology 38(1):69–77.

Toal, K.R. and J.T. Collins.
2003. *Gastrophryne carolinensis* (eastern narrow-mouthed toad). Maximum size. Herpetological Review 34(1):50.

Toal, K.R. and R.S. Reiserer.
1992. Geographic distribution *Eumeces obtusirostris*. Herpetological Review 23(3):89.

Topping, M.S. and C.A. Ingersol.
1981. Fecundity in the hellbender, *Cryptobranchus alleganiensis*. Copeia 1981(4):873–876.

Tracy-Smith, E., D.L. Galat, and R.B. Jacobson.
2012. Effects of flow dynamics on the aquatic-terrestrial transition zone (ATTZ) of lower Missouri River sandbars with implications for selected biota. River Research and Applications 28(7):793–813.

Trapp, M.M.
1956. Range and natural history of the ringed salamander, *Ambystoma annulatum* Cope (Ambystomidae). The Southwestern Naturalist 1(2):78–82.
1959. Studies on the life history of *Ambystoma annulatum* (Cope). MS thesis. University of Arkansas, Fayetteville. 36 pp.

Trauth, J.B., S.E. Trauth, and R.L. Johnson.
2006. Best management practices and drought combine to silence the Illinois Chorus Frog in Arkansas. Wildlife Society Bulletin 34(2): 514–518.

Trauth, S.E.
1982. *Cemophora coccinea* (scarletsnake). Reproduction. Herpetological Review 13(4):126.
1984. Seasonal incidence and reproduction in the western slender glass lizard, *Ophisaurus attenuatus attenuatus* (Reptilia, Anguidae), in Arkansas. The Southwestern Naturalist 29(3):271–275.
1989. Female reproductive traits of the southern leopard frog, *Rana sphenocephala* (Anura: Ranidae), from Arkansas. Journal of the Arkansas Academy of Science 43:105–108.
1990. Flooding as a factor in the decimation of a population of green water snakes (*Nerodia cyclopion cyclopion*) from Arkansas. Bulletin of Chicago Herpetological Society 25(1):1–3.
1993. Enlarged posterior maxillary teeth in the scarlet snake, *Cemophora coccinea* (Serpentes: Colubridae), using scanning electron microscopy. Journal of Arkansas Academy of Science 47:157–160.

Trauth, S.E, M.E. Cartwright, and W.E. Meshaka.
1989a. Winter breeding in the ringed salamander, *Ambystoma annulatum* (Caudata: Ambystomatidae) from Arkansas. The Southwestern Naturalist 34(1):145–146.

Trauth, S.E., B.G. Cochran, D.A. Saugey, W.R. Posey, II, and W.A. Stone.
1993. Distribution of the mole salamander, *Ambystoma talpoideum* (Urodela: Ambystomatidae), in Arkansas with notes on paedomorphic populations. Journal of the Arkansas Academy of Science 47:154–156.

Trauth, S.E., R.L. Cox, Jr., B.P. Butterfield, D.A. Saugey, and W.E. Meshaka, Jr.
1990. Reproduction phenophases and clutch characteristics of selected Arkansas amphibians. Journal of the Arkansas Academy of Science 44:107–113.

Trauth, S.E., R.L. Cox, W.E. Meshaka, Jr., B.P. Butterfield, and A. Holt.
1994. Female reproduction traits in selected Arkansas snakes. Journal of the Arkansas Academy of Sciences 48:196–209.

TRAUTH, S.E., R.L. COX JR., J.D. WILHIDE, and H.J. WORLEY.
1995. Egg mass characteristics of terrestrial morphs of the mole salamander, *Ambystoma talpoideum* (Caudata: Ambystomatidae), from northeastern Arkansas and clutch comparisons with other Ambystoma species. Journal of the Arkansas Academy of Science 49:193–196.
TRAUTH, S.E. and A. HOLT.
1993. Notes on the breeding biology of Hurter's spadefoot toad, *Scaphiopus holbrookii hurterii*, in Arkansas. Bulletin of Chicago Herpetological Society 28:236–239.
TRAUTH, S.E. and C.T. MCALLISTER.
1995. Vertebrate prey of selected Arkansas snakes. Journal of the Arkansas Academy of Science 49:188–192.
TRAUTH, S.E., M.L. MCCALLUM, R.R. JORDAN, and D.A. SAUGEY.
2006. Brooding postures and nest site fidelity in the western slimy salamander, *Plethodon albagula* (Caudata: Plethodontidae), from an abandoned mine shaft in Arkansas. Herpetological Natural History 9(2):141–149.
TRAUTH, S.E., W.E. MESHAKA JR., and B.P. BUTTERFIELD.
1989b. Reproduction and larval development in the marbled salamander, *Ambystoma opacum* (Caudata: Ambystomatidae), from Arkansas. Journal of Arkansas Academy of Science 43:109–111.
TRAUTH, S.E., H.W. ROBISON, and M.V. PLUMMER.
2004. The Amphibians and Reptiles of Arkansas. University of Arkansas Press. Fayetteville. 421 pp.
TRAUTH, S.E., J.D. WILHIDE, L.C. HUNT, A. HOLT, T.L. KLOTZ, and S.A. WOOLBRIGHT.
1996. *Cnemidophorus sexlineatus* (six-lined racerunner). Aquatic behavior. Herpetological Review 27(1):20–21.
TREMBLEY, F.J.
1948. The effects of predation on the fish population of Pocono Mountains lakes. Proceedings of the Pennsylvania Academy of Science 22:44–49.
TUCKER, A.D. and K.N. SLOAN.
1997. Growth and reproductive estimates from alligator snapping turtles, *Macroclemys temminckii*, taken by commercial harvest in Louisiana. Chelonian Conservation and Biology 2(4):587–592.
TUCKER, J.K.
1976. Observations on the birth of a brood of Kirtland's water snake, *Clonophis kirtlandii* (Kennicott) (Reptilia, Serpentes, Colubridae). Journal of Herpetology 10(1):53–54.
1995. Early post-transformational growth in the Illinois chorus frog (*Pseudacris streckeri illinoensis*). Journal of Herpetology 29(2):314–316.
1997. Food habits of the fossorial frog *Pseudacris streckeri illinoensis*. Herpetological Natural History 5(1):83–87.
2000. Growth and survivorship in the Illinois chorus frog (*Pseudacris streckeri illinoensis*). Transactions of the Illinois State Academy of Science 93:63–68.
TUCKER, J.K., D.W. SOERGEL, and J.B. HATCHER.
1995. Flood-associated activities of some reptiles and amphibians at Carlyle Lake, Fayette County, Illinois. Transactions of the Illinois State Academy of Science 88:73–81.
TUMA, M.W.
1993. *Kinosternon flavescens* (yellow mud turtle). Multiple nesting. Herpetological Review 24(1):31.
2006. Range, habitat use, and seasonal activity of the yellow mud turtle (*Kinosternon flavescens*) in northwestern Illinois: Implications for site-specific conservation and management. Chelonian Conservation and Biology 5(1):108–120.
TUMLISON, R., G.R. CLINE, and P. ZWANK.
1990a. Surface habitat associations of the Oklahoma salamander (*Eurycea tynerensis*). Herpetologica 46(2):169–175.
1990b. Prey selection in the Oklahoma salamander (*Eurycea tynerensis*). Journal of Herpetology 24(2):222–225.
1990c. Morphological discrimination between the Oklahoma salamander (*Eurycea tynerensis*) and the graybelly salamander (*Eurycea multiplicata griseogaster*). Copeia 1990:242–246.
TUMLISON, R. and S.E. TRAUTH.
2006. A novel facultative mutualistic relationship between bufonid tadpoles and flagellated green algae. Herpetological Conservation and Biology 1(1):51–55.
TURNER, L.K.
1995. Reproduction and diet of *Pseudemys concinna* inhabiting a cold water reservoir in southwest Missouri. MS thesis. Southwest Missouri State University, Springfield. 39 pp.
TYLER, M.J., T. BURTON, and A.M. BAUER.
2001. Parotoid or parotid: on the nomenclature of an amphibian skin gland. Herpetological Review 32(2):79–81.
UETZ, P., P. FREED, and J HOŠEK (eds.).
2020. The reptile database. http://www.reptile.database.org. Accessed on 13 March 2020.
ULTSCH, G.R. and S.J. ARCENEAUX.
1988. Gill loss in larvae *Amphiuma tridactylum*. Journal of Herpetology 22(3):347–348.
[USFWS] U.S. FISH AND WILDLIFE SERVICE.
2011. Endangered and threatened wildlife and plants: endangered status for the Ozark hellbender salamander. Federal Register 76:61956–61978.
VERMERSCH, T.G. and R.E. KUNTZ.
1986. Snakes of south-central Texas. Eakin Press. Austin, Texas. 137 pp.
VITT, L.J. and W.E. COOPER, JR.
1985. The relationship between reproduction and lipid cycling in *Eumeces laticeps* with comments on brooding ecology. Herpetologica 41(4):419–432.
1989. Maternal care in skinks (*Eumeces*). Journal of Herpetology 23(1):29–34.
VOGT, R.C.
1979. Cleaning/feeding symbiosis between grackles (*Quiscalus*: Icteridae) and map turtles (*Graptemys*: Emydidae). Auk 96(3):608–609.
1980. Natural history of the map turtles *Graptemys pseudogeographica* and *G. ouachitensis* in Wisconsin. Tulane Studies in Zoology and Botany 22(1):17–48.
1981. Natural History of Amphibians and Reptiles of Wisconsin. Milwaukee Pub. Mus., Milwaukee, Wisconsin. 205 pp.

1993. Systematics of the false map turtles (*Graptemys pseudogeographica* complex: Reptilia, Testudines, Emydidae). Annals of Carnegie Museum 62(1):1–46.
2018. *Graptemys ouachitensis* Cagle 1953—Ouachita Map Turtle. Conservation Biology of Freshwater Turtles and Tortoises. Chelonian Research Monographs, 5:103.1–103.13.

Vogt, R.C. and J.J. Bull.
1982. Temperature controlled sex determination in turtles: Ecological and behavioral aspects. Herpetologica 38(1):156–164.

Vogt, R.C., G. Bulté, and J.B. Iverson.
2018. *Graptemys geographica* (LeSueur 1917)—Northern Map Turtle, Common Map Turtle. Conservation Biology of Freshwater Turtles and Tortoises. Chelonian Research Monographs 5:104.1–104.18.

Wallace, J.E., W. Fratto, and V.A. Barko.
2007. A comparison of three sampling gears for capturing aquatic turtles in Missouri: The environmental variables related to species richness and diversity. Transactions of the Missouri Academy of Science 41:7–13.

Walley, H.D. and M.V. Plummer.
2000. *Opheodrys aestivus* (Linnaeus) rough green snake. Catalogue of American Amphibians and Reptiles 718.1–718.14.

Warner, D.A.
2000. Ecological observations on the six-lined racerunner (*Cnemidophorus sexlineatus*) in northwestern Illinois. Transactions of the Illinois State Academy of Science 93(3):239–248.

Watermolen, D.J.
1991. *Storeria occipitomaculata occipitomaculata* (northern red-belly snake). Behavior. Herpetological Review 22(2):61.

Watt, C.L., P.A. Tappe, and M.F. Roth.
2002. Concentrations of American alligator populations in Arkansas. Journal of Arkansas Academy of Science 56:243–249.

Weinstein, S.A., C.F. DeWitt, and L.A. Smith.
1992. Variability of venom-neutralizing properties of serum from snakes of the colubrid genus *Lampropeltis*. Journal of Herpetology 26(4):452–461.

Weinstein, S.A. and D.E. Keyler.
2009. Local envenoming by the western hognose snake (*Heterodon nasicus*): A case report and review of medically significant *Heterodon* bites. Toxicon 54(2009):354–360.

Welchert, D.M. and M.S. Mills.
2014. *Sistrurus catenatus tergeminus* (western massasauga). Reproduction. Herpetological Review 45(1):147–148.

Wells, K.D.
1976. Multiple egg clutches in the green frog (*Rana clamitans*). Herpetologica 32(1):85–87.
1977. Territoriality and male mating success in the green frog (*Rana clamitans*). Ecology 58(4):750–762.

Wells, P.H. and W. Gordon.
1958. Brooding slimy salamanders, *Plethodon glutinosus glutinosus* (Green). National Speleological Society Bulletin 20:23–24.

Werler, J.E. and J.R. Dixon.
2000. Texas snakes: Identification, distribution, and natural history. University of Texas Press. Austin, Texas. 544 pp.

Werner, J.K., B.A. Maxell, P. Hendricks, and D.L. Flath.
2004. Amphibians and Reptiles of Montana. Mountain Press Publishing Company. Missoula, Montana. 262 pp.

Wessels, J.L., E.T. Carter, C.L. Hively, L.E. Hayter, and B.M. Fitzpatrick.
2018. Population viability of nonnative Mediterranean house geckos (*Hemidactylus turcicus*) at an urban site near the northern invasion front. Journal of Herpetology 52(2):215–222.

Wharton, C.H.
1960. Birth and behavior of a brood of cottonmouths, *Agkistrodon piscivorus piscivorus* with notes on tail-luring. Herpetologica 16(2):125–129.

Wheeler, B.A., E. Prosen, A. Mathis, and R.F. Wilkinson.
2003. Population decline of a long-lived salamander: A 20+ year study of hellbenders, *Cryptobranchus alleganiensis*. Biological Conservation 109(1):151–156.

White, D., Jr. and D. Moll.
1991. Clutch size and annual reproductive potential of the turtle *Graptemys geographica* in a Missouri stream. Journal of Herpetology 25(4):493–494.
1992. Restricted diet of the common map turtle, *Graptemys geographica* in a Missouri stream. The Southwestern Naturalist 37(3):317–318.

Wickham, M.M.
1922. Notes on the migration of *Macrochelys lacertina*. Proceedings of Oklahoma Academy of Science 2:20–22.

Wilhoft, D.C., E. Hotaling, and P. Franks.
1983. Effects of temperature on sex determination in embryos of the snapping turtle, *Chelydra serpentina*. Journal of Herpetology 17(1):38–42.

Wilkinson, R.F., Jr.
1962. Reproductive cycle of the ring-neck snake, *Diadophis punctatus*. MS thesis. University of Missouri, Columbia. 28 pp.

Wilkinson, R.F., C.L. Peterson, D. Moll, and T. Holder.
1993. Reproductive biology of *Plethodon dorsalis* in northwestern Arkansas. Journal of Herpetology 27(1):85–87.

Williams, P.K.
1973. Seasonal movements and population dynamics of four sympatric mole salamanders, Genus *Ambystoma*. PhD dissertation. Indiana University, Bloomington, Indiana.

Williams, T.A. and J.L. Christiansen.
1981. The niches of two sympatric softshell turtles, *Trionyx muticus* and *Trionyx spiniferus*, in Iowa. Journal of Herpetology 15(3):303–308.

Willis, Y.L.
1954. Breeding, transformation, and determination of age of the bullfrog (*Rana catesbeiana* Shaw) in Missouri. MA thesis. University of Missouri, Columbia. 63 pp.

WILLIS, Y.L., D.L. MOYLE, and T.S. BASKETT.
1956. Emergence, breeding, hibernation, movements and transformation of the bullfrog, *Rana catesbeiana*, in Missouri. Copeia 1956(1):30–41.
WILSMANN, L.A. and M.A. SELLERS, JR.
1988. *Clonophis kirtlandii* rangewide survey. Unpublished report submitted to U.S.F.W.S. Region 3, Office of Endangered Species, Minneapolis-St. Paul, Minnesota.
WINTER, D.T. and R.F. WILKINSON, JR.
1994. A skeletochronological approach to determine the age of ringed salamanders (*Ambystoma annulatum*) by using growth rings found in phalanges. Missouri Herpetological Association Newsletter 7:4.
WITTENBERG, R.D.
2012. Foraging ecology of the timber rattlesnake (*Crotalus horridus*) in a fragmented agricultural landscape. Herpetological Conservation and Biology 7(3):449–461.
WOOD, J.T.
1948. *Microhyla c. carolinensis* in an ant nest. Herpetologica 4:226.
WRIGHT, A.H.
1931. Life-histories of the Frogs of Okefinokee Swamp, Georgia. The Macmillan Company. New York, New York. 497 pp.
WRIGHT, A.H. and A.A. WRIGHT.
1949. Handbook of Frogs and Toads of the U.S. and Canada. Comstock Publishing Company. Ithaca, New York. vi + 286 pp.
1957. Handbook of Snakes of the United States and Canada. Comstock Publishing Company. Ithaca, New York. Two volumes. xviii +1105 pp.
WYMAN, R.L.
1971. The courtship behavior of the small-mouthed salamander, *Ambystoma texanum*. Herpetologica 27(4):491–498.
YEARLY, C.M.
1979. Support for a mutualistic relationship between the western narrow-mouthed frog, *Gastrophryne olivacea*, and the tarantula, *Dugesiella hentzii*. MS thesis. University of Tulsa, Oklahoma.
ZIMMER-SHAFFER, S.A., J.T. BRIGGLER, and J.J. MILLSPAUGH.
2014. Modeling the effects of commercial harvest on population growth of river turtles. Chelonian Conservation and Biology 13(2):227–236.
ZWEIFEL, R.G.
1980. Aspects of the biology of a laboratory population of kingsnakes. Pp. 141–152. *In:* J.B. Murphy and J.T. Collins (eds). Reproductive biology and diseases of captive reptiles. Society for the Studies of Amphibians and Reptiles, Contributions to Herpetology No. 1. 277 pp.

Index